Student Solutions Manual

Intermediate Algebra with Applications
Fifth Edition

Richard N. Aufmann
Palomar College

Vernon C. Barker
Palomar College

Joanne S. Lockwood
Plymouth State College

Laurel Technical Services

Houghton Mifflin Company   Boston New York

Senior Sponsoring Editor: Maureen O'Connor
Senior Associate Editor: Dawn Nuttall
Editorial Assistant: Lauren M. Gagliardi
Senior Manufacturing Coordinator: Sally Culler
Marketing Manager: Ros Kane
Editorial Assistant: Amanda Bafaro

Copyright © 2000 by Houghton Mifflin Company. All rights reserved.

No part of this work may be reproduced or transmitted in any form or by any means, electronic or mechanical, including photocopying and recording, or by any information storage or retrieval system without the prior written permission of Houghton Mifflin Company unless such copying is expressly permitted by federal copyright law. Address inquiries to College Permissions, Houghton Mifflin Company, 222 Berkeley Street, Boston, MA 02116-3764.

Printed in the U.S.A.

ISBN: 0-395-96965-4

23456789-POO-03 02 01 00

Table of Contents

Chapter 1 Review of Real Numbers	1
Chapter 2 First-Degree Equations and Inequalities	13
Chapter 3 Linear Functions and Inequalities in Two Variables	52
Chapter 4 Systems of Equations and Inequalities	74
Chapter 5 Polynomials and Exponents	125
Chapter 6 Rational Expressions	158
Chapter 7 Rational Exponents and Radicals	200
Chapter 8 Quadratic Equations and Inequalities	230
Chapter 9 Functions and Relations	285
Chapter 10 Exponential and Logarithmic Functions	300
Chapter 11 Sequences and Series	325
Chapter 12 Conic Sections	350

Chapter 1: Review of Real Numbers

Section 1.1

Concept Review 1.1

1. Sometimes true
   The only exception to this statement is for zero.
   $|0| = 0$, which is not a positive number.

3. Sometimes true
   If x is a positive integer, then $-x$ is a negative integer.
   If x is a negative integer, then $-x$ is a positive integer.

5. Never true
   $-4 < -2$ but $(-4)^2$ is not less than $(-2)^2$.

7. Always true

Objective 1 Exercises

1. -14: c, e
   9: a, b, c, d
   0: b, c
   53: a, b, c, d
   7.8: none
   -626: c, e

3. $-\dfrac{15}{2}$: b, d
   0: a, b, d
   -3: a, b, d
   π: c, d
   $2.\overline{33}$: b, d
   4.232232223: c, d
   $\dfrac{\sqrt{5}}{4}$: c, d
   $\sqrt{7}$: c, d

9. -27

11. $-\dfrac{3}{4}$

13. 0

15. $\sqrt{33}$

17. 91

19. Replace x with each element in the set and determine whether the inequality is true.
    $x < 5$
    $-3 < 5$ True
    $0 < 5$ True
    $7 < 5$ False
    The inequality is true for -3 and 0.

21. Replace y with each element in the set and determine whether the inequality is true.
    $y > -4$
    $-6 > -4$ False
    $-4 > -4$ False
    $7 > -4$ True
    The inequality is true for 7.

23. Replace w with each element in the set and determine whether the inequality is true.
    $w \le -1$
    $-2 \le -1$ True
    $-1 \le -1$ True
    $0 \le -1$ False
    $1 \le -1$ False
    The inequality is true for -2 and -1.

25. Replace b with each element in the set and evaluate the expression.
    $-b$
    $-(-9) = 9$
    $-(0) = 0$
    $-(9) = -9$

27. Replace c with each element in the set and evaluate the expression.
    $|c|$
    $|-4| = 4$
    $|0| = 0$
    $|4| = 4$

29. Replace m with each element in the set and evaluate the expression.
    $-|m|$
    $-|-6| = -6$
    $-|-2| = -2$
    $-|0| = 0$
    $-|1| = -1$
    $-|4| = -4$

Objective 2 Exercises

33. $\{-2, -1, 0, 1, 2, 3, 4\}$

35. $\{2, 4, 6, 8, 10, 12\}$

37. $\{3, 6, 9, 12, 15, 18, 21, 24, 27, 30\}$

39. $\{-35, -30, -25, -20, -15, -10, -5\}$

41. $\{x \mid x > 4,\ x \text{ is an integer}\}$

43. $\{x \mid x \ge -2\}$

45. $\{x \mid 0 < x < 1\}$

47. $\{x \mid 1 \le x \le 4\}$

49. $A \cup B = \{1, 2, 4, 6, 9\}$

Chapter 1: Review of Real Numbers

51. $A \cup B = \{2, 3, 5, 8, 9, 10\}$
53. $A \cup B = \{-4, -2, 0, 2, 4, 8\}$
55. $A \cup B = \{1, 2, 3, 4, 5\}$
57. $A \cap B = \{6\}$
59. $A \cap B = \{5, 10, 20\}$
61. $A \cap B = \varnothing$
63. $A \cap B = \{4, 6\}$
65. $\{x | -1 < x < 5\}$
67. $\{x | 0 \leq x \leq 3\}$
69. $\{x | x < 2\}$
71. $\{x | x \geq 1\}$
73. $\{x | x > 1\} \cup \{x | x < -1\}$
75. $\{x | x \leq 2\} \cap \{x | x \geq 0\}$
77. $\{x | x > 1\} \cap \{x | x \geq -2\}$
79. $\{x | x > 2\} \cup \{x | x > 1\}$
81. $\{x | 0 < x < 8\}$
83. $\{x | -5 \leq x \leq 7\}$
85. $\{x | -3 \leq x < 6\}$
87. $\{x | x \leq 4\}$
89. $\{x | x > 5\}$
91. $(-2, 4)$
93. $[-1, 5]$
95. $(-\infty, 1)$
97. $[-2, 6)$
99. $(-\infty, \infty)$
101. $(-2, 5)$

103. $[-1, 2]$
105. $(-\infty, 3]$
107. $[3, \infty)$
109. $(-\infty, 2] \cup [4, \infty)$
111. $[-1, 2] \cap [0, 4]$
113. $(2, \infty) \cup (-2, 4]$

Applying Concepts 1.1

115. $A \cup B$ is
$\{x | -1 \leq x \leq 1\} \cup \{x | 0 \leq x \leq 1\} = \{x | -1 \leq x \leq 1\} = A$

117. $B \cap B$ is set B.

119. $A \cap R$ is $\{x | -1 \leq x \leq 1\}$, which is set A.

121. $B \cup R$ is the set of real numbers, R.

123. $R \cup R$ is the set R.

125. $B \cap C$ is $\{x | 0 \leq x \leq 1\} \cap \{x | -1 \leq x \leq 0\}$, which contains only the number 0.

127.
129.
131.
133.

135. $A \cup B = \{x | x > 0, x \text{ is an integer}\}$

137. $A \cap B = \{x | x \geq 15, x \text{ is an odd integer}\}$

139. The answer is b and c. For example:

　　a. $\dfrac{5-4}{3-2} \le 0$
　　　　$1 \le 0$
　　　　False

　　b. $\dfrac{2-3}{5-4} \le 0$
　　　　$-1 \le 0$
　　　　True

　　c. $\dfrac{5-4}{2-3} \le 0$
　　　　$-1 \le 0$
　　　　True

　　d. $\dfrac{4-5}{2-3} \le 0$
　　　　$1 \le 0$
　　　　False

Section 1.2

Concept Review 1.2

1. Sometimes true
   $(-2) + 4 = 2$, a positive number
   $(-8) + 4 = -4$, a negative number

3. Never true
   $\dfrac{1}{2} + \dfrac{2}{3} = \dfrac{7}{6}$
   $\dfrac{1+2}{2+3} = \dfrac{3}{5}$

5. Always true

7. Never true
   The Order of Operations says to work inside parentheses before doing exponents.

9. Always true

Objective 1 Exercises

5. $-18 + (-12) = -30$

7. $5 - 22 = 5 + (-22) = -17$

9. $3 \cdot 4(-8) = 12 \cdot (-8) = -96$

11. $18 \div (-3) = -6$

13. $-60 \div (-12) = 5$

15. $-20(35)(-16) = -700(-16) = 11{,}200$

17. $(-271)(-365) = 98{,}915$

19. $|12(-8)| = |-96| = 96$

21. $|15 - (-8)| = |15 + 8| = |23| = 23$

23. $|-56 \div 8| = |-7| = 7$

25. $|-153 \div (-9)| = |17| = 17$

27. $-|-8| + |-4| = -8 + 4 = -4$

29. $-30 + (-16) - 14 - 2 = -30 + (-16) + (-14) + (-2)$
   $= -46 + (-14) + (-2)$
   $= -60 + (-2)$
   $= -62$

31. $-2 + (-19) - 16 + 12 = -2 + (-19) + (-16) + 12$
   $= -21 + (-16) + 12$
   $= -37 + 12$
   $= -25$

33. $13 - |6 - 12| = 13 - |6 + (-12)|$
   $= 13 - |-6|$
   $= 13 - 6$
   $= 13 + (-6)$
   $= 7$

35. $738 - 46 + (-105) = 738 + (-46) + (-105)$
   $= 692 + (-105)$
   $= 587$

37. $-442 \div (-17) = 26$

39. $-4897 \div 59 = -83$

Objective 2 Exercises

43. $\dfrac{7}{12} + \dfrac{5}{16} = \dfrac{28}{48} + \dfrac{15}{48} = \dfrac{28+15}{48} = \dfrac{43}{48}$

45. $-\dfrac{5}{9} - \dfrac{14}{15} = -\dfrac{25}{45} - \dfrac{42}{45} = \dfrac{-25-42}{45} = -\dfrac{67}{45}$

47. $-\dfrac{1}{3} + \dfrac{5}{9} - \dfrac{7}{12} = -\dfrac{12}{36} + \dfrac{20}{36} - \dfrac{21}{36}$
   $= \dfrac{-12+20-21}{36}$
   $= -\dfrac{13}{36}$

49. $\dfrac{2}{3} - \dfrac{5}{12} + \dfrac{5}{24} = \dfrac{16}{24} - \dfrac{10}{24} + \dfrac{5}{24} = \dfrac{16-10+5}{24} = \dfrac{11}{24}$

51. $\dfrac{5}{8} - \dfrac{7}{12} + \dfrac{1}{2} = \dfrac{15}{24} - \dfrac{14}{24} + \dfrac{12}{24} = \dfrac{15-14+12}{24} = \dfrac{13}{24}$

53. $\left(\dfrac{6}{35}\right)\left(-\dfrac{5}{16}\right) = -\dfrac{6 \cdot 5}{35 \cdot 16} = -\dfrac{\overset{1}{2} \cdot 3 \cdot \overset{1}{5}}{5 \cdot 7 \cdot 2 \cdot 2 \cdot 2 \cdot 2} = -\dfrac{3}{56}$

55. $-\dfrac{8}{15} \div \dfrac{4}{5} = -\dfrac{8}{15} \cdot \dfrac{5}{4} = -\dfrac{8 \cdot 5}{15 \cdot 4} = -\dfrac{\overset{1}{2} \cdot \overset{1}{2} \cdot 2 \cdot \overset{1}{5}}{3 \cdot \underset{1}{5} \cdot \underset{1}{2} \cdot \underset{1}{2}} = -\dfrac{2}{3}$

57. $-\dfrac{11}{24} \div \dfrac{7}{12} = -\dfrac{11}{24} \cdot \dfrac{12}{7}$
$= -\dfrac{11 \cdot 12}{24 \cdot 7}$
$= -\dfrac{11 \cdot \cancel{2} \cdot \cancel{2} \cdot \cancel{3}}{\cancel{2} \cdot \cancel{2} \cdot 2 \cdot \cancel{3} \cdot 7}$
$= -\dfrac{11}{14}$

59. $\left(-\dfrac{5}{12}\right)\left(\dfrac{4}{35}\right)\left(\dfrac{7}{8}\right) = -\dfrac{5 \cdot 4 \cdot 7}{12 \cdot 35 \cdot 8}$
$= -\dfrac{\cancel{5} \cdot \cancel{2} \cdot \cancel{2} \cdot \cancel{7}}{2 \cdot 2 \cdot 3 \cdot \cancel{5} \cdot \cancel{7} \cdot \cancel{2} \cdot \cancel{2} \cdot 2}$
$= -\dfrac{1}{24}$

61. $\begin{array}{r} -14.270 \\ +\ 1.296 \\ \hline -12.974 \end{array}$

$-14.27 + 1.296 = -12.974$

63. $\begin{array}{r} -7.840 \\ +\ 1.832 \\ \hline -6.008 \end{array}$

$1.832 - 7.84 = -6.008$

65. $(0.03)(10.5)(6.1) = (0.315)(6.1) = 1.9215$

67. $0.9 \to 9$
$5.418 \to 54.18$

$\begin{array}{r} 6.02 \\ 9\overline{)\,54.18} \\ \underline{-54} \\ 0\ 1 \\ \underline{-\ \ 0} \\ 18 \\ \underline{-18} \\ 0 \end{array}$

$5.418 \div (-0.9) = -6.02$

69. $0.065 \to 65$
$0.4355 \to 435.5$

$\begin{array}{r} 6.7 \\ 65\overline{)\,435.5} \\ \underline{-390} \\ 455 \\ \underline{-455} \\ 0 \end{array}$

$-0.4355 \div 0.065 = -6.7$

71. $38.241 \div [-(-6.027)] - 7.453$
$= 38.241 \div 6.027 + (-7.453)$
$\approx 6.345 + (-7.453)$
≈ -1.11

73. $-287.3069 \div 0.1415 \approx -2030.44$

Objective 3 Exercises

75. $5^3 = 5 \cdot 5 \cdot 5 = 125$

77. $-2^3 = -(2 \cdot 2 \cdot 2) = -8$

79. $(-5)^3 = (-5)(-5)(-5) = -125$

81. $2^2 \cdot 3^4 = (2)(2) \cdot (3)(3)(3)(3) = 4 \cdot 81 = 324$

83. $-2^2 \cdot 3^2 = -(2)(2) \cdot (3)(3) = -4 \cdot 9 = -36$

85. $(-2)^3 \cdot (-3)^2 = (-2)(-2)(-2) \cdot (-3)(-3)$
$= -8 \cdot 9$
$= -72$

87. $4 \cdot 2^3 \cdot 3^3 = 4 \cdot (2)(2)(2) \cdot (3)(3)(3)$
$= 4 \cdot 8 \cdot 27$
$= 32 \cdot 27$
$= 864$

89. $2^2 \cdot (-10)(-2)^2 = (2)(2) \cdot (-10) \cdot (-2)(-2)$
$= 4 \cdot (-10)(4)$
$= -40(4)$
$= -160$

91. $(-3)^3 \cdot 15 \cdot (-2)^4 = (-27) \cdot 15 \cdot (16)$
$= -405 \cdot (16)$
$= -6480$

93. $2^5 \cdot (-3)^4 \cdot 4^5 = 32 \cdot (81) \cdot 1024$
$= 2592 \cdot 1024$
$= 2,654,208$

Objective 4 Exercises

97. $5 - 3(8 \div 4)^2 = 5 - 3(2)^2 = 5 - 3(4) = 5 - 12 = -7$

99. $16 - \dfrac{2^2 - 5}{3^2 + 2} = 16 - \dfrac{4 - 5}{9 + 2}$
$= 16 - \dfrac{-1}{11}$
$= 16 + \dfrac{1}{11}$
$= \dfrac{177}{11}$

101. $\dfrac{3 + \frac{2}{3}}{\frac{11}{16}} = \dfrac{\frac{11}{3}}{\frac{11}{16}} = \dfrac{11}{3} \cdot \dfrac{16}{11} = \dfrac{16}{3}$

103. $5[(2 - 4) \cdot 3 - 2] = 5[(-2) \cdot 3 - 2]$
$= 5[-6 - 2]$
$= 5[-8]$
$= -40$

105. $16 - 4\left(\dfrac{8-2}{3-6}\right) \div \dfrac{1}{2} = 16 - 4\left(\dfrac{6}{-3}\right) \div \dfrac{1}{2}$
$= 16 - 4(-2) \div \dfrac{1}{2}$
$= 16 - (-8) \div \dfrac{1}{2}$
$= 16 - (-8) \cdot 2$
$= 16 - (-16)$
$= 16 + 16$
$= 32$

107. $6[3 - (-4 + 2) \div 2] = 6[3 - (-2) \div 2]$
$= 6[3 - (-1)]$
$= 6[3 + 1]$
$= 6[4]$
$= 24$

109. $\dfrac{1}{2} - \left(\dfrac{2}{3} \div \dfrac{5}{9}\right) + \dfrac{5}{6} = \dfrac{1}{2} - \left(\dfrac{2}{3} \cdot \dfrac{9}{5}\right) + \dfrac{5}{6}$
$= \dfrac{1}{2} - \dfrac{6}{5} + \dfrac{5}{6}$
$= \dfrac{15}{30} - \dfrac{36}{30} + \dfrac{25}{30}$
$= \dfrac{15 - 36 + 25}{30}$
$= \dfrac{4}{30}$
$= \dfrac{2}{15}$

111. $\dfrac{1}{2} - \dfrac{\frac{17}{25}}{4 - \frac{3}{5}} + \dfrac{1}{5} = \dfrac{1}{2} - \dfrac{\frac{17}{25}}{\frac{17}{5}} + \dfrac{1}{5}$
$= \dfrac{1}{2} - \left(\dfrac{17}{25} \cdot \dfrac{5}{17}\right) + \dfrac{1}{5}$
$= \dfrac{1}{2} - \dfrac{1}{5} + \dfrac{1}{5}$
$= \dfrac{1}{2}$

113. $\dfrac{2}{3} - \left[\dfrac{3}{8} + \dfrac{5}{6}\right] \div \dfrac{3}{5} = \dfrac{2}{3} - \left[\dfrac{9}{24} + \dfrac{20}{24}\right] \div \dfrac{3}{5}$
$= \dfrac{2}{3} - \dfrac{29}{24} \div \dfrac{3}{5}$
$= \dfrac{2}{3} - \dfrac{29}{24} \cdot \dfrac{5}{3}$
$= \dfrac{2}{3} - \dfrac{145}{72}$
$= \dfrac{48}{72} - \dfrac{145}{72}$
$= -\dfrac{97}{72}$

115. $0.4(1.2 - 2.3)^2 + 5.8 = 0.4(-1.1)^2 + 5.8$
$= 0.4(1.21) + 5.8$
$= 0.484 + 5.8$
$= 6.284$

117. $1.75 \div 0.25 - (1.25)^2 = 1.75 \div 0.25 - 1.5625$
$= 7 - 1.5625$
$= 5.4375$

119. $25.76 \div (6.96 - 3.27)^2 = 25.76 \div (3.69)^2$
$= 25.76 \div 13.6161$
≈ 1.891878

Applying Concepts 1.2

121. 0

123. No, the multiplicative inverse of zero is undefined.

125. $7^{18} = 1,628,413,597,910,449$
The ones digit is 9.

127. 5^{234} has over 150 digits. The last three are 625.

129. The order of operations is $a^{(b^c)}$.

Section 1.3

Concept Review 1.3

1. Sometimes true
The reciprocal of 1 is 1, a whole number. The reciprocal of 2 is $\dfrac{1}{2}$, not a whole number.

3. Sometimes true
$2xy$ and $3xy$ are like terms with the same variables. $2xy$ and $2x^2y$ are unlike terms with same variables.

5. Always true

Objective 1 Exercises

1. $3 \cdot 4 = 4 \cdot 3$

3. $(3 + 4) + 5 = 3 + (4 + 5)$

5. $\dfrac{5}{0}$ is undefined.

7. $3(x + 2) = 3x + 6$

9. $\dfrac{0}{-6} = 0$

11. $\dfrac{1}{mn}(mn) = 1$

13. $2(3x) = (2 \cdot 3) \cdot x$

15. A Division Property of Zero

17. The Inverse Property of Multiplication

19. The Addition Property of Zero

21. A Division Property of Zero

23. The Distributive Property

25. The Associative Property of Multiplication

Objective 2 Exercises

29. $ab + dc$
    $(2)(3) + (-4)(-1) = 6 + 4 = 10$

31. $4cd \div a^2$
    $4(-1)(-4) \div (2)^2 = 4(-1)(-4) \div 4$
    $= (-4)(-4) \div 4$
    $= 16 \div 4$
    $= 4$

33. $(b - 2a)^2 + c$
    $[3 - 2(2)]^2 + (-1) = [3 - 4]^2 + (-1)$
    $= [-1]^2 + (-1)$
    $= 1 + (-1)$
    $= 0$

35. $(bc + a)^2 \div (d - b)$
    $[(3)(-1) + 2]^2 \div (-4 - 3) = [-3 + 2]^2 \div (-7)$
    $= [-1]^2 \div (-7)$
    $= -\dfrac{1}{7}$

37. $\dfrac{1}{4}a^4 - \dfrac{1}{6}bc$
    $\dfrac{1}{4}(2)^4 - \dfrac{1}{6}(3)(-1) = \dfrac{1}{4}(16) - \dfrac{1}{6}(3)(-1)$
    $= 4 - \dfrac{1}{6}(3)(-1)$
    $= 4 - \dfrac{1}{2}(-1)$
    $= 4 - \left(-\dfrac{1}{2}\right)$
    $= 4 + \dfrac{1}{2}$
    $= \dfrac{9}{2}$

39. $\dfrac{3ac}{-4} - c^2$
    $\dfrac{3(2)(-1)}{-4} - (-1)^2 = \dfrac{6(-1)}{-4} - (-1)^2$
    $= \dfrac{-6}{-4} - (-1)^2$
    $= \dfrac{3}{2} - (-1)^2$
    $= \dfrac{3}{2} - 1$
    $= \dfrac{1}{2}$

41. $\dfrac{3b - 5c}{3a - c}$
    $\dfrac{3(3) - 5(-1)}{3(2) - (-1)} = \dfrac{9 - (-5)}{6 - (-1)} = \dfrac{9 + 5}{6 + 1} = \dfrac{14}{7} = 2$

43. $\dfrac{a - d}{b + c}$
    $\dfrac{2 - (-4)}{3 + (-1)} = \dfrac{2 + 4}{3 + (-1)} = \dfrac{6}{2} = 3$

45. $-a|a + 2d|$
    $-2|2 + 2(-4)| = -2|2 + (-8)|$
    $= -2|-6|$
    $= -2(6)$
    $= -12$

47. $\dfrac{2a - 4d}{3b - c}$
    $\dfrac{2(2) - 4(-4)}{3(3) - (-1)} = \dfrac{4 - (-16)}{9 - (-1)}$
    $= \dfrac{4 + 16}{9 + 1}$
    $= \dfrac{20}{10}$
    $= 2$

49. $-3d \div \left|\dfrac{ab - 4c}{2b + c}\right|$
    $-3(-4) \div \left|\dfrac{2(3) - 4(-1)}{2(3) + (-1)}\right| = -3(-4) \div \left|\dfrac{6 - (-4)}{6 + (-1)}\right|$
    $= -3(-4) \div \left|\dfrac{6 + 4}{6 + (-1)}\right|$
    $= -3(-4) \div \left|\dfrac{10}{5}\right|$
    $= -3(-4) \div |2|$
    $= -3(-4) \div 2$
    $= 12 \div 2$
    $= 6$

51. $2(d - b) \div (3a - c)$
    $2(-4 - 3) \div [3(2) - (-1)] = 2(-7) \div [6 - (-1)]$
    $= 2(-7) \div [6 + 1]$
    $= 2(-7) \div 7$
    $= -14 \div 7$
    $= -2$

53. $-d^2 - c^3 a$
    $-(-4)^2 - (-1)^3(2) = -16 - (-1)(2)$
    $= -16 + 2$
    $= -14$

55. $-d^3 + 4ac$
    $-(-4)^3 + 4(2)(-1) = -(-64) + 8(-1)$
    $= 64 - 8$
    $= 56$

57. $4^{(a)^2}$
    $4^{(2^2)} = 4^4 = 256$

59. $V = LWH$
    $V = (14)(10)(6)$
    $V = 840$
    The volume is 840 in^3.

61. $V = \frac{1}{3}s^2 h$
$V = \frac{1}{3}(3^2)5$
$V = 15$
The volume is 15 ft^2.

63. $V = \frac{4}{3}\pi r^3 \qquad r = \frac{1}{2}d = \frac{1}{2}(3) = 1.5$
$V = \frac{4}{3}\pi(1.5)^3$
$V = 4.5\pi$
$V \approx 14.14$
The volume is 4.5π cm^3.
The volume is approximately 14.14 cm^3.

65. $SA = 2LW + 2LH + 2WH$
$SA = 2(5)(4) + 2(5)(3) + 2(4)(3)$
$SA = 40 + 30 + 24$
$SA = 94$
The surface area is 94 m^2.

67. $SA = s^2 + 4\left(\frac{1}{2}\right)bh$
$SA = 4^2 + 2(4)(5)$
$SA = 16 + 40 = 56$
The surface area is 56 m^2.

69. $SA = 2\pi r^2 + 2\pi rh$
$SA = 2\pi(6^2) + 2\pi(6)(2)$
$SA = 72\pi + 24\pi$
$SA = 96\pi$
$SA \approx 301.59$
The surface area is 96π in^2.
The surface area is approximately 301.59 in^2.

71. Density of statue = $\frac{\text{weight of statue}}{\text{volume of statue}}$
Density of statue = $\frac{15}{60.48}$
Density of statue = 0.25 lb / in^3
The statue has a density of 0.25 lb/in^3.

Objective 3 Exercises

73. $3x + 10x = 13x$

75. $-2x + 5x - 7x = 3x - 7x = -4x$

77. $-2a + 7b + 9a = 7a + 7b$

79. $12\left(\frac{1}{12}\right)x = x$

81. $-3(x - 2) = -3x + 6$

83. $(x + 2)5 = 5x + 10$

85. $-(-x - y) = x + y$

87. $3(x - 2y) - 5 = 3x - 6y - 5$

89. $-2a - 3(3a - 7) = -2a - 9a + 21 = -11a + 21$

91. $2x - 3(x - 2y) = 2x - 3x + 6y = -x + 6y$

93. $5[-2 - 6(a - 5)] = 5[-2 - 6a + 30]$
$= 5[28 - 6a]$
$= 140 - 30a$

95. $5[y - 3(y - 2x)] = 5[y - 3y + 6x]$
$= 5[-2y + 6x]$
$= -10y + 30x$

97. $4(-a - 2b) - 2(3a - 5b) = -4a - 8b - 6a + 10b$
$= -10a + 2b$

99. $-7(2a - b) + 2(-3b + a) = -14a + 7b - 6b + 2a$
$= -12a + b$

101. $2x - 4[x - 4(y - 2[5y + 3])]$
$= 2x - 4[x - 4(y - 10y - 6)]$
$= 2x - 4[x - 4(-9y - 6)]$
$= 2x - 4[x + 36y + 24]$
$= 2x - 4x - 144y - 96$
$= -2x - 144y - 96$

103. $3x + 8(x - 4) - 3(2x - y)$
$= 3x + 8x - 32 - 6x + 3y$
$= 5x - 32 + 3y$

105. $\frac{1}{4}[14x - 3(x - 8) - 7x] = \frac{1}{4}[14x - 3x + 24 - 7x]$
$= \frac{1}{4}[4x + 24]$
$= x + 6$

Applying Concepts 1.3

107. $4(3y + 1) = 12y + 4$
The statement is correct; it uses the Distributive Property.

109. $2 + 3x = (2 + 3)x = 5x$
The statement is not correct; it mistakenly uses the Distributive Property. It is in an irreducible statement. That is, the answer is $2 + 3x$.

111. $2(3y) = (2 \cdot 3)(2y) = 12y$
The statement is not correct; it incorrectly uses the Associative Property of Multiplication. The correct answer is $(2 \cdot 3)y = 6y$.

113. $-x^2 + y^2 = y^2 - x^2$
The statement is correct; it uses the Commutative Property of Addition.

115. $3a + 4(b + a)$

  a. $3a + (4b + 4a)$ Distributive Property

  b. $3a + (4a + 4b)$ Commutative Property of Addition

  c. $(3a + 4a) + 4b$ Associative Property of Addition

  d. $(3 + 4)a + 4b$ Distributive Property
     $7a + 4b$

117. $5(3a + 1)$

    a. $5(3a) + 5(1)$ Distributive Property

    b. $(5 \cdot 3)a + 5(1)$ Associative Property of
        $15a + 5(1)$ Multiplication

    c. $15a + 5$ Multiplication Property of One

Section 1.4

Concept Review 1.4

1. Never true
   The smaller number is represented by $12 - x$.

3. Never true
   The sum of twice x and 4 is represented by $2x + 4$.

5. Sometimes true
   The square of $-x$ is represented by $(-x)^2$. The only exception is for the number 0.
   $-0^2 = (-0)^2 = 0$

Objective 1 Exercises

1. the unknown number: n
   The sum of the number and two: $n + 2$
   $n - (n + 2) = n - n - 2 = -2$

3. the unknown number: n
   one-third of the number: $\frac{1}{3}n$
   four-fifths of the number: $\frac{4}{5}n$
   $\frac{1}{3}n + \frac{4}{5}n = \frac{5}{15}n + \frac{12}{15}n = \frac{17}{15}n$

5. the unknown number: n
   the product of eight and the number: $8n$
   $5(8n) = 40n$

7. the unknown number: n
   the product of seventeen and the number: $17n$
   twice the number: $2n$
   $17n - 2n = 15n$

9. the unknown number: n
   the square of the number: n^2
   the total of twelve and the square of the number: $12 + n^2$
   $n^2 - (12 + n^2) = n^2 - 12 - n^2 = -12$

11. the unknown number: n
    the sum of five times the number and 12: $5n + 12$
    the product of the number and fifteen: $15n$
    $15n + (5n + 12) = 15n + 5n + 12 = 20n + 12$

13. Let the smaller number be x.
    The larger number is $15 - x$.
    The sum of twice the smaller number and two more than the larger number
    $2x + (15 - x + 2) = 2x + (17 - x) = x + 17$

15. Let the larger number be x.
    Then the smaller number is $34 - x$.
    The difference between two more than the smaller number and twice the larger number
    $[(34 - x) + 2] - 2x = 34 - x + 2 - 2x = 36 - 3x$

Objective 2 Exercises

17. The population of Milan, Italy: P
    The population of San Paolo, Brazil: $4P$

19. Amount earned by Arnold Palmer: A
    Amount earned by Dennis Rodman: $\frac{2}{3}A$

21. The measure of angle B: x
    The measure of angle A is twice that of angle B: $2x$
    The measure of angle C is twice that of angle A: $2(2x) = 4x$

23. The flying time between Los Angeles and New York: t
    The flying time between New York and Los Angeles: y
    The total round-trip time: $t + y = 12$
    The trip from New York to Los Angeles can be expressed as $y = 12 - t$.

Applying Concepts 1.4

25. Three more than twice a number.

27. The product of two and three more than a number.

29. One-half the acceleration due to gravity: $\frac{1}{2}g$
    Time squared: t^2
    The product: $\frac{1}{2}gt^2$

31. The product of A and v^2: Av^2

Focus on Problem Solving

1. a. *Understand the problem.* We must determine the weight of water in the cup. To do this, we need the volume of the cup and the density (weight per unit volume) of water. The dimensions of the cup are in inches, so the volume will be in cubic inches. Therefore, the density must be found in ounces per cubic inch.

    b. *Devise a plan.* Consult a reference book to find the formula for the volume of a cone and the density of water. The formula for the volume of a cone is $V = \frac{1}{3}\pi r^2 h$. The density of water is 62.4 lb/ft^3. The plan is to convert the density to ounces per cubic inch and then use the formula $w = dv$ where w is the weight in ounces, d is the density of water in ounces per cubic inch, and v is the volume in cubic inches.

    c. *Carry out the plan.* Find the volume of the cone.
    $r = 1.5, h = 4$
    $V = \frac{1}{3}\pi r^2 h = \frac{1}{3}\pi(1.5)^2(4) \approx 9.425 \text{ in}^3$
    Convert 62.4 lb/ft^3 to ounces per cubic inch.
    $d = 62.4 \frac{\text{lb}}{\text{ft}^3}$
    $= 62.4 \frac{\text{lb}}{\text{ft}^3} \cdot \frac{1 \text{ ft}^3}{1728 \text{ in}^3} \cdot \frac{16 \text{ oz}}{\text{lb}}$
    $\approx 0.578 \frac{\text{oz}}{\text{in}^3}$
    Substitute the values in the formula $w = dv$.
    $w \approx 0.578 \frac{\text{oz}}{\text{in}^3} \cdot 9.425 \text{ in}^3 \approx 5.45 \text{ oz}$
    The cup will hold 5.45 oz of water.

    d. *Review the solution.* A cup 4 in. tall is a fairly large cup, so it seems reasonable that it would hold about one-third of a pound.

2. a. *Understand the problem.* We are to determine the dimensions of a 12-oz soft drink can. We are to use an approximation of the distance that a hand can reach around 75% of the circumference of the can. We need to know the formula for the volume of a right circular cylinder and the volume in cubic inches of 12 fl oz. We also need to make an approximation of the length of a hand.

    b. *Devise a plan.* From a resource book, we find that the volume of a right circular cylinder is $V = \pi r^2 h$. Approximate the length of a hand is 7 in. From this approximation, we can use the formula $C = 2\pi r$ to find the radius of the can. After finding the volume of 12 fl oz and the radius of the can, we find the height of the can.

    c. *Carry out the plan.* The length of the hand is 75% of the circumference.
    $7 = 0.75C$
    $9.33 \approx C$
    Use the formula $C = 2\pi r$ to find the radius.
    $C = 2\pi r$
    $9.333 = 2\pi r$
    $1.485 \approx r$
    Use the fact that 128 fl oz = 1 gal and 1 gal = 231 in^3 to find the volume of 12 fl oz.
    $V = 12 \text{ fl oz}$
    $= 12 \text{ fl oz} \cdot \frac{1 \text{ gal}}{128 \text{ fl oz}} \cdot \frac{231 \text{ in}^3}{1 \text{ gal}}$
    $\approx 21.656 \text{ in}^3$
    Use the formula for the volume of a right circular cylinder to find the height of the can.
    $V = \pi r^2 h$
    $21.656 = \pi(1.485)^2 h$
    $\frac{21.656}{\pi(1.495)^2} = h$
    $3.13 \approx h$
    The radius of the can is approximately 1.5 in., and the height is approximately 3.1 in.

    d. *Review the solution.* The diameter of the can is approximately the same as the height of the can. The diameter seems too large and the height seems too small. The approximation of the distance of the hand reaching around the can may be too large.

Projects and Group Activities

Water Displacement

1. Volume of the cylinder is $V = \pi r^2 h$, where $r = 2$ and $h = 10$.
    $V = \pi(2)^2(10)$
    $V = 40\pi$
    The volume of the water displaced is $V = LWH$, where $L = 30$, $W = 20$, and $H = x$.
    $40\pi = (30)(20)x$
    $\frac{2}{30}\pi = x$
    $0.21 \approx x$
    The water will rise approximately 0.21 cm.

2. The volume of $\frac{2}{3}$ of the sphere is
    $V = \frac{2}{3}\left(\frac{4}{3}\pi r^3\right)$, where $r = 6$.
    $V = \frac{8}{9}\pi(6)^3$
    $V = 192\pi$
    The volume of the water displaced is $V = LWH$, where $L = 20$, $W = 16$ and $H = x$.
    $192\pi = (20)(16)x$
    $\frac{3}{5}\pi = x$
    $1.88 \approx x$
    The water will rise approximately 1.88 in.

Chapter 1: Review of Real Numbers

3. Find the volume of the statue by finding the volume of the water displaced by the statue.
   $V = LWH$, where $L = 12$, $W = 12$ and $H = 0.42$.
   $V = (12)(12)(0.42) = 60.48$
   The volume of the statue is 60.48 cubic inches.
   density = weight ÷ volume
   density = $15 \div 60.48 \approx 0.25$
   The density of the statue is approximately 0.25 lb/in^3.

Chapter Review Exercises

1. $\frac{3}{4}; -\frac{3}{4} + \frac{3}{4} = 0$

2. Replace x with the elements in the set and determine whether the inequality is true.
   $x > -1$
   $-4 > -1$ False
   $-2 > -1$ False
   $0 > -1$ True
   $2 > -1$ True
   The inequality is true for 0 and 2.

3. $p \in \{-4, 0, 7\}$
   $-|p|$
   $-|-4| = -4$
   $-|0| = 0$
   $-|7| = -7$

4. $\{-2, -1, 0, 1, 2, 3\}$

5. $\{x \mid x < -3, x \in \text{real numbers}\}$

6. $\{x \mid -2 \leq x \leq 3\}$

7. $A \cup B = \{1, 2, 3, 4, 5, 6, 7, 8\}$

8. $A \cap B = \{2, 3\}$

9. $[-3, \infty)$

10. $\{x \mid x < 1\}$

11. $\{x \mid x \leq -3\} \cup \{x \mid x > 0\}$

12. $(-2, 4]$

13. $-10 - (-3) - 8 = -10 + 3 + (-8) = -7 + (-8) = -15$

14. $-204 \div (-17) = 12$

15. $18 - |-12 + 8| = 18 - |-4| = 18 - 4 = 14$

16. $-2 \cdot (4^2) \cdot (-3)^2 = -2 \cdot 16 \cdot 9 = -32 \cdot 9 = -288$

17. $-\frac{3}{8} + \frac{3}{5} - \frac{1}{6} = -\frac{45}{120} + \frac{72}{120} - \frac{20}{120}$
    $= \frac{-45 + 72 - 20}{120}$
    $= \frac{7}{120}$

18. $\frac{3}{5}\left(-\frac{10}{21}\right)\left(-\frac{7}{15}\right) = \frac{3 \cdot 10 \cdot 7}{5 \cdot 21 \cdot 15}$
    $= \frac{3 \cdot 2 \cdot 5 \cdot 7}{5 \cdot 3 \cdot 7 \cdot 3 \cdot 5}$
    $= \frac{2}{15}$

19. $-\frac{3}{8} \div \frac{3}{5} = -\frac{3}{8} \cdot \frac{5}{3}$
    $= -\frac{3 \cdot 5}{8 \cdot 3}$
    $= -\frac{5}{8}$

20. $-4.07 + 2.3 - 1.07 = -1.77 - 1.07 = -2.84$

21. $-3.286 \div (-1.06) = 3.1$

22. $20 \div \frac{3^2 - 2^2}{3^2 + 2^2} = 20 \div \frac{9 - 4}{9 + 4}$
    $= 20 \div \frac{5}{13}$
    $= 20 \cdot \frac{13}{5}$
    $= 52$

23. $2a^2 - \frac{3b}{a} = 2(-3)^2 - \frac{3(2)}{-3}$
    $= 2(-3)^2 - \frac{6}{-3}$
    $= 2(-3)^2 - (-2)$
    $= 2(9) - (-2)$
    $= 18 + 2$
    $= 20$

24. $(a - 2b^2) \div ab$
    $(4 - 2(-3)^2) \div (4)(-3) = (4 - 2(9)) \div (4)(-3)$
    $= (4 - 18) \div (4)(-3)$
    $= -14 \div [(4)(-3)]$
    $= -14 \div -12$
    $= \frac{-14}{-12}$
    $= \frac{7}{6}$

25. 3

26. y

27. (ab)

28. 4

29. The Inverse Property of Addition

30. The Associative Property of Multiplication

31. $-2(x-3) + 4(2-x) = -2x + 6 + 8 - 4x = -6x + 14$

32. $4y - 3[x - 2(3 - 2x) - 4y] = 4y - 3[x - 6 + 4x - 4y]$
    $= 4y - 3[5x - 6 - 4y]$
    $= 4y - 15x + 18 + 12y$
    $= 16y - 15x + 18$

33. The unknown number: x
    The sum of the number and four: $x + 4$
    $4(x + 4) = 4x + 16$

34. The unknown number: x
    The difference between the number and two: $x - 2$
    Twice the difference between the number and two: $2(x - 2)$
    $2(x - 2) + 8 = 2x - 4 + 8 = 2x + 4$

35. Let x be the smaller of the numbers. Then the larger number is $40 - x$.
    The sum of twice x and five more than $40 - x$.
    $2x + (40 - x + 5) = x + 45$

36. Let x be the larger number.
    Then the smaller number is $9 - x$.
    The difference between three more than twice $(9 - x)$ and one more than x.
    $[2(9 - x) + 3] - (x + 1) = 18 - 2x + 3 - x - 1$
    $= -3x + 20$

37. The width of the rectangle: W
    The length is 3 feet less than $3W$.
    The length is $3W - 3$.

38. Let the first integer be n.
    The second integer is five more than four times n.
    $4n + 5$ is the magnitude of the second integer.

Chapter Test

1. 12

2. Replace x with each element in the set and determine whether the inequality is true.
   $-1 > x$
   $-1 > -5$ True
   $-1 > 3$ False
   $-1 > 7$ False
   The inequality is true for -5.

3. $2 - (-12) + 3 - 5 = 2 + 12 + 3 + (-5)$
   $= 14 + 3 + (-5)$
   $= 17 + (-5)$
   $= 12$

4. $(-2)(-3)(-5) = (6)(-5) = -30$

5. $-180 \div 12 = -15$

6. $|-3 - (-5)| = |-3 + 5| = |2| = 2$

7. $-5^2 \cdot 4 = -25 \cdot 4 = -100$

8. $(-2)^3(-3)^2 = (-8)(9) = -72$

9. $\dfrac{2}{3} - \dfrac{5}{12} + \dfrac{4}{9} = \dfrac{24}{36} - \dfrac{15}{36} + \dfrac{16}{36}$
   $= \dfrac{24 - 15 + 16}{36}$
   $= \dfrac{25}{36}$

10. $\left(-\dfrac{2}{3}\right)\left(\dfrac{9}{15}\right)\left(\dfrac{10}{27}\right) = -\dfrac{2 \cdot 3 \cdot 3 \cdot 2 \cdot 5}{3 \cdot 3 \cdot 5 \cdot 3 \cdot 3 \cdot 3}$
    $= -\dfrac{4}{27}$

11. $4.27 - 6.98 + 1.3 = -2.71 + 1.3 = -1.41$

12. $-15.092 \div 3.08 = -4.9$

13. $12 - 4\left(\dfrac{5^2 - 1}{3}\right) \div 16 = 12 - 4\left(\dfrac{25 - 1}{3}\right) \div 16$
    $= 12 - 4\left(\dfrac{24}{3}\right) \div 16$
    $= 12 - 4(8) \div 16$
    $= 12 - 32 \div 16$
    $= 12 - 2$
    $= 10$

14. $8 - 4(2 - 3)^2 \div 2 = 8 - 4(-1)^2 \div 2$
    $= 8 - 4(1) \div 2$
    $= 8 - 4 \div 2$
    $= 8 - 2$
    $= 6$

15. $(a - b)^2 \div (2b + 1) = (2 - (-3))^2 \div (2(-3) + 1)$
    $= (5)^2 \div (-6 + 1)$
    $= (5)^2 \div (-5)$
    $= 25 \div (-5)$
    $= -5$

16. $\dfrac{b^2 - c^2}{a - 2c} = \dfrac{(3)^2 - (-1)^2}{2 - 2(-1)}$
    $= \dfrac{9 - 1}{2 - (-2)}$
    $= \dfrac{8}{4}$
    $= 2$

17. 4

18. The Distributive Property

19. $3x - 2(x - y) - 3(y - 4x)$
    $= 3x - 2x + 2y - 3y + 12x$
    $= 13x - y$

20. $2x - 4[2 - 3(x + 4y) - 2]$
    $= 2x - 4[2 - 3x - 12y - 2]$
    $= 2x - 4[-3x - 12y]$
    $= 2x + 12x + 48y$
    $= 14x + 48y$

21. the unknown number: n
    three less than the number: $n - 3$
    the product of three less than the number and nine: $(n-3)(9)$
    $13 - (n-3)(9) = 13 - 9n + 27 = 40 - 9n$

22. The unknown number: n
    The total of twelve times the number and twenty-seven: $12n + 27$
    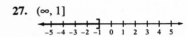
    $\frac{1}{3}(12n + 27) = 4n + 9$

23. $A \cup B = \{1, 2, 3, 4, 5, 7\}$

24. $A \cup B = \{-2, -1, 0, 1, 2, 3\}$

25. $A \cap B = \{5, 7\}$

26. $A \cap B = \{-1, 0, 1\}$

27. $(\infty, 1]$

28. $(3, \infty)$

29. $\{x | x \leq 3\} \cup \{x | x < -2\}$

30. $\{x | x < 3\} \cap \{x | x > -2\}$

Chapter 2: First-Degree Equations and Inequalities

Section 2.1

Concept Review 2.1

1. Never true
   An equation must have an equals sign.

3. Sometimes true
   $8 = 4 + 2$ is a false equation. $6 = 6$ is a true equation. Neither equation contains a variable.

5. Never true
   This equation has no solution. Any replacement value for x will result in a false equation.

7. Always true

Objective 1 Exercises

5. $x - 2 = 7$
   $x - 2 + 2 = 7 + 2$
   $x = 9$
   The solution is 9.

7. $a + 3 = -7$
   $a + 3 - 3 = -7 - 3$
   $a = -10$
   The solution is -10.

9. $3x = 12$
   $\dfrac{3x}{3} = \dfrac{12}{3}$
   $x = 4$
   The solution is 4.

11. $\dfrac{2}{7} + x = \dfrac{17}{21}$
    $\dfrac{2}{7} - \dfrac{2}{7} + x = \dfrac{17}{21} - \dfrac{2}{7}$
    $x = \dfrac{17}{21} - \dfrac{6}{21}$
    $x = \dfrac{11}{21}$
    The solution is $\dfrac{11}{21}$.

13. $\dfrac{5}{8} - y = \dfrac{3}{4}$
    $\dfrac{5}{8} - \dfrac{5}{8} - y = \dfrac{3}{4} - \dfrac{5}{8}$
    $-y = \dfrac{6}{8} - \dfrac{5}{8}$
    $-y = \dfrac{1}{8}$
    $(-1)(-y) = \dfrac{1}{8}(-1)$
    $y = -\dfrac{1}{8}$
    The solution is $-\dfrac{1}{8}$.

15. $\dfrac{3}{5}y = 12$
    $\dfrac{5}{3}\left(\dfrac{3}{5}y\right) = \dfrac{5}{3}(12)$
    $y = 20$
    The solution is 20.

17. $\dfrac{3a}{7} = -21$
    $\dfrac{7}{3}\left(\dfrac{3}{7}a\right) = \dfrac{7}{3}(-21)$
    $a = -49$
    The solution is -49.

19. $-\dfrac{5}{12}y = \dfrac{7}{16}$
    $-\dfrac{12}{5}\left(-\dfrac{5}{12}y\right) = -\dfrac{12}{5}\left(\dfrac{7}{16}\right)$
    $y = -\dfrac{21}{20}$
    The solution is $-\dfrac{21}{20}$.

21. $b - 14.72 = -18.45$
    $b - 14.72 + 14.72 = -18.45 + 14.72$
    $b = -3.73$
    The solution is -3.73.

23. $3x + 5x = 12$
    $8x = 12$
    $\dfrac{8x}{8} = \dfrac{12}{8}$
    $x = \dfrac{3}{2}$
    The solution is $\dfrac{3}{2}$.

25. $2x - 4 = 12$
    $2x - 4 + 4 = 12 + 4$
    $2x = 16$
    $\dfrac{2x}{2} = \dfrac{16}{2}$
    $x = 8$

27. $4x + 2 = 4x$
    $4x - 4x + 2 = 4x - 4x$
    $2 = 0$
    The equation has no solution.

29. $2x + 2 = 3x + 5$
    $2x - 3x + 2 = 3x - 3x + 5$
    $-x + 2 = 5$
    $-x + 2 - 2 = 5 - 2$
    $-x = 3$
    $(-1)(-x) = (-1)(3)$
    $x = -3$
    The solution is -3.

31.
$2 - 3t = 3t - 4$
$2 - 3t - 3t = 3t - 3t - 4$
$2 - 6t = -4$
$2 - 2 - 6t = -4 - 2$
$-6t = -6$
$\dfrac{-6t}{-6} = \dfrac{-6}{-6}$
$t = 1$
The solution is 1.

33.
$2a - 3a = 7 - 5a$
$-a = 7 - 5a$
$-a + 5a = 7 - 5a + 5a$
$4a = 7$
$\dfrac{4a}{4} = \dfrac{7}{4}$
$a = \dfrac{7}{4}$
The solution is $\dfrac{7}{4}$.

35.
$\dfrac{5}{8}b - 3 = 12$
$\dfrac{5}{8}b - 3 + 3 = 12 + 3$
$\dfrac{5}{8}b = 15$
$\dfrac{8}{5}\left(\dfrac{5}{8}b\right) = \dfrac{8}{5}(15)$
$b = 24$
The solution is 24.

37.
$b + \dfrac{1}{5}b = 2$
$\dfrac{6}{5}b = 2$
$\dfrac{5}{6}\left(\dfrac{6}{5}b\right) = \dfrac{5}{6}(2)$
$b = \dfrac{5}{3}$
The solution is $\dfrac{5}{3}$.

39.
$2x - 9x + 3 = 6 - 5x$
$-7x + 3 = 6 - 5x$
$-7x + 5x + 3 = 6 - 5x + 5x$
$-2x + 3 = 6$
$-2x + 3 - 3 = 6 - 3$
$-2x = 3$
$\dfrac{-2x}{-2} = \dfrac{3}{-2}$
$x = -\dfrac{3}{2}$
The solution is $-\dfrac{3}{2}$.

41.
$2y - 4 + 8y = 7y - 8 + 3y$
$10y - 4 = 10y - 8$
$10y - 10y - 4 = 10y - 10y - 8$
$-4 = -8$
The equation has no solution.

43.
$9 + 4x - 12 = -3x + 5x + 8$
$4x - 3 = 2x + 8$
$4x - 2x - 3 = 2x - 2x + 8$
$2x - 3 = 8$
$2x - 3 + 3 = 8 + 3$
$2x = 11$
$\dfrac{2x}{2} = \dfrac{11}{2}$
$x = \dfrac{11}{2}$
The solution is $\dfrac{11}{2}$.

45.
$5.3y + 0.35 = 5.02y$
$5.3y - 5.3y + 0.35 = 5.02y - 5.3y$
$0.35 = -0.28y$
$\dfrac{0.35}{-0.28} = \dfrac{-0.28y}{-0.28}$
$-1.25 = y$
The solution is -1.25.

Objective 2 Exercises

47.
$2x + 3(x - 5) = 15$
$2x + 3x - 15 = 15$
$5x - 15 = 15$
$5x = 30$
$\dfrac{5x}{5} = \dfrac{30}{5}$
$x = 6$
The solution is 6.

49.
$5(2 - b) = -3(b - 3)$
$10 - 5b = -3b + 9$
$10 = 2b + 9$
$1 = 2b$
$\dfrac{1}{2} = \dfrac{2b}{2}$
$\dfrac{1}{2} = b$
The solution is $\dfrac{1}{2}$.

51.
$3(y - 5) - 5y = 2y + 9$
$3y - 15 - 5y = 2y + 9$
$-2y - 15 = 2y + 9$
$-4y - 15 = 9$
$-4y = 24$
$\dfrac{-4y}{-4} = \dfrac{24}{-4}$
$y = -6$
The solution is -6.

53. $2x - 3(x-4) = 2(3-2x) + 2$
$2x - 3x + 12 = 6 - 4x + 2$
$-x + 12 = 8 - 4x$
$3x + 12 = 8$
$3x = -4$
$\dfrac{3x}{3} = \dfrac{-4}{3}$
$x = -\dfrac{4}{3}$
The solution is $-\dfrac{4}{3}$.

55. $-4(7y - 1) + 5y = -2(3y + 4) - 3y$
$-28y + 4 + 5y = -6y - 8 - 3y$
$-23y + 4 = -9y - 8$
$-14y + 4 = -8$
$-14y = -12$
$\dfrac{-14y}{-14} = \dfrac{-12}{-14}$
$y = \dfrac{6}{7}$
The solution is $\dfrac{6}{7}$.

57. $2[4 + 2(5 - x) - 2x] = 4x - 7$
$2[4 + 10 - 2x - 2x] = 4x - 7$
$2[14 - 4x] = 4x - 7$
$28 - 8x = 4x - 7$
$28 - 12x = -7$
$-12x = -35$
$\dfrac{-12x}{-12} = \dfrac{-35}{-12}$
$x = \dfrac{35}{12}$
The solution is $\dfrac{35}{12}$.

59. $-3[x + 4(x + 1)] = x + 4$
$-3(x + 4x + 4) = x + 4$
$-3(5x + 4) = x + 4$
$-15x - 12 = x + 4$
$-16x - 12 = 4$
$-16x = 16$
$\dfrac{-16x}{-16} = \dfrac{16}{-16}$
$x = -1$
The solution is -1.

61. $5 - 6[2t - 2(t + 3)] = 8 - t$
$5 - 6(2t - 2t - 6) = 8 - t$
$5 - 12t + 12t + 36 = 8 - t$
$41 = 8 - t$
$33 = -t$
$-33 = t$
The solution is -33.

63. $3[x - (2 - x) - 2x] = 3(4 - x)$
$3[x - 2 + x - 2x] = 12 - 3x$
$3(-2) = 12 - 3x$
$-6 = 12 - 3x$
$-18 = -3x$
$\dfrac{-18}{-3} = \dfrac{-3x}{-3}$
$6 = x$
The solution is 6.

65. $\dfrac{3}{4}t - \dfrac{7}{12}t = 1$
$12\left(\dfrac{3}{4}t - \dfrac{7}{12}t\right) = 12 \cdot 1$
$\dfrac{12 \cdot 3t}{4} - \dfrac{12 \cdot 7t}{12} = 12$
$9t - 7t = 12$
$2t = 12$
$\dfrac{2t}{2} = \dfrac{12}{2}$
$t = 6$
The solution is 6.

67. $\dfrac{1}{2}x - \dfrac{3}{4}x + \dfrac{5}{8} = \dfrac{3}{2}x - \dfrac{5}{2}$
$8\left(\dfrac{1}{2}x - \dfrac{3}{4}x + \dfrac{5}{8}\right) = 8\left(\dfrac{3}{2}x - \dfrac{5}{2}\right)$
$\dfrac{8 \cdot x}{2} - \dfrac{8 \cdot 3x}{4} + \dfrac{8 \cdot 5}{8} = \dfrac{8 \cdot 3x}{2} - \dfrac{8 \cdot 5}{2}$
$4x - 6x + 5 = 12x - 20$
$-2x + 5 = 12x - 20$
$-14x + 5 = -20$
$-14x = -25$
$\dfrac{-14x}{-14} = \dfrac{-25}{-14}$
$x = \dfrac{25}{14}$
The solution is $\dfrac{25}{14}$.

69. $\dfrac{2a - 9}{5} + 3 = 2a$
$5\left(\dfrac{2a - 9}{5} + 3\right) = 5 \cdot 2a$
$\dfrac{5(2a - 9)}{5} + 5 \cdot 3 = 10a$
$2a - 9 + 15 = 10a$
$2a + 6 = 10a$
$6 = 8a$
$\dfrac{6}{8} = \dfrac{8a}{8}$
$\dfrac{3}{4} = a$
The solution is $\dfrac{3}{4}$.

71.
$$\frac{2x-1}{4}+\frac{3x+4}{8}=\frac{1-4x}{12}$$
$$24\left(\frac{2x-1}{4}+\frac{3x+4}{8}\right)=24\left(\frac{1-4x}{12}\right)$$
$$\frac{24(2x-1)}{4}+\frac{24(3x+4)}{8}=\frac{24(1-4x)}{12}$$
$$6(2x-1)+3(3x+4)=2(1-4x)$$
$$12x-6+9x+12=2-8x$$
$$21x+6=2-8x$$
$$29x+6=2$$
$$29x=-4$$
$$\frac{29x}{29}=\frac{-4}{29}$$
$$x=-\frac{4}{29}$$
The solution is $-\frac{4}{29}$.

73.
$$\frac{1}{5}(20x+30)=\frac{1}{3}(6x+36)$$
$$15\left[\frac{1}{5}(20x+30)\right]=15\left[\frac{1}{3}(6x+36)\right]$$
$$3(20x+30)=5(6x+36)$$
$$60x+90=30x+180$$
$$30x+90=180$$
$$30x=90$$
$$\frac{30x}{30}=\frac{90}{30}$$
$$x=3$$
The solution is 3.

75. $2(y-4)+8=\frac{1}{2}(6y+20)$
$$2y-8+8=\frac{6y}{2}+\frac{20}{2}$$
$$2y=3y+10$$
$$-y=10$$
$$(-1)(-y)=(-1)(10)$$
$$y=-10$$
The solution is −10.

77. $\frac{1}{4}(7-x)=\frac{2}{3}(x+2)$
$$12\left[\frac{1}{4}(7-x)\right]=12\left[\frac{2}{3}(x+2)\right]$$
$$3(7-x)=8(x+2)$$
$$21-3x=8x+16$$
$$21-11x=16$$
$$-11x=-5$$
$$\frac{-11x}{-11}=\frac{-5}{-11}$$
$$x=\frac{5}{11}$$
The solution is $\frac{5}{11}$.

79. $-1.6(b-2.35)=-11.28$
$$-1.6b+3.76=-11.28$$
$$-1.6b=-15.04$$
$$\frac{-1.6b}{-1.6}=\frac{-15.04}{-1.6}$$
$$b=9.4$$
The solution is 9.4.

Objective 3 Exercises

81. Strategy
To find the number of bags purchased, write and solve an equation using b to represent the number of bags purchased.

Solution
The first bag purchased cost $10.90. That means that $b-1$ bags cost $10.50(b-1)$.
$$84.40=10.90+10.50(b-1)$$
$$84.40=10.90+10.50b-10.50$$
$$84.40=0.40+10.50b$$
$$84=10.50b$$
$$8=b$$
The customer purchased 8 bags of feed.

83. Strategy
To find the charge per hour for labor, write and solve an equation using L to represent the charge per hour for labor.

Solution
The total charge for labor was $4L$ dollars.
$$316.55=4L+148.55$$
$$168=4L$$
$$42=L$$
Labor cost $42 per hour.

85. Strategy
To find the employee's regular hourly rate, write and solve an equation using h to represent the hourly rate and $1.5h$ to represent the overtime rate.

Solution
The wages earned for the first forty hours plus the wages earned for overtime are $374.50.
$$40h+9(1.5h)=374.50$$
$$40h+13.5h=374.50$$
$$53.5h=374.50$$
$$h=7$$
The regular hourly rate is $7.

87. **Strategy**
To find the number of each type of ticket, write and solve an equation using g to represent the number of grandstand tickets and $12 - g$ to represent the number of upper field box tickets.

Solution
The total price of the tickets was $275.
$$20g + 27(12 - g) = 275$$
$$20g + 324 - 27g = 275$$
$$-7g + 324 = 275$$
$$-7g = -49$$
$$g = 7$$
The fraternity purchased 7 grandstand tickets and 5 upper field box tickets.

89. **Strategy**
To find the number of mezzanine tickets purchased, write and solve an equation using m to represent the number of mezzanine tickets and $7 - m$ to represent the number of balcony tickets.

Solution
The total cost of the tickets was $275.00.
$$50m + 35(7 - m) = 275$$
$$50m + 245 - 35m = 275$$
$$15m + 245 = 275$$
$$15m = 30$$
$$m = 2$$
Two mezzanine tickets were purchased.

Applying Concepts 2.1

91. $\dfrac{1}{\frac{1}{y}} = -9$
$$y = -9$$
The solution is -9.

93. $\dfrac{10}{\frac{3}{x}} - 5 = 4x$
$$10\left(\dfrac{x}{3}\right) - 5 = 4x$$
$$\dfrac{10}{3}x - 5 = 4x$$
$$\dfrac{10}{3}x - \dfrac{10}{3}x - 5 = 4x - \dfrac{10}{3}x$$
$$-5 = \dfrac{2}{3}x$$
$$\dfrac{3}{2}(-5) = \dfrac{3}{2}\left(\dfrac{2}{3}x\right)$$
$$-\dfrac{15}{2} = x$$
The solution is $-\dfrac{15}{2}$.

95. $2[3(x + 4) - 2(x + 1)] = 5x + 3(1 - x)$
$$2[3x + 12 - 2x - 2] = 5x + 3 - 3x$$
$$2(x + 10) = 2x + 3$$
$$2x + 20 = 2x + 3$$
$$2x - 2x + 20 = 2x - 2x + 3$$
$$20 = 3$$
There is no solution.

97. $\dfrac{4[(x - 3) + 2(1 - x)]}{5} = x + 1$
$$\dfrac{4(x - 3 + 2 - 2x)}{5} = x + 1$$
$$\dfrac{4(-x - 1)}{5} = x + 1$$
$$\dfrac{-4x - 4}{5} = x + 1$$
$$5\left(\dfrac{-4x - 4}{5}\right) = 5(x + 1)$$
$$-4x - 4 = 5x + 5$$
$$-4x + 4x - 4 = 5x + 4x + 5$$
$$-4 = 9x + 5$$
$$-4 - 5 = 9x + 5 - 5$$
$$-9 = 9x$$
$$\dfrac{-9}{9} = \dfrac{9x}{9}$$
$$-1 = x$$
The solution is -1.

99. $3(2x + 2) - 4(x - 3) = 2(x + 9)$
$$6x + 6 - 4x + 12 = 2x + 18$$
$$2x + 18 = 2x + 18$$
$$2x - 2x + 18 = 2x - 2x + 18$$
$$18 = 18$$
$$0 = 0$$
The solution is all real numbers.

Section 2.2

Concept Review 2.2

1. Always true

3. Always true

5. Sometimes true
If n is even, then $n + 1$, $n + 3$, and $n + 5$ represent consecutive odd integers.
If n is odd, then $n + 1$, $n + 3$, and $n + 5$ represent consecutive even integers.

Chapter 2: First-Degree Equations and Inequalities

Objective 1 Exercises

3. **Strategy**
   Number of nickels: x
   Number of dimes: $56 - x$

Coin	Number	Value	Total Value
Nickel	x	5	$5x$
Dime	$56 - x$	10	$10(56 - x)$

   The sum of the total values of each type of coin equals the total value of all the coins (400 cents).
   $5x + 10(56 - x) = 400$

   Solution
   $5x + 10(56 - x) = 400$
   $5x + 560 - 10x = 400$
   $-5x + 560 = 400$
   $-5x = -160$
   $x = 32$
   $56 - x = 56 - 32 = 24$
   There are 24 dimes in the bank.

5. **Strategy**
   Number of twenty-dollar bills: x
   Number of five-dollar bills: $68 - x$

Bill	Number	Value	Total Value
20-dollar	x	20	$20x$
5-dollar	$68 - x$	5	$5(68 - x)$

   The sum of the total values of each type of bill equals the total value of all the bills (730 dollars).
   $20x + 5(68 - x) = 730$

   Solution
   $20x + 5(68 - x) = 730$
   $20x + 340 - 5x = 730$
   $15x + 340 = 730$
   $15x = 390$
   $x = 26$
   The cashier has 26 twenty-dollar bills.

7. **Strategy**
   Number of 20¢ stamps: x
   Number of 15¢ stamps: $3x - 8$

Stamp	Number	Value	Total Value
20¢	x	20	$20x$
15¢	$3x - 8$	15	$15(3x - 8)$

   The sum of the total values of each type of stamp equals the total value of all the stamps (400 cents).
   $20x + 15(3x - 8) = 400$

   Solution
   $20x + 15(3x - 8) = 400$
   $20x + 45x - 120 = 400$
   $65x - 120 = 400$
   $65x = 520$
   $x = 8$
   $3x - 8 = 3(8) - 8 = 24 - 8 = 16$
   There are eight 20¢ stamps and sixteen 15¢ stamps.

9. **Strategy**
   Number of quarters: x
   Number of dimes: $4x$
   Number of nickels: $25 - 5x$

Coin	Number	Value	Total Value
Quarter	x	25	$25x$
Dime	$4x$	10	$10(4x)$
Nickel	$25 - 5x$	5	$5(25 - 5x)$

   The sum of the total values of each type of coin equals the total value of all the coins (205 cents).
   $25x + 10(4x) + 5(25 - 5x) = 205$

   Solution
   $25x + 10(4x) + 5(25 - 5x) = 205$
   $25x + 40x + 125 - 25x = 205$
   $40x + 125 = 205$
   $40x = 80$
   $x = 2$
   $4x = 4 \cdot 2 = 8$
   There are 8 dimes in the bank.

11. **Strategy**
    Number of 3¢ stamps: x
    Number of 8¢ stamps: $2x - 3$
    Number of 13¢ stamps: $2(2x - 3)$

Stamp	Number	Value	Total Value
3¢	x	3	$3x$
8¢	$2x - 3$	8	$8(2x - 3)$
13¢	$2(2x - 3)$	13	$13(2)(2x - 3)$

    The sum of the total values of each type of stamp equals the total value of all the stamps (253 cents).
    $3x + 8(2x - 3) + 26(2x - 3) = 253$

    Solution
    $3x + 8(2x - 3) + 26(2x - 3) = 253$
    $3x + 16x - 24 + 52x - 78 = 253$
    $71x - 102 = 253$
    $71x = 355$
    $x = 5$
    There are five 3¢ stamps in the collection.

13. **Strategy**
    Number of 18¢ stamps: x
    Number of 8¢ stamps: $2x$
    Number of 11¢ stamps: $x + 3$

Stamp	Number	Value	Total Value
18¢	x	18	$18x$
8¢	$2x$	8	$8(2x)$
11¢	$x+3$	11	$11(x+3)$

    The sum of the total values of each type of stamp equals the total value of all the stamps (348 cents).
    $18x + 8(2x) + 11(x + 3) = 348$

    Solution
    $$18x + 8(2x) + 11(x+3) = 348$$
    $$18x + 16x + 11x + 33 = 348$$
    $$45x + 33 = 348$$
    $$45x = 315$$
    $$x = 7$$
    There are seven 18¢ stamps in the collection.

Objective 2 Exercises

17. **Strategy**
    The smaller integer: n
    The larger integer: $10 - n$
    Three times the larger integer is three less than eight times the smaller integer.
    $3(10 - n) = 8n - 3$

    Solution
    $$3(10 - n) = 8n - 3$$
    $$30 - 3n = 8n - 3$$
    $$-11n = -33$$
    $$n = 3$$
    $10 - n = 10 - 3 = 7$
    The integers are 3 and 7.

19. **Strategy**
    The larger integer: n
    The smaller integer: $n - 8$
    The sum of the two integers is fifty.
    $n + (n - 8) = 50$

    Solution
    $$n + (n - 8) = 50$$
    $$2n - 8 = 50$$
    $$2n = 58$$
    $$n = 29$$
    $n - 8 = 29 - 8 = 21$
    The two integers are 21 and 29.

21. **Strategy**
    The first number: n
    The second number: $2n + 2$
    The third number: $3n - 5$
    The sum of the three numbers is 123.

    Solution
    $$n + (2n + 2) + (3n - 5) = 123$$
    $$6n - 3 = 123$$
    $$6n = 126$$
    $$n = 21$$
    $2n + 2 = 2(21) + 2 = 42 + 2 = 44$
    $3n - 5 = 3(21) - 5 = 63 - 5 = 58$
    The numbers are 21, 44, and 58.

23. **Strategy**
    The first integer: n
    The second consecutive integer: $n + 1$
    The third consecutive integer: $n + 2$
    The sum of the integers is –57.
    $n + (n + 1) + (n + 2) = -57$

    Solution
    $$n + (n + 1) + (n + 2) = -57$$
    $$3n + 3 = -57$$
    $$3n = -60$$
    $$n = -20$$
    $n + 1 = -20 + 1 = -19$
    $n + 2 = -20 + 2 = -18$
    The integers are –20, –19, and –18.

25. **Strategy**
    The first odd integer: n
    The second consecutive odd integer: $n + 2$
    The third consecutive odd integer: $n + 4$
    Five times the smallest of the three integers is ten more than twice the largest.
    $5n = 2(n + 4) + 10$

    Solution
    $$5n = 2(n + 4) + 10$$
    $$5n = 2n + 8 + 10$$
    $$5n = 2n + 18$$
    $$3n = 18$$
    $$n = 6$$
    Since 6 is not an odd integer, there is no solution.

27. **Strategy**
    The first odd integer: n
    The second consecutive odd integer: $n + 2$
    The third consecutive odd integer: $n + 4$
    Three times the middle integer is seven more than the sum of the first and third integers.
    $3(n + 2) = [n + (n + 4)] + 7$

    Solution
    $$3(n + 2) = [n + (n + 4)] + 7$$
    $$3n + 6 = 2n + 11$$
    $$n = 5$$
    $n + 2 = 5 + 2 = 7$
    $n + 4 = 5 + 4 = 9$
    The odd integers are 5, 7, and 9.

Chapter 2: First-Degree Equations and Inequalities

Applying Concepts 2.2

29. Strategy
    Number of nickels: x
    Number of dimes: $2x - 2$
    The number of nickels plus the number of dimes is the total number of coins in the bank (52).

    Solution
    $$x + (2x - 2) = 52$$
    $$3x - 2 = 52$$
    $$3x = 54$$
    $$\frac{3x}{3} = \frac{54}{3}$$
    $$x = 18$$
    $2x - 2 = 2(18) - 2 = 36 - 2 = 34$
    There are 18 nickels in the bank and 34 dimes in the bank.
    The value of the coins in the bank is equal to the value of the dimes $(\$.10)(34)$ plus the value of the nickels $(\$.05)(18)$.
    $\$.10(34) + \$.05(18) = \$3.40 + \$.90 = \$4.30$
    The total value of the coins in the bank is $4.30.

31. Strategy
    Number of 5¢ stamps: x
    Number of 3¢ stamps: $x + 6$
    Number of 7¢ stamps: $(x + 6) + 2 = x + 8$

Stamp	Number	Value	Total Value
5¢	x	5	$5x$
3¢	$x + 6$	3	$3(x + 6)$
7¢	$x + 8$	7	$7(x + 8)$

    The sum of the total value of each type of stamp equals the total value of all the stamps (194 cents.)
    $5x + 3(x + 6) + 7(x + 8) = 194$

    Solution
    $$5x + 3(x + 6) + 7(x + 8) = 194$$
    $$5x + 3x + 18 + 7x + 56 = 194$$
    $$15x + 74 = 194$$
    $$15x = 120$$
    $$x = 8$$
    $x + 6 = 8 + 6 = 14$
    There are fourteen 3¢ stamps on the collection.

33. Strategy
    First odd integer: n
    Second consecutive odd integer: $n + 2$
    Third consecutive odd integer: $n + 4$
    Fourth consecutive odd integer: $n + 6$
    The sum of the four integers is –64.
    $n + (n + 2) + (n + 4) + (n + 6) = -64$

    Solution
    $$n + (n + 2) + (n + 4) + (n + 6) = -64$$
    $$4n + 12 = -64$$
    $$4n = -76$$
    $$n = -19$$
    $n + 6 = -19 + 6 = -13$
    The smallest of the four integers is –19.
    The largest of the four integers is –13.
    The sum of the smallest and largest integers is $-19 + (-13)$ or –32.

35. Strategy
    Units digit: x
    Tens digit: $x - 1$
    Hundreds digit: $6 - (x + x - 1)$
    The value of the number is 12 more than 100 times the hundreds digit.

    Solution
    $$x + 10(x - 1) + 100[6 - (x + x - 1)] = 100[6 - (x + x - 1)] + 12$$
    $$x + 10x - 10 + 100[6 - 2x + 1)] = 100[6 - 2x + 1] + 12$$
    $$11x - 10 + 600 - 200x + 100 = 600 - 200x + 100 + 12$$
    $$-189x + 690 = -200x + 712$$
    $$11x = 22$$
    $$x = 2$$
    $x - 1 = 2 - 1 = 1$
    $6 - (x + x - 1) = 6 - (2 + 2 - 1) = 6 - 3 = 3$
    The number is 312.

Section 2.3

Concept Review 2.3

1. Sometimes true
   This is true only when the same amounts of gold at each price are mixed.

3. Never true
   It takes a time of $t - \frac{1}{2}$ for the cyclist to overtake the runner.

5. Never true
   The car travels 3 h at a rate of 40 mph and 2 h at a rate of 60 mph for a total travel time of 5 h. The total distance traveled is 240 mi. The average speed is $240 \div 5 = 48$. The average speed is 48 mph.

Objective 1 Exercises

3. Strategy
   Cost of mixture: x

	Amount	Cost	Value
Snow peas	20	1.99	39.80
Petite onions	14	1.19	16.66
Mixture	34	x	$34x$

   The sum of the values before mixing equals the value after mixing.
   $39.80 + 16.66 = 34x$

   Solution
   $56.46 = 34x$
   $1.66 = x$
   The cost per pound of the mixture is $1.66.

5. Strategy
   Number of adult tickets: x
   Number of child tickets: $460 - x$

	Amount	Cost	Value
Adult tickets	x	5.00	$5x$
Child tickets	$460-x$	2.00	$2(460-x)$

   The sum of the values of each type of ticket sold equals the total value of all the tickets sold (1880 dollars).
   $5x + 2(460 - x) = 1880$

   Solution
   $5x + 2(460 - x) = 1880$
   $5x + 920 - 2x = 1880$
   $3x + 920 = 1880$
   $3x = 960$
   $x = 320$
   There were 320 adult tickets sold.

7. Strategy
   Liters of imitation maple syrup: x

	Amount	Cost	Value
Imitation Syrup	x	4.00	$4x$
Maple Syrup	50	9.50	9.50(50)
Mixture	$50+x$	5.00	$5(50+x)$

   The sum of the values before mixing equals the value after mixing.
   $4x + 9.50(50) = 5(50 + x)$

   Solution
   $4x + 9.50(50) = 5(50 + x)$
   $4x + 475 = 250 + 5x$
   $-x + 475 = 250$
   $-x = -225$
   $x = 225$
   The mixture must contain 225 L of imitation maple syrup.

9. Strategy
   Number of pounds of nuts used x
   Number of pounds of pretzels used $20 - x$

	Amount	Cost	Value
Nuts	x	3.99	$3.99x$
Pretzels	$20-x$	1.29	$1.29(20-x)$
Mixture	20	2.37	47.40

   The sum of the values before mixing equals the value after mixing.
   $3.99x + 1.29(20 - x) = 47.40$
   $3.99x + 25.80 - 1.29x = 47.40$
   $2.70x + 25.80 = 47.40$
   $2.70x = 21.60$
   $x = 8$
   The mixture must contain 8 pounds of nuts.

11. Strategy
    Cost per pound of mixture: x

	Amount	Cost	Value
$6.00 tea	30	6.00	6.00(30)
$3.20 tea	70	3.20	3.20(70)
Mixture	100	x	$100x$

    The sum of the values before mixing equals the value after mixing.
    $6.00(30) + 3.20(70) = 100x$

    Solution
    $6.00(30) + 3.20(70) = 100x$
    $180 + 224 = 100x$
    $404 = 100x$
    $4.04 = x$
    The cost of the mixture is $4.04 per pound.

13. Strategy
Gallons of cranberry juice: x

	Amount	Cost	Value
Cranberry	x	4.20	$4.20x$
Apple	50	2.10	$50(2.10)$
Mixture	$50 + x$	3.00	$3(50 + x)$

The sum of the values before mixing equals the value after mixing.
$4.20x + 50(2.10) = 3(50 + x)$

Solution
$$4.20x + 50(2.10) = 3(50 + x)$$
$$4.20x + 105 = 150 + 3x$$
$$1.20x + 105 = 150$$
$$1.20x = 45$$
$$x = 37.5$$
The mixture must contain 37.5 gal of cranberry juice.

Objective 2 Exercises

15. Strategy
t = time for bicyclist
$t + 0.5$ = time for in-line skater

	Rate	Time	Distance
Bicyclist	18	t	$18t$
In-line skater	10	$t + 0.5$	$10(t + 0.5)$

The bicyclist and in-line skater travel the same distance.
$18t = 10(t + 0.5)$
$18t = 10t + 5$
$8t = 5$
$t = 0.625$
$d = 18t = 11.25$
The bicyclist will overtake the in-line skater after 11.25 miles.

17. Strategy
Rate of the first plane: r
Rate of the second plane: $r + 80$

	Rate	Time	Distance
1st plane	r	1.5	$1.5r$
2nd plane	$r + 80$	1.5	$1.5(r + 80)$

The total distance traveled by the two planes is 1380 mi.
$1.5r + 1.5(r + 80) = 1380$

Solution
$$1.5r + 1.5(r + 80) = 1380$$
$$1.5r + 1.5r + 120 = 1380$$
$$3r + 120 = 1380$$
$$3r = 1260$$
$$r = 420$$
$r + 80 = 420 + 80 = 500$
The speed of the first plane is 420 mph.
The speed of the second plane is 500 mph.

19. Strategy
Time to the island: t
Time returning from the island: $6 - t$

	Rate	Time	Distance
Going	18	t	$18t$
Returning	12	$6 - t$	$12(6 - t)$

The distance to the island is the same as the distance returning.
$18t = 12(6 - t)$

Solution
$$18t = 12(6 - t)$$
$$18t = 72 - 12t$$
$$30t = 72$$
$$t = 2.4$$
$d = rt = 18(2.4) = 43.2$
The distance to the island is 43.2 mi.

21. Strategy
Rate of the second plane: r
Rate of the first plane: $r - 50$

	Rate	Time	Distance
2nd plane	r	2.5	$2.5r$
1st plane	$r - 50$	2.5	$2.5(r - 50)$

The total distance traveled by the two planes is 1400 mi.
$2.5r + 2.5(r - 50) = 1400$

Solution
$$\begin{aligned} 2.5r + 2.5(r - 50) &= 1400 \\ 2.5r + 2.5r - 125 &= 1400 \\ 5r - 125 &= 1400 \\ 5r &= 1525 \\ r &= 305 \\ r - 50 = 305 - 50 &= 255 \end{aligned}$$
The rate of the first plane is 255 mph.
The rate of the second plane is 305 mph.

23. Strategy
Time to the repair shop: t
Time walking home: $1 - t$

	Rate	Time	Distance
To repair shop	14	t	$14t$
Walking home	3.5	$1-t$	$3.5(1-t)$

The distance to the repair shop is the same as the distance walking home.
$14t = 3.5(1 - t)$

Solution
$$\begin{aligned} 14t &= 3.5(1 - t) \\ 14t &= 3.5 - 3.5t \\ 17.5t &= 3.5 \\ t &= 0.2 \end{aligned}$$
$d = rt = 14(0.2) = 2.8$
The distance between the student's home and the bicycle shop is 2.8 mi.

25. Strategy
Time Washington to Pittsburgh train travels: t
Time Pittsburgh to Washington train travels: $t - 1$

	r	t	d
Washington to Pittsburgh	60	t	$60t$
Pittsburgh to Washington	40	$t - 1$	$40(t - 1)$

The trains together cover the distance from Washington to Pittsburgh, 260 miles.
$60t + 40(t - 1) = 260$

Solution
$$\begin{aligned} 60t + 40t - 40 &= 260 \\ 100t - 40 &= 260 \\ 100t &= 300 \\ t &= 3 \end{aligned}$$
$t - 1 = 2$
The two trains will pass each other after 2 hours.

Applying Concepts 2.3

27. The time it will take for the jogger to reach the end of the parade is the time it takes to run 2 mi at an apparent rate of 9 mph (the rate of the parade + the rate of the jogger.)
$d = rt$
$2 = 9t$
$t = \dfrac{2}{9}$

It will take the jogger $\dfrac{2}{9}$ h to reach the end of the parade.

24   *Chapter 2: First-Degree Equations and Inequalities*

29. The distance between them 2 min before impact is equal to the sum of the distances each one can travel during 2 min.

$2 \text{ minutes} \cdot \dfrac{1 \text{ hour}}{60 \text{ minutes}} = 0.03\overline{3}$ hour

Distance between cars = rate of first car $\cdot$ $0.03\overline{3}$ + rate of second car $\cdot$ $0.03\overline{3}$

Distance between cars = $40 \cdot 0.03\overline{3} + 60 \cdot 0.03\overline{3} = 3.3\overline{3}$

The cars are $3.3\overline{3}$ $\left(\text{or } 3\dfrac{1}{3}\right)$ miles apart 2 min before impact.

31. Rate during the second mile: x

	Rate	Distance	Time
1st mile	30	1	$\dfrac{1}{30}$
2nd mile	x	1	$\dfrac{1}{x}$
Both miles	60	2	$\dfrac{2}{60} = \dfrac{1}{30}$

The time traveled during the first mile plus the time traveled during the second mile is equal to the total time traveled during both miles.

$\dfrac{1}{30} + \dfrac{1}{x} = \dfrac{1}{30}$

Solution

$\dfrac{1}{30} + \dfrac{1}{x} = \dfrac{1}{30}$

$\dfrac{1}{x} = 0$

$x\left(\dfrac{1}{x}\right) = 0 \cdot x$

$1 = 0$

There is no solution to the equation. No, it is not possible to increase the speed enough.

Section 2.4

Concept Review 2.4

1. Never true
   The amount invested in the other account is $10,000 - x$.

3. Never true
   A mixture can never have a greater concentration of an ingredient than the concentrations of both substances going into the mixture.

Objective 1 Exercises

3. Strategy
Amount invested at 6.75%: x
Amount invested at 7.25%: $40,000 - x$

	Principal	Rate	Interest
Amount at 6.75%	x	0.0675	$0.0675x$
Amount at 7.25%	$40,000 - x$	0.0725	$0.0725(40,000 - x)$

The sum of the interest earned by the two investments equals the total annual interest earned ($2825)
$0.0675x + 0.0725(40,000 - x) = 2825$

Solution
$0.0675x + 0.0725(40,000 - x) = 2825$
$0.0675x + 2900 - 0.0725x = 2825$
$-0.005x = -75$
$x = 15,000$

The amount invested in the certificate of deposit is $15,000.

5. Strategy
Amount invested at 10.5%: x

	Principal	Rate	Interest
Amount at 8.4%	5000	0.084	$0.084(5000)$
Amount at 10.5%	x	0.105	$0.105x$
Amount at 9%	$5000 + x$	0.09	$0.09(5000 + x)$

The sum of the interest earned by the two investments equals the interest earned by the total investment.
$0.084(5000) + 0.105x = 0.09(5000 + x)$

Solution
$0.084(5000) + 0.105x = 0.09(5000 + x)$
$420 + 0.105x = 450 + 0.09x$
$0.015x = 30$
$x = 2000$

$2000 more must be invested at 10.5%.

7. Strategy
Amount invested at 8.5%: x
Amount invested at 6.4%: $8000 - x$

	Principal	Rate	Interest
Amount at 8.5%	x	0.085	$0.085x$
Amount at 6.4%	$8000 - x$	0.064	$0.064(8000 - x)$

The sum of the interest earned by the two investments equals the total annual interest earned ($575).
$0.085x + 0.064(8000 - x) = 575$

Solution
$0.085x + 0.064(8000 - x) = 575$
$0.085x + 512 - 0.064x = 575$
$0.021x = 63$
$x = 3000$
$8000 - x = 8000 - 3000 = 5000$

The amount invested at 8.5% is $3000.
The amount invested at 6.4% is $5000.

9. Strategy
Amount of additional money at 10%: x

	Principal	Rate	Interest
Amount at 5.5%	6000	0.055	330
Amount at 10%	x	0.10	$0.10x$
Combined Investment	$6000 + x$	0.07	$0.07(6000 + x)$

The interest from the combined investment is the sum of the interests from each investment.
$330 + 0.10x = 0.07(6000 + x)$
$330 + 0.10x = 420 + 0.07x$
$0.03x = 90$
$x = 3000$
The additional amount that should be invested at 10% is $3000.

11. Strategy
Amount invested at 4.2%: x
Amount invested at 6%: $13{,}600 - x$

	Principal	Rate	Interest
Amount 4.2%	x	0.042	$0.042x$
Amount at 6%	$13{,}600 - x$	0.006	$0.006(13{,}600 - x)$

The interest earned on one investment is equal to the interest earned on the other investment.
$0.042x = 0.06(13{,}600 - x)$

Solution
$0.042x = 0.06(13{,}600 - x)$
$0.042x = 816 - 0.06x$
$0.102x = 816$
$x = 8000$
$13{,}600 - x = 13{,}600 - 8000 = 5600$
The amount that should be invested at 4.2% is $8000.
the amount that should be invested at 6% is $5600.

Objective 2 Exercises

15. Strategy
Percent concentration of resulting alloy: x

	Amount	Percent	Quantity
60% alloy	15	0.60	9
20% alloy	45	0.20	9
Mixture	60	x	$60x$

The sum of the quantities before mixing is equal to the quantity after mixing.
$9 + 9 = 60x$

Solution
$9 + 9 = 60x$
$18 = 60x$
$0.30 = x$
The resulting alloy is 30% silver.

17. Strategy
Percent concentration of the resulting alloy: x

	Amount	Percent	Quantity
70%	25	0.70	$0.70(25)$
15%	50	0.15	$0.15(50)$
Mixture	75	x	$75x$

The sum of the quantities before mixing is equal to the quantity after mixing.
$0.70(25) + 0.15(50) = 75x$

Solution
$0.70(25) + 0.15(50) = 75x$
$17.5 + 7.5 = 75x$
$25 = 75x$
$0.3\overline{3} = x$
the resulting alloy is $33\frac{1}{3}\%$ silver.

19. Strategy
Pounds of 12% aluminum alloy: x

	Amount	Percent	Quantity
12%	x	0.12	$0.12x$
30%	400	0.30	$0.30(400)$
20%	$400 + x$	0.20	$0.20(400 + x)$

The sum of the quantities before mixing is equal to the quantity after mixing.
$0.12x + 0.30(400) = 0.20(400 + x)$

Solution
$0.12x + 0.30(400) = 0.20(400 + x)$
$0.12x + 120 = 80 + 0.20x$
$-0.08x + 120 = 80$
$-0.08x = -40$
$x = 500$
500 lb of the 12% aluminum alloy must be used.

21. Strategy
Liters of 65% solution: x
Liters of 15% solution: $50 - x$

	Amount	Percent	Quantity
65% solution	x	65%	$0.65x$
15% solution	$50 - x$	15%	$0.15(50 - x)$
Mixture	50	40%	$0.40(50)$

The sum of the quantities before mixing is equal to the quantity after mixing.
$0.65x + 0.15(50 - x) = 0.40(50)$

Solution
$0.65x + 0.15(50 - x) = 0.40(50)$
$0.65x + 7.5 - 0.15x = 20$
$0.5x = 12.5$
$x = 25$
25 L of 65% disinfectant solution and 25 L of 15% disinfectant solution were used.

23. Strategy
Number of quarts of water: x

	Amount	Percent	Quantity
Water	x	0	0
80% antifreeze	5	0.80	4
50% antifreeze	$5 + x$	0.50	$0.50(5 + x)$

The sum of the quantities before mixing is equal to the quantity after mixing.
$0 + 4 = 0.50(5 + x)$

Solution
$4 = 2.5 + 0.50x$
$1.5 = 0.5x$
$x = 3$
3 quarts of water should be added.

25. Strategy
Ounces of water to be added: x
Ounces of 5% solution: $60 + x$

	Amount	Percent	Quantity
Water	x	0	0
7.5% solution	60	0.075	4.5
5% solution	$60 + x$	0.05	$0.05(60 + x)$

The sum of the quantities before mixing is equal to the quantity after mixing.
$0 + 4.5 = 0.05(60 + x)$

Solution
$0 + 4.5 = 0.05(60 + x)$
$4.5 = 3 + 0.05x$
$1.5 = 0.05x$
$x = 30$
30 oz of water should be added.

27. Strategy
Percent concentration of result: x

	Amount	Percent	Quantity
5% fruit juice	12	0.05	$0.05(12)$
Water	2	0	0
Result	10	x	$10x$

The sum of the quantities before mixing is equal to the quantity after mixing.
$0.05(12) + 0 = 10x$

Solution
$0.05(12) + 0 = 10x$
$0.60 = 10x$
$x = 0.06$
The result is 6% fruit juice.

29. Strategy
Percent concentration of the resulting alloy: x

	Amount	Percent	Quantity
54%	80	0.54	$0.54(80)$
22%	200	0.22	$0.22(200)$
Mixture	280	x	$280x$

The sum of the quantities before mixing is equal to the quantity after mixing.
$0.54(80) + 0.22(200) = 280x$

Solution
$0.54(80) + 0.22(200) = 280x$
$43.2 + 44 = 280x$
$87.2 = 280x$
$0.3114 \approx x$
The resulting alloy is about 31.1% copper.

28 *Chapter 2: First-Degree Equations and Inequalities*

Applying Concepts 2.4

31. Strategy
Total amount invested: x
Amount invested at 9%: $0.25x$
Amount invested at 8%: $0.30x$
Amount invested at 9.5%: $0.45x$

	Amount	Percent	Quantity
Amount at 9%	$0.25x$	0.09	$0.09(0.25x)$
Amount at 8%	$0.30x$	0.08	$0.08(0.30x)$
Amount at 9.5%	$0.45x$	0.095	$0.095(0.45x)$

The total annual interest earned is $1785.

Solution
$$0.09(0.25x) + 0.08(0.3x) + 0.095(0.45x) = 1785$$
$$0.0225x + 0.024x + 0.04275x = 1785$$
$$0.08925x = 1785$$
$$x = 20,000$$

$0.25x = 0.25(20,000) = 5000$
$0.3x = 0.3(20,000) = 6000$
$0.45x = 0.45(20,000) = 9000$
The amount invested at 9% was $5000.
The amount invested at 8% was $6000.
The amount invested at 9.5% was $9000.

33. Strategy
Cost per pound of mixture: x

	Amount	Cost	Value
$5.50 tea	50	550	550(50)
$4.40 tea	75	440	440(75)
Mixture	125	x	$125x$

The sum of the quantities before mixing is equal to the quantity after mixing.
$550(50) + 440(75) = 125x$

Solution
$$550(50) + 440(75) = 125x$$
$$27,500 + 33,000 = 125x$$
$$60,500 = 125x$$
$$484 = x$$
The tea mixture would cost $4.84 per pound.

35. Strategy
Grams of water: x

	Amount	Percent	Quantity
Pure water	x	0	0
Pure acid	20	1.00	1.00(20)
25% acid	20 + x	0.25	0.25(20 + x)

The sum of the quantities before mixing is equal to the quantity after mixing.
$0 + 1.00(20) = 0.25(20 + x)$

Solution
$0 + 1.00(20) = 0.25(20 + x)$
$20 = 5 + 0.25x$
$15 = 0.25x$
$60 = x$

60 g of pure water were in the beaker.

37. a. The percent increases for '91 and '96 are very close but the percent increase for '96 is slightly higher.

  b. Since there was a positive percent increase every year, the costs were highest in 1997.

Section 2.5

Concept Review 2.5

1. Always true

3. Sometimes true
The rule states that when dividing an inequality by a negative integer, we must reverse the inequality.

5. Sometimes true
This is not true for $a = 0$.

Objective 1 Exercises

3. $x - 3 < 2$
$x < 5$
$\{x | x < 5\}$

5. $4x \leq 8$
$\dfrac{4x}{4} \leq \dfrac{8}{4}$
$x \leq 2$
$\{x | x \leq 2\}$

7. $-2x > 8$
$\dfrac{-2x}{-2} < \dfrac{8}{-2}$
$x < -4$
$\{x | x < -4\}$

9. $3x - 1 > 2x + 2$
$x - 1 > 2$
$x > 3$
$\{x | x > 3\}$

11. $2x - 1 > 7$
$2x > 8$
$\dfrac{2x}{2} > \dfrac{8}{2}$
$x > 4$
$\{x | x > 4\}$

13. $6x + 3 > 4x - 1$
$2x + 3 > -1$
$2x > -4$
$\dfrac{2x}{2} > \dfrac{-4}{2}$
$x > -2$
$\{x | x > -2\}$

15. $8x + 1 \geq 2x + 13$
$6x + 1 \geq 13$
$6x \geq 12$
$\dfrac{6x}{6} \geq \dfrac{12}{6}$
$x \geq 2$
$\{x | x \geq 2\}$

17. $7 - 2x \geq 1$
$-2x \geq -6$
$\dfrac{-2x}{-2} \leq \dfrac{-6}{-2}$
$x \leq 3$
$\{x | x \leq 3\}$

19. $4x - 2 < x - 11$
$3x - 2 < -11$
$3x < -9$
$\dfrac{3x}{3} < \dfrac{-9}{3}$
$x < -3$
$\{x | x < -3\}$

21. $x + 7 \geq 4x - 8$
$-3x + 7 \geq -8$
$-3x \geq -15$
$\dfrac{-3x}{-3} \leq \dfrac{-15}{-3}$
$x \leq 5$
$(-\infty, 5]$

23. $6 - 2(x - 4) \leq 2x + 10$
$6 - 2x + 8 \leq 2x + 10$
$14 - 2x \leq 2x + 10$
$14 - 4x \leq 10$
$-4x \leq -4$
$\dfrac{-4x}{-4} \geq \dfrac{-4}{-4}$
$x \geq 1$
$[1, \infty)$

25. $2(1 - 3x) - 4 > 10 + 3(1 - x)$
$2 - 6x - 4 > 10 + 3 - 3x$
$-6x - 2 > 13 - 3x$
$-3x - 2 > 13$
$-3x > 15$
$\dfrac{-3x}{-3} < \dfrac{15}{-3}$
$x < -5$
$(-\infty, -5)$

27. $\dfrac{3}{5}x - 2 < \dfrac{3}{10} - x$
$10\left(\dfrac{3}{5}x - 2\right) < 10\left(\dfrac{3}{10} - x\right)$
$6x - 20 < 3 - 10x$
$16x - 20 < 3$
$16x < 23$
$\dfrac{16x}{16} < \dfrac{23}{16}$
$x < \dfrac{23}{16}$
$\left(-\infty, \dfrac{23}{16}\right)$

29. $\dfrac{1}{3}x - \dfrac{3}{2} \geq \dfrac{7}{6} - \dfrac{2}{3}x$
$6\left(\dfrac{1}{3}x - \dfrac{3}{2}\right) \geq 6\left(\dfrac{7}{6} - \dfrac{2}{3}x\right)$
$2x - 9 \geq 7 - 4x$
$6x - 9 \geq 7$
$6x \geq 16$
$\dfrac{6x}{6} \geq \dfrac{16}{6}$
$x \geq \dfrac{8}{3}$
$\left[\dfrac{8}{3}, \infty\right)$

31. $\dfrac{1}{2}x - \dfrac{3}{4} > \dfrac{7}{4}x - 2$
$4\left(\dfrac{1}{2}x - \dfrac{3}{4}\right) > 4\left(\dfrac{7}{4}x - 2\right)$
$2x - 3 > 7x - 8$
$-5x - 3 > -8$
$-5x > -5$
$\dfrac{-5x}{-5} < \dfrac{-5}{-5}$
$x < 1$
$(-\infty, 1)$

33. $2 - 2(7 - 2x) < 3(3 - x)$
$2 - 14 + 4x < 9 - 3x$
$-12 + 4x < 9 - 3x$
$-12 + 7x < 9$
$7x < 21$
$\dfrac{7x}{7} < \dfrac{21}{7}$
$x < 3$
$(-\infty, 3)$

Objective 2 Exercises

35. $3x < 6$ and $x + 2 > 1$
$x < 2 \qquad x > -1$
$\{x | x < 2\} \quad \{x | x > -1\}$
$\{x | x < 2\} \cap \{x | x > -1\} = (-1, 2)$

37. $x + 2 \geq 5$ or $3x \leq 3$
$x \geq 3 \qquad x \leq 1$
$\{x | x \geq 3\} \quad \{x | x \leq 1\}$
$\{x | x \geq 3\} \cup \{x | x \leq 1\} = (-\infty, 1] \cup [3, \infty)$

39. $-2x > -8$ and $-3x < 6$
$x < 4 \qquad x > -2$
$\{x | x < 4\} \quad \{x | x > -2\}$
$\{x | x < 4\} \cap \{x | x > -2\} = (-2, 4)$

41. $\dfrac{1}{3}x < -1$ or $2x > 0$
$x < -3 \qquad x > 0$
$\{x | x < -3\} \quad \{x | x > 0\}$
$\{x | x < -3\} \cup \{x | x > 0\} = (-\infty, -3) \cup (0, \infty)$

43. $x+4 \geq 5$ and $2x \geq 6$
    $x \geq 1 \qquad x \geq 3$
    $\{x|x \geq 1\} \qquad \{x|x \geq 3\}$
    $\{x|x \geq 1\} \cap \{x|x \geq 3\} = [3, \infty)$

45. $-5x > 10$ and $x+1 > 6$
    $x < -2 \qquad x > 5$
    $\{x|x < -2\} \qquad \{x|x > 5\}$
    $\{x|x < -2\} \cap \{x|x > 5\} = \emptyset$

47. $2x-3 > 1$ and $3x-1 < 2$
    $2x > 4 \qquad 3x < 3$
    $x > 2 \qquad x < 1$
    $\{x|x > 2\} \qquad \{x|x < 1\}$
    $\{x|x > 2\} \cap \{x|x < 1\} = \emptyset$

49. $3x+7 < 10$ or $2x-1 > 5$
    $3x < 3 \qquad 2x > 6$
    $x < 1 \qquad x > 3$
    $\{x|x < 1\} \qquad \{x|x > 3\}$
    $\{x|x < 1\} \cup \{x|x > 3\} = (-\infty, 1) \cup (3, \infty)$

51. $-5 < 3x+4 < 16$
    $-5-4 < 3x+4-4 < 16-4$
    $-9 < 3x < 12$
    $\frac{-9}{3} < \frac{3x}{3} < \frac{12}{3}$
    $-3 < x < 4$
    $\{x|-3 < x < 4\}$

53. $0 < 2x-6 < 4$
    $0+6 < 2x-6+6 < 4+6$
    $6 < 2x < 10$
    $\frac{6}{2} < \frac{2x}{2} < \frac{10}{2}$
    $3 < x < 5$
    $\{x|3 < x < 5\}$

55. $4x-1 > 11$ or $4x-1 \leq -11$
    $4x > 12 \qquad 4x \leq -10$
    $x > 3 \qquad x \leq -\frac{5}{2}$
    $\{x|x > 3\} \qquad \{x|x \leq -\frac{5}{2}\}$
    $\{x|x > 3\} \cup \{x|x \leq -\frac{5}{2}\}$
    $= \{x|x > 3 \text{ or } x \leq -\frac{5}{2}\}$

57. $2x+3 \geq 5$ and $3x-1 > 11$
    $2x \geq 2 \qquad 3x > 12$
    $x \geq 1 \qquad x > 4$
    $\{x|x \geq 1\} \qquad \{x|x > 4\}$
    $\{x|x \geq 1\} \cap \{x|x > 4\} = \{x|x > 4\}$

59. $9x-2 < 7$ and $3x-5 > 10$
    $9x < 9 \qquad 3x > 15$
    $x < 1 \qquad x > 5$
    $\{x|x < 1\} \qquad \{x|x > 5\}$
    $\{x|x < 1\} \cap \{x|x > 5\} = \emptyset$

61. $3x-11 < 4$ or $4x+9 \geq 1$
    $3x < 15 \qquad 4x \geq -8$
    $x < 5 \qquad x \geq -2$
    $\{x|x < 5\} \qquad \{x|x \geq -2\}$
    $\{x|x < 5\} \cup \{x|x \geq -2\}$
    $= \{x|x \text{ is a real number}\}$

63. $3-2x > 7$ and $5x+2 > -18$
    $-2x > 4 \qquad 5x > -20$
    $x < -2 \qquad x > -4$
    $\{x|x < -2\} \qquad \{x|x > -4\}$
    $\{x|x < -2\} \cap \{x|x > -4\} = \{x|-4 < x < -2\}$

65. $5-4x > 21$ or $7x-2 > 19$
    $-4x > 16 \qquad 7x > 21$
    $x < -4 \qquad x > 3$
    $\{x|x < -4\} \qquad \{x|x > 3\}$
    $\{x|x < -4\} \cup \{x|x > 3\} = \{x|x < -4 \text{ or } x > 3\}$

67. $3-7x \leq 31$ and $5-4x > 1$
    $-7x \leq 28 \qquad -4x > -4$
    $x \geq -4 \qquad x < 1$
    $\{x|x \geq -4\} \qquad \{x|x < 1\}$
    $\{x|x \geq -4\} \cap \{x|x < 1\} = \{x|-4 \leq x < 1\}$

69. $\frac{2}{3}x-4 > 5$ or $x+\frac{1}{2} < 3$
    $\frac{2}{3}x > 9 \qquad x < \frac{5}{2}$
    $x > \frac{27}{2}$
    $\{x|x > \frac{27}{2}\} \qquad \{x|x < \frac{5}{2}\}$
    $\{x|x > \frac{27}{2}\} \cup \{x|x < \frac{5}{2}\}$
    $= \{x|x > \frac{27}{2} \text{ or } x < \frac{5}{2}\}$

71. $-\frac{3}{8} \leq 1-\frac{1}{4}x \leq \frac{7}{2}$
    $-3 \leq 8-2x \leq 28$
    $-11 \leq -2x \leq 20$
    $\frac{11}{2} \geq x \geq -10$
    $\{x|-10 \leq x \leq \frac{11}{2}\}$

Objective 3 Exercises

73. Strategy
the unknown number: x
five times the difference between the number and two: $5(x-2)$
the quotient of two times the number and three: $2x \div 3$

Solution
five times the difference between a number and two > the quotient of two times the number and three

$$5(x-2) \geq \frac{2x}{3}$$
$$5x - 10 \geq \frac{2x}{3}$$
$$3(5x - 10) \geq 2x$$
$$15x - 30 \geq 2x$$
$$13x - 30 \geq 0$$
$$13x \geq 30$$
$$x \geq \frac{30}{13}$$
$$x \geq 2\frac{4}{13}$$

The smallest integer is 3.

75. Strategy
the width of the rectangle: x
the length of the rectangle: $4x + 2$
To find the maximum width, substitute the given values in the inequality $2L + 2W < 34$ and solve.

Solution
$$2L + 2W < 34$$
$$2(4x + 2) + 2x < 34$$
$$8x + 4 + 2x < 34$$
$$10x + 4 < 34$$
$$10x < 30$$
$$x < 3$$

The maximum width of the rectangle is 2 ft.

77. Strategy
To find the four consecutive integers, write and solve a compound inequality using x to represent the first integer.

Solution
Lower limit of the sum < sum < Upper limit of the sum
$$62 < x + (x+1) + (x+2) + (x+3) < 78$$
$$62 < 4x + 6 < 78$$
$$62 - 6 < 4x + 6 - 6 < 78 - 6$$
$$56 < 4x < 72$$
$$\frac{56}{4} < \frac{4x}{4} < \frac{72}{4}$$
$$14 < x < 18$$

The four integers are 15, 16, 17, and 18; or 16, 17, 18, and 19; or 17, 18, 19, and 20.

79. Strategy
First side of the triangle: $x + 1$
Second side of the triangle: x
Third side of the triangle: $x + 2$

Solution
The perimeter of the triangle is more than 15 in. and less than 25 in.
$$15 < P < 25$$
$$15 < (x+1) + x + (x+2) < 25$$
$$15 < 3x + 3 < 25$$
$$15 - 3 < 3x + 3 - 3 < 25 - 3$$
$$12 < 3x < 22$$
$$\frac{12}{3} < \frac{3x}{3} < \frac{22}{3}$$
$$4 < x < 7\frac{1}{3}$$

If $x = 5$: $5 + 1 = 6$; $5 + 2 = 7$;
$5 + 6 + 7 = 18 =$ perimeter
If $x = 6$: $6 + 1 = 7$; $6 + 2 = 8$;
$6 + 7 + 8 = 21 =$ perimeter
If $x = 7$: $7 + 1 = 8$; $7 + 2 = 9$;
$7 + 8 + 9 = 24 =$ perimeter
The lengths of the second side could be 5 in., 6 in., or 7 in.

81. Strategy
To find the number of minutes, write and solve an inequality using N to represent the number of minutes of cellular phone time.

Solution
cost of second option $\leq$ cost of first option
$$0.4N + 35 \leq 99$$
$$0.4N \leq 64$$
$$N \leq 160$$

A customer can use a cellular phone 160 min before the charges exceed the first option.

83. Strategy
To find the number of pages for which the AirTouch plan is less expensive, solve an inequality using N to represent the number of pages.

Cost of AirTouch < Cost of TopPage
$$6.95 + 0.10(x - 400) < 3.95 + 0.15(x - 400)$$
$$6.95 + 0.10x - 40 < 3.95 + 0.15x - 60$$
$$33.05 + 0.10x < -56.05 + 0.15x$$
$$23 < 0.05x$$
$$460 < x$$

AirTouch is less expensive for jobs more than 460 pages.

85. Strategy
To find the number of minutes for which a call will be cheaper to pay with coins, solve an inequality using N to represent the number of minutes.
Coins < Calling card

Solution
$(0.70) + 0.15(N-3) < 0.35 + 0.196 + 0.126(N-1)$
$0.70 + 0.15N - 0.45 < 0.35 + 0.196 + 0.126N - 0.126$
$0.25 + 0.15N < 0.42 + 0.126N$
$0.024N < .17$
$N < 7.08$
Using coins will be cheaper for 7 minutes or less.

87. Strategy
To find the number of checks that have to be written for Glendale Federal to cost less than the competitor, solve an inequality using N to represent the number of checks.
Glendale account < Other account
$8 + 0.12(N-100) < 5 + 0.15(N-100)$
$8 + 0.12N - 12 < 5 + 0.15N - 15$
$0.12N - 4 < 0.15N - 10$
$-0.03N < -6$
$N > 200$
The Glendale Federal account will cost less for more than 200 checks.

89. Strategy
To find the range of miles that a car can travel, write and solve an inequality using N to represent the range of miles.

Solution
$22(19.5) < N < 27.5(19.5)$
$429 < N < 536.25$
The range of miles is between 429 mi and 536.25 mi.

Applying Concepts 2.5

91. $8x - 7 < 2x + 9$
$6x - 7 < 9$
$6x < 16$
$x < \frac{16}{6}$
$x < \frac{8}{3}$
$\{1, 2\}$

93. $5 + 3(2 + x) > 8 + 4(x - 1)$
$5 + 6 + 3x > 8 + 4x - 4$
$11 + 3x > 4 + 4x$
$3x > -7 + 4x$
$-x > -7$
$x < 7$
$\{1, 2, 3, 4, 5, 6\}$

95. $-3x < 15$ and $x + 2 < 7$
$x > -5 \qquad x < 5$
$\{x | x > -5\} \cap \{x | x < 5\} = \{x | -5 < x < 5\}$
$\{1, 2, 3, 4\}$

97. $\qquad -4 \leq 3x + 8 < 16$
$-4 + (-8) \leq 3x + 8 + (-8) < 16 + (-8)$
$\qquad -12 \leq 3x < 8$
$\qquad \frac{-12}{3} \leq \frac{3x}{3} < \frac{8}{3}$
$\qquad -4 \leq x < \frac{8}{3}$
$\{1, 2\}$

99. Strategy
To find the temperature range in degrees Celsius, write and solve a compound inequality.

Solution
$77 < \frac{9}{5}C + 32 < 86$
$77 - 32 < \frac{9}{5}C + 32 - 32 < 86 - 32$
$45 < \frac{9}{5}C < 54$
$\frac{5}{9}(45) < \frac{5}{9}\left(\frac{9}{5}C\right) < \frac{5}{9}(54)$
$25 < C < 30$
The temperature is between 25°C and 30°C.

101. Strategy
To find the largest whole number of minutes the call could last, set up an inequality with N representing the number of minutes and $N-3$ representing the number of minutes after the first 3 minutes.

Solution
$$156 + 52(N-3) < 540$$
$$156 + 52N - 156 < 540$$
$$52N < 540$$
$$N < 10.38$$
The largest whole number of minutes the call could last is 10 min.

Section 2.6

Concept Review 2.6

1. Always true

3. Never true
   The absolute value of a number is always positive.

5. Never true
   $|x+b| < c$ is equivalent to $-c < x+b < c$.

Objective 1 Exercises

3. $|x| = 7$
   $x = 7 \qquad x = -7$
   The solutions are 7 and −7.

5. $|-t| = 3$
   $-t = 3 \qquad -t = -3$
   $t = -3 \qquad t = 3$
   The solutions are −3 and 3.

7. $|-t| = -3$
   There is no solution to this equation because the absolute value of a number must be non-negative.

9. $|x+2| = 3$
   $x+2 = 3 \qquad x+2 = -3$
   $x = 1 \qquad x = -5$
   The solutions are 1 and −5.

11. $|y-5| = 3$
    $y-5 = 3 \qquad y-5 = -3$
    $y = 8 \qquad y = 2$
    The solutions are 8 and 2.

13. $|a-2| = 0$
    $a-2 = 0$
    $a = 2$
    The solution is 2.

15. $|x-2| = -4$
    There is no solution to this equation because the absolute value of a number must be non-negative.

17. $|2x-5| = 4$
    $2x-5 = 4 \qquad 2x-5 = -4$
    $2x = 9 \qquad 2x = 1$
    $x = \dfrac{9}{2} \qquad x = \dfrac{1}{2}$
    The solutions are $\dfrac{9}{2}$ and $\dfrac{1}{2}$.

19. $|2-5x| = 2$
    $2-5x = 2 \qquad 2-5x = -2$
    $-5x = 0 \qquad -5x = -4$
    $x = 0 \qquad x = \dfrac{4}{5}$
    The solutions are 0 and $\dfrac{4}{5}$.

21. $|5x+5| = 0$
    $5x+5 = 0$
    $5x = -5$
    $x = -1$
    The solution is −1.

23. $|2x+5| = -2$
    There is no solution to this equation because the absolute value of a number must be non-negative.

25. $|x-9| - 3 = 2$
    $|x-9| = 5$
    $x-9 = 5 \qquad x-9 = -5$
    $x = 14 \qquad x = 4$
    The solutions are 14 and 4.

27. $|8-y| - 3 = 1$
    $|8-y| = 4$
    $8-y = 4 \qquad 8-y = -4$
    $-y = -4 \qquad -y = -12$
    $y = 4 \qquad y = 12$
    The solutions are 4 and 12.

29. $|4x-7| - 5 = -5$
    $|4x-7| = 0$
    $4x-7 = 0$
    $4x = 7$
    $x = \dfrac{7}{4}$
    The solution is $\dfrac{7}{4}$.

31. $|3x-2| + 1 = -1$
    $|3x-2| = -2$
    There is no solution to this equation because the absolute value of a number must be non-negative.

33. $|4b+3|-2=7$
$|4b+3|=9$

$4b+3=9 \qquad 4b+3=-9$
$4b=6 \qquad 4b=-12$
$b=\dfrac{3}{2} \qquad b=-3$

The solutions are $\dfrac{3}{2}$ and -3.

35. $|5x-2|+5=7$
$|5x-2|=2$

$5x-2=2 \qquad 5x-2=-2$
$5x=4 \qquad 5x=0$
$x=\dfrac{4}{5} \qquad x=0$

The solutions are $\dfrac{4}{5}$ and 0.

37. $2-|x-5|=4$
$-|x-5|=2$
$|x-5|=-2$
There is no solution to this equation because the absolute value of a number must be non-negative.

39. $|3x-4|+8=3$
$|3x-4|=-5$
There is no solution to this equation because the absolute value of a number must be non-negative.

41. $5+|2x+1|=8$
$|2x+1|=3$

$2x+1=3 \qquad 2x+1=-3$
$2x=2 \qquad 2x=-4$
$x=1 \qquad x=-2$

The solutions are 1 and -2.

43. $3-|5x+3|=3$
$-|5x+3|=0$
$|5x+3|=0$
$5x+3=0$
$5x=-3$
$x=-\dfrac{3}{5}$

The solution is $-\dfrac{3}{5}$.

Objective 2 Exercises

47. $|x|>3$
$x>3 \quad \text{or} \quad x<-3$
$\{x|x>3\} \qquad \{x|x<-3\}$
$\{x|x>3\} \cup \{x|x<-3\} = \{x|x>3 \text{ or } x<-3\}$

49. $|x+1|>2$
$x+1>2 \quad \text{or} \quad x+1<-2$
$x>1 \qquad x<-3$
$\{x|x>1\} \qquad \{x|x<-3\}$
$\{x|x>1\} \cup \{x|x<-3\} = \{x|x>1 \text{ or } x<-3\}$

51. $|x-5|\le 1$
$-1\le x-5\le 1$
$-1+5\le x-5+5\le 1+5$
$4\le x\le 6$
$\{x|4\le x\le 6\}$

53. $|2-x|\ge 3$
$2-x\le -3 \quad \text{or} \quad 2-x\ge 3$
$-x\le -5 \qquad -x\ge 1$
$x\ge 5 \qquad x\le -1$
$\{x|x\ge 5\} \qquad \{x|x\le -1\}$
$\{x|x\ge 5\} \cup \{x|x\le -1\} = \{x|x\ge 5 \text{ or } x\le -1\}$

55. $|2x+1|<5$
$-5<2x+1<5$
$-5-1<2x+1-1<5-1$
$-6<2x<4$
$\dfrac{-6}{2}<\dfrac{2x}{2}<\dfrac{4}{2}$
$-3<x<2$
$\{x|-3<x<2\}$

57. $|5x+2|>12$
$5x+2>12 \quad \text{or} \quad 5x+2<-12$
$5x>10 \qquad 5x<-14$
$x>2 \qquad x<-\dfrac{14}{5}$
$\{x|x>2\} \qquad \left\{x\middle|x<-\dfrac{14}{5}\right\}$
$\{x|x>2\} \cup \left\{x\middle|x<-\dfrac{14}{5}\right\} = \left\{x\middle|x>2 \text{ or } x<-\dfrac{14}{5}\right\}$

59. $|4x-3|\le 2$
The absolute value of a number must be non-negative. The solution set is the empty set, $\varnothing$.

61. $|2x+7|>-5$
$2x+7>-5 \quad \text{or} \quad 2x+7<5$
$2x>-12 \qquad 2x<-2$
$x>-6 \qquad x<-1$
$\{x|x>-6\} \qquad \{x|x<-1\}$
$\{x|x>-6\} \cup \{x|x<-1\} = \{x|x \text{ is a real number}\}$

63. $|4-3x| \geq 5$

$\quad 4-3x \geq 5 \quad$ or $\quad 4-3x \leq -5$
$\quad -3x \geq 1 \qquad\qquad -3x \leq -9$
$\quad x \leq -\dfrac{1}{3} \qquad\qquad x \geq 3$

$\left\{x \mid x \leq -\dfrac{1}{3}\right\} \quad \{x \mid x \geq 3\}$

$\left\{x \mid x \leq -\dfrac{1}{3}\right\} \cup \{x \mid x \geq 3\} = \left\{x \mid x \leq -\dfrac{1}{3} \text{ or } x \geq 3\right\}$

65. $|5-4x| \leq 13$

$\quad -13 \leq 5-4x \leq 13$
$\quad -13-5 \leq 5-5-4x \leq 13-5$
$\quad -18 \leq -4x \leq 8$
$\quad \dfrac{-18}{-4} \geq \dfrac{-4x}{-4} \geq \dfrac{8}{-4}$
$\quad \dfrac{9}{2} \geq x \geq -2$

$\left\{x \mid -2 \leq x \leq \dfrac{9}{2}\right\}$

67. $|6-3x| \leq 0$

$\quad 6-3x = 0$
$\quad -3x = -6$
$\quad x = 2$

$\{x \mid x = 2\}$

69. $|2-9x| > 20$

$\quad 2-9x > 20 \quad$ or $\quad 2-9x < -20$
$\quad -9x > 18 \qquad\qquad -9x < -22$
$\quad x < -2 \qquad\qquad x > \dfrac{22}{9}$

$\{x \mid x < -2\} \cup \left\{x \mid x > \dfrac{22}{9}\right\} = \left\{x \mid x < -2 \text{ or } x > \dfrac{22}{9}\right\}$

Objective 3 Exercises

71. Strategy
Let b represent the diameter of the bushing, T the tolerance, and d the lower and upper limits of the diameter. Solve the absolute value inequality $|d-b| \leq T$ for d.

Solution
$|d-b| \leq T$
$|d-1.75| \leq 0.008$
$\quad -0.008 \leq d-1.75 \leq 0.008$
$-0.008 + 1.75 \leq d-1.75+1.75 \leq 0.008+1.75$
$\quad 1.742 \leq d \leq 1.758$

The lower and upper limits of the diameter of the bushing are 1.742 in. and 1.758 in.

73. Strategy
Let p represent the prescribed amount of medication, T the tolerance, and m the lower and upper limits of the amount of medication. Solve the absolute value inequality $|m-p| \leq T$ for m.

Solution
$|m-p| \leq T$
$|m-2.5| \leq 0.2$
$\quad -0.2 \leq m-2.5 \leq 0.2$
$-0.2+2.5 \leq m-2.5+2.5 \leq 0.2+2.5$
$\quad 2.3 \leq m \leq 2.7$

The lower and upper limits of the amount of medicine to be given to the patient are 2.3 cc and 2.7 cc.

75. Strategy
Let v represent the prescribed number of volts, T the tolerance, and m the upper and lower limits of the amount of voltage. Solve the absolute value inequality $|m-v| \leq T$ for m.

Solution
$|m-v| \leq T$
$|m-220| \leq 25$
$\quad -25 \leq m-220 \leq 25$
$-25+220 \leq m-220+220 \leq 25+220$
$\quad 195 \leq m \leq 245$

The lower and upper limits of the amount of voltage on which the motor will run are 195 volts and 245 volts.

77. Strategy
Let r represent the diameter of the piston rod, T the tolerance, and L the lower and upper limits of the diameter. Solve the absolute value inequality $|L-r| \leq T$ for L.

Solution
$|L-r| \leq T$
$\left|L-3\dfrac{5}{16}\right| \leq \dfrac{1}{64}$

$\quad -\dfrac{1}{64} \leq L-3\dfrac{5}{16} \leq \dfrac{1}{64}$

$-\dfrac{1}{64}+3\dfrac{5}{16} \leq L-3\dfrac{5}{16}+3\dfrac{5}{16} \leq \dfrac{1}{64}$

$-\dfrac{1}{64}+3\dfrac{5}{16} \leq L-3\dfrac{5}{16}+3\dfrac{5}{16} \leq \dfrac{1}{64}+3\dfrac{5}{16}$

$\quad 3\dfrac{19}{64} \leq L \leq 3\dfrac{21}{64}$

The lower and upper limits of the length of the piston rod are $3\dfrac{19}{64}$ in. and $3\dfrac{21}{64}$ in.

79. Strategy
Let M represent the amount of ohms, T the tolerance, and r the given amount of the resistor. Find the tolerance and solve $|M - r| \leq T$ for M.

Solution
$T = (0.10)(15,000) = 1500$ ohms
$$|M - r| \leq T$$
$$|M - 15,000| \leq 1,500$$
$$-1,500 \leq M - 15,000 \leq 1,500$$
$$-1,500 + 15,000 \leq M - 15,000 + 15,000 \leq 1,500 + 15,000$$
$$13,500 \leq M \leq 16,500$$
The lower and upper limits of the resistor are 13,500 ohms and 16,500 ohms.

81. Strategy
Let M represent the amount of ohms, T the tolerance, and r the given amount of the resistor. Find the tolerance and solve $|M - r| \leq T$ for M.

Solution
$T = (.05)(56) = 2.8$
$$|M - r| \leq T$$
$$|M - 56| \leq 2.8$$
$$-2.8 \leq M - 56 \leq 2.8$$
$$-2.8 + 56 \leq M - 56 + 56 \leq 2.8 + 56$$
$$53.2 \leq M \leq 58.8$$
The lower and upper limits of the resistor are 53.2 ohms and 58.8 ohms.

Applying Concepts 2.6

83. $\left|\dfrac{3x-2}{4}\right| + 5 = 6$

$\left|\dfrac{3x-2}{4}\right| = 1$

$\dfrac{3x-2}{4} = 1 \qquad \dfrac{3x-2}{4} = -1$

$3x - 2 = 4 \qquad 3x - 2 = -4$

$3x = 6 \qquad 3x = -2$

$x = 2 \qquad x = -\dfrac{2}{3}$

The solutions are 2 and $-\dfrac{2}{3}$.

85. $\left|\dfrac{2x-1}{5}\right| \leq 3$

$-3 \leq \dfrac{2x-1}{5} \leq 3$

$5(-3) \leq 5\left(\dfrac{2x-1}{5}\right) \leq 5(3)$

$-15 \leq 2x - 1 \leq 15$

$-15 + 1 \leq 2x - 1 + 1 \leq 15 + 1$

$-14 \leq 2x \leq 16$

$\dfrac{-14}{2} \leq \dfrac{2x}{2} \leq \dfrac{16}{2}$

$-7 \leq x \leq 8$

$\{x | -7 \leq x \leq 8\}$

87. $|y + 6| = y + 6$
Any value of y that makes $y + 6$ negative will result in a false equation because the left side of the equation will be positive and the right side of the equation will be negative. Therefore, the equation is true if $y + 6$ is greater than or equal to zero.
$y + 6 \geq 0$
$y \geq -6$
$\{y | y \geq -6\}$

89. $|b - 7| = 7 - b$
Any value of b that makes $7 - b$ negative will result in a false equation because the left side of the equation will be positive and the right side of the equation will be negative. Therefore, the equation is true if $7 - b$ is greater than or equal to zero.
$7 - b \geq 0$
$b \leq 7$
$\{b | b \leq 7\}$

91. $|x - 5| < 2$

93. **a.** $|x + y| \leq |x| + |y|$

b. $|x - y| \geq |x| - |y|$

c. $||x| - |y|| \geq |x| - |y|$

d. $\left|\dfrac{x}{y}\right| = \dfrac{|x|}{|y|}, y \neq 0$

e. $|xy| = |x||y|$

Chapter 2: First-Degree Equations and Inequalities

Focus on Problem Solving

1. **a.** The goal is to find the cost of iced tea.

   b. The information that Johanna spent one-third of her allowance on a book is unnecessary.

   c. Johanna spent $5 for a sandwich and tea, and one-fifth of this cost was for the iced tea. This is sufficient information for us to find the cost of iced tea.

2. **a.** The goal is to find the speed of the plane.

   b. To find the speed, we need to know the time of travel and the distance. We do not know the distance. There is not enough information to find the speed of the plane.

3. **a.** The goal is to find the amount of cowhide that will cover 10 baseballs.

   b. The formula for the surface area of a sphere is needed. It can be found in a reference book.

   c. The radius of the baseball is all the dimensions that are needed to find the surface area of the cowhide.

4. **a.** The number in a baker's dozen can be found in a reference book. Multiply this number by 7 and you will find the number of donuts.

5. **a.** The goal is to find the size of the smallest prime number.

   b. The first line indicates that the smallest prime number is 2. The rest of the information is unnecessary.

Projects and Group Activities

Venn Diagrams

1. **a.** 9

   b. 11

   c. 2

2.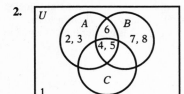

3.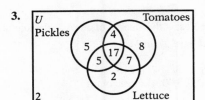

   a. 5

   b. 7

   c. 2

Absolute Value Equations and Inequalities

1. $|3x-4|=|5x-8|$

   $3x-4 = 5x-8$ or $3x-4 = -(5x-8)$
   $-2x = -4$ $3x-4 = -5x+8$
   $x = 2$ $8x = 12$
   $\qquad\qquad\qquad\qquad x = \dfrac{3}{2}$

   The solutions are 2 and $\dfrac{3}{2}$.

2. $|5x-3|=|3x-3|$

   $5x-3 = 3x-3$ or $5x-3 = -(3x-3)$
   $2x = 0$ $5x-3 = -3x+3$
   $x = 0$ $8x = 6$
   $\qquad\qquad\qquad\qquad x = \dfrac{3}{4}$

   The solutions are 0 and $\dfrac{3}{4}$.

3. $|2x-1|=|x|$

   $2x-1 = x$ or $2x-1 = -x$
   $x = 1$ $3x = 1$
   $\qquad\qquad\quad x = \dfrac{1}{3}$

   The solutions are 1 and $\dfrac{1}{3}$.

4. $|x+1|=|x|$

   $x+1 = x$ or $x+1 = -x$
   $1 = 0$ $2x = -1$
   no solution $x = -\dfrac{1}{2}$

   The solution is $-\dfrac{1}{2}$.

5. $\left|\dfrac{2x-3}{5}\right| = |2-x|$

$\dfrac{2x-3}{5} = 2-x$ or $\dfrac{2x-3}{5} = -(2-x)$

$5\left(\dfrac{2x-3}{5}\right) = 5(2-x)$ $\dfrac{2x-3}{5} = -2+x$

$2x-3 = 10-5x$ $5\left(\dfrac{2x-3}{5}\right) = 5(-2+x)$

$7x = 13$

$x = \dfrac{13}{7}$ $2x-3 = -10+5x$

$-3x = -7$

$x = \dfrac{7}{3}$

The solutions are $\dfrac{13}{7}$ and $\dfrac{7}{3}$.

6. $\left|\dfrac{3-2x}{3}\right| = \left|\dfrac{x-4}{2}\right|$

$\dfrac{3-2x}{3} = \dfrac{x-4}{2}$ or $\dfrac{3-2x}{3} = -\left(\dfrac{x-4}{2}\right)$

$6\left(\dfrac{3-2x}{3}\right) = 6\left(\dfrac{x-4}{2}\right)$ $\dfrac{3-2x}{3} = \dfrac{-x+4}{2}$

$2(3-2x) = 3(x-4)$

$6-4x = 3x-12$ $6\left(\dfrac{3-2x}{3}\right) = 6\left(\dfrac{-x+4}{2}\right)$

$-7x = -18$

$x = \dfrac{18}{7}$ $2(3-2x) = 3(-x+4)$

$6-4x = -3x+12$

$-x = 6$

$x = -6$

The solutions are $\dfrac{18}{7}$ and -6.

7. $|2x-3| > |2x-5|$

$(2x-3)^2 > (2x-5)^2$

$4x^2 - 12x + 9 > 4x^2 - 20x + 25$

$4x^2 - 4x^2 - 12x + 9 > 4x^2 - 4x^2 - 20x + 25$

$-12x + 9 > -20x + 25$

$8x + 9 > 25$

$8x > 16$

$x > 2$

The solution is $\{x | x > 2\}$.

8. $|x+4| < |3-x|$

$(x+4)^2 < (3-x)^2$

$x^2 + 8x + 16 < 9 - 6x + x^2$

$x^2 - x^2 + 8x + 16 < 9 - 6x + x^2 - x^2$

$8x + 16 < 9 - 6x$

$14x + 16 < 9$

$14x < -7$

$x < -\dfrac{1}{2}$

The solution is $\left\{x \Big| x < -\dfrac{1}{2}\right\}$.

9. $|3x-2| < |3x+2|$

$(3x-2)^2 < (3x+2)^2$

$9x^2 - 12x + 4 < 9x^2 + 12x + 4$

$9x^2 - 9x^2 - 12x + 4 < 9x^2 - 9x^2 + 12x + 4$

$-12x + 4 < 12x + 4$

$-24x + 4 < 4$

$-24x < 0$

$x > 0$

The solution is $\{x | x > 0\}$.

Chapter Review Exercises

1. $x + 4 = -5$

$x + 4 - 4 = -5 - 4$

$x = -9$

The solution is -9.

2. $\dfrac{2}{3} = x + \dfrac{3}{4}$

$\dfrac{2}{3} - \dfrac{3}{4} = x + \dfrac{3}{4} - \dfrac{3}{4}$

$\dfrac{8}{12} - \dfrac{9}{12} = x$

$-\dfrac{1}{12} = x$

The solution is $-\dfrac{1}{12}$.

3. $-3x = -21$

$\dfrac{-3x}{-3} = \dfrac{-21}{-3}$

$x = 7$

The solution is 7.

4. $\dfrac{2}{3}x = \dfrac{4}{9}$

$\dfrac{3}{2}\left(\dfrac{2}{3}x\right) = \dfrac{3}{2}\left(\dfrac{4}{9}\right)$

$x = \dfrac{2}{3}$

The solution is $\dfrac{2}{3}$.

5. $3y - 5 = 3 - 2y$

$3y + 2y - 5 = 3 - 2y + 2y$

$5y - 5 = 3$

$5y - 5 + 5 = 3 + 5$

$5y = 8$

$\dfrac{5y}{5} = \dfrac{8}{5}$

$y = \dfrac{8}{5}$

The solution is $\dfrac{8}{5}$.

6.
$$3x - 3 + 2x = 7x - 15$$
$$5x - 3 = 7x - 15$$
$$5x - 3 - 7x = 7x - 15 - 7x$$
$$-2x - 3 = -15$$
$$-2x - 3 + 3 = -15 + 3$$
$$-2x = -12$$
$$\frac{-2x}{-2} = \frac{-12}{-2}$$
$$x = 6$$
The solution is 6.

7.
$$2(x - 3) = 5(4 - 3x)$$
$$2x - 6 = 20 - 15x$$
$$2x - 6 + 15x = 20 - 15x + 15x$$
$$17x - 6 = 20$$
$$17x - 6 + 6 = 20 + 6$$
$$17x = 26$$
$$\frac{17x}{17} = \frac{26}{17}$$
$$x = \frac{26}{17}$$
The solution is $\frac{26}{17}$.

8.
$$2x - (3 - 2x) = 4 - 3(4 - 2x)$$
$$2x - 3 + 2x = 4 - 12 + 6x$$
$$4x - 3 = -8 + 6x$$
$$4x - 3 - 6x = -8 + 6x - 6x$$
$$-2x - 3 = -8$$
$$-2x - 3 + 3 = -8 + 3$$
$$-2x = -5$$
$$\frac{-2x}{-2} = \frac{-5}{-2}$$
$$x = \frac{5}{2}$$
The solution is $\frac{5}{2}$.

9.
$$\frac{1}{2}x - \frac{5}{8} = \frac{3}{4}x + \frac{3}{2}$$
$$8\left(\frac{1}{2}x - \frac{5}{8}\right) = 8\left(\frac{3}{4}x + \frac{3}{2}\right)$$
$$8\left(\frac{1}{2}x\right) - 8\left(\frac{5}{8}\right) = 8\left(\frac{3}{4}x\right) + 8\left(\frac{3}{2}\right)$$
$$4x - 5 = 6x + 12$$
$$4x - 5 - 6x = 6x + 12 - 6x$$
$$-2x - 5 = 12$$
$$-2x - 5 + 5 = 12 + 5$$
$$-2x = 17$$
$$\frac{-2x}{-2} = \frac{17}{-2}$$
$$x = -\frac{17}{2}$$
The solution is $-\frac{17}{2}$.

10.
$$\frac{2x - 3}{3} + 2 = \frac{2 - 3x}{5}$$
$$15\left(\frac{2x - 3}{3} + 2\right) = 15\left(\frac{2 - 3x}{5}\right)$$
$$\frac{15(2x - 3)}{3} + 15(2) = \frac{15(2 - 3x)}{5}$$
$$5(2x - 3) + 30 = 3(2 - 3x)$$
$$10x - 15 + 30 = 6 - 9x$$
$$10x + 15 = 6 - 9x$$
$$10x + 15 + 9x = 6 - 9x + 9x$$
$$19x + 15 = 6$$
$$19x + 15 - 15 = 6 - 15$$
$$19x = -9$$
$$\frac{19x}{19} = \frac{-9}{19}$$
$$x = -\frac{9}{19}$$
The solution is $-\frac{9}{19}$.

11.
$$3x - 7 > -2$$
$$3x > 5$$
$$\frac{3x}{3} > \frac{5}{3}$$
$$x > \frac{5}{3}$$
The solution is $\left(\frac{5}{3}, \infty\right)$.

12.
$$2x - 9 < 8x + 15$$
$$2x - 8x - 9 < 8x - 8x + 15$$
$$-6x - 9 < 15$$
$$-6x - 9 + 9 < 15 + 9$$
$$-6x < 24$$
$$\frac{-6x}{-6} > \frac{24}{-6}$$
$$x > -4$$
The solution is $(-4, \infty)$.

13.
$$\frac{2}{3}x - \frac{5}{8} \geq \frac{5}{4}x + 3$$
$$24\left(\frac{2}{3}x - \frac{5}{8}\right) \geq 24\left(\frac{5}{4}x + 3\right)$$
$$16x - 15 \geq 30x + 72$$
$$16x - 30x - 15 \geq 30x - 30x + 72$$
$$-14x - 15 \geq 72$$
$$-14x - 15 + 15 \geq 72 + 15$$
$$-14x \geq 87$$
$$\frac{-14x}{-14} \leq \frac{87}{-14}$$
$$x \leq -\frac{87}{14}$$
The solution is $\left\{x \mid x \leq -\frac{87}{14}\right\}$.

14.
$$2 - 3(x-4) \leq 4x - 2(1-3x)$$
$$2 - 3x + 12 \leq 4x - 2 + 6x$$
$$-3x + 14 \leq 10x - 2$$
$$-3x - 10x + 14 \leq 10x - 10x - 2$$
$$-13x + 14 \leq -2$$
$$-13x + 14 - 14 \leq -2 - 14$$
$$-13x \leq -16$$
$$\frac{-13x}{-13} \geq \frac{-16}{-13}$$
$$x \geq \frac{16}{13}$$
The solution is $\left\{x \mid x \geq \frac{16}{13}\right\}$.

15.
$$-5 < 4x - 1 < 7$$
$$-5 + 1 < 4x - 1 + 1 < 7 + 1$$
$$-4 < 4x < 8$$
$$\frac{-4}{4} < \frac{4x}{4} < \frac{8}{4}$$
$$-1 < x < 2$$
The solution is $(-1, 2)$.

16. $5x - 2 > 8$ or $3x + 2 < -4$
$\quad 5x > 10 \qquad\qquad 3x < -6$
$\quad x > 2 \qquad\qquad\quad x < -2$
$\{x \mid x > 2\} \qquad \{x \mid x < -2\}$
$\{x \mid x > 2\} \cup \{x \mid x < -2\} = \{x \mid x > 2 \text{ or } x < -2\}$
The solution is $(-\infty, -2) \cup (2, \infty)$.

17. $3x < 4$ and $x + 2 > -1$
$\quad x < \frac{4}{3} \qquad\qquad x > -3$
$\left\{x \mid x < \frac{4}{3}\right\} \quad \{x \mid x > -3\}$
$\left\{x \mid x < \frac{4}{3}\right\} \cap \{x \mid x > -3\} = \left\{x \mid -3 < x < \frac{4}{3}\right\}$

18. $3x - 2 > -4$ or $7x - 5 < 3x + 3$
$\quad 3x > -2 \qquad\qquad 4x - 5 < 3$
$\quad \frac{3x}{3} > \frac{-2}{3} \qquad\qquad 4x < 8$
$\quad x > -\frac{2}{3} \qquad\qquad \frac{4x}{4} < \frac{8}{4}$
$\qquad\qquad\qquad\qquad x < 2$
$\left\{x \mid x > -\frac{2}{3}\right\} \quad \{x \mid x < 2\}$
$\left\{x \mid x > -\frac{2}{3}\right\} \cup \{x \mid x < 2\} = \{x \mid x \text{ is any real number}\}$

19. $|2x - 3| = 8$
$2x - 3 = 8$ or $2x - 3 = -8$
$2x = 11 \qquad\qquad 2x = -5$
$x = \frac{11}{2} \qquad\qquad x = -\frac{5}{2}$
The solutions are $\frac{11}{2}$ and $-\frac{5}{2}$.

20. $|5x + 8| = 0$
$5x + 8 = 0$
$5x = -8$
$x = -\frac{8}{5}$
The solution is $-\frac{8}{5}$.

21. $6 + |3x - 3| = 2$
$|3x - 3| = -4$
There is no solution to this equation because the absolute value of a number must be non-negative.

22. $|2x - 5| \leq 3$
$\quad -3 \leq 2x - 5 \leq 3$
$-3 + 5 \leq 2x - 5 + 5 \leq 3 + 5$
$\quad 2 \leq 2x \leq 8$
$\quad \frac{2}{2} \leq \frac{2x}{2} \leq \frac{8}{2}$
$\quad 1 \leq x \leq 4$
The solution is $\{x \mid 1 \leq x \leq 4\}$.

23. $|4x - 5| \geq 3$
$4x - 5 \geq 3$ or $4x - 5 \leq -3$
$4x \geq 8 \qquad\qquad 4x \leq 2$
$x \geq 2 \qquad\qquad\quad x \leq \frac{1}{2}$
$\{x \mid x \geq 2\} \quad \left\{x \mid x < \frac{1}{2}\right\}$
$\{x \mid x \geq 2\} \cup \left\{x \mid x \leq \frac{1}{2}\right\} = \left\{x \mid x \leq \frac{1}{2} \text{ or } x \geq 2\right\}$

24. $|5x - 4| < -2$
There is no solution to this equation because the absolute value of a number must be non-negative.

25. **Strategy**
Let b represent the diameter of the bushing, T the tolerance, and d the lower and upper limits of the diameter. Solve the absolute value inequality $|d - b| \leq T$ for d.

Solution
$$|d - b| \leq T$$
$$|d - 2.75| \leq 0.003$$
$$-0.003 \leq d - 2.75 \leq 0.003$$
$$-0.003 + 2.75 \leq d - 2.75 + 2.75 \leq 0.003 + 2.75$$
$$2.747 \leq d \leq 2.753$$
The lower and upper limits of the diameter of the bushing are 2.747 in. and 2.753 in.

Chapter 2: First-Degree Equations and Inequalities

26. **Strategy**
    Let p represent the prescribed amount of medication, T the tolerance, and m the lower and upper limits of the amount of medication. Solve the absolute value inequality $|m - p| \leq T$ for m.

    Solution
    $|m - p| \leq T$
    $|m - 2| \leq 0.25$
    $-0.25 \leq m - 2 \leq 0.25$
    $-0.25 + 2 \leq m - 2 + 2 \leq 0.25 + 2$
    $1.75 \leq m \leq 2.25$
    The lower and upper limits of the amount of medicine to be given to the patient are 1.75 cc and 2.25 cc.

27. **Strategy**
    The smaller integer: n
    The larger integer: $20 - n$
    Five times the smaller integer is two more than twice the larger integer.
    $5n = 2 + 2(20 - n)$

    Solution
    $5n = 2 + 2(20 - n)$
    $5n = 2 + 40 - 2n$
    $5n = 42 - 2n$
    $7n = 42$
    $n = 6$
    $20 - n = 20 - 6 = 14$
    The integers are 6 and 14.

28. **Strategy**
    First consecutive integer: x
    Second consecutive integer: $x + 1$
    Third consecutive integer: $x + 2$
    Five times the middle integer is twice the sum of the other two integers.
    $5(x + 1) = 2[x + (x + 2)]$

    Solution
    $5(x + 1) = 2[x + (x + 2)]$
    $5x + 5 = 2(2x + 2)$
    $5x + 5 = 4x + 4$
    $x + 5 = 4$
    $x = -1$
    $x + 1 = -1 + 1 = 0$
    $x + 2 = -1 + 2 = 1$
    The integers are -1, 0, and 1.

29. **Strategy**
    Number of nickels: x
    Number of dimes: $x + 3$
    Number of quarters: $30 - (2x + 3) = 27 - 2x$

Coin	Number	Value	Total Value
Nickel	x	5	$5x$
Dime	$x + 3$	10	$10(x + 3)$
Quarter	$27 - 2x$	25	$25(27 - 2x)$

    The sum of the total values of each type of coin equals the total value of all the coins (355 cents).
    $5x + 10(x + 3) + 25(27 - 2x) = 355$

    Solution
    $5x + 10(x + 3) + 25(27 - 2x) = 355$
    $5x + 10x + 30 + 675 - 50x = 355$
    $-35x + 705 = 355$
    $-35x = -350$
    $x = 10$
    $27 - 2x = 27 - 2(10) = 27 - 20 = 7$
    There are 7 quarters in the collection.

30. **Strategy**
    Cost per ounce of the mixture: x

	Amount	Cost	Value
Pure silver	40	$8.00	8(40)
Alloy	200	$3.50	3.50(200)
Mixture	240	x	$240x$

    The sum of the values before mixing equals the value after mixing.
    $8(40) + 3.50(200) = 240x$

    Solution
    $8(40) + 3.50(200) = 240x$
    $320 + 700 = 240x$
    $1020 = 240x$
    $4.25 = x$
    The mixture costs $4.25 per ounce.

31. **Strategy**
 Gallons of apple juice: x

	Amount	Cost	Value
Apple juice	x	3.20	$3.20x$
Cranberry juice	40	5.50	$40(5.50)$
Mixture	$40 + x$	4.20	$4.20(40 + x)$

The sum of the values before mixing equals the value after mixing.
$3.20x + 40(5.50) = 4.20(40 + x)$

Solution
$3.20x + 40(5.50) = 4.20(40 + x)$
$3.20x + 220 = 168 + 4.20x$
$-x = -52$
$x = 52$
The mixture must contain 52 gal of apple juice.

32. **Strategy**
 Rate of the first plane: r
 Rate of the second plane: $r + 80$

	Rate	Time	Distance
1st plane	r	1.75	$1.75r$
2nd plane	$r + 80$	1.75	$1.75(r + 80)$

The total distance traveled by the two planes is 1680 mi.
$1.75r + 1.75(r + 80) = 1680$

Solution
$1.75r + 1.75(r + 80) = 1680$
$1.75r + 1.75r + 140 = 1680$
$3.5r + 140 = 1680$
$3.5r = 1540$
$r = 440$
$r + 80 = 440 + 80 = 520$
The speed of the first plane is 440 mph. The speed of the second plane is 520 mph.

33. **Strategy**
 Amount invested at 10.5%: x
 Amount invested at 6.4%: $8000 - x$

	Principal	Rate	Interest
Amount at 10.5%	x	0.105	$0.105x$
Amount at 6.4%	$8000 - x$	0.064	$0.064(8000 - x)$

The sum of the interest earned by the two investments equals the total annual interest earned ($635).
$0.105x + 0.064(8000 - x) = 635$

Solution
$0.105x + 0.064(8000 - x) = 635$
$0.105x + 512 - 0.064x = 635$
$0.041x + 512 = 635$
$0.041x = 123$
$x = 3000$
$8000 - x = 8000 - 3000 = 5000$
The amount invested at 10.5% was $3000.
The amount invested at 6.4% was $5000.

34. **Strategy**
 Pounds of 30% tin: x
 Pounds of 70% tin: $500 - x$

	Amount	Percent	Quantity
30%	x	0.30	$0.30x$
70%	$500 - x$	0.70	$0.70(500 - x)$
40%	500	0.40	$0.40(500)$

The sum of the quantities before mixing is equal to the quantity after mixing.
$0.30x + 0.70(500 - x) = 0.40(500)$

Solution
$0.30x + 0.70(500 - x) = 0.40(500)$
$0.30x + 350 - 0.70x = 200$
$-0.40x + 350 = 200$
$-0.40x = -150$
$x = 375$
$500 - x = 500 - 375 = 125$
375 lb of 30% tin and 125 lb of 70% tin were used.

35. **Strategy**
 To find the minimum amount of sales, write and solve an inequality using N to represent the amount of sales.

Solution
$800 + 0.04N \geq 3000$
$0.04N \geq 2200$
$N \geq 55{,}000$
The executive's amount of sales must be $55,000 or more.

44 *Chapter 2: First-Degree Equations and Inequalities*

36. **Strategy**
To find the range of scores, write and solve an inequality using N to represent the score on the last test.

Solution
$$80 \le \frac{92+66+72+88+N}{5} \le 90$$
$$80 \le \frac{318+N}{5} \le 90$$
$$5 \cdot 80 \le 5 \cdot \frac{318+N}{5} \le 5 \cdot 90$$
$$400 \le 318+N \le 450$$
$$400-318 \le 318+N-318 \le 450-318$$
$$82 \le N \le 132$$
Since 100 is the maximum score, the range of scores to receive a B grade is $82 \le x \le 100$.

Chapter Test

1. $x-2=-4$
$x-2+2=-4+2$
$x=-2$
The solution is –2.

2. $x+\dfrac{3}{4}=\dfrac{5}{8}$
$x+\dfrac{3}{4}-\dfrac{3}{4}=\dfrac{5}{8}-\dfrac{3}{4}$
$x=\dfrac{5}{8}-\dfrac{6}{8}$
$x=-\dfrac{1}{8}$
The solution is $-\dfrac{1}{8}$.

3. $-\dfrac{3}{4}y=-\dfrac{5}{8}$
$-\dfrac{4}{3}\left(-\dfrac{3}{4}y\right)=-\dfrac{4}{3}\left(-\dfrac{5}{8}\right)$
$y=\dfrac{5}{6}$
The solution is $\dfrac{5}{6}$.

4. $3x-5=7$
$3x-5+5=7+5$
$3x=12$
$\dfrac{3x}{3}=\dfrac{12}{3}$
$x=4$
The solution is 4.

5. $\dfrac{3}{4}y-2=6$
$\dfrac{3}{4}y-2+2=6+2$
$\dfrac{3}{4}y=8$
$\dfrac{4}{3}\left(\dfrac{3}{4}y\right)=\dfrac{4}{3}(8)$
$y=\dfrac{32}{3}$
The solution is $\dfrac{32}{3}$.

6. $2x-3-5x=8+2x-10$
$-3x-3=-2+2x$
$-3x-3+3x=-2+2x+3x$
$-3=-2+5x$
$-3+2=-2+5x+2$
$-1=5x$
$\dfrac{-1}{5}=\dfrac{5x}{5}$
$-\dfrac{1}{5}=x$
$x=-\dfrac{1}{5}$
The solution is $-\dfrac{1}{5}$.

7. $2[x-(2-3x)-4]=x-5$
$2[x-2+3x-4]=x-5$
$2[4x-6]=x-5$
$8x-12=x-5$
$8x-x-12=x-x-5$
$7x-12=-5$
$7x-12+12=-5+12$
$7x=7$
$\dfrac{7x}{7}=\dfrac{7}{7}$
$x=1$
The solution is 1.

8. $\dfrac{2}{3}x-\dfrac{5}{6}x=4$
$\dfrac{4}{6}x-\dfrac{5}{6}x=4$
$-\dfrac{1}{6}x=4$
$-6\left(-\dfrac{1}{6}x\right)=-6(4)$
$x=-24$
The solution is –24.

9. $$\frac{2x+1}{3} - \frac{3x+4}{6} = \frac{5x-9}{9}$$
$$18\left(\frac{2x+1}{3} - \frac{3x+4}{6}\right) = 18\left(\frac{5x-9}{9}\right)$$
$$18\left(\frac{2x+1}{3}\right) - 18\left(\frac{3x+4}{6}\right) = 18\left(\frac{5x-9}{9}\right)$$
$$6(2x+1) - 3(3x+4) = 2(5x-9)$$
$$12x + 6 - 9x - 12 = 10x - 18$$
$$3x - 6 = 10x - 18$$
$$3x - 6 - 10x = 10x - 18 - 10x$$
$$-7x - 6 = -18$$
$$-7x - 6 + 6 = -18 + 6$$
$$-7x = -12$$
$$\frac{-7x}{-7} = \frac{-12}{-7}$$
$$x = \frac{12}{7}$$

The solution is $\frac{12}{7}$.

10. $2x - 5 \geq 5x + 4$
$-3x - 5 \geq 4$
$-3x \geq 9$
$\frac{-3x}{-3} \leq \frac{9}{-3}$
$x \leq -3$
$(-\infty, -3]$

11. $4 - 3(x+2) < 2(2x+3) - 1$
$4 - 3x - 6 < 4x + 6 - 1$
$-2 - 3x < 4x + 5$
$-7x < 7$
$\frac{-7x}{-7} > \frac{7}{-7}$
$x > -1$
$(-1, \infty)$

12. $3x - 2 > 4$ or $4 - 5x < 14$
$3x > 6 \qquad\quad -5x < 10$
$x > 2 \qquad\quad \frac{-5x}{-5} > \frac{10}{-5}$
$\qquad\qquad\qquad x > -2$
$\{x | x > 2\} \quad \{x | x > -2\}$
$\{x | x > 2\} \cup \{x | x > -2\} = \{x | x > -2\}$

13. $4 - 3x \geq 7$ and $2x + 3 \geq 7$
$-3x \geq 3 \qquad\quad 2x \geq 4$
$\frac{-3x}{-3} \leq \frac{3}{-3} \qquad \frac{2x}{2} \geq \frac{4}{2}$
$x \leq -1 \qquad\qquad x \geq 2$
$\{x | x \leq -1\} \cap \{x | x \geq 2\} = \varnothing$

14. $|3 - 5x| = 12$
$3 - 5x = 12 \qquad 3 - 5x = -12$
$-5x = 9 \qquad\quad -5x = -15$
$x = -\frac{9}{5} \qquad\qquad x = 3$

The solutions are $-\frac{9}{5}$ and 3.

15. $2 - |2x - 5| = -7$
$-|2x - 5| = -9$
$|2x - 5| = 9$
$2x - 5 = 9 \qquad 2x - 5 = -9$
$2x = 14 \qquad\quad 2x = -4$
$x = 7 \qquad\qquad x = -2$

The solutions are 7 and –2.

16. $|3x - 1| \leq 2$
$-2 \leq 3x - 1 \leq 2$
$-2 + 1 \leq 3x - 1 + 1 \leq 2 + 1$
$-1 \leq 3x \leq 3$
$\frac{-1}{3} \leq \frac{3x}{3} \leq \frac{3}{3}$
$-\frac{1}{3} \leq x \leq 1$
$\left\{x \mid -\frac{1}{3} \leq x \leq 1\right\}$

17. $|2x - 1| > 3$
$2x - 1 > 3$ or $2x - 1 < -3$
$2x > 4 \qquad\quad 2x < -2$
$x > 2 \qquad\qquad x < -1$
$\{x | x > 2\} \cup \{x | x < -1\} = \{x | x > 2 \text{ or } x < -1\}$

18. $4 + |2x - 3| = 1$
$|2x - 3| = -3$

There is no solution because the absolute value of a number is always non-negative.

19. **Strategy**
To find the number of miles, write and solve an inequality using N to represent the number of miles.

Solution
cost of car A < cost of car B
$12 + 0.10N < 24$
$0.10N < 12$
$N < 120$

It costs less to rent from Agency A if the car is driven less than 120 mi.

20. **Strategy**
Let p represent the prescribed amount of medication, T the tolerance, and m the lower and upper limits of the given amount of medication.

Solve the absolute value inequality $|m - p| \leq T$ form m.

Solution
$|m - p| \leq T$
$|m - 3| \leq 0.1$
$-0.1 \leq m - 3 \leq 0.1$
$-0.1 + 3 \leq m - 3 + 3 \leq 0.1 + 3$
$2.9 \leq m \leq 3.1$

The lower and upper limits of the amount of medication to be given to the patient are 2.9 cc and 3.1 cc.

21. Strategy
    Number of 15¢ stamps: x
    Number of 11¢ stamps: $2x$
    Number of 24¢ stamps: $30 - 3x$

Stamp	Number	Value	Total Value
15¢	x	15	$15x$
11¢	$2x$	11	$11(2x)$
24¢	$30-3x$	24	$24(30-3x)$

    The sum of the total values of each type of stamp equals the total value of all the stamps (440 cents).
    $15x + 11(2x) + 24(30 - 3x) = 440$

    Solution
    $15x + 11(2x) + 24(30 - 3x) = 440$
    $15x + 22x + 720 - 72x = 440$
    $-35x + 720 = 440$
    $-35x = -280$
    $x = 8$
    $30 - 3x = 30 - 3(8) = 30 - 24 = 6$
    There are six 24¢ stamps.

22. Strategy
    Price of hamburger mixture: x

	Amount	Cost	Value
$1.60 hamburger	100	1.60	1.60(100)
$3.20 hamburger	60	3.20	3.20(60)
Mixture	160	x	$160x$

    The sum of the values before mixing equals the value after mixing.
    $1.60(100) + 3.20(60) = 160x$

    Solution
    $1.60(100) + 3.20(60) = 160x$
    $160 + 192 = 160x$
    $352 = 160x$
    $2.20 = x$
    The price of the hamburger mixture is $2.20/lb.

23. Strategy
Time jogger runs a distance: t

Time jogger returns same distance: $1\frac{45}{60} - t$

	Rate	Time	Distance
Jogger runs a distance	8	t	$8t$
Jogger returns same distance	6	$\frac{7}{4} - t$	$6\left(\frac{7}{4} - t\right)$

The jogger runs a distance and returns the same distance.

$$8t = 6\left(\frac{7}{4} - t\right)$$

Solution

$$8t = 6\left(\frac{7}{4} - t\right)$$
$$8t = \frac{21}{2} - 6t$$
$$14t = \frac{21}{2}$$
$$\frac{1}{14}(14t) = \frac{1}{14}\left(\frac{21}{2}\right)$$
$$t = \frac{3}{4}$$

The jogger ran for $\frac{3}{4}$ hour.

$$8t = 8 \cdot \frac{3}{4} = 6$$

The jogger ran a distance of 6 mi one way. The jogger ran a total distance of 12 mi.

24. Strategy
Amount invested at 7.8%: x
Amount invested at 9%: $12{,}000 - x$

	Principal	Rate	Interest
Amount invested at 7.8%	x	0.078	$0.078x$
Amount invested at 9%	$12{,}000 - x$	0.09	$0.09(12{,}000 - x)$

The sum of the interest earned by the two investments equals the total annual interest earned ($1020).
$0.078x + 0.09(12{,}000 - x) = 1020$

Solution
$0.078x + 0.09(12{,}000 - x) = 1020$
$0.078x + 1080 - 0.09x = 1020$
$-0.012x = -60$
$x = 5000$
$12{,}000 - x = 12{,}000 - 50000 = 7000$
The amount invested at 7.8% was $5000.
The amount invested at 9% was $7000.

25. Strategy
Ounces of pure water: x

	Amount	Percent	Quantity
Pure water	x	0	0
8% salt	60	0.08	0.08(60)
3% salt	$60 + x$	0.03	$0.03(60 + x)$

The sum of the quantities before mixing is equal to the quantity after mixing.
$0 + 0.08(60) = 0.03(60 + x)$

Solution
$$0 + 0.08(60) = 0.03(60 + x)$$
$$4.8 = 1.8 + 0.03x$$
$$3 = 0.03x$$
$$100 = x$$
There are 100 oz of pure water.

Cumulative Review Exercises

1. $-2^2 \cdot 3^3 = -(2 \cdot 2)(3 \cdot 3 \cdot 3) = -(4)(27) = -108$

2. $4 - (2-5)^2 \div 3 + 2 = 4 - (-3)^2 \div 3 + 2$
$= 4 - 9 \div 3 + 2$
$= 4 - 3 + 2$
$= 1 + 2$
$= 3$

3. $4 \div \dfrac{\frac{3}{8} - 1}{5} \cdot 2 = 4 \div \dfrac{-\frac{5}{8}}{5} \cdot 2$
$= 4 \div \left(-\dfrac{5}{8} \cdot \dfrac{1}{5}\right) \cdot 2$
$= 4 \div \left(-\dfrac{1}{8}\right) \cdot 2$
$= 4 \cdot (-8) \cdot 2$
$= -32 \cdot 2$
$= -64$

4. $2a^2 - (b-c)^2 = 2(2)^2 - (3-(-1))^2$
$= 2 \cdot 4 - (3+1)^2$
$= 2 \cdot 4 - 4^2$
$= 2 \cdot 4 - 16$
$= 8 - 16$
$= -8$

5. The Commutative Property of Addition

6. $A \cap B = \{3, 9\}$

7. $3x - 2[x - 3(2 - 3x) + 5] = 3x - 2[x - 6 + 9x + 5]$
$= 3x - 2[10x - 1]$
$= 3x - 20x + 2$
$= -17x + 2$

8. $5[y - 2(3 - 2y) + 6] = 5[y - 6 + 4y + 6]$
$= 5[5y]$
$= 25y$

9. $4 - 3x = -2$
$4 - 3x - 4 = -2 - 4$
$-3x = -6$
$\dfrac{-3x}{-3} = \dfrac{-6}{-3}$
$x = 2$
The solution is 2.

10. $-\dfrac{5}{6}b = -\dfrac{5}{12}$
$\left(-\dfrac{6}{5}\right)\left(-\dfrac{5}{6}\right)b = \left(-\dfrac{6}{5}\right)\left(-\dfrac{5}{12}\right)$
$b = \dfrac{1}{2}$
The solution is $\dfrac{1}{2}$.

11. $2x + 5 = 5x + 2$
$2x + 5 - 5x = 5x + 2 - 5x$
$-3x + 5 = 2$
$-3x + 5 - 5 = 2 - 5$
$-3x = -3$
$\dfrac{-3x}{-3} = \dfrac{-3}{-3}$
$x = 1$
The solution is 1.

12. $\dfrac{5}{12}x - 3 = 7$
$\dfrac{5}{12}x - 3 + 3 = 7 + 3$
$\dfrac{5}{12}x = 10$
$\left(\dfrac{12}{5}\right)\left(\dfrac{5}{12}\right)x = \left(\dfrac{12}{5}\right)10$
$x = 24$
The solution is 24.

13. $2[3 - 2(3 - 2x)] = 2(3 + x)$
$2[3 - 6 + 4x] = 6 + 2x$
$2[-3 + 4x] = 6 + 2x$
$-6 + 8x = 6 + 2x$
$-6 + 8x - 2x = 6 + 2x - 2x$
$-6 + 6x = 6$
$-6 + 6x + 6 = 6 + 6$
$6x = 12$
$\dfrac{6x}{6} = \dfrac{12}{6}$
$x = 2$
The solution is 2.

14. $3[2x - 3(4 - x)] = 2(1 - 2x)$
    $3[2x - 12 + 3x] = 2 - 4x$
    $3[5x - 12] = 2 - 4x$
    $15x - 36 = 2 - 4x$
    $15x - 36 + 4x = 2 - 4x + 4x$
    $19x - 36 = 2$
    $19x - 36 + 36 = 2 + 36$
    $19x = 38$
    $\dfrac{19x}{19} = \dfrac{38}{19}$
    $x = 2$
    The solution is 2.

15. $\dfrac{3x-1}{4} - \dfrac{4x-1}{12} = \dfrac{3+5x}{8}$
    $24\left(\dfrac{3x-1}{4} - \dfrac{4x-1}{12}\right) = 24\left(\dfrac{3+5x}{8}\right)$
    $\dfrac{24(3x-1)}{4} - \dfrac{24(4x-1)}{12} = \dfrac{24(3+5x)}{8}$
    $6(3x-1) - 2(4x-1) = 3(3+5x)$
    $18x - 6 - 8x + 2 = 9 + 15x$
    $10x - 4 = 9 + 15x$
    $10x - 4 - 15x = 9 + 15x - 15x$
    $-5x - 4 = 9$
    $-5x - 4 + 4 = 9 + 4$
    $-5x = 13$
    $\dfrac{-5x}{-5} = \dfrac{13}{-5}$
    $x = -\dfrac{13}{5}$
    The solution is $-\dfrac{13}{5}$.

16. $3x - 2 \geq 6x + 7$
    $3x \geq 6x + 9$
    $-3x \geq 9$
    $x \leq -3$
    $\{x | x \leq -3\}$

17. $5 - 2x \geq 6$ and $3x + 2 \geq 5$
    $-2x \geq 1$ $\quad\quad$ $3x \geq 3$
    $x \leq -\dfrac{1}{2}$ $\quad\quad$ $x \geq 1$
    $\{x | x \leq -\dfrac{1}{2}\}$ $\quad$ $\{x | x \geq 1\}$
    $\{x | x \leq -\dfrac{1}{2}\} \cap \{x | x \geq 1\} = \varnothing$

18. $4x - 1 > 5$ or $2 - 3x < 8$
    $4x > 6$ $\quad\quad$ $-3x < 6$
    $x > \dfrac{3}{2}$ $\quad\quad$ $x > -2$
    $\{x | x > \dfrac{3}{2}\}$ $\quad$ $\{x | x > -2\}$
    $\{x | x > \dfrac{3}{2}\} \cup \{x | x > -2\} = \{x | x > -2\}$

19. $|3 - 2x| = 5$
    $3 - 2x = 5$ $\quad\quad$ $3 - 2x = -5$
    $-2x = 2$ $\quad\quad$ $-2x = -8$
    $x = -1$ $\quad\quad$ $x = 4$
    The solutions are -1 and 4.

20. $3 - |2x - 3| = -8$
    $-|2x - 3| = -11$
    $|2x - 3| = 11$
    $2x - 3 = 11$ $\quad\quad$ $2x - 3 = -11$
    $2x = 14$ $\quad\quad$ $2x = -8$
    $x = 7$ $\quad\quad$ $x = -4$
    The solutions are 7 and -4.

21. $|3x - 5| \leq 4$
    $-4 \leq 3x - 5 \leq 4$
    $-4 + 5 \leq 3x - 5 + 5 \leq 4 + 5$
    $1 \leq 3x \leq 9$
    $\dfrac{1}{3} \leq \dfrac{3x}{3} \leq \dfrac{9}{3}$
    $\dfrac{1}{3} \leq x \leq 3$
    $\{x | \dfrac{1}{3} \leq x \leq 3\}$

22.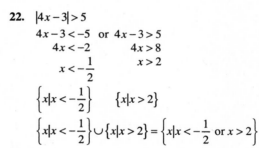

23. $\{x | x \geq -2\}$
    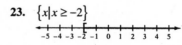

24. (number line graph)

25. The unknown number: n
    three times the number: $3n$
    the sum of three times the number and six: $3n + 6$
    $(3n + 6) + 3n = 6n + 6$

Chapter 2: First-Degree Equations and Inequalities

26. Strategy
The first integer: n
Second consecutive odd integer: $n + 2$
Third consecutive odd integer: $n + 4$
Three times the sum of the first and third integers is fifteen more than the second integer.

Solution
$$3[n+(n+4)] = (n+2)+15$$
$$3(2n+4) = n+17$$
$$6n+12 = n+17$$
$$5n+12 = 17$$
$$5n = 5$$
$$n = 1$$
The first integer is 1.

27. Strategy
Number of 11¢ stamps: n
Number of 9¢ stamps: $2n - 5$

Stamp	Number	Value	Total Value
9¢	$2n - 5$	9	$9(2n - 5)$
11¢	n	11	$11n$

The sum of the total values of each denomination of stamp equals the total value of all the stamps (187 cents).
$9(2n - 5) + 11n = 187$

Solution
$$9(2n-5)+11n = 187$$
$$18n-45+11n = 187$$
$$29n-45 = 187$$
$$29n = 232$$
$$n = 8$$
$2n - 5 = 2(8) - 5 = 16 - 5 = 11$
There are eleven 9¢ stamps.

28. Strategy
Number of adult tickets: n
Number of children's tickets: $75 - n$

	Number	Value	Total Value
Adult tickets	n	2.25	$2.25n$
Children's tickets	$75 - n$	0.75	$0.75(75 - n)$

The sum of the total values of each denomination of ticket equals the total value of all the tickets ($128.25).
$2.25n + 0.75(75 - n) = 128.25$

Solution
$$2.25n + 0.75(75-n) = 128.25$$
$$2.25n + 56.25 - 0.75n = 128.25$$
$$1.5n + 56.25 = 128.25$$
$$1.5n = 72$$
$$n = 48$$
48 adult tickets were sold.

29. Strategy
Slower plane: x
Faster plane: $x + 120$

	Rate	Time	Distance
Slower plane	x	2.5	$2.5x$
Faster plane	$x + 120$	2.5	$2.5(x + 120)$

The two planes travel a total distance of 1400 mi.
$2.5x + 2.5(x + 120) = 1400$

Solution
$$2.5x + 2.5(x+120) = 1400$$
$$2.5x + 2.5x + 300 = 1400$$
$$5x + 300 = 1400$$
$$5x = 1100$$
$$x = 220$$
$x + 120 = 220 + 120 = 340$
The speed of the faster plane is 340 mph.

30. Strategy
Liters of 12% acid solution: x

Solution	Amount	Percent	Quantity
12%	x	0.12	$0.12x$
5%	4	0.05	$0.05(4)$
8%	$x + 4$	0.08	$0.08(x + 4)$

The sum of the quantities before mixing equals the quantity after mixing.
$0.12x + 0.05(4) = 0.08(x + 4)$

Solution
$$0.12x + 0.05(4) = 0.08(x+4)$$
$$0.12x + 0.2 = 0.08x + 0.32$$
$$0.04x + 0.2 = 0.32$$
$$0.04x = 0.12$$
$$x = 3$$
3 L of 12% acid solution must be in the mixture.

31. Strategy
Amount invested at 9.8%: x
Amount invested at 12.8%: $10,000 - x$

	Principal	Rate	Interest
Amount at 9.8%	x	0.098	$0.098x$
Amount at 12.8%	$10,000 - x$	0.128	$0.128(10,000 - x)$

The sum of the interest earned by the two investments is equal to the total annual interest earned ($1085).
$0.098x + 0.128(10,000 - x) = 1085$

Solution
$$0.098x + 0.128(10,000 - x) = 1085$$
$$0.098x + 1280 - 0.128x = 1085$$
$$-0.03x + 1280 = 1085$$
$$-0.03x = -195$$
$$x = 6500$$
$6500 was invested at 9.8%.

Chapter 3: Linear Functions and Inequalities in Two Variables

Section 3.1

Concept Review 3.1

1. Always true

3. Never true
   The first number is the *x*-coordinate and the second number is the *y*-coordinate.

5. Never true
   The point (−2, −4) is in the third quadrant.

Objective 1 Exercises

3.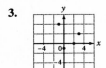

5. A(0, 3)
   B(1, 1)
   C(3, −4)
   D(−4, 4)

7.

9.

11. $y = x^2$
    Ordered pairs: (−2, 4)
    (−1, 1)
    (0, 0)
    (1, 1)
    (2, 4)

13. $y = |x + 1|$
    Ordered pairs: (−5, 4)
    (−3, 2)
    (0, 1)
    (3, 4)
    (5, 6)

15. $y = -x^2 + 2$
    Ordered pairs: (−2, −2)
    (−1, 1)
    (0, 2)
    (1, 1)
    (2, −2)

17. $y = x^3 - 2$
    Ordered pairs: (−1, −3)
    (0, −2)
    (1, −1)
    (2, 6)

Objective 2 Exercises

21. $d = \sqrt{(4-3)^2 + (1-5)^2}$
    $= \sqrt{1 + 16}$
    $d = \sqrt{17}$
    $x_m = \dfrac{3+4}{2} = \dfrac{7}{2}$
    $y_m = \dfrac{1+5}{2} = 3$
    The length is $\sqrt{17}$ and the midpoint is $\left(\dfrac{7}{2}, 3\right)$.

23. $d = \sqrt{(-2-0)^2 + (4-3)^2}$
    $= \sqrt{4 + 1}$
    $d = \sqrt{5}$
    $x_m = \dfrac{0 + (-2)}{2} = -1$
    $y_m = \dfrac{3+4}{2} = \dfrac{7}{2}$
    The length is $\sqrt{5}$ and the midpoint is $\left(-1, \dfrac{7}{2}\right)$.

25. $d = \sqrt{[2-(-3)]^2 + [-4-(-5)]^2}$
$= \sqrt{5^2 + 1^2}$
$d = \sqrt{26}$
$x_m = \dfrac{-3+2}{2} = -\dfrac{1}{2}$
$y_m = \dfrac{-5+(-4)}{2} = -\dfrac{9}{2}$

The length is $\sqrt{26}$ and the midpoint is $\left(-\dfrac{1}{2}, -\dfrac{9}{2}\right)$.

27. $d = \sqrt{(-1-5)^2 + [5-(-2)]^2}$
$= \sqrt{(-6)^2 + 7^2}$
$d = \sqrt{85}$
$x_m = \dfrac{5+(-1)}{2} = \dfrac{4}{2} = 2$
$y_m = \dfrac{-2+5}{2} = \dfrac{3}{2}$

The length is $\sqrt{85}$ and the midpoint is $\left(2, \dfrac{3}{2}\right)$.

29. $d = \sqrt{(2-5)^2 + [-5-(-5)]^2}$
$= \sqrt{(-3)^2 + 0^2}$
$d = 3$
$x_m = \dfrac{5+2}{2} = \dfrac{7}{2}$
$y_m = \dfrac{-5+(-5)}{2} = -5$

The length is 3 and the midpoint is $\left(\dfrac{7}{2}, -5\right)$.

Objective 3 Exercises

31. **a.** The number of calories can be read by looking at the *y*-axis. Each dotted line represents an increment of 250 The number of calories for a hamburger is 275.

    b. The number of milligrams can be read by looking at the *x*-axis. Each dotted line on the *x*-axis represents an increment of 100. The number of milligrams of sodium in a Big Mac is 1100.

33. **a.** Look for 200 on the *y*-axis. Then look for the first point that lies above the 200 horizontal line. This point has an *x*-coordinate of 1994.

    b. Strategy
    To find the percent increase, you apply the formula $\dfrac{\text{increase in number}}{\text{original number}}$.
    Then apply this percentage to the 1996 usage.

    Solution
    $\dfrac{440 - 338}{338} = \dfrac{102}{338} = 30.2\%$
    $(1 + 0.302)440 = 572.88$
    573 million people would be using cell phones in 1997.

35. Label the years on the *x*-axis. The *y*-values represent percent so the values on the *y*-axis should go from 0 to 100. Since some of the values are between multiples of 10, the graph would be easier to read with increments of 5.

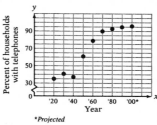
*Projected

Applying Concepts 3.1

37. Ordered pairs: (−2, 4)
    (−1, 1)
    (0, 0)
    (1, 1)
    (2, 4)

39.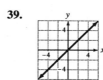

Section 3.2

Concept Review 3.2

1. Sometimes true
   By definition, a function cannot have different second coordinates with the same first coordinate.

3. Sometimes true
   The function $f(x) = \dfrac{2}{x-3}$ is not defined for $x = 3$.

5. Always true

Objective 1 Exercises

9. Function

11. Function

13. Function

15. Not a function

17. Yes, this table defines a function because no weight occurs more than once.

21. $f(x) = 5x - 4$
$f(3) = 5(3) - 4$
$f(3) = 15 - 4$
$f(3) = 11$

23. $f(x) = 5x - 4$
$f(0) = 5(0) - 4$
$f(0) = -4$

25. $G(t) = 4 - 3t$
$G(0) = 4 - 3(0)$
$G(0) = 4$

27. $G(t) = 4 - 3t$
$G(-2) = 4 - 3(-2)$
$G(-2) = 4 + 6$
$G(-2) = 10$

29. $q(r) = r^2 - 4$
$q(3) = 3^2 - 4$
$q(3) = 9 - 4$
$q(3) = 5$

31. $q(r) = r^2 - 4$
$q(-2) = (-2)^2 - 4$
$q(-2) = 4 - 4$
$q(-2) = 0$

33. $F(x) = x^2 + 3x - 4$
$F(4) = 4^2 + 3(4) - 4$
$F(4) = 16 + 12 - 4$
$F(4) = 24$

35. $F(x) = x^2 + 3x - 4$
$F(-3) = (-3)^2 + 3(-3) - 4$
$F(-3) = 9 - 9 - 4$
$F(-3) = -4$

37. $H(p) = \dfrac{3p}{p+2}$
$H(1) = \dfrac{3(1)}{1+2}$
$H(1) = \dfrac{3}{3}$
$H(1) = 1$

39. $H(p) = \dfrac{3p}{p+2}$
$H(t) = \dfrac{3t}{t+2}$

41. $s(t) = t^3 - 3t + 4$
$s(-1) = (-1)^3 - 3(-1) + 4$
$s(-1) = -1 + 3 + 4$
$s(-1) = 6$

43. $s(t) = t^3 - 3t + 4$
$s(a) = a^3 - 3a + 4$

45. **a.** To find the tax for an income of $40,135, find the interval where $40,135 lies. 40,135 is between 40,100 and 40,150. The tax for this interval is $2387.

b. The income $40,249 is between 40,200 and 40,250. The tax for this interval is $2393.

47. **a.** $3000

b. $950

49. Domain = {1, 2, 3, 4, 5}
Range = {1, 4, 7, 10, 13}

51. Domain = {0, 2, 4, 6}
Range = {1, 2, 3, 4}

53. Domain = {1, 3, 5, 7, 9}
Range = {0}

55. Domain = {−2, −1, 0, 1, 2}
Range = {0, 1, 2}

57. $x = 1$

59. $x = -8$

61. No values are excluded.

63. No values are excluded.

65. $x = 0$

67. No values are excluded.

69. No values are excluded.

71. $x = -2$

73. No values are excluded.

75. $f(x) = 4x - 3$
$f(0) = 4(0) - 3 = -3$
$f(1) = 4(1) - 3 = 1$
$f(2) = 4(2) - 3 = 5$
$f(3) = 4(3) - 3 = 9$
Range = {−3, 1, 5, 9}

77. $g(x) = 5x - 8$
$g(-3) = 5(-3) - 8 = -23$
$g(-1) = 5(-1) - 8 = -13$
$g(0) = 5(0) - 8 = -8$
$g(1) = 5(1) - 8 = -3$
$g(3) = 5(3) - 8 = 7$
Range = $\{-23, -13, -8, -3, 7\}$

79. $h(x) = x^2$
$h(-2) = (-2)^2 = 4$
$h(-1) = (-1)^2 = 1$
$h(0) = 0^2 = 0$
$h(1) = 1^2 = 1$
$h(2) = 2^2 = 4$
Range = $\{0, 1, 4\}$

81. $f(x) = 2x^2 - 2x + 2$
$f(-4) = 2(-4)^2 - 2(-4) + 2$
$\quad = 32 + 8 + 2 = 42$
$f(-2) = 2(-2)^2 - 2(-2) + 2$
$\quad = 8 + 4 + 2 = 14$
$f(0) = 2(0)^2 - 2(0) + 2 = 2$
$f(4) = 2(4)^2 - 2(4) + 2$
$\quad = 32 - 8 + 2 = 26$
Range = $\{2, 14, 26, 42\}$

83. $H(x) = \dfrac{5}{1-x}$
$H(-2) = \dfrac{5}{1-(-2)} = \dfrac{5}{3}$
$H(0) = \dfrac{5}{1-0} = 5$
$H(2) = \dfrac{5}{1-2} = -5$
Range = $\left\{-5, \dfrac{5}{3}, 5\right\}$

85. $f(x) = \dfrac{2}{x-4}$
$f(-2) = \dfrac{2}{-2-4} = -\dfrac{1}{3}$
$f(0) = \dfrac{2}{0-4} = -\dfrac{1}{2}$
$f(2) = \dfrac{2}{2-4} = -1$
$f(6) = \dfrac{2}{6-4} = 1$
Range = $\left\{-1, -\dfrac{1}{2}, -\dfrac{1}{3}, 1\right\}$

87. $H(x) = 2 - 3x - x^2$
$H(-5) = 2 - 3(-5) - (-5)^2$
$\quad = 2 + 15 - 25 = -8$
$H(0) = 2 - 3(0) - 0^2 = 2$
$H(5) = 2 - 3(5) - 5^2$
$\quad = 2 - 15 - 25 = -38$
Range = $\{-38, -8, 2\}$

Applying Concepts 3.2

89. $P(x) = 4x + 7$; $P(-2 + h) - P(-2)$
$P(-2+h) - P(-2) = 4(-2+h) + 7 - [4(-2) + 7]$
$\quad = -8 + 4h + 7 + 8 - 7$
$\quad = 4h$

91. $f(z) = 8 - 3z$; $f(-3+h) - f(-3)$
$f(-3+h) - f(-3) = 8 - 3(-3+h) - [8 - 3(-3)]$
$\quad = 8 + 9 - 3h - 8 - 9$
$\quad = -3h$

93. Find the value of the function
$s = f(v) = 0.017v^2$ when $v = 60$.
$s = 0.017v^2$
$s = 0.017(60)^2$
$s = 61.2$
A car will skid 61.2 feet.

95. a. 20 ft/s

b. 28 ft/s

97. a. 60°F

b. 50°F

99. a.

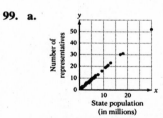

b. The 435 representatives to the House are apportioned among the states according to their populations.

Section 3.3

Concept Review 3.3

1. Never true
   The equation of a linear function is a first-degree equation. This equation has a variable in the denominator, so it is not a first-degree equation.

3. Always true

5. Sometimes true
   The only time that this occurs is when the line passes through the point (0, 0).

Objective 1 Exercises

3.

5.

7.

9.

11.

Objective 2 Exercises

15. $2x + y = -3$
    $y = -2x - 3$

17. $x - 4y = 8$
    $-4y = -x + 8$
    $y = \dfrac{1}{4}x - 2$

19. $x - 3y = 0$
    $-3y = -x$
    $y = \dfrac{1}{3}x$

21. $3x - y = -2$
    $-y = -3x - 2$
    $y = 3x + 2$

23. x-intercept: y-intercept
    $x - 2y = -4$ $x - 2y = -4$
    $x - 2(0) = -4$ $0 - 2y = -4$
    $x = -4$ $-2y = -4$
                        $y = 2$
    $(-4, 0)$ $(0, 2)$

25. x-intercept: y-intercept:
    $2x - 3y = 9$ $2x - 3y = 9$
    $2x - 3(0) = 9$ $2(0) - 3y = 9$
    $2x = 9$ $-3y = 9$
    $x = \dfrac{9}{2}$ $y = -3$

    $\left(\dfrac{9}{2}, 0\right)$ $(0, -3)$

27. x-intercept:
$2x - y = 4$
$2x - 0 = 4$
$2x = 4$
$x = 2$
$(2, 0)$

y-intercept:
$2x - y = 4$
$2(0) - y = 4$
$-y = 4$
$y = -4$
$(0, -4)$

29. x-intercept:
$3x + 2y = 5$
$3x + 2(0) = 5$
$3x = 5$
$x = \frac{5}{3}$
$\left(\frac{5}{3}, 0\right)$

y-intercept:
$3x + 2y = 5$
$3(0) + 2y = 5$
$2y = 5$
$y = \frac{5}{2}$
$\left(0, \frac{5}{2}\right)$

31. x-intercept:
$3x + 2y = 4$
$3x + 2(0) = 4$
$3x = 4$
$x = \frac{4}{3}$
$\left(\frac{4}{3}, 0\right)$

y-intercept:
$3x + 2y = 4$
$3(0) + 2y = 4$
$2y = 4$
$y = 2$
$(0, 2)$

33. x-intercept:
$3x - 5y = 9$
$3x - 5(0) = 9$
$3x = 9$
$x = 3$
$(3, 0)$

y-intercept:
$3x - 5y = 9$
$3(0) - 5y = 9$
$-5y = 9$
$y = -\frac{9}{5}$
$\left(0, -\frac{9}{5}\right)$

35. The laborer will earn $169.50 for working 30 hours.

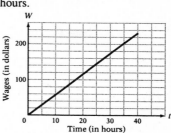

37. The realtor will earn $4000 for selling $60,000 worth of property.

39. The caterer will charge $614 for 120 hot appetizers.

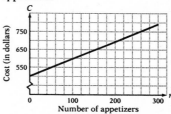

Applying Concepts 3.3

41. $s(p) = 0.80p$
$s(p) = 0.80(200)$
$s(p) = 160$
The sale price is $160.

43.

45.

47.

49. a.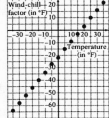

b. Yes

c. Domain: {−35, −30, −25, −20, −15, −10, −5, 0, 5, 10, 15, 20, 25, 30, 35}
Range: {−64, −58, −52, −46, −40, −34, −27, −22, −15, −9, −3, 3, 10, 16, 22}

d. $f(0) \approx -21 \neq -22$ (This is the only ordered pair that does not satisfy the function.) The ordered pair $(0, -22)$ does not satisfy the function.

e. $f(x) = 1.2321x - 21.2667$
$y = 1.2321x - 21.2667$
$0 = 1.2321x - 21.2667$
$21.2667 = 1.2321x$
$17 \approx x$
The x-intercept is $(17, 0)$.
Given a wind speed of 10 mph, the wind-chill factor is 0°F when the air temperature is 17°F.
$f(x) = 1.2321x - 21.2667$
$y = 1.2321x - 21.2667$
$y = 1.2321(0) - 21.2667$
$y = -21.2667$
$y \approx -21.3$
The y-intercept is $(0, -21.3)$.
Given a wind speed of 10 mph, the wind-chill factor is −21.3°F when the air temperature is 0°F.

Section 3.4

Concept Review 3.4

1. Always true

3. Never true
   The slope of a vertical line is undefined.

Objective 1 Exercises

3. $P_1(1, 3)$, $P_2(3, 1)$
$m = \dfrac{y_2 - y_1}{x_2 - x_1} = \dfrac{1-3}{3-1} = \dfrac{-2}{2} = -1$
The slope is −1.

5. $P_1(-1, 4)$, $P_2(2, 5)$
$m = \dfrac{y_2 - y_1}{x_2 - x_1} = \dfrac{5-4}{2-(-1)} = \dfrac{1}{3}$
The slope is $\dfrac{1}{3}$.

7. $P_1(-1, 3)$, $P_2(-4, 5)$
$m = \dfrac{y_2 - y_1}{x_2 - x_1} = \dfrac{5-3}{-4-(-1)} = \dfrac{2}{-3} = -\dfrac{2}{3}$
The slope is $-\dfrac{2}{3}$.

9. $P_1(0, 3)$, $P_2(4, 0)$
$m = \dfrac{y_2 - y_1}{x_2 - x_1} = \dfrac{0-3}{4-0} = \dfrac{-3}{4} = -\dfrac{3}{4}$
The slope is $-\dfrac{3}{4}$.

11. $P_1(2, 4)$, $P_2(2, -2)$
$m = \dfrac{y_2 - y_1}{x_2 - x_1} = \dfrac{-2-4}{2-2} = \dfrac{-6}{0}$
The slope is undefined.

13. $P_1(2, 5)$, $P_2(-3, -2)$
$m = \dfrac{y_2 - y_1}{x_2 - x_1} = \dfrac{-2-5}{-3-2} = \dfrac{-7}{-5} = \dfrac{7}{5}$
The slope is $\dfrac{7}{5}$.

15. $P_1(2, 3)$, $P_2(-1, 3)$
$m = \dfrac{y_2 - y_1}{x_2 - x_1} = \dfrac{3-3}{-1-2} = \dfrac{0}{-3} = 0$
The line has zero slope.

17. $P_1(0, 4)$, $P_2(-2, 5)$
$m = \dfrac{y_2 - y_1}{x_2 - x_1} = \dfrac{5-4}{-2-0} = \dfrac{1}{-2} = -\dfrac{1}{2}$
The slope is $-\dfrac{1}{2}$.

19. $P_1(-3, -1)$, $P_2(-3, 4)$
$m = \dfrac{y_2 - y_1}{x_2 - x_1} = \dfrac{4-(-1)}{-3-(-3)} = \dfrac{5}{0}$
The slope is undefined.

21. $m = \dfrac{1250 - 800}{8 - 3} = \dfrac{450}{5} = 90$
The price of the Flutter baby was increasing at a rate of $90/month.

23. $m = \dfrac{13 - 6}{40 - 180} = \dfrac{7}{-140} = -0.05$
For each mile driven, the number of gallons in the tank is reduced by 0.05 gallons.

25. **Strategy**
Find lines that have slope that match the rates of the runners.

Solution
Lois has the highest rate so the line with the steepest slope represents Lois. Line A has the steepest slope so it represents Lois's distance. The line which represents Tanya's distance must have a slope of $\frac{6}{1}$. Line B goes through the points (1, 6) and (0, 0) so its slope is $\frac{6}{1}$. So line B represents Tanya's distance.
In one hour the difference between Lois's and Tanya's distances is 3. Line C goes through the point (1, 3) so it represents the distance between Lois and Tanya.

27. **a.** The slope for a ramp that is 6 inches high and 5 feet long is $\frac{6 \text{ inches}}{60 \text{ inches}} = \frac{1}{10}$ or 0.1.
$\frac{1}{12} = 0.0833$. Since $0.1 > 0.0833$, the ramp does not meet the ANSI requirements.

   b. The slope for a ramp that is 12 inches high and 170 inches long is $\frac{12}{170}$ or 0.07. Since $0.07 < 0.08$, the ramp does meet the ANSI requirements.

Objective 2 Exercises

29.

31.

33.

35. $x - 3y = 3$
$-3y = -x + 3$
$y = \frac{1}{3}x - 1$

37. $4x + y = 2$
$y = -4x + 2$

39.

41.

Applying Concepts 3.4

43. increases by 2

45. increases by $\frac{1}{2}$

47. $P_1 = (3, 2)$
$P_2 = (4, 6)$
$P_3 = (5, k)$

P_1 to P_2: $m = \frac{6-2}{4-3} = 4$

The slope from P_1 to P_3 and that from P_2 to P_3 must also be 4.

Set the slope from P_1 to P_3 equal to 4.

$\frac{2-k}{3-5} = 4$

$\frac{2-k}{-2} = 4$

$2 - k = -8$

$k = 10$

This checks out against P_2 to P_3, so $k = 10$.

49. a.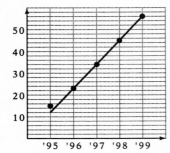

c. $y = 11x - 21933$
$y = 11(2000) - 21933$
$y = 67$
The estimated number of on-line households in 2000 is 67 million.

d. Increase between 1997 and 1998
$= 45 - 34$
$= 11$ million.

e. The number of on-line households is increasing at a rate of 11 million households per year.

f. $y = 11x - 21,933$
$y = 11(1994) - 21,933$
$y = 1$
The estimated number of on-line households in 1994 is 1 million.

g. Sample answer: The rate of new household users will slow down as most households become on-line.

Section 3.5

Concept Review 3.5

1. Always true

3. Never true
The point-slope formula is given by
$y - y_1 = m(x - x_1)$.

5. Never true
The line represented by the equation $y = 2x - \frac{1}{2}$ has slope 2 and y-intercept $-\frac{1}{2}$.

7. Never true
A horizontal line has zero slope.

Objective 1 Exercises

3. $m = 2, b = 5$
$y = mx + b$
$y = 2x + 5$
The equation of the line is $y = 2x + 5$.

5. $m = \frac{1}{2}, (x_1, y_1) = (2, 3)$
$y - y_1 = m(x - x_1)$
$y - 3 = \frac{1}{2}(x - 2)$
$y - 3 = \frac{1}{2}x - 1$
$y = \frac{1}{2}x + 2$
The equation of the line is $y = \frac{1}{2}x + 2$.

7. $m = -\frac{5}{3}, (x_1, y_1) = (3, 0)$
$y - y_1 = m(x - x_1)$
$y - 0 = -\frac{5}{3}(x - 3)$
$y = -\frac{5}{3}x + 5$
The equation of the line is $y = -\frac{5}{3}x + 5$.

9. $m = -3, (x_1, y_1) = (-1, 7)$
$y - y_1 = m(x - x_1)$
$y - 7 = -3[x - (-1)]$
$y - 7 = -3(x + 1)$
$y - 7 = -3x - 3$
$y = -3x + 4$
The equation of the line is $y = -3x + 4$.

11. $m = \frac{1}{2}, (x_1, y_1) = (0, 0)$
$y - y_1 = m(x - x_1)$
$y - 0 = \frac{1}{2}(x - 0)$
$y = \frac{1}{2}x$
The equation of the line is $y = \frac{1}{2}x$.

13. $m = 3, (x_1, y_1) = (2, -3)$
$y - y_1 = m(x - x_1)$
$y - (-3) = 3(x - 2)$
$y + 3 = 3x - 6$
$y = 3x - 9$
The equation of the line is $y = 3x - 9$.

15. $m = -\frac{2}{3}, (x_1, y_1) = (3, 5)$
$y - y_1 = m(x - x_1)$
$y - 5 = -\frac{2}{3}(x - 3)$
$y - 5 = -\frac{2}{3}x + 2$
$y = -\frac{2}{3}x + 7$
The equation of the line is $y = -\frac{2}{3}x + 7$.

17. $m = -1$, $(x_1, y_1) = (0, -3)$
$$y - y_1 = m(x - x_1)$$
$$y - (-3) = -1(x - 0)$$
$$y + 3 = -x + 0$$
$$y = -x - 3$$
The equation of the line is $y = -x - 3$.

19. The slope is undefined; $(x_1, y_1) = (3, -4)$.
The line is a vertical line. All points on the line have an abscissa of 3. The equation of the line is $x = 3$.

21. $m = 0$, $(x_1, y_1) = (-2, -3)$
$$y - y_1 = m(x - x_1)$$
$$y - (-3) = 0[x - (-2)]$$
$$y + 3 = 0$$
$$y = -3$$
The equation of the line is $y = -3$.

23. $m = -2$, $(x_1, y_1) = (4, -5)$
$$y - y_1 = m(x - x_1)$$
$$y - (-5) = -2(x - 4)$$
$$y + 5 = -2x + 8$$
$$y = -2x + 3$$
The equation of the line is $y = -2x + 3$.

25. The slope is undefined; $(x_1, y_1) = (-5, -1)$.
The line is a vertical line. All points on the line have an abscissa of -5. The equation of the line is $x = -5$.

Objective 2 Exercises

27. $P_1(0, 2)$, $P_2(3, 5)$
$$m = \frac{y_2 - y_1}{x_2 - x_1} = \frac{5 - 2}{3 - 0} = \frac{3}{3} = 1$$
$$y - y_1 = m(x - x_1)$$
$$y - 2 = 1(x - 0)$$
$$y - 2 = x$$
$$y = x + 2$$
The equation of the line is $y = x + 2$.

29. $P_1(0, -3)$, $P_2(-4, 5)$
$$m = \frac{y_2 - y_1}{x_2 - x_1} = \frac{5 - (-3)}{-4 - 0} = \frac{8}{-4} = -2$$
$$y - y_1 = m(x - x_1)$$
$$y - (-3) = -2(x - 0)$$
$$y + 3 = -2x$$
$$y = -2x - 3$$
The equation of the line is $y = -2x - 3$.

31. $P_1(-1, 3)$, $P_2(2, 4)$
$$m = \frac{y_2 - y_1}{x_2 - x_1} = \frac{4 - 3}{2 - (-1)} = \frac{1}{3}$$
$$y - y_1 = m(x - x_1)$$
$$y - 3 = \frac{1}{3}[x - (-1)]$$
$$y - 3 = \frac{1}{3}(x + 1)$$
$$y - 3 = \frac{1}{3}x + \frac{1}{3}$$
$$y = \frac{1}{3}x + \frac{10}{3}$$
The equation of the line is $y = \frac{1}{3}x + \frac{10}{3}$.

33. $P_1(0, 3)$, $P_2(2, 0)$
$$m = \frac{y_2 - y_1}{x_2 - x_1} = \frac{0 - 3}{2 - 0} = \frac{-3}{2} = -\frac{3}{2}$$
$$y - y_1 = m(x - x_1)$$
$$y - 3 = -\frac{3}{2}(x - 0)$$
$$y - 3 = -\frac{3}{2}x$$
$$y = -\frac{3}{2}x + 3$$
The equation of the line is $y = -\frac{3}{2}x + 3$.

35. $P_1(-2, -3)$, $P_2(-1, -2)$
$$m = \frac{y_2 - y_1}{x_2 - x_1} = \frac{-2 - (-3)}{-1 - (-2)} = \frac{1}{1} = 1$$
$$y - y_1 = m(x - x_1)$$
$$y - (-3) = 1[x - (-2)]$$
$$y + 3 = x + 2$$
$$y = x - 1$$
The equation of the line is $y = x - 1$.

37. $P_1(2, 3)$, $P_2(5, 5)$
$$m = \frac{y_2 - y_1}{x_2 - x_1} = \frac{5 - 3}{5 - 2} = \frac{2}{3}$$
$$y - y_1 = m(x - x_1)$$
$$y - 3 = \frac{2}{3}(x - 2)$$
$$y - 3 = \frac{2}{3}x - \frac{4}{3}$$
$$y = \frac{2}{3}x + \frac{5}{3}$$
The equation of the line is $y = \frac{2}{3}x + \frac{5}{3}$.

39. $P_1(2, 0)$, $P_2(0, -1)$

$m = \dfrac{y_2 - y_1}{x_2 - x_1} = \dfrac{-1-0}{0-2} = \dfrac{-1}{-2} = \dfrac{1}{2}$

$y - y_1 = m(x - x_1)$

$y - 0 = \dfrac{1}{2}(x - 2)$

$y = \dfrac{1}{2}x - 1$

The equation of the line is $y = \dfrac{1}{2}x - 1$.

41. $P_1(3, -4)$, $P_2(-2, -4)$

$m = \dfrac{y_2 - y_1}{x_2 - x_1} = \dfrac{-4-(-4)}{-2-3} = \dfrac{0}{-5} = 0$

$y - y_1 = m(x - x_1)$

$y - (-4) = 0(x - 3)$

$y + 4 = 0$

$y = -4$

The equation of the line is $y = -4$.

43. $P_1(0, 0)$, $P_2(4, 3)$

$m = \dfrac{y_2 - y_1}{x_2 - x_1} = \dfrac{3-0}{4-0} = \dfrac{3}{4}$

$y - y_1 = m(x - x_1)$

$y - 0 = \dfrac{3}{4}(x - 0)$

$y = \dfrac{3}{4}x$

The equation of the line is $y = \dfrac{3}{4}x$.

45. $P_1(-2, 5)$, $P_2(-2, -5)$

$m = \dfrac{y_2 - y_1}{x_2 - x_1} = \dfrac{-5-5}{-2-(-2)} = \dfrac{-10}{0}$

The slope is undefined. The line is a vertical line. All points on the line have an abscissa of –2. The equation of the line is $x = -2$.

47. $P_1(2, 1)$, $P_2(-2, -3)$

$m = \dfrac{y_2 - y_1}{x_2 - x_1} = \dfrac{-3-1}{-2-2} = \dfrac{-4}{-4} = 1$

$y - y_1 = m(x - x_1)$

$y - 1 = 1(x - 2)$

$y - 1 = x - 2$

$y = x - 1$

The equation of the line is $y = x - 1$.

49. $P_1(0, 3)$, $P_2(3, 0)$

$m = \dfrac{y_2 - y_1}{x_2 - x_1} = \dfrac{0-3}{3-0} = \dfrac{-3}{3} = -1$

$y - y_1 = m(x - x_1)$

$y - 3 = -1(x - 0)$

$y - 3 = -x$

$y = -x + 3$

The equation of the line is $y = -x + 3$.

Objective 3 Exercises

51. Strategy
Use the slope-intercept form to determine the equation of the line.

Solution
y-intercept $= 0$
$m = 1200$
$y = mx + b$
$y = 1200x + 0$
The equation that represents the ascent of the plane is $y = 1200x$ where x stands for the number of minutes after take-off. To find the height of the plane when $x = 11$
$y = 1200(11)$
$y = 13{,}200$
The plane will be 13,200 feet in the air after 11 minutes.

53. Strategy
Use the slope-intercept form to find the equation.

Solution
The y-intercept is 4.95 because it is the charge for 0 minutes.
$b = 4.95$
$m = 0.59$
$y = mx + b$
$y = 0.59x + 4.95$
The equation for the monthly cost of the phone is $y = 0.59x + 4.95$.
To find the cost for 13 minutes of use,
$y = 0.59(13) + 4.95$
$y = 12.62$
It costs $12.62 to use the cellular phone for 13 minutes.

55. Strategy
Use the point-slope formula.

Solution
$(x_1, y_1) = (2, 126)$, $(x_2, y_2) = (3, 189)$

$m = \dfrac{189 - 126}{3 - 2} = 63$

$y - y_1 = m(x - x_1)$
$y - y_1 = 63(x - x_1)$
$y - 126 = 63(x - 2)$
$y - 126 = 63x - 126$
$y = 63x$

The equation to approximate the number of calories in a hamburger is $y = 63x$ where x represents the ounces in the hamburger. To predict the number of calories in a 5-ounce serving of hamburger,
$y = 63 \cdot 5$
$y = 315$
A 5-ounce serving of hamburger will contain 315 calories.

57. **Strategy**
Use the point-slope formula.

Solution
$(x_1, y_1) = (1927, 33.5)$, $(x_2, y_2) = (1997, 3.3)$
$m = \dfrac{y_2 - y_1}{x_2 - x_1} = \dfrac{3.3 - 33.5}{1997 - 1927} = \dfrac{-30.2}{70} = -0.4$
$y - y_1 = m(x - x_1)$
$y - 33.5 = -0.4(x - 1927)$
$y - 33.5 = -0.4x + 770.8$
$\quad\quad y = -0.4x + 804.3$
The equation to predict how long a flight took is $y = -0.4x + 804.3$ where x represents the year. To predict how long a flight between the two cities would have taken in 1967,
$y = -0.4(1967) + 804.3$
$\quad = -786.8 + 804.3 = 17.5$
It would have taken 17.5 hours in 1967 for a plane to cross the Atlantic.

59. **Strategy**
Use the slope-intercept form of the equation.

Solution
The y-intercept is 16 gallons because 0 miles have been driven at the beginning of the trip. The slope is -0.032.
$y = mx + b$
$y = -0.032x + 16$
The equation to find the number of gallons in the tank is $y = -0.032x + 16$ where x represents the number of miles traveled. To predict the number of gallons in the tank after 150 miles,
$y = -0.032(150) + 16$
$y = 11.2$
11.2 gallons will be in the tank when 150 miles have been driven.

Applying Concepts 3.5

61. To find the x-intercept, set y to 0.
$0 = mx + b$
$-b = mx$
$-\dfrac{b}{m} = x$
The x-intercept is $\left(-\dfrac{b}{m}, 0\right)$.

63. Find the equation of the line. The two points are $(1, 3)$ and $(-1, 5)$.
$m = \dfrac{5 - 3}{-1 - 1} = \dfrac{2}{-2} = -1$
$y - 3 = -1(x - 1)$
$y - 3 = -x + 1$
$\quad y = -x + 4$
The function is $f(x) = -x + 4$
$\quad\quad\quad\quad\quad\quad f(4) = -4 + 4 = 0$

65. If m is a given constant, changing b causes the graph of the line to move up or down.

Section 3.6

Concept Review 3.6

1. Sometimes true
The only time perpendicular lines have the same y-intercept is when the point of intersection of the two lines is on the y-axis.

3. Never true
The product of the slopes of two perpendicular lines is equal to -1.
$\left(\dfrac{3}{2}\right)\left(-\dfrac{3}{2}\right) = -\dfrac{9}{4}$; the lines are not perpendicular.

5. Always true

7. Always true

Objective 1 Exercises

3. $x = -2$ is a vertical line.
$y = 3$ is a horizontal line.
The lines are perpendicular.

5. $x = -3$ is a vertical line.
$y = \dfrac{1}{3}$ is a horizontal line.
The lines are not parallel.

7. $y = \dfrac{2}{3}x - 4$, $m_1 = \dfrac{2}{3}$
$y = -\dfrac{3}{2}x - 4$, $m_2 = -\dfrac{3}{2}$
$m_1 \neq m_2$
The lines are not parallel.

9. $y = \dfrac{4}{3}x - 2$, $m_1 = \dfrac{4}{3}$
$y = -\dfrac{3}{4}x + 2$, $m_2 = -\dfrac{3}{4}$
$m_1 \cdot m_2 = \dfrac{4}{3}\left(-\dfrac{3}{4}\right) = -1$
The lines are perpendicular.

11. $2x + 3y = 2$
$\quad 3y = -2x + 2$
$\quad\quad y = -\dfrac{2}{3}x + \dfrac{2}{3}$, $m_1 = -\dfrac{2}{3}$
$2x + 3y = -4$
$\quad 3y = -2x - 4$
$\quad\quad y = -\dfrac{2}{3}x - \dfrac{4}{3}$, $m_2 = -\dfrac{2}{3}$
$m_1 = m_2 = -\dfrac{2}{3}$
The lines are parallel.

13. $x - 4y = 2$
    $-4y = -x + 2$
    $y = \frac{1}{4}x - \frac{1}{2}$, $m_1 = \frac{1}{4}$
    $4x + y = 8$
    $y = -4x + 8$, $m_2 = -4$
    $m_1 \cdot m_2 = \frac{1}{4}(-4) = -1$
    The lines are perpendicular.

15. $m_1 = \frac{6-2}{1-3} = \frac{4}{-2} = -2$
    $m_2 = \frac{-1-3}{-1-(-1)} = \frac{-4}{0}$
    $m_1 \neq m_2$
    The lines are not parallel.

17. $m_1 = \frac{-1-2}{4-(-3)} = \frac{-3}{7} = -\frac{3}{7}$
    $m_2 = \frac{-4-3}{-2-1} = \frac{-7}{-3} = \frac{7}{3}$
    $m_1 \cdot m_2 = -\frac{3}{7}\left(\frac{7}{3}\right) = -1$
    The lines are perpendicular.

19. $m_1 = \frac{2-0}{0-(-5)} = \frac{2}{5}$
    $m_2 = \frac{-1-1}{0-5} = \frac{-2}{-5} = \frac{2}{5}$
    $m_1 = m_2 = \frac{2}{5}$
    The lines are parallel.

21. $2x - 3y = 2$
    $-3y = -2x + 2$
    $y = \frac{2}{3}x - \frac{2}{3}$
    $m = \frac{2}{3}$
    $y - y_1 = m(x - x_1)$
    $y - (-4) = \frac{2}{3}[x - (-2)]$
    $y + 4 = \frac{2}{3}(x + 2)$
    $y + 4 = \frac{2}{3}x + \frac{4}{3}$
    $y = \frac{2}{3}x - \frac{8}{3}$
    The equation of the line is $y = \frac{2}{3}x - \frac{8}{3}$.

23. $y = -3x + 4$
    $m_1 = -3$
    $m_1 \cdot m_2 = -1$
    $-3 \cdot m_2 = -1$
    $m_2 = \frac{1}{3}$
    $y - y_1 = m(x - x_1)$
    $y - 1 = \frac{1}{3}(x - 4)$
    $y - 1 = \frac{1}{3}x - \frac{4}{3}$
    $y = \frac{1}{3}x - \frac{1}{3}$
    The equation of the line is $y = \frac{1}{3}x - \frac{1}{3}$.

25. $3x - 5y = 2$
    $-5y = -3x + 2$
    $y = \frac{3}{5}x - \frac{2}{5}$
    $m_1 = \frac{3}{5}$
    $m_1 \cdot m_2 = -1$
    $\frac{3}{5} \cdot m_2 = -1$
    $m_2 = -\frac{5}{3}$
    $y - y_1 = m(x - x_1)$
    $y - (-3) = -\frac{5}{3}[x - (-1)]$
    $y + 3 = -\frac{5}{3}(x + 1)$
    $y + 3 = -\frac{5}{3}x - \frac{5}{3}$
    $y = -\frac{5}{3}x - \frac{14}{3}$
    The equation of the line is $y = -\frac{5}{3}x - \frac{14}{3}$.

Applying Concepts 3.6

27. Write the equations of the lines in slope-intercept form.
    (1) $A_1 x + B_1 y = C_1$
    $B_1 y = C_1 - A_1 x$
    $y = \frac{C_1}{B_1} - \frac{A_1}{B_1}x$
    (2) $A_2 x + B_2 y = C_2$
    $B_2 y = C_2 - A_2 x$
    $y = \frac{C_2}{B_2} - \frac{A_2}{B_2}x$
    The slope of the second line must be the negative reciprocal of the first for the lines to be perpendicular, so $\frac{A_1}{B_1} = -\frac{B_2}{A_2}$.

29. Strategy
Find a line perpendicular to either of the given lines. The line must form a triangle with the other two lines, so the line cannot go through the point (6, –1).

Solution
Start with the line $y = -\frac{1}{2}x + 2$.
The slope of a line perpendicular to this line is 2, and the line has the form $y = 2x + b$. Since the line cannot go through the point (6, –1), the value of b cannot be –13. So one possible solution is any equation of the form $y = 2x + b$, where $b \ne -13$.

Now consider the line $y = \frac{2}{3}x + 5$. The slope of a line perpendicular to this line is $-\frac{3}{2}$, and the line has the form $y = -\frac{3}{2}x + c$. Since the line cannot go through the point (6, –1), the value of c cannot be 8. So the other possible solution is any equation of the form $y = -\frac{3}{2}x + c$, where $c \ne 8$.

31. The equation of the tangent line is $y = -x + 9$.
y-intercept: (0, 9)
x-intercept: (9, 0)

Section 3.7

Concept Review 3.7

1. Never true
   The solution of a linear inequality is a half-plane.

3. Always true

5. Always true

Objective 1 Exercises

3. $y \le \frac{3}{2}x - 3$

5. $y < \frac{4}{5}x - 2$

7. $x < -\frac{1}{3}y + 2$

9. $x + 3y < 4$
$3y < -x + 4$
$y < -\frac{1}{3}x + \frac{4}{3}$

11. $2x + 3y \ge 6$
$3y \ge -2x + 6$
$y \ge -\frac{2}{3}x + 2$

13. $-x + 2y > -8$
$2y > x - 8$
$y > \frac{1}{2}x - 4$

15. $y - 4 < 0$
$y < 4$

17. $6x + 5y < 15$
$5y < -6x + 15$
$y < -\frac{6}{5}x + 3$

19. $-5x + 3y \geq -12$
$3y \geq 5x - 12$
$y \geq \dfrac{5}{3}x - 4$

Focus on Problem Solving

1. Note that $2 + 100 = 102$, $4 + 98 = 102$, and $6 + 96 = 102$. There are 25 sums of 102. Thus the sum of the first 50 even integers is
$2 + 4 + 6 + \ldots + 100 = 25 \cdot 102 = 2550$

2. Note that $1 + 101 = 102$, $3 + 99 = 102$, and $5 + 97 = 102$. There is an odd number of integers, so all the numbers will not pair. The sum will be 25 pairs of 102 plus 51.
$1 + 3 + 5 + \ldots + 101 = 25 \cdot 102 + 51 = 2601$

3. The sequence $1, 3, 5, \ldots, 101$ is called an arithmetic sequence. From a reference book, we determine that the sum of n terms on an arithmetic sequence is $S = \dfrac{n}{2}(a_1 + a_n)$. Here a_1 is the first term of the sequence and a_n is the last term of the sequence. $n = 51$, $a_1 = 1$, and $a_n = 101$.
$S = \dfrac{n}{2}(a_1 + a_2)$
$S = \dfrac{51}{2}(1 + 101)$
$S = 2601$

4. From the pattern shown,
$\dfrac{1}{1 \cdot 2} + \dfrac{1}{2 \cdot 3} + \dfrac{1}{3 \cdot 4} + \ldots + \dfrac{1}{49 \cdot 50} = \dfrac{49}{50}$.

5.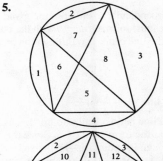

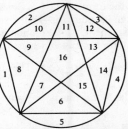

With two points, the chord divides the circle into 2 regions. With three points, the chords divide the circle into 4 or 2^2 regions. With four points, the chords divide the circle into 8 or 2^3 regions. With five points, the chords divide the circle into 16 or 2^4 regions.

Conjecture: When n points are chosen, the chords will divide the circle into 2^{n-1} regions.

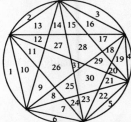

With six points, the chords should divide the circle into $2^5 = 32$ regions. As you can see from the diagram, the circle is divided into 31 regions when 6 points are used. Our conjecture is not true.

6.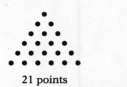

21 points   28 points

The next two triangular numbers are 21 and 28.

Projects and Group Activities

Introduction to Graphing Calculators

1. $y = 2x + 1$

   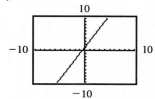

2. $y = -x + 2$

   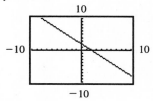

3. $3x + 2y = 6$
   $2y = -3x + 6$
   $y = -\dfrac{3}{2}x + 3$

   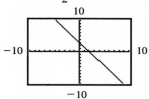

4. $y = 50x$

   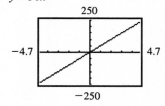

5. $y = \dfrac{2}{3}x - 3$

   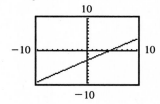

6. $4x + 3y = 75$
   $3y = -4x + 75$
   $y = -\dfrac{4}{3}x + 25$

   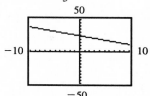

Chapter Review Exercises

1. $y = \dfrac{x}{x-2}$
   $y = \dfrac{4}{4-2}$
   $y = 2$
   The ordered pair is (4, 2).

2. $x_m = \dfrac{x_1 + x_2}{2} = \dfrac{-2+3}{2} = \dfrac{1}{2}$
   $y_m = \dfrac{y_1 + y_2}{2} = \dfrac{4+5}{2} = \dfrac{9}{2}$
   Length $= \sqrt{(x_2 - x_1)^2 + (y_2 - y_1)^2}$
   $= \sqrt{[3-(-2)]^2 + (5-4)^2}$
   $= \sqrt{5^2 + 1^2}$
   $= \sqrt{26}$
   The midpoint is $\left(\dfrac{1}{2}, \dfrac{9}{2}\right)$ and the length is $\sqrt{26}$.

3. $y = x^2 - 2$
   Ordered pairs: (−2, 2)
   (−1, −1)
   (0, −2)
   (1, −1)
   (2, 2)

4.

5. $P(x) = 3x + 4$
   $P(-2) = 3(-2) + 4 = -2$

   $P(a) = 3(a) + 4$
   $P(a) = 3a + 4$

6. Domain = {−1, 0, 1, 2, 5}
   Range = {0, 2, 3}

68 Chapter 3: Linear Functions and Inequalities in Two Variables

7. $f(x) = x^2 - 2$
   $f(-2) = (-2)^2 - 2 = 2$
   $f(-1) = (-1)^2 - 2 = -1$
   $f(0) = 0^2 - 2 = -2$
   $f(1) = 1^2 - 2 = -1$
   $f(2) = 2^2 - 2 = 2$
   Range = $\{-2, -1, 2\}$

8. Division by zero is undefined. Therefore, -4 must be excluded from the domain.

9. To find the x-intercept, let $y = 0$.
   $4x - 6(0) = 12$
   $4x = 12$
   $x = 3$
   The x-intercept is $(3, 0)$.
   To find the y-intercept, let $x = 0$.
   $4(0) - 6y = 12$
   $-6y = 12$
   $y = -2$
   The y-intercept is $(0, -2)$.

10.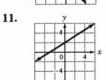

11.

12. $m = \dfrac{y_2 - y_1}{x_2 - x_1}$
    $m = \dfrac{2 - (-2)}{-1 - 3} = \dfrac{4}{-4} = -1$

13. x-intercept y-intercept
    $3x + 2y = -4$ $3x + 2y = -4$
    $3x + 2(0) = -4$ $3(0) + 2y = -4$
    $3x = -4$ $2y = -4$
    $x = -\dfrac{4}{3}$ $y = -2$
    $\left(-\dfrac{4}{3}, 0\right)$ $(0, -2)$

14.

15. Use the point-slope form to find the equation of the line.
    $y - y_1 = m(x - x_1)$
    $y - 4 = \dfrac{5}{2}[x - (-3)]$
    $y - 4 = \dfrac{5}{2}(x + 3)$
    $y - 4 = \dfrac{5}{2}x + \dfrac{15}{2}$
    $y = \dfrac{5}{2}x + \dfrac{23}{2}$
    The equation of the line is $y = \dfrac{5}{2}x + \dfrac{23}{2}$.

16. $P_1(-2, 4),\ P_2(4, -3)$
    $m = \dfrac{y_2 - y_1}{x_2 - x_1} = \dfrac{-3 - 4}{4 - (-2)} = \dfrac{-7}{6} = -\dfrac{7}{6}$
    $y - y_1 = m(x - x_1)$
    $y - 4 = -\dfrac{7}{6}[x - (-2)]$
    $y - 4 = -\dfrac{7}{6}(x + 2)$
    $y - 4 = -\dfrac{7}{6}x - \dfrac{7}{3}$
    $y = -\dfrac{7}{6}x + \dfrac{5}{3}$
    The equation of the line is $y = -\dfrac{7}{6}x + \dfrac{5}{3}$.

17. $y = -3x + 4$
    $m = -3$
    $y - y_1 = m(x - x_1)$
    $y - (-2) = -3(x - 3)$
    $y + 2 = -3x + 9$
    $y = -3x + 7$
    The equation of the line is $y = -3x + 7$.

18. $2x - 3y = 4$
    $-3y = -2x + 4$
    $y = \dfrac{2}{3}x - \dfrac{4}{3}$
    $m = \dfrac{2}{3}$
    $y - y_1 = m(x - x_1)$
    $y - (-4) = \dfrac{2}{3}[x - (-2)]$
    $y + 4 = \dfrac{2}{3}(x + 2)$
    $y + 4 = \dfrac{2}{3}x + \dfrac{4}{3}$
    $y = \dfrac{2}{3}x - \dfrac{8}{3}$
    The equation of the line is $y = \dfrac{2}{3}x - \dfrac{8}{3}$.

19. $y = -\dfrac{2}{3}x + 6$

$m_1 = -\dfrac{2}{3}$

$m_1 \cdot m_2 = -1$

$-\dfrac{2}{3}m_2 = -1$

$m_2 = \dfrac{3}{2}$

$y - y_1 = m(x - x_1)$

$y - 5 = \dfrac{3}{2}(x - 2)$

$y - 5 = \dfrac{3}{2}x - 3$

$y = \dfrac{3}{2}x + 2$

The equation of the line is $y = \dfrac{3}{2}x + 2$.

20. $4x - 2y = 7$

$-2y = -4x + 7$

$y = 2x - \dfrac{7}{2}$

$m_1 = 2$

$m_1 \cdot m_2 = -1$

$2m_2 = -1$

$m_2 = -\dfrac{1}{2}$

$y - y_1 = m(x - x_1)$

$y - (-1) = -\dfrac{1}{2}[x - (-3)]$

$y + 1 = -\dfrac{1}{2}(x + 3)$

$y + 1 = -\dfrac{1}{2}x - \dfrac{3}{2}$

$y = -\dfrac{1}{2}x - \dfrac{5}{2}$

The equation of the line is $y = -\dfrac{1}{2}x - \dfrac{5}{2}$.

21. $y \geq 2x - 3$

22. $3x - 2y < 6$

$-2y < -3x + 6$

$y > \dfrac{3}{2}x - 3$

23.

After 4 hours the car will have traveled 220 miles.

24. $m = \dfrac{y_2 - y_1}{x_2 - x_1} = \dfrac{12{,}000 - 6000}{500 - 200} = \dfrac{6000}{300} = 20$

The slope is 20. The slope represents the cost per calculator manufactured. The cost of manufacturing one calculator is $20.

25. The y-intercept is $(0, 25{,}000)$.
The slope is 80.
$y = mx + b$
$y = 80x + 25{,}000$
The linear function is $y = 80x + 25{,}000$.
Predict the cost of building a house with 2000 ft^2.
$y = 80(2000) + 25{,}000$
$y = 185{,}000$
The house will cost $185,000 to build.

Chapter Test

1. $P(x) = 2 - x^2$
Ordered pairs: $(-2, -2)$
$(-1, 1)$
$(0, 2)$
$(1, 1)$
$(2, -2)$

2. $y = 2x + 6$
$y = 2(-3) + 6$
$y = -6 + 6$
$y = 0$
The ordered-pair solution is $(-3, 0)$.

3.

4. $2x + 3y = -3$
$3y = -2x - 3$
$y = -\dfrac{2}{3}x - 1$

5. The equation of the vertical line that contains $(-2, 3)$ is $x = -2$.

6. $x_m = \dfrac{x_1 + x_2}{2} = \dfrac{4 + (-5)}{2} = -\dfrac{1}{2}$

   $y_m = \dfrac{y_1 + y_2}{2} = \dfrac{2 + 8}{2} = 5$

   Length $= \sqrt{(x_2 - x_1)^2 + (y_2 - y_1)^2}$
   $= \sqrt{(-5 - 4)^2 + (8 - 2)^2}$
   $= \sqrt{81 + 36}$
   $= \sqrt{117}$

   The midpoint is $\left(-\dfrac{1}{2}, 5\right)$ and the length is $\sqrt{117}$.

7. $P_1(-2, 3)$, $P_2(4, 2)$

   $m = \dfrac{y_2 - y_1}{x_2 - x_1} = \dfrac{2 - 3}{4 - (-2)} = -\dfrac{1}{6}$

   The slope of the line is $-\dfrac{1}{6}$.

8. $P(x) = 3x^2 - 2x + 1$
   $P(2) = 3(2)^2 - 2(2) + 1$
   $P(2) = 9$

9. x-intercept: y-intercept:
   $2x - 3y = 6$ $2x - 3y = 6$
   $2x - 3(0) = 6$ $2(0) - 3y = 6$
   $2x = 6$ $-3y = 6$
   $x = 3$ $y = -2$
   $(3, 0)$ $(0, -2)$

10.

11. $m = \dfrac{2}{5}$, $(x_1, y_1) = (-5, 2)$

    $y - 2 = \dfrac{2}{5}[x - (-5)]$

    $y - 2 = \dfrac{2}{5}(x + 5)$

    $y - 2 = \dfrac{2}{5}x + 2$

    $y = \dfrac{2}{5}x + 4$

    The equation of the line is $y = \dfrac{2}{5}x + 4$.

12. $x = 0$

13. $P_1(3, -4)$, $P_2(-2, 3)$

    $m = \dfrac{y_2 - y_1}{x_2 - x_1} = \dfrac{3 - (-4)}{-2 - 3} = \dfrac{3 + 4}{-5} = -\dfrac{7}{5}$

    $y - y_1 = m(x - x_1)$

    $y - (-4) = -\dfrac{7}{5}(x - 3)$

    $y + 4 = -\dfrac{7}{5}x + \dfrac{21}{5}$

    $y = -\dfrac{7}{5}x + \dfrac{1}{5}$

    The equation of the line is $y = -\dfrac{7}{5}x + \dfrac{1}{5}$.

14. A horizontal line has a slope of 0.
    $y - y_1 = m(x - x_1)$
    $y - (-3) = 0(x - 4)$
    $y + 3 = 0$
    $y = -3$
    The equation of the line is $y = -3$.

15. Domain = $\{-4, -2, 0, 3\}$
    Range = $\{0, 2, 5\}$

16. $y = -\dfrac{3}{2}x - 6$

    $m = -\dfrac{3}{2}$

    $y - y_1 = m(x - x_1)$

    $y - 2 = -\dfrac{3}{2}(x - 1)$

    $y - 2 = -\dfrac{3}{2}x + \dfrac{3}{2}$

    $y = -\dfrac{3}{2}x + \dfrac{7}{2}$

    The equation of the line is $y = -\dfrac{3}{2}x + \dfrac{7}{2}$.

17. $y = -\dfrac{1}{2}x - 3$

    $m_1 = -\dfrac{1}{2}$

    $m_1 \cdot m_2 = -1$

    $-\dfrac{1}{2}m_2 = -1$

    $m_2 = 2$

    $y - y_1 = m(x - x_1)$

    $y - (-3) = 2[x - (-2)]$

    $y + 3 = 2(x + 2)$

    $y + 3 = 2x + 4$

    $y = 2x + 1$

    The equation of the line is $y = 2x + 1$.

18. $3x - 4y > 8$
$\quad -4y > -3x + 8$
$\quad y < \dfrac{3}{4}x - 2$

19. Dependent variable: number of students: (y)
Independent variable: tuition cost (x)
$m = \dfrac{\text{change in } y}{\text{change in } x} = \dfrac{-6}{20} = -\dfrac{3}{10}$
$P_1(250, 100)$
Use the point-slope formula to find the equation.
$y - y_1 = m(x - x_1)$
$y - 100 = -\dfrac{3}{10}(x - 250)$
$y - 100 = -\dfrac{3}{10}x + 75$
$y = -\dfrac{3}{10}x + 175$
The equation that predicts the number of students for a certain tuition is $y = -\dfrac{3}{10}x + 175$. Predict the number of students when the tuition is $300.
$y = -\dfrac{3}{10}x + 175$
$y = -\dfrac{3}{10}(300) + 175$
$y = 85$
When the tuition is $300, 85 students will enroll.

20. **Strategy**
 • Use two points on the graph to find the slope of the line.

Solution
$(x_1, y_1) = (3, 40,000), \ (x_2, y_2) = (12, 10,000)$
$m = \dfrac{y_2 - y_1}{x_2 - x_1} = \dfrac{10,000 - 40,000}{12 - 3} = -\dfrac{10,000}{3}$
The value of the house decreases by $3333.33 per year.

Cumulative Review Exercises

1. The Commutative Property of Multiplication

2. $3 - \dfrac{x}{2} = \dfrac{3}{4}$
$4\left(3 - \dfrac{x}{2}\right) = \dfrac{3}{4}(4)$
$12 - 2x = 3$
$12 - 2x - 12 = 3 - 12$
$-2x = -9$
$x = \dfrac{9}{2}$
The solution is $\dfrac{9}{2}$.

3. $2[y - 2(3 - y) + 4] = 4 - 3y$
$2(y - 6 + 2y + 4) = 4 - 3y$
$2(3y - 2) = 4 - 3y$
$6y - 4 = 4 - 3y$
$9y - 4 = 4$
$9y = 8$
$y = \dfrac{8}{9}$
The solution is $\dfrac{8}{9}$.

4. $\dfrac{1 - 3x}{2} + \dfrac{7x - 2}{6} = \dfrac{4x + 2}{9}$
$18\left(\dfrac{1 - 3x}{2} + \dfrac{7x - 2}{6}\right) = 18\left(\dfrac{4x + 2}{9}\right)$
$9(1 - 3x) + 3(7x - 2) = 2(4x + 2)$
$9 - 27x + 21x - 6 = 8x + 4$
$-6x + 3 = 8x + 4$
$-14x = 1$
$x = -\dfrac{1}{14}$
The solution is $-\dfrac{1}{14}$.

5. $x - 3 < -4$ or $2x + 2 > 3$
$\quad x < -1$ $2x > 1$
$\{x \mid x < -1\}$ $x > \dfrac{1}{2}$
$\qquad\qquad\qquad\qquad \left\{x \mid x > \dfrac{1}{2}\right\}$
$\{x \mid x < -1\} \cup \left\{x \mid x > \dfrac{1}{2}\right\} = \left\{x \mid x < -1 \text{ or } x > \dfrac{1}{2}\right\}$

6. $8 - |2x - 1| = 4$
$-|2x - 1| = -4$
$|2x - 1| = 4$
$2x - 1 = 4 \qquad\qquad 2x - 1 = -4$
$2x = 5 \qquad\qquad\quad 2x = -3$
$x = \dfrac{5}{2} \qquad\qquad\quad x = -\dfrac{3}{2}$
The solutions are $\dfrac{5}{2}$ and $-\dfrac{3}{2}$.

7. $|3x - 5| < 5$
$-5 < 3x - 5 < 5$
$-5 + 5 < 3x - 5 + 5 < 5 + 5$
$0 < 3x < 10$
$\dfrac{1}{3}(0) < \dfrac{1}{3}(3x) < 10\left(\dfrac{1}{3}\right)$
$0 < x < \dfrac{10}{3}$
$\left\{x \mid 0 < x < \dfrac{10}{3}\right\}$

8. $4 - 2(4 - 5)^3 + 2 = 4 - 2(-1)^3 + 2$
$\qquad\qquad\qquad\quad = 4 + 2 + 2$
$\qquad\qquad\qquad\quad = 8$

9. $(a-b)^2 \div ab$ for $a = 4$ and $b = -2$
   $[4-(-2)]^2 \div 4(-2) = 6^2 \div 4(-2)$
   $ = 36 \div 4(-2)$
   $ = 9 \cdot (-2)$
   $ = -18$

10. $\{x|x < -2\} \cup \{x|x > 0\}$

    (number line from -5 to 5 with open circles at -2 and 0)

11. Solve each inequality.
    $3x - 1 < 4 \qquad x - 2 > 2$
    $3x < 5 \qquad\quad x > 4$
    $x < \dfrac{5}{3}$

    The solution is $\left\{x \middle| x < \dfrac{5}{3} \text{ and } x > 4\right\}$, and there is no such value, so the solution is the null set.

12. $P(x) = x^2 + 5$
    $P(-3) = (-3)^2 + 5$
    $P(-3) = 14$

13. $y = -\dfrac{5}{4}x + 3$
    $y = -\dfrac{5}{4}(-8) + 3$
    $y = 10 + 3$
    $y = 13$
    The ordered-pair solution is $(-8, 13)$.

14. $P_1(-1, 3)$, $P_2(3, -4)$
    $m = \dfrac{y_2 - y_1}{x_2 - x_1} = \dfrac{-4 - 3}{3 - (-1)} = \dfrac{-7}{4} = -\dfrac{7}{4}$

15. $m = \dfrac{3}{2}$, $(x_1, y_1) = (-1, 5)$
    $y - y_1 = m(x - x_1)$
    $y - 5 = \dfrac{3}{2}[x - (-1)]$
    $y - 5 = \dfrac{3}{2}(x + 1)$
    $y - 5 = \dfrac{3}{2}x + \dfrac{3}{2}$
    $y = \dfrac{3}{2}x + \dfrac{13}{2}$

    The equation of the line is $y = \dfrac{3}{2}x + \dfrac{13}{2}$.

16. $(x_1, y_1) = (4, -2)$, $(x_2, y_2) = (0, 3)$
    $m = \dfrac{y_2 - y_1}{x_2 - x_1} = \dfrac{3 - (-2)}{0 - 4} = \dfrac{3 + 2}{-4} = -\dfrac{5}{4}$
    $y - y_1 = m(x - x_1)$
    $y - (-2) = -\dfrac{5}{4}(x - 4)$
    $y + 2 = -\dfrac{5}{4}x + 5$
    $y = -\dfrac{5}{4}x + 3$

    The equation of the line is $y = -\dfrac{5}{4}x + 3$.

17. $y = -\dfrac{3}{2}x + 2$, $m = -\dfrac{3}{2}$
    $y - y_1 = m(x - x_1)$
    $y - 4 = -\dfrac{3}{2}(x - 2)$
    $y - 4 = -\dfrac{3}{2}x + 3$
    $y = -\dfrac{3}{2}x + 7$

    The equation of the line is $y = -\dfrac{3}{2}x + 7$.

18. $3x - 2y = 5$
    $-2y = -3x + 5$
    $y = \dfrac{3}{2}x - \dfrac{5}{2}$
    $m_1 = \dfrac{3}{2}$
    $m_1 \cdot m_2 = -1$
    $\dfrac{3}{2}m_2 = -1$
    $m_2 = -\dfrac{2}{3}$
    $y - y_1 = m(x - x_1)$
    $y - 0 = -\dfrac{2}{3}(x - 4)$
    $y = -\dfrac{2}{3}x + \dfrac{8}{3}$

    The equation of the line is $y = -\dfrac{2}{3}x + \dfrac{8}{3}$.

19. x-intercept: $\qquad\qquad$ y-intercept:
    $3x - 5y = 15 \qquad\quad 3x - 5y = 15$
    $3x - 5(0) = 15 \qquad 3(0) - 5y = 15$
    $3x = 15 \qquad\qquad\quad -5y = 15$
    $x = 5 \qquad\qquad\qquad y = -3$
    $(5, 0) \qquad\qquad\qquad (0, -3)$

    (graph showing line through $(5, 0)$ and $(0, -3)$)

20.

21. $3x - 2y \geq 6$
$-2y \geq -3x + 6$
$y \leq \dfrac{3}{2}x - 3$

22. Strategy
• Number of nickels: $3x$
Number of quarters: x

Coin	Number	Value	Total Value
Nickels	$3x$	5	$3x(5)$
Quarters	x	25	$25x$

• The sum of the total values of each denomination of coin equals the total value of all the coins (160).
$3(x)(5) + 25x = 160$

Solution
$15x + 25x = 160$
$40x = 160$
$x = 4$
$3x = 12$
There are 12 nickels in the purse.

23. Strategy
• Rate of first plane: x
Rate of second plane: $2x$

	Rate	Time	Distance
First plane	x	3	$3x$
Second plane	$2x$	3	$3(2x)$

• The two planes travel a total distance of 1800 miles.
$3x + 3(2x) = 1800$

Solution
$3x + 3(2x) = 1800$
$3x + 6x = 1800$
$9x = 1800$
$x = 200$
$2x = 400$
The rate of the first plane is 200 mph and the rate of the second plane is 400 mph.

24. Strategy
• Pounds of coffee costing $3.00: x
Pounds of coffee costing $8.00: $80 - x$

	Amount	Cost	Value
$3.00 coffee	x	3.00	$3x$
$8.00 coffee	$80 - x$	8.00	$8(80 - x)$
$5.00 mixture	80	5.00	$5(80)$

• The sum of the values of each part of the mixture equals the value of the mixture.
$3x + 8(80 - x) = 5(80)$

Solution
$3x + 640 - 8x = 400$
$-5x + 640 = 400$
$-5x = -240$
$x = 48$
$80 - x = 32$
The mixture consists of 48 lb of $3 coffee and 32 lb of $8 coffee.

25. Strategy
To write the equation
• Use points on the graph to find the slope of the line.
• Locate the y-intercept of the line on the graph.
• Use the slope-intercept form of an equation to write the equation of the line.

Solution
$(x_1, y_1) = (0, 15{,}000)$, $(x_2, y_2) = (6, 0)$
The y-intercept is $(0, 15{,}000)$.
$m = \dfrac{y_2 - y_1}{x_2 - x_1} = \dfrac{0 - 15{,}000}{6 - 0} = -2500$
$y = mx + b$
$y = -2500x + 15{,}000$
The value of the truck decreases by $2500 each year.

Chapter 4: Systems of Equations and Inequalities

Section 4.1

Concept Review 4.1

1. Always true

3. Sometimes true
   A system of equations with two unknowns has one solution, an infinite number of solutions, or no solution.

5. Always true

Objective 1 Exercises

3.
   The solution is (3, −1).

5.
   The solution is (2, 4).

7.
   The solution is (4, 3).

9.
   The solution is (4, −1).

11.
    The solution is (3, −2).

13.
    The lines are parallel and therefore do not intersect. The system of equations has no solution. The system is inconsistent.

15.
    The system of equations is dependent. The solutions are the ordered pairs $\left(x, \dfrac{2}{5}x - 2\right)$.

17.
    The solution is (0, −3).

Objective 2 Exercises

21. (1) $3x - 2y = 4$
    (2) $\quad\quad x = 2$
    Substitute the value of x into equation (1).
    $$3x - 2y = 4$$
    $$3(2) - 2y = 4$$
    $$6 - 2y = 4$$
    $$-2y = -2$$
    $$y = 1$$
    The solution is (2, 1).

23. (1) $y = 2x - 1$
    (2) $x + 2y = 3$
    Substitute $2x - 1$ for y in equation (2).
    $$x + 2y = 3$$
    $$x + 2(2x - 1) = 3$$
    $$x + 4x - 2 = 3$$
    $$5x - 2 = 3$$
    $$5x = 5$$
    $$x = 1$$
    Substitute into equation (1).
    $$y = 2x - 1$$
    $$y = 2(1) - 1$$
    $$y = 2 - 1$$
    $$y = 1$$
    The solution is (1, 1).

25. (1) $4x - 3y = 5$
   (2) $y = 2x - 3$
   Substitute $2x - 3$ for y in equation (1).
   $$4x - 3y = 5$$
   $$4x - 3(2x - 3) = 5$$
   $$4x - 6x + 9 = 5$$
   $$-2x + 9 = 5$$
   $$-2x = -4$$
   $$x = 2$$
   Substitute into equation (2).
   $$y = 2x - 3$$
   $$y = 2(2) - 3$$
   $$y = 4 - 3$$
   $$y = 1$$
   The solution is (2, 1).

27. (1) $x = 2y + 4$
   (2) $4x + 3y = -17$
   Substitute $2y + 4$ for x in equation (2).
   $$4x + 3y = -17$$
   $$4(2y + 4) + 3y = -17$$
   $$8y + 16 + 3y = -17$$
   $$11y + 16 = -17$$
   $$11y = -33$$
   $$y = -3$$
   Substitute into equation (1).
   $$x = 2y + 4$$
   $$x = 2(-3) + 4$$
   $$x = -6 + 4$$
   $$x = -2$$
   The solution is (−2, −3).

29. (1) $5x + 4y = -1$
   (2) $y = 2 - 2x$
   Substitute $2 - 2x$ for y in equation (1).
   $$5x + 4y = -1$$
   $$5x + 4(2 - 2x) = -1$$
   $$5x + 8 - 8x = -1$$
   $$-3x + 8 = -1$$
   $$-3x = -9$$
   $$x = 3$$
   Substitute into equation (2).
   $$y = 2 - 2x$$
   $$y = 2 - 2(3)$$
   $$y = 2 - 6$$
   $$y = -4$$
   The solution is (3, −4).

31. (1) $7x - 3y = 3$
   (2) $x = 2y + 2$
   Substitute $2y + 2$ for x in equation (1).
   $$7x - 3y = 3$$
   $$7(2y + 2) - 3y = 3$$
   $$14y + 14 - 3y = 3$$
   $$11y + 14 = 3$$
   $$11y = -11$$
   $$y = -1$$
   Substitute into equation (2).
   $$x = 2y + 2$$
   $$x = 2(-1) + 2$$
   $$x = -2 + 2$$
   $$x = 0$$
   The solution is (0, −1).

33. (1) $2x + 2y = 7$
   (2) $y = 4x + 1$
   Substitute $4x + 1$ for y in equation (1).
   $$2x + 2y = 7$$
   $$2x + 2(4x + 1) = 7$$
   $$2x + 8x + 2 = 7$$
   $$10x + 2 = 7$$
   $$10x = 5$$
   $$x = \frac{1}{2}$$
   Substitute into equation (2).
   $$y = 4x + 1$$
   $$y = 4\left(\frac{1}{2}\right) + 1$$
   $$y = 2 + 1$$
   $$y = 3$$
   The solution is $\left(\frac{1}{2}, 3\right)$.

35. (1) $3x + y = 5$
   (2) $2x + 3y = 8$
   Solve equation (1) for y.
   $$3x + y = 5$$
   $$y = -3x + 5$$
   Substitute into equation (2).
   $$2x + 3y = 8$$
   $$2x + 3(-3x + 5) = 8$$
   $$2x - 9x + 15 = 8$$
   $$-7x + 15 = 8$$
   $$-7x = -7$$
   $$x = 1$$
   Substitute into equation (1).
   $$3x + y = 5$$
   $$3(1) + y = 5$$
   $$3 + y = 5$$
   $$y = 2$$
   The solution is (1, 2).

37. (1) $x + 3y = 5$
(2) $2x + 3y = 4$
Solve equation (1) for x.
$x + 3y = 5$
$x = -3y + 5$
Substitute into equation (2).
$2x + 3y = 4$
$2(-3y + 5) + 3y = 4$
$-6y + 10 + 3y = 4$
$-3y + 10 = 4$
$-3y = -6$
$y = 2$
Substitute into equation (1).
$x + 3y = 5$
$x + 3(2) = 5$
$x + 6 = 5$
$x = -1$
The solution is $(-1, 2)$.

39. (1) $3x + 4y = 14$
(2) $2x + y = 1$
Solve equation (2) for y.
$2x + y = 1$
$y = -2x + 1$
Substitute into equation (1).
$3x + 4y = 14$
$3x + 4(-2x + 1) = 14$
$3x - 8x + 4 = 14$
$-5x + 4 = 14$
$-5x = 10$
$x = -2$
Substitute into equation (2).
$2x + y = 1$
$2(-2) + y = 1$
$-4 + y = 1$
$y = 5$
The solution is $(-2, 5)$.

41. (1) $3x + 5y = 0$
(2) $x - 4y = 0$
Solve equation (2) for x.
$x - 4y = 0$
$x = 4y$
Substitute into equation (1).
$3x + 5y = 0$
$3(4y) + 5y = 0$
$12y + 5y = 0$
$17y = 0$
$y = 0$
Substitute into equation (2).
$x - 4y = 0$
$x - 4(0) = 0$
$x = 0$
The solution is $(0, 0)$.

43. (1) $5x - 3y = -2$
(2) $-x + 2y = -8$
Solve equation (2) for x.
$-x + 2y = -8$
$-x = -2y - 8$
$x = 2y + 8$
Substitute $2y + 8$ for x in equation (1).
$5x - 3y = -2$
$5(2y + 8) - 3y = -2$
$10y + 40 - 3y = -2$
$7y + 40 = -2$
$7y = -42$
$y = -6$
Substitute into equation (2).
$-x + 2y = -8$
$-x + 2(-6) = -8$
$-x - 12 = -8$
$-x = 4$
$x = -4$
The solution is $(-4, -6)$.

45. (1) $y = 3x + 2$
(2) $y = 2x + 3$
Substitute $2x + 3$ for y in equation (1).
$y = 3x + 2$
$2x + 3 = 3x + 2$
$2x = 3x - 1$
$-x = -1$
$x = 1$
Substitute into equation (2).
$y = 2x + 3$
$y = 2(1) + 3$
$y = 2 + 3$
$y = 5$
The solution is $(1, 5)$.

47. (1) $x = 2y + 1$
(2) $x = 3y - 1$
Substitute $3y - 1$ for x in equation (1).
$x = 2y + 1$
$3y - 1 = 2y + 1$
$3y = 2y + 2$
$y = 2$
Substitute into equation (2).
$x = 3y - 1$
$x = 3(2) - 1$
$x = 6 - 1$
$x = 5$
The solution is $(5, 2)$.

49. (1) $y = 5x - 1$

(2) $y = 5 - x$

Substitute $5 - x$ for y in equation (1).
$$y = 5x - 1$$
$$5 - x = 5x - 1$$
$$-x = 5x - 6$$
$$-6x = -6$$
$$x = 1$$
Substitute into equation (2).
$$y = 5 - x$$
$$y = 5 - 1$$
$$y = 4$$
The solution is $(1, 4)$.

Applying Concepts 4.1

51. Inconsistent equations have the same slope but different y-intercepts. Solve the two equations for y, and then set the slopes equal to each other.

(1) $2x - 2y = 5$
$$-2y = -2x + 5$$
$$y = x - \frac{5}{2}$$

(2) $kx - 2y = 3$
$$-2y = -kx + 3$$
$$y = \frac{k}{2}x - \frac{3}{2}$$
$$\frac{k}{2} = 1$$
$$k = 2$$
The value of k is 2.

53. Inconsistent equations have the same slope but different y-intercepts. Solve the two equations for y, and then set the slopes equal to each other.

(1) $x = 6y + 6$
$$x - 6 = 6y$$
$$\frac{1}{6}x - 1 = y$$

(2) $kx - 3y = 6$
$$-3y = -kx + 6$$
$$y = \frac{k}{3}x - 2$$
$$\frac{k}{3} = \frac{1}{6}$$
$$k = \frac{1}{2}$$
The value of k is $\frac{1}{2}$.

55. Strategy

Solve a system of equations using x to represent one number and y to represent the second number.

Solution

(1) $x + y = 44$

(2) $x - 8 = y$

Solve equation (1) for y.
$$x + y = 44$$
$$y = -x + 44$$
Substitute $-x + 44$ for y in equation (2).
$$x - 8 = y$$
$$x - 8 = -x + 44$$
$$2x = 52$$
$$x = 26$$
Substitute into equation (1).
$$x + y = 44$$
$$26 + y = 44$$
$$y = 18$$
The numbers are 26 and 18.

57. Strategy

Solve a system of equations using x to represent one number and y to represent the second number.

Solution

(1) $x + y = 19$

(2) $2x - 5 = y$

Solve equation (1) for y.
$$x + y = 19$$
$$y = -x + 19$$
Substitute $-x + 19$ for y in equation (2).
$$2x - 5 = y$$
$$2x - 5 = -x + 19$$
$$3x = 24$$
$$x = 8$$
Substitute into equation (1).
$$x + y = 19$$
$$8 + y = 19$$
$$y = 11$$
The numbers are 8 and 11.

59.
$$\frac{2}{a} + \frac{3}{b} = 4$$
$$2\left(\frac{1}{a}\right) + 3\left(\frac{1}{b}\right) = 4$$
(1) $\quad 2x + 3y = 4$
$$\frac{4}{a} + \frac{1}{b} = 3$$
$$4\left(\frac{1}{a}\right) + \frac{1}{b} = 3$$
(2) $\quad 4x + y = 3$

Solve equation (1) for y.
$$2x + 3y = 4$$
$$3y = -2x + 4$$
$$y = \left(-\frac{2}{3}\right)x + \frac{4}{3}$$

Substitute $-\frac{2}{3}x + \frac{4}{3}$ for y in equation (2).
$$4x + y = 3$$
$$4x + \left(-\frac{2}{3}x + \frac{4}{3}\right) = 3$$
$$\frac{10}{3}x + \frac{4}{3} = 3$$
$$10x + 4 = 9$$
$$10x = 5$$
$$x = \frac{1}{2}$$

Substitute into equation (1).
$$2x + 3y = 4$$
$$2\left(\frac{1}{2}\right) + 3y = 4$$
$$1 + 3y = 4$$
$$3y = 3$$
$$y = 1$$

Replace x by $\frac{1}{a}$.
$$x = \frac{1}{2}$$
$$\frac{1}{a} = \frac{1}{2}$$
$$2a\left(\frac{1}{a}\right) = 2a\left(\frac{1}{2}\right)$$
$$2 = a$$

Replace y by $\frac{1}{b}$.
$$y = 1$$
$$\frac{1}{b} = 1$$
$$b\left(\frac{1}{b}\right) = b(1)$$
$$1 = b$$

The solution is (2, 1).

61.
$$\frac{1}{a} + \frac{3}{b} = 2$$
$$\frac{1}{a} + 3\left(\frac{1}{b}\right) = 2$$
(1) $\quad x + 3y = 2$
$$\frac{4}{a} - \frac{1}{b} = 3$$
$$4\left(\frac{1}{a}\right) - \frac{1}{b} = 3$$
(2) $\quad 4x - y = 3$

Solve equation (1) for y.
$$x + 3y = 2$$
$$3y = -x + 2$$
$$y = -\frac{1}{3}x + \frac{2}{3}$$

Substitute $-\frac{1}{3}x + \frac{2}{3}$ for y in equation (2).
$$4x - y = 3$$
$$4x - \left(-\frac{1}{3}x + \frac{2}{3}\right) = 3$$
$$4x + \frac{1}{3}x - \frac{2}{3} = 3$$
$$\frac{13}{3}x = \frac{11}{3}$$
$$x = \frac{11}{13}$$

Substitute into equation (1).
$$x + 3y = 2$$
$$\frac{11}{13} + 3y = 2$$
$$3y = \frac{15}{13}$$
$$y = \frac{15}{39} = \frac{5}{13}$$

Replace x by $\frac{1}{a}$.
$$x = \frac{11}{13}$$
$$\frac{1}{a} = \frac{11}{13}$$
$$13a\left(\frac{1}{a}\right) = 13a\left(\frac{11}{13}\right)$$
$$13 = 11a$$
$$\frac{13}{11} = a$$

Replace y by $\frac{1}{b}$.
$$y = \frac{5}{13}$$
$$\frac{1}{b} = \frac{5}{13}$$
$$13b\left(\frac{1}{b}\right) = 13b\left(\frac{5}{13}\right)$$
$$13 = 5b$$
$$\frac{13}{5} = b$$

The solution is $\left(\frac{13}{11}, \frac{13}{5}\right)$.

63. $y = -\dfrac{1}{2}x + 2$
$y = 2x - 1$

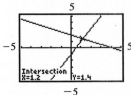

The solution is (1.20, 1.40).

65. $y = \sqrt{2}x - 1$
$y = -\sqrt{3}x + 1$

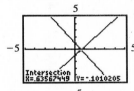

The solution is (0.64, −0.10).

67. a. $y_1 = -0.0175x + 35.853$
$= -0.0175(2000) + 35.853 = 0.853$
$y_2 = 0.0146x - 28.658$
$= 0.0146(2000) - 28.658 = 0.542$
Since $0.853 > 0.542$ the population of Phoenix will not exceed the population of Detroit in the year 2000.

b. $y_1 = y_2$
$-0.0175x + 35.853 = 0.0146x - 28.658$
$64.511 = 0.0321x$
$2009.6885 = x$
In 2009 the population of Phoenix will first exceed the population of Detroit.

c. The slope of y_1 is
$-0.0175(1,000,000) = 17,500$. The slope indicates that the population of Detroit is decreasing at the rate of 17,500 people per year.

d. The slope of y_2 is
$0.0146(1,000,000) = 14,600$. The slope indicates that the population of Phoenix is increasing at the rate of 14,600 people per year.

Section 4.2

Concept Review 4.2

1. Always true

3. Sometimes true
The solution of a system of three equations in three variables may be a point, a line, or a plane, or the system may not have a solution.

5. Never true
The system is inconsistent and has no solutions.

Objective 1 Exercises

3. (1) $x - y = 5$
(2) $x + y = 7$
Eliminate y. Add the equations.
$2x = 12$
$x = 6$
Replace x in equation (1).
$x - y = 5$
$6 - y = 5$
$-y = -1$
$y = 1$
The solution is (6, 1).

5. (1) $3x + y = 4$
(2) $x + y = 2$
Eliminate y.
$3x + y = 4$
$-1(x + y) = -1(2)$

$3x + y = 4$
$-x - y = -2$
Add the equations.
$2x = 2$
$x = 1$
Replace x in equation (2).
$x + y = 2$
$1 + y = 2$
$y = 1$
The solution is (1, 1).

7. (1) $3x + y = 7$
(2) $x + 2y = 4$
Eliminate y.
$-2(3x + y) = -2(7)$
$x + 2y = 4$

$-6x - 2y = -14$
$x + 2y = 4$
Add the equations.
$-5x = -10$
$x = 2$
Replace x in equation (2).
$x + 2y = 4$
$2 + 2y = 4$
$2y = 2$
$y = 1$
The solution is (2, 1).

9. (1) $3x - y = 4$
   (2) $6x - 2y = 8$
   Eliminate y.
   $-2(3x - y) = -2(4)$
   $6x - 2y = 8$

   $-6x + 2y = -8$
   $6x - 2y = 8$
   Add the equations.
   $0 = 0$
   This is a true equation. The equations are dependent. The solutions are the ordered pairs $(x, 3x - 4)$.

11. (1) $2x + 5y = 9$
    (2) $4x - 7y = -16$
    Eliminate x.
    $-2(2x + 5y) = -2(9)$
    $4x - 7y = -16$

    $-4x - 10y = -18$
    $4x - 7y = -16$
    Add the equations.
    $-17y = -34$
    $y = 2$
    Replace y in equation (1).
    $2x + 5y = 9$
    $2x + 5(2) = 9$
    $2x + 10 = 9$
    $2x = -1$
    $x = -\dfrac{1}{2}$
    The solution is $\left(-\dfrac{1}{2}, 2\right)$.

13. (1) $4x - 6y = 5$
    (2) $2x - 3y = 7$
    Eliminate y.
    $4x - 6y = 5$
    $-2(2x - 3y) = -2(7)$

    $4x - 6y = 5$
    $-4x + 6y = -14$
    Add the equations.
    $0 = -9$
    This is not a true equation. The system of equations is inconsistent and therefore has no solution.

15. (1) $3x - 5y = 7$
    (2) $x - 2y = 3$
    Eliminate x.
    $3x - 5y = 7$
    $-3(x - 2y) = -3(3)$

    $3x - 5y = 7$
    $-3x + 6y = -9$
    Add the equations.
    $y = -2$
    Replace y in equation (2).
    $x - 2y = 3$
    $x - 2(-2) = 3$
    $x + 4 = 3$
    $x = -1$
    The solution is $(-1, -2)$.

17. (1) $3x + 2y = 16$
    (2) $2x - 3y = -11$
    Eliminate y.
    $3(3x + 2y) = 3(16)$
    $2(2x - 3y) = 2(-11)$

    $9x + 6y = 48$
    $4x - 6y = -22$
    Add the equations.
    $13x = 26$
    $x = 2$
    Replace x in equation (1).
    $3x + 2y = 16$
    $3(2) + 2y = 16$
    $6 + 2y = 16$
    $2y = 10$
    $y = 5$
    The solution is $(2, 5)$.

19. (1) $4x + 4y = 5$
    (2) $2x - 8y = -5$
    Eliminate y.
    $2(4x + 4y) = 2(5)$
    $2x - 8y = -5$

    $8x + 8y = 10$
    $2x - 8y = -5$
    Add the equations.
    $10x = 5$
    $x = \dfrac{1}{2}$
    Replace x in equation (1).
    $4x + 4y = 5$
    $4\left(\dfrac{1}{2}\right) + 4y = 5$
    $2 + 4y = 5$
    $4y = 3$
    $y = \dfrac{3}{4}$
    The solution is $\left(\dfrac{1}{2}, \dfrac{3}{4}\right)$.

21. (1) $5x + 4y = 0$
(2) $3x + 7y = 0$
Eliminate x.
$-3(5x + 4y) = -3(0)$
$5(3x + 7y) = 5(0)$

$-15x - 12y = 0$
$15x + 35y = 0$
Add the equations.
$23y = 0$
$y = 0$
Replace y in equation (1).
$5x + 4y = 0$
$5x + 4(0) = 0$
$5x = 0$
$x = 0$
The solution is $(0, 0)$.

23. (1) $3x - 6y = 6$
(2) $9x - 3y = 8$
Eliminate y.
$3x - 6y = 6$
$-2(9x - 3y) = -2(8)$

$3x - 6y = 6$
$-18x + 6y = -16$
Add the equations.
$-15x = -10$
$x = \dfrac{2}{3}$
Replace x in the equation (1).
$3x - 6y = 6$
$3\left(\dfrac{2}{3}\right) - 6y = 6$
$2 - 6y = 6$
$-6y = 4$
$y = -\dfrac{2}{3}$
The solution is $\left(\dfrac{2}{3}, -\dfrac{2}{3}\right)$.

25. (1) $5x + 2y = 2x + 1$
(2) $2x - 3y = 3x + 2$
Write the equations in the form $Ax + By = C$.
$5x + 2y = 2x + 1$
$3x + 2y = 1$

$2x - 3y = 3x + 2$
$-x - 3y = 2$
Solve the system.
$3x + 2y = 1$
$-x - 3y = 2$
Eliminate x.
$3x + 2y = 1$
$3(-x - 3y) = 3(2)$

$3x + 2y = 1$
$-3x - 9y = 6$
Add the equations.
$-7y = 7$
$y = -1$
Replace y in the equation $-x - 3y = 2$.
$-x - 3y = 2$
$-x - 3(-1) = 2$
$-x + 3 = 2$
$-x = -1$
$x = 1$
The solution is $(1, -1)$.

27. (1) $\dfrac{2}{3}x - \dfrac{1}{2}y = 3$
(2) $\dfrac{1}{3}x - \dfrac{1}{4}y = \dfrac{3}{2}$
Clear the fractions.
$6\left(\dfrac{2}{3}x - \dfrac{1}{2}y\right) = 6(3)$
$12\left(\dfrac{1}{3}x - \dfrac{1}{4}y\right) = 12\left(\dfrac{3}{2}\right)$

$4x - 3y = 18$
$4x - 3y = 18$
Eliminate x.
$-1(4x - 3y) = -1(18)$
$4x - 3y = 18$

$-4x + 3y = -18$
$4x - 3y = 18$
Add the equations.
$0 = 0$
This is a true equation. The equations are dependent. The solutions are the ordered pairs $\left(x, \dfrac{4}{3}x - 6\right)$.

29. (1) $\frac{2}{5}x - \frac{1}{3}y = 1$
   (2) $\frac{3}{5}x + \frac{2}{3}y = 5$
Clear the fractions.
$15\left(\frac{2}{5}x - \frac{1}{3}y\right) = 15(1)$
$15\left(\frac{3}{5}x + \frac{2}{3}y\right) = 15(5)$

$6x - 5y = 15$
$9x + 10y = 75$
Eliminate y.
$2(6x - 5y) = 2(15)$
$9x + 10y = 75$

$12x - 10y = 30$
$9x + 10y = 75$
Add the equations.
$21x = 105$
$x = 5$
Replace x in equation (1).
$\frac{2}{5}x - \frac{1}{3}y = 1$
$\frac{2}{5}(5) - \frac{1}{3}y = 1$
$2 - \frac{1}{3}y = 1$
$-\frac{1}{3}y = -1$
$y = 3$
The solution is (5, 3).

31. (1) $\frac{3}{4}x + \frac{2}{5}y = -\frac{3}{20}$
   (2) $\frac{3}{2}x - \frac{1}{4}y = \frac{3}{4}$
Clear the fractions.
$20\left(\frac{3}{4}x + \frac{2}{5}y\right) = 20\left(-\frac{3}{20}\right)$
$4\left(\frac{3}{2}x - \frac{1}{4}y\right) = 4\left(\frac{3}{4}\right)$

$15x + 8y = -3$
$6x - y = 3$
Eliminate y.
$15x + 8y = -3$
$8(6x - y) = 8(3)$

$15x + 8y = -3$
$48x - 8y = 24$
Add the equations.
$63x = 21$
$x = \frac{1}{3}$
Replace x in equation (2).
$\frac{3}{2}x - \frac{1}{4}y = \frac{3}{4}$
$\frac{3}{2}\left(\frac{1}{3}\right) - \frac{1}{4}y = \frac{3}{4}$
$\frac{1}{2} - \frac{1}{4}y = \frac{3}{4}$
$-\frac{1}{4}y = \frac{1}{4}$
$y = -1$
The solution is $\left(\frac{1}{3}, -1\right)$.

33. (1) $4x - 5y = 3y + 4$
    (2) $2x + 3y = 2x + 1$
    Write the equations in the form $Ax + By = C$.
    $4x - 5y = 3y + 4$
    $4x - 8y = 4$

    $2x + 3y = 2x + 1$
    $3y = 1$
    Solve the system.
    $4x - 8y = 4$
    $3y = 1$
    Solve the equation $3y = 1$ for y.
    $3y = 1$
    $y = \frac{1}{3}$
    Replace y in the equation $4x - 8y = 4$.
    $4x - 8y = 4$
    $4x - 8\left(\frac{1}{3}\right) = 4$
    $4x - \frac{8}{3} = 4$
    $4x = \frac{20}{3}$
    $x = \frac{5}{3}$
    The solution is $\left(\frac{5}{3}, \frac{1}{3}\right)$.

35. (1) $2x + 5y = 5x + 1$
    (2) $3x - 2y = 3y + 3$
    Write the equations in the form $Ax + By = C$.
    $2x + 5y = 5x + 1$
    $-3x + 5y = 1$

    $3x - 2y = 3y + 3$
    $3x - 5y = 3$
    Solve the system.
    $-3x + 5y = 1$
    $3x - 5y = 3$
    Add the equations.
    $0 = 4$
    This is not a true equation. The system of equations is inconsistent and therefore has no solution.

Objective 2 Exercises

41. (1) $x + 3y + z = 6$
    (2) $3x + y - z = -2$
    (3) $2x + 2y - z = 1$
    Eliminate z. Add equations (1) and (2).
    $x + 3y + z = 6$
    $3x + y - z = -2$

    $4x + 4y = 4$

    Multiply both sides of the equation by $\frac{1}{4}$.

    (4) $x + y = 1$
    Add equations (1) and (3).
    $x + 3y + z = 6$
    $2x + 2y - z = 1$

    (5) $3x + 5y = 7$
    Multiply equation (4) by -3 and add to equation (5).
    $-3(x + y) = -3(1)$
    $3x + 5y = 7$

    $-3x - 3y = -3$
    $3x + 5y = 7$

    $2y = 4$
    $y = 2$
    Replace y by 2 in equation (4).
    $x + y = 1$
    $x + 2 = 1$
    $x = -1$
    Replace x by -1 and y by 2 in equation (1).
    $x + 3y + z = 6$
    $-1 + 3(2) + z = 6$
    $-1 + 6 + z = 6$
    $5 + z = 6$
    $z = 1$
    The solution is $(-1, 2, 1)$.

43. (1) $x - 2y + z = 6$
(2) $x + 3y + z = 16$
(3) $3x - y - z = 12$
Eliminate z. Add equations (1) and (3).
$x - 2y + z = 6$
$3x - y - z = 12$

(4) $4x - 3y = 18$
Add equations (2) and (3).
$x + 3y + z = 16$
$3x - y - z = 12$

$4x + 2y = 28$
Multiply both sides of the equation by $\frac{1}{2}$.
(5) $2x + y = 14$
Multiply equation (5) by 3 and add to equation (4).
$3(2x + y) = 3(14)$
$4x - 3y = 18$

$6x + 3y = 42$
$4x - 3y = 18$

$10x = 60$
$x = 6$
Replace x by 6 in equation (5).
$2x + y = 14$
$2(6) + y = 14$
$12 + y = 14$
$y = 2$
Replace x by 6 and y by 2 in equation (1).
$x - 2y + z = 6$
$6 - 2(2) + z = 6$
$6 - 4 + z = 6$
$2 + z = 6$
$z = 4$
The solution is (6, 2, 4).

45. (1) $2y + z = 7$
(2) $2x - z = 3$
(3) $x - y = 3$
Eliminate z. Add equations (1) and (2).
$2y + z = 7$
$2x - z = 3$

$2x + 2y = 10$
Multiply both sides of the equation by $\frac{1}{2}$.
(4) $x + y = 5$
Add equations (3) and (4).
$x - y = 3$
$x + y = 5$

$2x = 8$
$x = 4$
Replace x by 4 in equation (4).
$x + y = 5$
$4 + y = 5$
$y = 1$
Replace y by 1 in equation (1).
$2y + z = 7$
$2(1) + z = 7$
$2 + z = 7$
$z = 5$
The solution is (4, 1, 5).

47. (1) $2x + y - 3z = 7$
(2) $x - 2y + 3z = 1$
(3) $3x + 4y - 3z = 13$
Eliminate z. Add equations (1) and (2).
$2x + y - 3z = 7$
$x - 2y + 3z = 1$

(4) $3x - y = 8$
Add equations (2) and (3).
$x - 2y + 3z = 1$
$3x + 4y - 3z = 13$

$4x + 2y = 14$
Multiply each side of the equation by $\frac{1}{2}$.
(5) $2x + y = 7$
Add equations (4) and (5).
$3x - y = 8$
$2x + y = 7$

$5x = 15$
$x = 3$
Replace x by 3 in equation (5).
$2x + y = 7$
$2(3) + y = 7$
$6 + y = 7$
$y = 1$
Replace x by 3 and y by 1 in equation (1).
$2x + y - 3z = 7$
$2(3) + 1 - 3z = 7$
$6 + 1 - 3z = 7$
$7 - 3z = 7$
$-3z = 0$
$z = 0$
The solution is (3, 1, 0).

49. (1) $3x + 4z = 5$
(2) $2y + 3z = 2$
(3) $2x - 5y = 8$
Eliminate z. Multiply equation (1) by -3 and equation (2) by 4.
Then add the equations.
$-3(3x + 4z) = -3(5)$
$4(2y + 3z) = 4(2)$

$-9x - 12z = -15$
$8y + 12z = 8$

(4) $-9x + 8y = -7$
Multiply equation (3) by 9 and equation (4) by 2.
Then add the equations.
$9(2x - 5y) = 9(8)$
$2(-9x + 8y) = 2(-7)$

$18x - 45y = 72$
$-18x + 16y = -14$

$-29y = 58$
$y = -2$
Replace y by -2 in equation (3).
$2x - 5y = 8$
$2x - 5(-2) = 8$
$2x + 10 = 8$
$2x = -2$
$x = -1$
Replace x by -1 in equation (1).
$3x + 4z = 5$
$3(-1) + 4z = 5$
$-3 + 4z = 5$
$4z = 8$
$z = 2$
The solution is $(-1, -2, 2)$.

51. (1) $x - 3y + 2z = 1$
(2) $x - 2y + 3z = 5$
(3) $2x - 6y + 4z = 3$

Eliminate x. Multiply equation (1) by -1 and add to equation (2).

$-1(x - 3y + 2z) = -1(1)$
$x - 2y + 3z = 5$

$-x + 3y - 2z = -1$
$x - 2y + 3z = 5$

(4) $y + z = 4$

Multiply equation (1) by -2 and add to equation (3).

$-2(x - 3y + 2z) = -2(1)$
$2x - 6y + 4z = 3$

$-2x + 6y - 4z = -2$
$2x - 6y + 4z = 3$

$0 = 1$

This is not a true equation. The system of equations is inconsistent and therefore has no solution.

53. (1) $3x - y - 2z = 11$
(2) $2x + y - 2z = 11$
(3) $x + 3y - z = 8$

Eliminate z. Multiply equation (1) by -1 and add to equation (2).

$-1(3x - y - 2z) = -1(11)$
$2x + y - 2z = 11$

$-3x + y + 2z = -11$
$2x + y - 2z = 11$

(4) $-x + 2y = 0$

Multiply equation (3) by -2 and add to equation (1).

$3x - y - 2z = 11$
$-2(x + 3y - z) = -2(8)$

$3x - y - 2z = 11$
$-2x - 6y + 2z = -16$

(5) $x - 7y = -5$

Add equations (4) and (5).

$-x + 2y = 0$
$x - 7y = -5$

$-5y = -5$
$y = 1$

Replace y by 1 in equation (4).

$-x + 2y = 0$
$-x + 2(1) = 0$
$-x + 2 = 0$
$-x = -2$
$x = 2$

Replace x by 2 and y by 1 in equation (3).

$x + 3y - z = 8$
$2 + 3(1) - z = 8$
$2 + 3 - z = 8$
$5 - z = 8$
$-z = 3$
$z = -3$

The solution is $(2, 1, -3)$.

55. (1) $\quad 4x + 5y + z = 6$
(2) $\quad 2x - y + 2z = 11$
(3) $\quad x + 2y + 2z = 6$
Eliminate z. Multiply equation (1) by -2 and add to equation (2).
$-2(4x + 5y + z) = -2(6)$
$\quad 2x - y + 2z = 11$

$-8x - 10y - 2z = -12$
$\quad 2x - y + 2z = 11$

(4) $\quad -6x - 11y = -1$
Multiply equation (2) by -1 and add to equation (3).
$-1(2x - y + 2z) = -1(11)$
$\quad x + 2y + 2z = 6$

$-2x + y - 2z = -11$
$\quad x + 2y + 2z = 6$

(5) $\quad -x + 3y = -5$
Multiply equation (5) by -6 and add to equation (4).
$-6x - 11y = -1$
$-6(-x + 3y) = -6(-5)$

$-6x - 11y = -1$
$\quad 6x - 18y = 30$

$\quad -29y = 29$
$\quad y = -1$
Replace y by -1 in equation (5).
$-x + 3y = -5$
$-x + 3(-1) = -5$
$-x - 3 = -5$
$-x = -2$
$x = 2$
Replace x by 2 and y by -1 in equation (1).
$4x + 5y + z = 6$
$4(2) + 5(-1) + z = 6$
$8 - 5 + z = 6$
$3 + z = 6$
$z = 3$
The solution is $(2, -1, 3)$.

57. (1) $\quad 3x + 2y - 3z = 8$
(2) $\quad 2x + 3y + 2z = 10$
(3) $\quad x + y - z = 2$
Eliminate z. Multiply equation (1) by 2 and equation (2) by 3.
Then add the equations.
$2(3x + 2y - 3z) = 2(8)$
$3(2x + 3y + 2z) = 3(10)$

$\quad 6x + 4y - 6z = 16$
$\quad 6x + 9y + 6z = 30$

(4) $\quad 12x + 13y = 46$
Multiply equation (3) by 2 and add to equation (2).
$2x + 3y + 2z = 10$
$2(x + y - z) = 2(2)$

$2x + 3y + 2z = 10$
$2x + 2y - 2z = 4$

(5) $\quad 4x + 5y = 14$
Multiply equation (5) by -3 and add to equation (4).
$12x + 13y = 46$
$-3(4x + 5y) = -3(14)$

$\quad 12x + 13y = 46$
$-12x - 15y = -42$

$\quad -2y = 4$
$\quad y = -2$
Replace y by -2 in equation (5).
$4x + 5y = 14$
$4x + 5(-2) = 14$
$4x - 10 = 14$
$4x = 24$
$x = 6$
Replace x by 6 and y by -2 in equation (3).
$x + y - z = 2$
$6 + (-2) - z = 2$
$4 - z = 2$
$-z = -2$
$z = 2$
The solution is $(6, -2, 2)$.

59. (1) $3x - 3y + 4z = 6$
(2) $4x - 5y + 2z = 10$
(3) $x - 2y + 3z = 4$
Eliminate x. Multiply equation (3) by -3 and add to equation (1).
$3x - 3y + 4z = 6$
$-3(x - 2y + 3z) = -3(4)$

$3x - 3y + 4z = 6$
$-3x + 6y - 9z = -12$

(4) $3y - 5z = -6$
Multiply equation (3) by -4 and add to equation (2).
$4x - 5y + 2z = 10$
$-4(x - 2y + 3z) = -4(4)$

$4x - 5y + 2z = 10$
$-4x + 8y - 12z = -16$

(5) $3y - 10z = -6$
Multiply equation (4) by -1 and add to equation (5).
$-1(3y - 5z) = -1(-6)$
$3y - 10z = -6$

$-3y + 5z = 6$
$3y - 10z = -6$

$-5z = 0$
$z = 0$
Replace z by 0 in equation (4).
$3y - 5z = -6$
$3y - 5(0) = -6$
$3y = -6$
$y = -2$
Replace y by -2 and z by 0 in equation (3).
$x - 2y + 3z = 4$
$x - 2(-2) + 3(0) = 4$
$x + 4 = 4$
$x = 0$
The solution is $(0, -2, 0)$.

61. (1) $2x + 2y + 3z = 13$
(2) $-3x + 4y - z = 5$
(3) $5x - 3y + z = 2$
Eliminate z. Multiply equation (2) by 3 and add to equation (1).
$2x + 2y + 3z = 13$
$3(-3x + 4y - z) = 3(5)$

$2x + 2y + 3z = 13$
$-9x + 12y - 3z = 15$

$-7x + 14y = 28$
Multiply each side of the equation by $\frac{1}{7}$.

(4) $-x + 2y = 4$
Add equations (2) and (3).
$-3x + 4y - z = 5$
$5x - 3y + z = 2$

(5) $2x + y = 7$
Multiply equation (4) by 2 and add to equation (5).
$2(-x + 2y) = 2(4)$
$2x + y = 7$

$-2x + 4y = 8$
$2x + y = 7$

$5y = 15$
$y = 3$
Replace y by 3 in equation (5).
$2x + y = 7$
$2x + 3 = 7$
$2x = 4$
$x = 2$
Replace x by 2 and y by 3 in equation (3).
$5x - 3y + z = 2$
$5(2) - 3(3) + z = 2$
$10 - 9 + z = 2$
$1 + z = 2$
$z = 1$
The solution is $(2, 3, 1)$.

63. (1) $\quad 5x+3y-z=5$
(2) $\quad 3x-2y+4z=13$
(3) $\quad 4x+3y+5z=22$
Eliminate z. Multiply equation (1) by 4 and add to equation (2).
$4(5x+3y-z)=4(5)$
$3x-2y+4z=13$

$20x+12y-4z=20$
$3x-2y+4z=13$

(4) $\quad 23x+10y=33$
Multiply equation (1) by 5 and add to equation (3).
$5(5x+3y-z)=5(5)$
$4x+3y+5z=22$

$25x+15y-5z=25$
$4x+3y+5z=22$

(5) $\quad 29x+18y=47$
Multiply equation (4) by -18 and equation (5) by 10. Then add the equations.
$-18(23x+10y)=-18(33)$
$10(29x+18y)=10(47)$

$-414x-180y=-594$
$290x+180y=470$

$-124x=-124$
$x=1$
Replace x by 1 in equation (4).
$23x+10y=33$
$23(1)+10y=33$
$23+10y=33$
$10y=10$
$y=1$
Replace x by 1 and y by 1 in equation (1).
$5x+3y-z=5$
$5(1)+3(1)-z=5$
$5+3-z=5$
$8-z=5$
$-z=-3$
$z=3$
The solution is (1, 1, 3).

Applying Concepts 4.2

65. (1) $\quad 0.4x-0.9y=-0.1$
(2) $\quad 0.3x+0.2y=0.8$
Multiply both sides of each equation by 10.
(1) $\quad 4x-9y=-1$
(2) $\quad 3x+2y=8$
Eliminate y.
$2(4x-9y)=2(-1)$
$9(3x+2y)=9(8)$

$8x-18y=-2$
$27x+18y=72$
Add the equations.
$35x=70$
$x=2$
Replace x in equation (1).
$4x-9y=-1$
$4(2)-9y=-1$
$8-9y=-1$
$-9y=-9$
$y=1$
The solution is (2, 1).

67. (1) $\quad 2.25x+1.5y=3$
(2) $\quad 1.75x+2.25y=1.25$
Multiply both sides of each equation by 100.
(1) $\quad 225x+150y=300$
(2) $\quad 175x+225y=125$
Eliminate y.
$3(225x+150y)=3(300)$
$-2(175x+225y)=-2(125)$

$675x+450y=900$
$-350x-450y=-250$
Add the equations.
$325x=650$
$x=2$
Replace x in equation (1).
$225x+150y=300$
$225(2)+150y=300$
$450+150y=300$
$150y=-150$
$y=-1$
The solution is (2, -1).

69.
(1) $1.6x - 0.9y + 0.3z = 2.9$
(2) $1.6x + 0.5y - 0.1z = 3.3$
(3) $0.8x - 0.7y + 0.1z = 1.5$

Multiply both sides of each equation by 10.
(1) $16x - 9y + 3z = 29$
(2) $16x + 5x - z = 33$
(3) $8x - 7y + z = 15$

Eliminate z. Add equations (2) and (3).
$16x + 5y - z = 33$
$8x - 7y + z = 15$

(4) $24x - 2y = 48$

Multiply equation (2) by 3 and add to equation (1).
$3(16x + 5y - z) = 3(33)$
$16x - 9y + 3z = 29$

$48x + 15y - 3z = 99$
$16x - 9y + 3z = 29$

(5) $64x + 6y = 128$

Multiply equation (4) by 3 and add to equation (5).
$3(24x - 2y) = 3(48)$
$64x + 6y = 128$

$72x - 6y = 144$
$64x + 6y = 128$

$136x = 272$
$x = 2$

Replace x by 2 in equation (4).
$24x - 2y = 48$
$24(2) - 2y = 48$
$48 - 2y = 48$
$-2y = 0$
$y = 0$

Replace x by 2 and y by 0 in equation (1).
$16x - 9y + 3z = 29$
$16(2) - 9(0) + 3z = 29$
$32 - 0 + 3z = 29$
$3z = -3$
$z = -1$

The solution is $(2, 0, -1)$.

71. Strategy
Substitute 3 for x, -2 for y, and 4 for z in the equations. Solve for A, B, and C.

Solution
$Ax + 3y + 2z = 8$
$A(3) + 3(-2) + 2(4) = 8$
$3A - 6 + 8 = 8$
$3A + 2 = 8$
$3A = 6$
$A = 2$

$2x + By - 3z = -12$
$2(3) + B(-2) - 3(4) = -12$
$6 - 2B - 12 = -12$
$-2B - 6 = -12$
$-2B = -6$
$B = 3$

$3x - 2y + Cz = 1$
$3(3) - 2(-2) + C(4) = 1$
$9 + 4 + 4C = 1$
$4C + 13 = 1$
$4C = -12$
$C = -3$

The value of A is 2. The value of B is 3. The value of C is -3.

73. Strategy
Solve a system of equations using x to represent the number of nickels, y to represent the number of dimes, and z to represent the number of quarters.

Solution
(1) $x + y + z = 30$
(2) $5x + 10y + 25z = 325$

Eliminate x by multiplying equation (1) by (-5) and adding equation (2).
$-5x - 5y - 5z = -150$
$5x + 10y + 25z = 325$
(3) $5y + 20z = 175$

Solve equation (3) for y in terms of z.
(3) $5y + 20z = 175$
$5y = -20z + 175$
$y = -4z + 35$

Replace y by $-4z + 35$ in equation (1) and solve for x in terms of z.
$x + y + z = 30$
$x + (-4z + 35) + z = 30$
$x - 4z + 35 + z = 30$
$x - 3z + 35 = 30$
$x = 3z - 5$

The number of nickels is $3z - 5$, the number of dimes is $-4z + 35$, and the number of quarters is z, where $z = 2, 3, 4, 5, 6, 7,$ or 8. All other values of z make the number of nickels or dimes negative, which is not possible.

75. (1) $\dfrac{1}{x} + \dfrac{2}{y} = 3$

(2) $\dfrac{1}{x} - \dfrac{3}{y} = -2$

Clear the fractions.

$xy\left(\dfrac{1}{x} + \dfrac{2}{y}\right) = xy \cdot 3$

$xy\left(\dfrac{1}{x} - \dfrac{3}{y}\right) = xy \cdot (-2)$

$y + 2x = 3xy$
$y - 3x = -2xy$

Eliminate y.

$y + 2x = 3xy$
$-y + 3x = 2xy$
$5x = 5xy$
$y = 1$

Substitute y into equation (2).

$\dfrac{1}{x} - \dfrac{3}{y} = -2$

$\dfrac{1}{x} - \dfrac{3}{1} = -2$

$\dfrac{1}{x} = 1$

$x = 1$

The solution is (1, 1).

77. (1) $\dfrac{3}{x} - \dfrac{5}{y} = -\dfrac{3}{2}$

(2) $\dfrac{1}{x} - \dfrac{2}{y} = -\dfrac{2}{3}$

Clear fractions.

$2xy\left(\dfrac{3}{x} - \dfrac{5}{y}\right) = 2xy\left(-\dfrac{3}{2}\right)$

$3xy\left(\dfrac{1}{x} - \dfrac{2}{y}\right) = 3xy\left(-\dfrac{2}{3}\right)$

$6y - 10x = -3xy$
$3y - 6x = -2xy$

Eliminate y.

$6y - 10x = -3xy$
$-6y + 12x = 4xy$
$2x = xy$
$y = 2$

Substitute y into equation (2).

$\dfrac{1}{x} - \dfrac{2}{y} = -\dfrac{2}{3}$

$\dfrac{1}{x} - \dfrac{2}{2} = -\dfrac{2}{3}$

$\dfrac{1}{x} = \dfrac{1}{3}$

$x = 3$

The solution is (3, 2).

Section 4.3

Concept Review 4.3

1. Always true

3. Sometimes true
   A square matrix has the same number of rows and columns. A matrix may have different numbers of rows and columns.

5. Always true

Objective 1 Exercises

3. $\begin{vmatrix} 2 & -1 \\ 3 & 4 \end{vmatrix} = 2(4) - 3(-1) = 8 + 3 = 11$

5. $\begin{vmatrix} 6 & -2 \\ -3 & 4 \end{vmatrix} = 6(4) - (-3)(-2) = 24 - 6 = 18$

7. $\begin{vmatrix} 3 & 6 \\ 2 & 4 \end{vmatrix} = 3(4) - 2(6) = 12 - 12 = 0$

9. $\begin{vmatrix} 1 & -1 & 2 \\ 3 & 2 & 1 \\ 1 & 0 & 4 \end{vmatrix} = 1\begin{vmatrix} 2 & 1 \\ 0 & 4 \end{vmatrix} + 1\begin{vmatrix} 3 & 1 \\ 1 & 4 \end{vmatrix} + 2\begin{vmatrix} 3 & 2 \\ 1 & 0 \end{vmatrix}$

$= 1(8 - 0) + 1(12 - 1) + 2(0 - 2)$
$= 8 + 11 - 4 = 15$

11. $\begin{vmatrix} 3 & -1 & 2 \\ 0 & 1 & 2 \\ 3 & 2 & -2 \end{vmatrix} = 3\begin{vmatrix} 1 & 2 \\ 2 & -2 \end{vmatrix} + 1\begin{vmatrix} 0 & 2 \\ 3 & -2 \end{vmatrix} + 2\begin{vmatrix} 0 & 1 \\ 3 & 2 \end{vmatrix}$

$= 3(-2 - 4) + 1(0 - 6) + 2(0 - 3)$
$= 3(-6) + 1(-6) + 2(-3)$
$= -18 - 6 - 6 = -30$

13. $\begin{vmatrix} 4 & 2 & 6 \\ -2 & 1 & 1 \\ 2 & 1 & 3 \end{vmatrix} = 4\begin{vmatrix} 1 & 1 \\ 1 & 3 \end{vmatrix} - 2\begin{vmatrix} -2 & 1 \\ 2 & 3 \end{vmatrix} + 6\begin{vmatrix} -2 & 1 \\ 2 & 1 \end{vmatrix}$

$= 4(3 - 1) - 2(-6 - 2) + 6(-2 - 2)$
$= 4(2) - 2(-8) + 6(-4)$
$= 8 + 16 - 24 = 0$

Objective 2 Exercises

17. $2x - 5y = 26$
$5x + 3y = 3$

$D = \begin{vmatrix} 2 & -5 \\ 5 & 3 \end{vmatrix} = 31$, $D_x = \begin{vmatrix} 26 & -5 \\ 3 & 3 \end{vmatrix} = 93$,

$D_y = \begin{vmatrix} 2 & 26 \\ 5 & 3 \end{vmatrix} = -124$

$x = \dfrac{D_x}{D} = \dfrac{93}{31} = 3$ $y = \dfrac{D_y}{D} = \dfrac{-124}{31} = -4$

The solution is (3, -4).

19. $x - 4y = 8$
    $3x + 7y = 5$

    $D = \begin{vmatrix} 1 & -4 \\ 3 & 7 \end{vmatrix} = 19$, $D_x = \begin{vmatrix} 8 & -4 \\ 5 & 7 \end{vmatrix} = 76$,

    $D_y = \begin{vmatrix} 1 & 8 \\ 3 & 5 \end{vmatrix} = -19$

    $x = \dfrac{D_x}{D} = \dfrac{76}{19} = 4$ $y = \dfrac{D_y}{D} = \dfrac{-19}{19} = -1$

    The solution is $(4, -1)$.

21. $2x + 3y = 4$
    $6x - 12y = -5$

    $D = \begin{vmatrix} 2 & 3 \\ 6 & -12 \end{vmatrix} = -42$, $D_x = \begin{vmatrix} 4 & 3 \\ -5 & -12 \end{vmatrix} = -33$,

    $D_y = \begin{vmatrix} 2 & 4 \\ 6 & -5 \end{vmatrix} = -34$

    $x = \dfrac{D_x}{D} = \dfrac{-33}{-42} = \dfrac{11}{14}$ $y = \dfrac{D_y}{D} = \dfrac{-34}{-42} = \dfrac{17}{21}$

    The solution is $\left(\dfrac{11}{14}, \dfrac{17}{21}\right)$.

23. $2x + 5y = 6$
    $6x - 2y = 1$

    $D = \begin{vmatrix} 2 & 5 \\ 6 & -2 \end{vmatrix} = -34$, $D_x = \begin{vmatrix} 6 & 5 \\ 1 & -2 \end{vmatrix} = -17$,

    $D_y = \begin{vmatrix} 2 & 6 \\ 6 & 1 \end{vmatrix} = -34$

    $x = \dfrac{D_x}{D} = \dfrac{-17}{-34} = \dfrac{1}{2}$ $y = \dfrac{D_y}{D} = \dfrac{-34}{-34} = 1$

    The solution is $\left(\dfrac{1}{2}, 1\right)$.

25. $-2x + 3y = 7$
    $4x - 6y = 9$

    $D = \begin{vmatrix} -2 & 3 \\ 4 & -6 \end{vmatrix} = 0$

    Since $D = 0$, $\dfrac{D_x}{D}$ is undefined. The system cannot be solved by Cramer's Rule.

27. $2x - 5y = -2$
    $3x - 7y = -3$

    $D = \begin{vmatrix} 2 & -5 \\ 3 & -7 \end{vmatrix} = 1$, $D_x = \begin{vmatrix} -2 & -5 \\ -3 & -7 \end{vmatrix} = -1$,

    $D_y = \begin{vmatrix} 2 & -2 \\ 3 & -3 \end{vmatrix} = 0$

    $x = \dfrac{D_x}{D} = \dfrac{-1}{1} = -1$ $y = \dfrac{D_y}{D} = \dfrac{0}{1} = 0$

    The solution is $(-1, 0)$.

29. $2x - y + 3z = 9$
    $x + 4y + 4z = 5$
    $3x + 2y + 2z = 5$

    $D = \begin{vmatrix} 2 & -1 & 3 \\ 1 & 4 & 4 \\ 3 & 2 & 2 \end{vmatrix} = -40$,

    $D_x = \begin{vmatrix} 9 & -1 & 3 \\ 5 & 4 & 4 \\ 5 & 2 & 2 \end{vmatrix} = -40$,

    $D_y = \begin{vmatrix} 2 & 9 & 3 \\ 1 & 5 & 4 \\ 3 & 5 & 2 \end{vmatrix} = 40$,

    $D_z = \begin{vmatrix} 2 & -1 & 9 \\ 1 & 4 & 5 \\ 3 & 2 & 5 \end{vmatrix} = -80$

    $x = \dfrac{D_x}{D} = \dfrac{-40}{-40} = 1$ $y = \dfrac{D_y}{D} = \dfrac{40}{-40} = -1$

    $z = \dfrac{D_z}{D} = \dfrac{-80}{-40} = 2$

    The solution is $(1, -1, 2)$.

31. $3x - y + z = 11$
    $x + 4y - 2z = -12$
    $2x + 2y - z = -3$

    $D = \begin{vmatrix} 3 & -1 & 1 \\ 1 & 4 & -2 \\ 2 & 2 & -1 \end{vmatrix} = -3$,

    $D_x = \begin{vmatrix} 11 & -1 & 1 \\ -12 & 4 & -2 \\ -3 & 2 & -1 \end{vmatrix} = -6$,

    $D_y = \begin{vmatrix} 3 & 11 & 1 \\ 1 & -12 & -2 \\ 2 & -3 & -1 \end{vmatrix} = 6$,

    $D_z = \begin{vmatrix} 3 & -1 & 11 \\ 1 & 4 & -12 \\ 2 & 2 & -3 \end{vmatrix} = -9$

    $x = \dfrac{D_x}{D} = \dfrac{-6}{-3} = 2$ $y = \dfrac{D_y}{D} = \dfrac{6}{-3} = -2$

    $z = \dfrac{D_z}{D} = \dfrac{-9}{-3} = 3$

    The solution is $(2, -2, 3)$.

33. $4x - 2y + 6z = 1$
    $3x + 4y + 2z = 1$
    $2x - y + 3z = 2$

    $D = \begin{vmatrix} 4 & -2 & 6 \\ 3 & 4 & 2 \\ 2 & -1 & 3 \end{vmatrix} = 0$

    Since $D = 0$, $\dfrac{D_x}{D}$ is undefined. The system cannot be solved by Cramer's Rule.

35. $5x - 4y + 2z = 4$
$3x - 5y + 3z = -4$
$3x + y - 5z = 12$

$D = \begin{vmatrix} 5 & -4 & 2 \\ 3 & -5 & 3 \\ 3 & 1 & -5 \end{vmatrix} = 50,$

$D_x = \begin{vmatrix} 4 & -4 & 2 \\ -4 & -5 & 3 \\ 12 & 1 & -5 \end{vmatrix} = 136,$

$D_y = \begin{vmatrix} 5 & 4 & 2 \\ 3 & -4 & 3 \\ 3 & 12 & -5 \end{vmatrix} = 112,$

$D_z = \begin{vmatrix} 5 & -4 & 4 \\ 3 & -5 & -4 \\ 3 & 1 & 12 \end{vmatrix} = -16$

$x = \dfrac{D_x}{D} = \dfrac{136}{50} = \dfrac{68}{25} \quad y = \dfrac{D_y}{D} = \dfrac{112}{50} = \dfrac{56}{25}$

$z = \dfrac{D_z}{D} = \dfrac{-16}{50} = -\dfrac{8}{25}$

The solution is $\left(\dfrac{68}{25}, \dfrac{56}{25}, -\dfrac{8}{25}\right).$

Objective 3 Exercises

39. $3x + y = 6$
$2x - y = -1$

$\begin{vmatrix} 3 & 1 & 6 \\ 2 & -1 & -1 \end{vmatrix}$

$\begin{vmatrix} 1 & \frac{1}{3} & 2 \\ 2 & -1 & -1 \end{vmatrix}$ Multiply row 1 by $\dfrac{1}{3}$.

$\begin{vmatrix} 1 & \frac{1}{3} & 2 \\ 0 & -\frac{5}{3} & -5 \end{vmatrix}$ Multiply row 1 by -2 and add to row 2.

$\begin{vmatrix} 1 & \frac{1}{3} & 2 \\ 0 & 1 & 3 \end{vmatrix}$ Multiply row 2 by $-\dfrac{3}{5}$.

$x + \left(\dfrac{1}{3}\right)y = 2$
$y = 3$
$x + \left(\dfrac{1}{3}\right)(3) = 2$
$x + 1 = 2$
$x = 1$

The solution is (1, 3).

41. $x - 3y = 8$
$3x - y = 0$

$\begin{vmatrix} 1 & -3 & 8 \\ 3 & -1 & 0 \end{vmatrix}$

$\begin{vmatrix} 1 & -3 & 8 \\ 0 & 8 & -24 \end{vmatrix}$ Multiply row 1 by -3 and add to row 2.

$\begin{vmatrix} 1 & -3 & 8 \\ 0 & 1 & -3 \end{vmatrix}$ Multiply row 2 by $\dfrac{1}{8}$.

$x - 3y = 8$
$y = -3$
$x - 3(-3) = 8$
$x + 9 = 8$
$x = -1$

The solution is (−1, −3).

43. $y = 4x - 10$
$2y = 5x - 11$

$4x - y = 10$
$5x - 2y = 11$

$\begin{vmatrix} 4 & -1 & 10 \\ 5 & -2 & 11 \end{vmatrix}$

$\begin{vmatrix} 1 & -\frac{1}{4} & \frac{5}{2} \\ 5 & -2 & 11 \end{vmatrix}$ Multiply row 1 by $\dfrac{1}{4}$.

$\begin{vmatrix} 1 & -\frac{1}{4} & \frac{5}{2} \\ 0 & -\frac{3}{4} & -\frac{3}{2} \end{vmatrix}$ Multiply row 1 by -5 and add to row 2.

$\begin{vmatrix} 1 & -\frac{1}{4} & \frac{5}{2} \\ 0 & 1 & 2 \end{vmatrix}$ Multiply row 2 by $-\dfrac{4}{3}$.

$x - \left(\dfrac{1}{4}\right)y = \dfrac{5}{2}$
$y = 2$
$x - \left(\dfrac{1}{4}\right)(2) = \dfrac{5}{2}$
$x - \dfrac{1}{2} = \dfrac{5}{2}$
$x = 3$

The solution is (3, 2).

45. $2x - y = -4$
$y = 2x - 8$

$2x - y = -4$
$2x - y = 8$

$\begin{vmatrix} 2 & -1 & -4 \\ 2 & -1 & 8 \end{vmatrix}$

$\begin{vmatrix} 1 & -\frac{1}{2} & -2 \\ 2 & -1 & 8 \end{vmatrix}$ Multiply row 1 by $\frac{1}{2}$.

$\begin{vmatrix} 1 & -\frac{1}{2} & -2 \\ 0 & 0 & 12 \end{vmatrix}$ Multiply row 1 by -2 and add to row 2.

$x - \frac{1}{2}y = -2$
$0 = 12$

This is not a true equation. The system of equations is inconsistent.

47. $4x - 3y = -14$
$3x + 4y = 2$

$\begin{vmatrix} 4 & -3 & -14 \\ 3 & 4 & 2 \end{vmatrix}$

$\begin{vmatrix} 1 & -\frac{3}{4} & -\frac{7}{2} \\ 3 & 4 & 2 \end{vmatrix}$ Multiply row 1 by $\frac{1}{4}$.

$\begin{vmatrix} 1 & -\frac{3}{4} & -\frac{7}{2} \\ 0 & \frac{25}{4} & \frac{25}{2} \end{vmatrix}$ Multiply row 1 by -3 and add to row 2.

$\begin{vmatrix} 1 & -\frac{3}{4} & -\frac{7}{2} \\ 0 & 1 & 2 \end{vmatrix}$ Multiply row 2 by $\frac{4}{25}$.

$x - \left(\frac{3}{4}\right)y = -\frac{7}{2}$
$y = 2$

$x - \left(\frac{3}{4}\right)(2) = -\frac{7}{2}$
$x - \frac{3}{2} = -\frac{7}{2}$
$x = -2$

The solution is $(-2, 2)$.

49. $5x + 4y + 3z = -9$
$x - 2y + 2z = -6$
$x - y - z = 3$

$\begin{vmatrix} 5 & 4 & 3 & -9 \\ 1 & -2 & 2 & -6 \\ 1 & -1 & -1 & 3 \end{vmatrix}$

$\begin{vmatrix} 1 & -2 & 2 & -6 \\ 5 & 4 & 3 & -9 \\ 1 & -1 & -1 & 3 \end{vmatrix}$ Interchange rows 1 and 2.

$\begin{vmatrix} 1 & -2 & 2 & -6 \\ 0 & 14 & -7 & 21 \\ 0 & 1 & -3 & 9 \end{vmatrix}$ Multiply row 1 by -5 and add to row 2. Multiply row 1 by -1 and add to row 3.

$\begin{vmatrix} 1 & -2 & 2 & -6 \\ 0 & 1 & -\frac{1}{2} & \frac{3}{2} \\ 0 & 1 & -3 & 9 \end{vmatrix}$ Multiply row 2 by $\frac{1}{14}$.

$\begin{vmatrix} 1 & -2 & 2 & -6 \\ 0 & 1 & -\frac{1}{2} & \frac{3}{2} \\ 0 & 0 & -\frac{5}{2} & \frac{15}{2} \end{vmatrix}$ Multiply row 2 by -1 and add to row 3.

$\begin{vmatrix} 1 & -2 & 2 & -6 \\ 0 & 1 & -\frac{1}{2} & \frac{3}{2} \\ 0 & 0 & 1 & -3 \end{vmatrix}$ Multiply row 3 by $-\frac{2}{5}$.

$x - 2y + 2z = -6$
$y - \left(\frac{1}{2}\right)z = \frac{3}{2}$
$z = -3$

$y - \left(\frac{1}{2}\right)(-3) = \frac{3}{2}$
$y + \frac{3}{2} = \frac{3}{2}$
$y = 0$

$x - 2(0) + 2(-3) = -6$
$x - 6 = -6$
$x = 0$

The solution is $(0, 0, -3)$.

51. $5x - 5y + 2z = 8$
$2x + 3y - z = 0$
$x + 2y - z = 0$

$\begin{vmatrix} 5 & -5 & 2 & 8 \\ 2 & 3 & -1 & 0 \\ 1 & 2 & -1 & 0 \end{vmatrix}$

$\begin{vmatrix} 1 & 2 & -1 & 0 \\ 2 & 3 & -1 & 0 \\ 5 & -5 & 2 & 8 \end{vmatrix}$ Interchange rows 1 and 3.

$\begin{vmatrix} 1 & 2 & -1 & 0 \\ 0 & -1 & 1 & 0 \\ 0 & -15 & 7 & 8 \end{vmatrix}$ Multiply row 1 by -2 and add to row 2. Multiply row 1 by -5 and add to row 3.

$\begin{vmatrix} 1 & 2 & -1 & 0 \\ 0 & 1 & -1 & 0 \\ 0 & -15 & 7 & 8 \end{vmatrix}$ Multiply row 2 by -1.

$\begin{vmatrix} 1 & 2 & -1 & 0 \\ 0 & 1 & -1 & 0 \\ 0 & 0 & -8 & 8 \end{vmatrix}$ Multiply row 2 by 15 and add to row 3.

$\begin{vmatrix} 1 & 2 & -1 & 0 \\ 0 & 1 & -1 & 0 \\ 0 & 0 & 1 & -1 \end{vmatrix}$ Multiply row 3 by $-\dfrac{1}{8}$.

$x + 2y - z = 0$
$y - z = 0$
$z = -1$

$y - (-1) = 0$
$y + 1 = 0$
$y = -1$

$x + 2(-1) - (-1) = 0$
$x - 2 + 1 = 0$
$x - 1 = 0$
$x = 1$

The solution is $(1, -1, -1)$.

53. $2x + 3y + z = 5$
$3x + 3y + 3z = 10$
$4x + 6y + 2z = 5$

$\begin{vmatrix} 2 & 3 & 1 & 5 \\ 3 & 3 & 3 & 10 \\ 4 & 6 & 2 & 5 \end{vmatrix}$

$\begin{vmatrix} 1 & \frac{3}{2} & \frac{1}{2} & \frac{5}{2} \\ 3 & 3 & 3 & 10 \\ 4 & 6 & 2 & 5 \end{vmatrix}$ Multiply row 1 by $\dfrac{1}{2}$.

$\begin{vmatrix} 1 & \frac{3}{2} & \frac{1}{2} & \frac{5}{2} \\ 0 & -\frac{3}{2} & \frac{3}{2} & \frac{5}{2} \\ 0 & 0 & 0 & -5 \end{vmatrix}$ Multiply row 1 by -3 and add to row 2. Multiply row 1 by -4 and add to row 3.

$x + \left(\dfrac{3}{2}\right)y + \left(\dfrac{1}{2}\right)z = \dfrac{5}{2}$
$-\left(\dfrac{3}{2}\right)y + \left(\dfrac{3}{2}\right)z = \dfrac{5}{2}$
$0 = -5$

This is not a true equation. The system of equations is inconsistent.

55. $3x + 2y + 3z = 2$
$6x - 2y + z = 1$
$3x + 4y + 2z = 3$

$\begin{vmatrix} 3 & 2 & 3 & 2 \\ 6 & -2 & 1 & 1 \\ 3 & 4 & 2 & 3 \end{vmatrix}$

$\begin{vmatrix} 1 & \frac{2}{3} & 1 & \frac{2}{3} \\ 6 & -2 & 1 & 1 \\ 3 & 4 & 2 & 3 \end{vmatrix}$ Multiply row 1 by $\frac{1}{3}$.

$\begin{vmatrix} 1 & \frac{2}{3} & 1 & \frac{2}{3} \\ 0 & -6 & -5 & -3 \\ 0 & 2 & -1 & 1 \end{vmatrix}$ Multiply row 1 by -6 and add to row 2. Multiply row 1 by -3 and add to row 3.

$\begin{vmatrix} 1 & \frac{2}{3} & 1 & \frac{2}{3} \\ 0 & 1 & \frac{5}{6} & \frac{1}{2} \\ 0 & 2 & -1 & 1 \end{vmatrix}$ Multiply row 2 by $-\frac{1}{6}$.

$\begin{vmatrix} 1 & \frac{2}{3} & 1 & \frac{2}{3} \\ 0 & 1 & \frac{5}{6} & \frac{1}{2} \\ 0 & 0 & -\frac{8}{3} & 0 \end{vmatrix}$ Multiply row 2 by -2 and add to row 3.

$\begin{vmatrix} 1 & \frac{2}{3} & 1 & \frac{2}{3} \\ 0 & 1 & \frac{5}{6} & \frac{1}{2} \\ 0 & 0 & 1 & 0 \end{vmatrix}$ Multiply row 3 by $-\frac{3}{8}$.

$x + \left(\frac{2}{3}\right)y + z = \frac{2}{3}$
$y + \left(\frac{5}{6}\right)z = \frac{1}{2}$
$z = 0$

$y + \left(\frac{5}{6}\right)(0) = \frac{1}{2}$
$y = \frac{1}{2}$

$x + \left(\frac{2}{3}\right)\left(\frac{1}{2}\right) + 0 = \frac{2}{3}$
$x + \frac{1}{3} = \frac{2}{3}$
$x = \frac{1}{3}$

The solution is $\left(\frac{1}{3}, \frac{1}{2}, 0\right)$.

57. $5x - 5y - 5z = 2$
$5x + 5y - 5z = 6$
$10x + 10y + 5z = 3$

$\begin{vmatrix} 5 & -5 & -5 & 2 \\ 5 & 5 & -5 & 6 \\ 10 & 10 & 5 & 3 \end{vmatrix}$

$\begin{vmatrix} 1 & -1 & -1 & \frac{2}{5} \\ 5 & 5 & -5 & 6 \\ 10 & 10 & 5 & 3 \end{vmatrix}$ Multiply row 1 by $\frac{1}{5}$.

$\begin{vmatrix} 1 & -1 & -1 & \frac{2}{5} \\ 0 & 10 & 0 & 4 \\ 0 & 20 & 15 & -1 \end{vmatrix}$ Multiply row 1 by -5 and add to row 2. Multiply row 1 by -10 and add to row 3.

$\begin{vmatrix} 1 & -1 & -1 & \frac{2}{5} \\ 0 & 1 & 0 & \frac{2}{5} \\ 0 & 20 & 15 & -1 \end{vmatrix}$ Multiply row 2 by $\frac{1}{10}$.

$\begin{vmatrix} 1 & -1 & -1 & \frac{2}{5} \\ 0 & 1 & 0 & \frac{2}{5} \\ 0 & 0 & 15 & -9 \end{vmatrix}$ Multiply row 2 by -20 and add to row 3.

$\begin{vmatrix} 1 & -1 & -1 & \frac{2}{5} \\ 0 & 1 & 0 & \frac{2}{5} \\ 0 & 0 & 1 & -\frac{3}{5} \end{vmatrix}$ Multiply row 3 by $\frac{1}{15}$.

$x - y - z = \frac{2}{5}$
$y = \frac{2}{5}$
$z = -\frac{3}{5}$

$x - \frac{2}{5} - \left(-\frac{3}{5}\right) = \frac{2}{5}$
$x - \frac{2}{5} + \frac{3}{5} = \frac{2}{5}$
$x + \frac{1}{5} = \frac{2}{5}$
$x = \frac{1}{5}$

The solution is $\left(\frac{1}{5}, \frac{2}{5}, -\frac{3}{5}\right)$.

59. $4x + 4y - 3z = 3$
$8x + 2y + 3z = 0$
$4x - 4y + 6z = -3$

$\begin{vmatrix} 4 & 4 & -3 & 3 \\ 8 & 2 & 3 & 0 \\ 4 & -4 & 6 & -3 \end{vmatrix}$

$\begin{vmatrix} 1 & 1 & -\frac{3}{4} & \frac{3}{4} \\ 8 & 2 & 3 & 0 \\ 4 & -4 & 6 & -3 \end{vmatrix}$ Multiply row 1 by $1/4$.

$\begin{vmatrix} 1 & 1 & -\frac{3}{4} & \frac{3}{4} \\ 0 & -6 & 9 & -6 \\ 0 & -8 & 9 & -6 \end{vmatrix}$ Multiply row 1 by -8 and add to row 2. Multiply row 1 by -4 and add to row 3.

$\begin{vmatrix} 1 & 1 & -\frac{3}{4} & \frac{3}{4} \\ 0 & 1 & -\frac{3}{2} & 1 \\ 0 & -8 & 9 & -6 \end{vmatrix}$ Multiply row 2 by $-\frac{1}{6}$.

$\begin{vmatrix} 1 & 1 & -\frac{3}{4} & \frac{3}{4} \\ 0 & 1 & -\frac{3}{2} & 1 \\ 0 & 0 & -3 & 2 \end{vmatrix}$ Multiply row 2 by 8 and add to row 3.

$\begin{vmatrix} 1 & 1 & -\frac{3}{4} & \frac{3}{4} \\ 0 & 1 & -\frac{3}{2} & 1 \\ 0 & 0 & 1 & -\frac{2}{3} \end{vmatrix}$ Multiply row 3 by $-\frac{1}{3}$.

$x + y - \left(\frac{3}{4}\right)z = \frac{3}{4}$
$y - \left(\frac{3}{2}\right)z = 1$
$z = -\frac{2}{3}$

$y - \left(\frac{3}{2}\right)\left(-\frac{2}{3}\right) = 1$
$y + 1 = 1$
$y = 0$

$x + 0 - \left(\frac{3}{4}\right)\left(-\frac{2}{3}\right) = \frac{3}{4}$
$x + \frac{1}{2} = \frac{3}{4}$
$x = \frac{1}{4}$

The solution is $\left(\frac{1}{4}, 0, -\frac{2}{3}\right)$.

Applying Concepts 4.3

61. $\begin{vmatrix} 1 & 0 & 2 \\ 4 & 3 & -1 \\ 0 & 2 & x \end{vmatrix} = -24$

Expand the minors of the first row.
$1(3x + 2) - 0 + 2(8 - 0) = -24$
$3x + 2 + 16 = -24$
$3x + 18 = -24$
$3x = -42$
$x = -14$

The solution is -14.

63. If all the elements in one row or one column of a 2×2 matrix are zeros, the value of the determinant of the matrix is 0.
For example,
$\begin{vmatrix} a_1 & 0 \\ b_1 & 0 \end{vmatrix} = a_1(0) - (0)b_1 = 0 - 0 = 0$

65. a. $\begin{vmatrix} x & x & a \\ y & y & b \\ z & z & c \end{vmatrix}$

Expand by minors of the first row.
$x(cy - bz) - x(cy - bz) + a(yz - yz)$
$= cxy - bxz - cxy + bxz + ayz - ayz$
$= 0$

b. If two columns of a determinant contain identical elements, the value of the determinant is 0.

67. $A = \dfrac{1}{2}\left\{\begin{vmatrix} x_1 & x_2 \\ y_1 & y_2 \end{vmatrix} + \begin{vmatrix} x_2 & x_3 \\ y_2 & y_3 \end{vmatrix} + \begin{vmatrix} x_3 & x_4 \\ y_3 & y_4 \end{vmatrix} + \ldots + \begin{vmatrix} x_n & x_1 \\ y_n & y_1 \end{vmatrix}\right\}$

$A = \dfrac{1}{2}\left\{\begin{vmatrix} 9 & 26 \\ -3 & 6 \end{vmatrix} + \begin{vmatrix} 26 & 18 \\ 6 & 21 \end{vmatrix} + \begin{vmatrix} 18 & 16 \\ 21 & 10 \end{vmatrix} + \begin{vmatrix} 16 & 1 \\ 10 & 11 \end{vmatrix} + \begin{vmatrix} 1 & 9 \\ 11 & -3 \end{vmatrix}\right\}$

$A = \dfrac{1}{2}(132 + 438 - 156 + 166 - 102) = 239$

The area of the polygon is 239 ft^2.

Section 4.4

Concept Review 4.4

1. Never true
   The rate up the river is $(y - x)$ mph.

3. Always true

Objective 1 Exercises

1. Strategy
Rate of the plane in calm air: p
Rate of the wind: w

	Rate	Time	Distance
With wind	$p + w$	2	$2(p + w)$
Against wind	$p - w$	2	$2(p - w)$

• The distance traveled with the wind is 320 mi.
The distance traveled against the wind is 280 mi.
$2(p + w) = 320$
$2(p - w) = 280$

Solution
$2(p + w) = 320$
$2(p - w) = 280$

$\dfrac{1}{2} \cdot 2(p + w) = \dfrac{1}{2} \cdot 320$
$\dfrac{1}{2} \cdot 2(p - w) = \dfrac{1}{2} \cdot 280$

$p + w = 160$
$p - w = 140$

$2p = 300$
$p = 150$

$p + w = 160$
$150 + w = 160$
$w = 10$

The rate of the plane in calm air is 150 mph. The rate of the wind is 10 mph.

3. Strategy
Rate of the cabin cruiser in calm water: x
Rate of the current: y

	Rate	Time	Distance
With current	$x + y$	3	$3(x + y)$
Against current	$x - y$	4	$4(x - y)$

• The distance traveled with the current is 48 mi.
The distance traveled against the current is 48 mi.
$3(x + y) = 48$
$4(x - y) = 48$

Solution
$3(x + y) = 48$
$4(x - y) = 48$

$\dfrac{1}{3} \cdot 3(x + y) = \dfrac{1}{3} \cdot 48$
$\dfrac{1}{4} \cdot 4(x - y) = \dfrac{1}{4} \cdot 48$

$x + y = 16$
$x - y = 12$

$2x = 28$
$x = 14$

$x + y = 16$
$14 + y = 16$
$y = 2$

The rate of the cabin cruiser in calm water is 14 mph. The rate of the current is 2 mph.

5. **Strategy**
Rate of the plane in calm air: x
Rate of the wind: y

	Rate	Time	Distance
With wind	$x + y$	2.5	$2.5(x + y)$
Against wind	$x - y$	3	$3(x - y)$

• The distance traveled with the wind is 450 mi. The distance traveled against the wind is 450 mi.
$2.5(x + y) = 450$
$3(x - y) = 450$

Solution
$2.5(x + y) = 450$
$3(x - y) = 450$

$\dfrac{1}{2.5} \cdot 2.5(x + y) = \dfrac{1}{2.5} \cdot 450$
$\dfrac{1}{3} \cdot 3(x - y) = \dfrac{1}{3} \cdot 450$

$x + y = 180$
$x - y = 150$

$2x = 330$
$x = 165$

$x + y = 180$
$165 + y = 180$
$y = 15$

The rate of the plane in calm air is 165 mph.
The rate of the wind is 15 mph.

7. **Strategy**
The rate of the boat in calm water: x
The rate of the current: y

	Rate	Time	Distance
With current	$x + y$	4	$4(x + y)$
Against current	$x - y$	4	$4(x - y)$

• The distance traveled with the current is 88 km. The distance traveled against the current is 64 km.
$4(x + y) = 88$
$4(x - y) = 64$

Solution
$4(x + y) = 88$
$4(x - y) = 64$

$\dfrac{1}{4} \cdot 4(x + y) = \dfrac{1}{4} \cdot 88$
$\dfrac{1}{4} \cdot 4(x - y) = \dfrac{1}{4} \cdot 64$

$x + y = 22$
$x - y = 16$

$2x = 38$
$x = 19$

$x + y = 22$
$19 + y = 22$
$y = 3$

The rate of the boat in calm water is 19 km/h. The rate of the current is 3 km/h.

9. **Strategy**
Rate of the plane in calm air: x
Rate of the wind: y

	Rate	Time	Distance
With wind	$x + y$	3	$3(x + y)$
Against wind	$x - y$	4	$4(x - y)$

• The distance traveled with the wind is 360 mi. The distance traveled against the wind is 360 mi.
$3(x + y) = 360$
$4(x - y) = 360$

Solution
$3(x + y) = 360$
$4(x - y) = 360$

$\frac{1}{3} \cdot 3(x + y) = \frac{1}{3} \cdot 360$
$\frac{1}{4} \cdot 4(x - y) = \frac{1}{4} \cdot 360$

$x + y = 120$
$x - y = 90$

$2x = 210$
$x = 105$

$x + y = 120$
$105 + y = 120$
$y = 15$

The rate of the plane in calm air is 105 mph. The rate of the wind is 15 mph.

11. **Strategy**
Rate of the boat in calm water: x
Rate of the current: y

	Rate	Time	Distance
With current	$x + y$	3	$3(x + y)$
Against current	$x - y$	3.6	$3.6(x - y)$

• The distance traveled with the current is 54 mi. The distance traveled against the current is 54 mi.
$3(x + y) = 54$
$3.6(x - 7) = 54$

Solution
$3(x + y) = 54$
$3.6(x - y) = 54$

$\frac{1}{3} \cdot 3(x + y) = \frac{1}{3} \cdot 54$
$\frac{1}{3.6} \cdot 3.6(x - y) = \frac{1}{3.6} \cdot 54$

$x + y = 18$
$x - y = 15$

$2x = 33$
$x = 16.5$

$x + y = 18$
$16.5 + y = 18$
$y = 1.5$

The rate of the boat in calm water is 16.5 mph. The rate of the current is 1.5 mph.

Objective 2 Exercises

13. Strategy
Cost of redwood: x
Cost of pine: y
First purchase:

	Amount	Cost	Total Value
Redwood	50	x	$50x$
Pine	90	y	$90y$

Second Purchase:

	Amount	Cost	Total Value
Redwood	200	x	$200x$
Pine	100	y	$100y$

• The first purchase cost $31.20.
The second purchase cost $78.
$$50x + 90y = 31.20$$
$$200x + 100y = 78$$

Solution
$$50x + 90y = 31.20$$
$$200x + 100y = 78$$

$$-4(50x + 90y) = -4(31.20)$$
$$200x + 100y = 78$$

$$-200x - 360y = -124.80$$
$$200x + 100y = 78$$

$$-260y = -46.80$$
$$y = 0.18$$

$$50x + 90y = 31.20$$
$$50x + 90(0.18) = 31.20$$
$$50x + 16.20 = 31.20$$
$$50x = 15$$
$$x = 0.30$$

The cost of the pine is $.18 per foot.
The cost of the redwood is $.30 per foot.

15. Strategy
Cost per unit of electricity: x
Cost per unit of gas: y
First month:

	Amount	Rate	Total Value
Electricity	400	x	$400x$
Gas	120	y	$120y$

Second month:

	Amount	Rate	Total Value
Electricity	350	x	$350x$
Gas	200	y	$200y$

• The total cost for the first month was $73.60.
The total cost for the second month was $72.
$$400x + 120y = 73.60$$
$$350x + 200y = 72$$

Solution
$$400x + 120y = 73.60$$
$$350x + 200y = 72$$

$$-5(400x + 120y) = -5(73.60)$$
$$3(350x + 200y) = 3(72)$$

$$-2000x - 600y = -368$$
$$1050x + 600y = 216$$

$$-950x = -152$$
$$x = 0.16$$

$$400x + 120y = 73.60$$
$$400(0.16) + 120y = 73.60$$
$$64 + 120y = 73.60$$
$$120y = 9.6$$
$$y = 0.08$$

The cost per unit of gas is $.08.

17. Strategy
Number of quarters in the bank: q
Number of dimes in the bank: d
Coins in the bank now:

Coin	Number	Value	Total Value
Quarter	q	25	$25q$
Dime	d	10	$10d$

Coins in the bank if the quarters were dimes and the dimes were quarters:

Coin	Number	Value	Total Value
Quarter	d	25	$25d$
Dime	q	10	$10q$

• The value of the quarters and dimes in the bank is $6.90.
The value of the quarters and dimes in the bank would be $7.80.

$25q + 10d = 690$
$10q + 25d = 780$

Solution
$25q + 10d = 690$
$10q + 25d = 780$

$5(25q + 10d) = 5(690)$
$-2(10q + 25d) = -2(780)$

$125q + 50d = 3450$
$-20q - 50d = -1560$

$105q = 1890$
$q = 18$

There are 18 quarters in the bank.

19. Strategy
Number of black-and-white TVs to be manufactured: b
Number of color TVs to be manufactured: c
Cost of materials:

Type of TV	Number	Cost	Total Cost
B & W	b	25	$25b$
Color	c	75	$75c$

Cost of labor:

Type of TV	Number	Cost	Total Cost
B & W	b	40	$40b$
Color	c	65	$65c$

• The company has budgeted $4800 for materials. The company has budgeted $4380 for labor.
$25b + 75c = 4800$
$40b + 65c = 4380$

Solution
$25b + 75c = 4800$
$40b + 65c = 4380$

$8(25b + 75c) = 8(4800)$
$-5(40b + 65c) = -5(4380)$

$200b + 600c = 38400$
$-200b - 325c = -21900$

$275c = 16500$
$c = 60$

The company plans to manufacture 60 color TVs during the week.

21. Strategy
Amount of the first powder to be used: x
Amount of the second powder to be used: y
Vitamin B_1:

	Amount	Percent	Quantity
1st powder	x	0.25	$0.25x$
2nd powder	y	0.15	$0.15y$

Vitamin B_2:

	Amount	Percent	Quantity
1st powder	x	0.15	$0.15x$
2nd powder	y	0.20	$0.20y$

• The mixture contains 117.5 mg of vitamin B_1.
The mixture contains 120 mg of vitamin B_2.
$0.25x + 0.15y = 117.5$
$0.15x + 0.20y = 120$

Solution
$0.25x + 0.15y = 117.5$
$0.15x + 0.20y = 120$

$-4(0.25x + 0.15y) = -4(117.5)$
$3(0.15x + 0.20y) = 3(120)$

$-1.0x - 0.60y = -470$
$0.45x + 0.60y = 360$

$-0.55x = -110$
$x = 200$

$0.25x + 0.15y = 117.5$
$0.25(200) + 0.15y = 117.5$
$50 + 0.15y = 117.5$
$0.15y = 67.5$
$y = 450$

The pharmacist should use 200 mg of the first powder and 450 mg of the second powder.

23. Strategy
Cost of a Model II computer: x
Cost of a Model VI computer: y
Cost of a Model IX computer: z
First shipment:

Computer	Number	Cost	Total Cost
Model II	4	x	$4x$
Model VI	6	y	$6y$
Model IX	10	z	$10z$

Second shipment:

Computer	Number	Cost	Total Cost
Model II	8	x	$8x$
Model VI	3	y	$3y$
Model IX	5	z	$5z$

Third Shipment:

Computer	Number	Cost	Total Cost
Model II	2	x	$2x$
Model VI	9	y	$9y$
Model IX	5	z	$5z$

• The bill for the first shipment was $114,000.
The bill for the second shipment was $72,000.
The bill for the third shipment was $81,000.
$4x + 6y + 10z = 114,000$
$8x + 3y + 5z = 72,000$
$2x + 9y + 5z = 81,000$

Chapter 4: Systems of Equations and Inequalities

Solution
$$4x + 6y + 10z = 114,000$$
$$8x + 3y + 5z = 72,000$$
$$2x + 9y + 5z = 81,000$$

$$4x + 6y + 10z = 114,000$$
$$-2(8x + 3y + 5z) = -2(72,000)$$

$$4x + 6y + 10z = 114,000$$
$$-16x - 6y - 10z = -144,000$$

$$-12x = -30,000$$
$$x = 2500$$

$$4x + 6y + 10z = 114,000$$
$$-2(2x + 9y + 5z) = -2(81,000)$$

$$4x + 6y + 10z = 114,000$$
$$-4x - 18y - 10z = 162,000$$

$$-12y = -48,000$$
$$y = 4000$$

$$4x + 6y + 10z = 114,000$$
$$4(2500) + 6(4000) + 10z = 114,000$$
$$10,000 + 24,000 + 10z = 114,000$$
$$34,000 + 10z = 114,000$$
$$10z = 80,000$$
$$z = 8000$$

The manufacturer charges $2500 for a Model II, $4000 for a Model VI, and $8000 for a Model IX computer.

25. Strategy
Amount earning 9% interest: x
Amount earning 7% interest: y
Amount earning 5% interest: z

Interest Rate	Amount deposited	Amount earned
9%	x	$0.09x$
7%	y	$0.07y$
5%	z	$0.05z$

• The total amount deposited is $18,000.
The total interest earned is $1340.
$$x + y + z = 18,000$$
$$x = 2z$$
$$0.09x + 0.07y + 0.05z = 1340$$

Solution
$$x + y + z = 18,000$$
$$x = 2z$$
$$0.09x + 0.07y + 0.05z = 1340$$

$$x + y + z = 18,000$$
$$2z + y + z = 18,000$$
$$3z + y = 18,000$$

$$0.09x + 0.07y + 0.05z = 1340$$
$$0.09(2z) + 0.07y + 0.05z = 1340$$
$$0.18z + 0.07y + 0.05z = 1340$$
$$0.23z + 0.07y = 1340$$

$$0.23z + 0.07y = 1340$$
$$-0.07(3z + y) = -0.07(18,000)$$

$$0.23z + 0.07y = 1340$$
$$-0.21z - 0.07y = -1260$$
$$0.02z = 80$$
$$z = 4000$$

$$x = 2z$$
$$x = 2(4000)$$
$$x = 8000$$

$$x + y + z = 18,000$$
$$8000 + y + 4000 = 18,000$$
$$12,000 + y = 18,000$$
$$y = 6000$$

The amounts deposited in each account are $8000 at 9% interest, $6000 at 7% interest, and $4000 at 5% interest.

27. Strategy
The sum of d_1 and d_3 is equal to the length of the rod, which is 15 in.
$w_1 = 5$, $w_2 = 1$, and $w_3 = 3$ so the equation that ensures the mobile will balance is $5d_1 = d_2 + 3d_3$.

$d_1 + d_3 = 15$
$5d_1 = d_2 + 3d_3$
$d_3 = 3d_2$

Solution
$d_1 + d_3 = 15$
$5d_1 = d_2 + 3d_3$
$d_3 = 3d_2$

$5d_1 = d_2 + 3d_3$
$5d_1 = d_2 + 3(3d_2)$
$5d_1 = d_2 + 9d_2$
$5d_1 = 10d_2$
$5d_1 - 10d_2 = 0$

$d_1 + d_3 = 15$
$d_1 + 3d_2 = 15$

$5d_1 - 10d_2 = 0$
$-5(d_1 + 3d_2) = -5(15)$

$5d_1 - 10d_2 = 0$
$-5d_1 - 15d_2 = -75$
$-25d_2 = -75$
$d_2 = 3$

$d_3 = 3d_2$
$d_3 = 3(3)$
$d_3 = 9$

$d_1 + d_3 = 15$
$d_1 + 9 = 15$
$d_1 = 6$

The distances are
$d_1 = 6$ in., $d_2 = 3$ in., and $d_3 = 9$ in.

29. Strategy
Number of nickels in the bank: n
Number of dimes in the bank: d
Number of quarters in the bank: q
Coins in the bank:

Coin	Number	Value	Total Value
Nickel	n	5	$5n$
Dime	d	10	$10d$
Quarter	q	25	$25q$

• The value of the nickels, dimes, and quarters in the bank is 200 cents.
$n + d + q = 19$
$5n + 10d + 25q = 200$
$n = 2d$

Solution
$n + d + q = 19$
$5n + 10d + 25q = 200$
$n = 2d$

$n + d + q = 19$
$2d + d + q = 19$
$3d + q = 19$

$5n + 10d + 25q = 200$
$5(2d) + 10d + 25q = 200$
$10d + 10d + 25q = 200$
$20d + 25q = 200$

$-25(3d + q) = -25(19)$
$20d + 25q = 200$

$-75d - 25q = -475$
$20d + 25q = 200$

$-55d = -275$
$d = 5$

$n = 2d$
$n = 10$

$n + d + q = 19$
$10 + 5 + q = 19$
$15 + q = 19$
$q = 4$

There are 10 nickels, 5 dimes, and 4 quarters in the bank.

31. Strategy
Amount invested at 9%: x
Amount invested at 12%: y
Amount invested at 8%: z

Interest Rate	Amount deposited	Amount earned
9%	x	$0.09x$
12%	y	$0.12y$
8%	z	$0.08z$

• The total amount invested is $33,000.
$$x + y + z = 33{,}000$$
$$0.09x + 0.12y + 0.08z = 3290$$
$$y = x + z - 5000$$

Solution
$$x + y + z = 33{,}000$$
$$x + (x + z - 5000) + z = 33{,}000$$
$$2x + 2z - 5000 = 33{,}000$$
$$2x + 2z = 38{,}000$$

$$0.09x + 0.12y + 0.08z = 3290$$
$$0.09x + 0.12(x + z - 5000) + 0.08z = 3290$$
$$0.09x + 0.12x + 0.12z - 600 + 0.08z = 3290$$
$$0.21x + 0.2z - 600 = 3290$$
$$0.21x + 0.2z = 3890$$

$$2x + 2z = 38{,}000$$
$$-10(0.21x + 0.2z) = -10(3890)$$

$$2x + 2z = 38{,}000$$
$$-2.1x - 2z = -38{,}900$$
$$-0.1x = -900$$
$$x = 9000$$

$$2x + 2z = 38{,}000$$
$$2(9000) + 2z = 38{,}000$$
$$18{,}000 + 2z = 38{,}000$$
$$2z = 20{,}000$$
$$z = 10{,}000$$

$$y = x + z - 5000$$
$$y = 9000 + 10{,}000 - 5000$$
$$y = 14{,}000$$

The amounts invested are $9000 at 9% interest, $14,000 at 12% interest, and $10,000 at 8% interest.

33. Strategy
Amount earned by Michael Jordan: m
Amount earned by Shaquille O'Neal: s
Amount earned by Arnold Palmer: a
$$m + s + a = 77$$
$$m = s + a - 1$$
$$s = a + 7$$

Solution
$$m + s + a = 77$$
$$m = s + a - 1$$
$$s = a + 7$$

$$m = s + a - 1$$
$$m = a + 7 + a - 1$$
$$m = 2a + 6$$

$$m + s + a = 77$$
$$(2a + 6) + (a + 7) + a = 77$$
$$4a + 13 = 77$$
$$4a = 64$$
$$a = 16$$

$$s = a + 7$$
$$s = 16 + 7$$
$$s = 23$$

$$m + s + a = 77$$
$$m + 23 + 16 = 77$$
$$m + 39 = 77$$
$$m = 38$$

Michael Jordan earned $38 million, Shaquille O'Neal earned $23 million, and Arnold Palmer earned $16 million.

Applying Concepts 4.4

35. Strategy
Write and solve a system of equations using x and y to represent the measures of the two angles.

Solution
$$x + y = 90$$
$$y = 8x + 9$$

$$x + (8x + 9) = 90$$
$$9x + 9 = 90$$
$$9x = 81$$
$$x = 9$$

$$x + y = 90$$
$$9 + y = 90$$
$$y = 81$$

The measures of the two angles are 9 and 81 degrees.

37. Strategy
• Number of nickels in the bank: n
Number of dimes in the bank: d
Number of quarters in the bank: q
Coins in the bank now:

Coin	Number	Value	Total Value
Nickel	n	5	$5n$
Dime	d	10	$10d$
Quarter	q	25	$25q$

Coins in the bank if nickels were dimes and dimes were nickels:

Coin	Number	Value	Total Value
Nickel	d	5	$5d$
Dime	n	10	$10n$
Quarter	q	25	$25q$

Coins in the bank if quarters were dimes and dimes were quarters:

Coin	Number	Value	Total Value
Nickel	n	5	$5n$
Dime	q	10	$10q$
Quarter	d	25	$25d$

• The value of the nickels, dimes, and quarters in the bank is 350 cents.
• The value of the nickels, dimes, and quarters would be 425 cents if the nickels were dimes and the dimes were nickels.
• The value of the nickels, dimes, and quarters would be 425 cents if the dimes were quarters and the quarters were dimes.

Solution
$5n + 10d + 25q = 350$
$10n + 5d + 25q = 425$
$5n + 25d + 10q = 425$

$5n + 10d + 25q = 350$
$-(5n + 25d + 10q) = -(425)$

$5n + 10d + 25q = 350$
$-5n - 25d - 10q = -425$

$-15d + 15q = -75$

$-2(5n + 10d + 25q) = -2(350)$
$10n + 5d + 25q = 425$

$-10n - 20d - 50q = -700$
$10n + 5d + 25q = 425$

$-15d - 25q = -275$

$-15d - 25q = -275$
$-(-15d + 15q) = -(-75)$

$-15d - 25q = -275$
$15d - 15q = 75$
$-40q = -200$
$q = 5$

$-15d + 15q = -75$
$-15d + 15(5) = -75$
$-15d + 75 = -75$
$-15d = -150$
$d = 10$

$5n + 10d + 25q = 350$
$5n + 10(10) + 25(5) = 350$
$5n + 100 + 125 = 350$
$5n + 225 = 350$
$5n = 125$
$n = 25$

There are 25 nickels, 10 dimes, and 5 quarters in the bank.

Section 4.5

Concept Review 4.5

1. Sometimes true
   The solution set of a system of inequalities can be a portion of the plane or the empty set.

3. Always true

108 Chapter 4: Systems of Equations and Inequalities

Objective 1 Exercises

3. Solve each inequality for y.
$$x - y \geq 3 \qquad x + y \leq 5$$
$$-y \geq -x + 3 \qquad y \leq -x + 5$$
$$y \leq x - 3$$

5. $y > 3x - 3 \qquad 2x + y \geq 2$
$$y \geq -2x + 2$$

7. Solve each inequality for y.
$$2x + y \geq -2 \qquad 6x + 3y \leq 6$$
$$y \geq -2x - 2 \qquad 3y \leq -6x + 6$$
$$y \leq -2x + 2$$

9. Solve each inequality for y.
$$3x - 2y < 6 \qquad y \leq 3$$
$$-2y < -3x + 6$$
$$y > \frac{3}{2}x - 3$$

11. Solve the inequality for y.
$$y > 2x - 6 \qquad x + y < 0$$
$$y < -x$$

13. Solve each inequality for the variable.
$$x + 1 \geq 0 \qquad y - 3 \leq 0$$
$$x \geq -1 \qquad y \leq 3$$

15. Solve each inequality for y.
$$2x + y \geq 4 \qquad 3x - 2y < 6$$
$$y \geq -2x + 4 \qquad -2y < -3x + 6$$
$$y > \frac{3}{2}x - 3$$

17. Solve each inequality for y.
$$x - 2y \leq 6 \qquad 2x + 3y \leq 6$$
$$-2y \leq -x + 6 \qquad 3y \leq -2x + 6$$
$$y \geq \frac{1}{2}x - 3 \qquad y \leq -\frac{2}{3}x + 2$$

19. Solve each inequality for y.
$$x - 2y \leq 4 \qquad 3x + 2y \leq 8$$
$$-2y \leq -x + 4 \qquad 2y \leq -3x + 8$$
$$y \geq \frac{1}{2}x - 2 \qquad y \leq -\frac{3}{2}x + 4$$
$$y \geq \frac{1}{2}x - 2$$
$$y \leq -\frac{3}{2}x + 4$$
$$x > -1$$

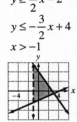

Applying Concepts 4.5

21. Solve each inequality for y.
$$2x + 3y \leq 15 \qquad 3x - y \leq 6$$
$$3y \leq -2x + 15 \qquad -y \leq -3x + 6$$
$$y \leq -\frac{2}{3}x + 5 \qquad y \geq 3x - 6$$
$$y \leq -\frac{2}{3}x + 5$$
$$y \geq 3x - 6$$
$$y \geq 0$$

23. Solve each inequality for y.
$x - y \leq 5$ $2x - y \geq 6$ $y \geq 0$
$-y \leq -x + 5$ $-y \geq -2x + 6$
$y \geq x - 5$ $y \leq 2x - 6$

25. Solve each inequality for y.
$2x - y \leq 4$ $3x + y < 1$ $y \leq 0$
$-y \leq -2x + 4$ $y < 1 - 3x$
$y \geq 2x - 4$

Focus on Problem Solving

1.

From the diagram, 5 teams will require 10 games in which every team plays every other team once.

2. The table of results is reproduced here for up to five teams.

Number of Teams	Number of Games	Possible Pattern
1	0	1
2	1	0 + 1
3	3	1 + 2
4	6	1 + 2 + 3
5	10	1 + 2 + 3 + 4

Note the pattern. For n teams, the number of games is the sum of the first $n - 1$ whole numbers.

3. Assuming that the pattern holds, the number of games scheduled for 15 teams is
Number of games = 1 + 2 + 3 + 4 + 5 + 6 + 7 + 8 + 9 + 10 + 11 + 12 + 13 + 14 = 105
If 15 teams are to play each other once, 105 games must be scheduled.

4. By continuing this reasoning, Team C had already played Team A and Team B, so there are 12 teams left for Team C to play. Consequently, you must schedule at least 14 + 13 + 12 games. Continuing with this argument, when we get to the two last teams, there will be 1 game left to play. Therefore, the total number of games to be played is 14 + 13 + 12 + 11 + 10 + 9 + 8 + 7 + 6 + 5 + 4 + 3 + 2 + 1 = 105. This is the same result obtained in Exercise 3.

Projects and Group Activities

Solving a First-Degree Equation with a Graphing Calculator

1. $3x - 1 = 5x + 1$
$y = 3x - 1$
$y = 5x + 1$

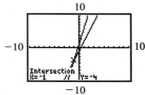

The solution is -1.

2. $3x + 2 = 4$
$y = 3x + 2$
$y = 4$

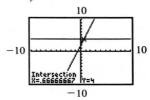

The solution is $\dfrac{2}{3}$.

3. $3 + 2(2x - 4) = 5(x - 3)$
$y = 3 + 2(2x - 4)$
$y = 5(x - 3)$

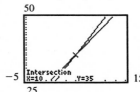

The solution is 10.

4. $2x - 4 = 5x - 3(x + 2) + 2$
$y = 2x - 4$
$y = 5x - 3(x + 2) + 2$

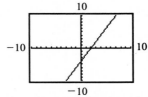

The solution is all real numbers.

5. Simplify the right side of the equation.
$5x - 3(x + 2) + 2 = 5x - 3x - 6 + 2$
$= 2x - 4$
Problem 4 relates to an identity since any replacement for x will result in a true equation.

6. $2x - 10 = 1.8x + 3.6$
$y = 2x - 10$
$y = 1.8x + 3.6$
Use the viewing window
[67.5, 68.5] by [123, 127].

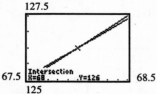

The solution is 68.

Chapter Review Exercises

1. (1) $2x - 6y = 15$
(2) $x = 3y + 8$
Substitute $3y + 8$ for x in equation (1).
$2(3y + 8) - 6y = 15$
$6y + 16 - 6y = 15$
$16 = 15$
This is not a true equation. The lines are parallel and the system is inconsistent.

2. (1) $3x + 12y = 18$
(2) $x + 4y = 6$
Solve equation (2) for x.
$x + 4y = 6$
$x = -4y + 6$
Substitute into equation (1).
$3x + 12y = 18$
$3(-4y + 6) + 12y = 18$
$-12y + 18 + 12y = 18$
$18 = 18$
This is a true equation. The equations are dependent. The solutions are the ordered pairs $\left(x, -\frac{1}{4}x + \frac{3}{2}\right)$.

3. (1) $3x + 2y = 2$
(2) $x + y = 3$
Eliminate y. Multiply equation (2) by -2 and add to equation (1).
$3x + 2y = 2$
$-2(x + y) = 3(-2)$

$3x + 2y = 2$
$-2x - 2y = -6$
Add the equations.
$x = -4$
Replace x in equation (2).
$x + y = 3$
$-4 + y = 3$
$y = 7$
The solution is $(-4, 7)$.

4. (1) $5x - 15y = 30$
(2) $x - 3y = 6$
Eliminate x. Multiply equation (2) by -5 and add to equation (1).
$5x - 15y = 30$
$-5(x - 3y) = 6(-5)$

$5x - 15y = 30$
$-5x + 15y = -30$
Add the equations.
$0 = 0$
This is a true equation. The equations are dependent. The solutions are the ordered pairs $\left(x, \frac{1}{3}x - 2\right)$.

5. (1) $3x + y = 13$
(2) $2y + 3z = 5$
(3) $x + 2z = 11$
Eliminate y. Multiply equation (1) by -2 and add to equation (2).
$-2(3x + y) = 13(-2)$
$2y + 3z = 5$

$-6x - 2y = -26$
$2y + 3z = 5$

(4) $-6x + 3z = -21$
Multiply equation (3) by 6 and add to equation (4).
$6(x + 2z) = 6(11)$
$-6x + 3z = -21$

$6x + 12z = 66$
$-6x + 3z = -21$

$15z = 45$
$z = 3$
Replace z by 3 in equation (3).
$x + 2z = 11$
$x + 2(3) = 11$
$x + 6 = 11$
$x = 5$
Replace x by 5 in equation (1).
$3x + y = 13$
$3(5) + y = 13$
$15 + y = 13$
$y = -2$
The solution is $(5, -2, 3)$.

6. (1) $3x - 4y - 2z = 17$
   (2) $4x - 3y + 5z = 5$
   (3) $5x - 5y + 3z = 14$
   Eliminate z. Multiply equation (1) by 5 and equation (2) by 2 and add the new equations.
   $5(3x - 4y - 2z) = 17(5)$
   $2(4x - 3y + 5z) = 5(2)$

   $15x - 20y - 10z = 85$
   $8x - 6y + 10z = 10$

   (4) $23x - 26y = 95$
   Multiply equation (1) by 3 and equation (3) by 2 and add the new equations.
   $3(3x - 4y - 2z) = (17)3$
   $2(5x - 5y + 3z) = (14)2$

   $9x - 12y - 6z = 51$
   $10x - 10y + 6z = 28$

   (5) $19x - 22y = 79$
   Multiply equation (4) by –11 and equation (5) by 13 and add the new equations.
   $-11(23x - 26y) = 95(-11)$
   $13(19x - 22y) = 79(13)$

   $-253x + 286y = -1045$
   $247x - 286y = 1027$

   $-6x = -18$
   $x = 3$
   Substitute x for 3 in equation (4).
   $23x - 26y = 95$
   $23(3) - 26y = 95$
   $69 - 26y = 95$
   $-26y = 26$
   $y = -1$
   Substitute x by 3 and y by –1 in equation (1).
   $3x - 4y - 2z = 17$
   $3(3) - 4(-1) - 2z = 17$
   $9 + 4 - 2z = 17$
   $-2z = 4$
   $z = -2$
   The solution is $(3, -1, -2)$.

7. $\begin{vmatrix} 6 & 1 \\ 2 & 5 \end{vmatrix} = 6(5) - 2(1) = 30 - 2 = 28$

8. $\begin{vmatrix} 1 & 5 & -2 \\ -2 & 1 & 4 \\ 4 & 3 & -8 \end{vmatrix} = 1\begin{vmatrix} 1 & 4 \\ 3 & -8 \end{vmatrix} - 5\begin{vmatrix} -2 & 4 \\ 4 & -8 \end{vmatrix} - 2\begin{vmatrix} -2 & 1 \\ 4 & 3 \end{vmatrix}$
   $= 1(-8 - 12) - 5(16 - 16) - 2(-6 - 4)$
   $= 1(-20) - 5(0) - 2(-10)$
   $= -20 - 0 + 20$
   $= 0$

9. $2x - y = 7$
   $3x + 2y = 7$
   $D = \begin{vmatrix} 2 & -1 \\ 3 & 2 \end{vmatrix} = 7$
   $D_x = \begin{vmatrix} 7 & -1 \\ 7 & 2 \end{vmatrix} = 21$
   $D_y = \begin{vmatrix} 2 & 7 \\ 3 & 7 \end{vmatrix} = -7$
   $x = \dfrac{D_x}{D} = \dfrac{21}{7} = 3$
   $y = \dfrac{D_y}{D} = \dfrac{-7}{7} = -1$
   The solution is $(3, -1)$.

10. $3x - 4y = 10$
    $2x + 5y = 15$
    $D = \begin{vmatrix} 3 & -4 \\ 2 & 5 \end{vmatrix} = 23$
    $D_x = \begin{vmatrix} 10 & -4 \\ 15 & 5 \end{vmatrix} = 110$
    $D_y = \begin{vmatrix} 3 & 10 \\ 2 & 15 \end{vmatrix} = 25$
    $x = \dfrac{D_x}{D} = \dfrac{110}{23}$
    $y = \dfrac{D_y}{D} = \dfrac{25}{23}$
    The solution is $\left(\dfrac{110}{23}, \dfrac{25}{23}\right)$.

11. $x + y + z = 0$
    $x + 2y + 3z = 5$
    $2x + y + 2z = 3$
    $D = \begin{vmatrix} 1 & 1 & 1 \\ 1 & 2 & 3 \\ 2 & 1 & 2 \end{vmatrix} = 2$
    $D_x = \begin{vmatrix} 0 & 1 & 1 \\ 5 & 2 & 3 \\ 3 & 1 & 2 \end{vmatrix} = -2$
    $D_y = \begin{vmatrix} 1 & 0 & 1 \\ 1 & 5 & 3 \\ 2 & 3 & 2 \end{vmatrix} = -6$
    $D_z = \begin{vmatrix} 1 & 1 & 0 \\ 1 & 2 & 5 \\ 2 & 1 & 3 \end{vmatrix} = 8$
    $x = \dfrac{D_x}{D} = \dfrac{-2}{2} = -1$
    $y = \dfrac{D_y}{D} = \dfrac{-6}{2} = -3$
    $z = \dfrac{D_z}{D} = \dfrac{8}{2} = 4$
    The solution is $(-1, -3, 4)$.

12. $x + 3y + z = 6$
$2x + y - z = 12$
$x + 2y - z = 13$

$D = \begin{vmatrix} 1 & 3 & 1 \\ 2 & 1 & -1 \\ 1 & 2 & -1 \end{vmatrix} = 7$

$D_x = \begin{vmatrix} 6 & 3 & 1 \\ 12 & 1 & -1 \\ 13 & 2 & -1 \end{vmatrix} = 14$

$D_y = \begin{vmatrix} 1 & 6 & 1 \\ 2 & 12 & -1 \\ 1 & 13 & -1 \end{vmatrix} = 21$

$D_z = \begin{vmatrix} 1 & 3 & 6 \\ 2 & 1 & 12 \\ 1 & 2 & 13 \end{vmatrix} = -35$

$x = \dfrac{D_x}{D} = \dfrac{14}{7} = 2$

$y = \dfrac{D_y}{D} = \dfrac{21}{7} = 3$

$z = \dfrac{D_z}{D} = \dfrac{-35}{7} = -5$

The solution is $(2, 3, -5)$.

13. (1) $x - 2y + z = 7$
(2) $3x - z = -1$
(3) $3y + z = 1$

Eliminate z. Add equations (2) and (3).
$3x - z = -1$
$3y + z = 1$

(4) $3x + 3y = 0$

Multiply equation (1) by -1 and add to equation (3).
$-(x - 2y + z) = -7$
$3y + z = 1$

$-x + 2y - z = -7$
$3y + z = 1$

(5) $-x + 5y = -6$

Multiply equation (4) by $\dfrac{1}{3}$ and add to equation (5).

$\dfrac{1}{3}(3x + 3y) = \dfrac{1}{3}(0)$
$-x + 5y = -6$

$x + y = 0$
$-x + 5y = -6$

$6y = -6$
$y = -1$

Replace y in equation (5).
$-x + 5y = -6$
$-x + 5(-1) = -6$
$-x - 5 = -6$
$-x = -1$
$x = 1$

Replace x in equation (2).
$3x - z = -1$
$3(1) - z = -1$
$3 - z = -1$
$-z = -4$
$z = 4$

The solution is $(1, -1, 4)$.

14. $3x - 2y = 2$
$-2x + 3y = 1$

$D = \begin{vmatrix} 3 & -2 \\ -2 & 3 \end{vmatrix} = 5$

$D_x = \begin{vmatrix} 2 & -2 \\ 1 & 3 \end{vmatrix} = 8$

$D_y = \begin{vmatrix} 3 & 2 \\ -2 & 1 \end{vmatrix} = 7$

$x = \dfrac{D_x}{D} = \dfrac{8}{5}$

$y = \dfrac{D_y}{D} = \dfrac{7}{5}$

The solution is $\left(\dfrac{8}{5}, \dfrac{7}{5}\right)$.

15. $2x - 2y - 6z = 1$
$4x + 2y + 3z = 1$
$2x - 3y - 3z = 3$

$\begin{vmatrix} 2 & -2 & -6 & 1 \\ 4 & 2 & 3 & 1 \\ 2 & -3 & -3 & 3 \end{vmatrix}$

$\begin{vmatrix} 1 & -1 & -3 & \frac{1}{2} \\ 4 & 2 & 3 & 1 \\ 2 & -3 & -3 & 3 \end{vmatrix}$ Multiply row 1 by $\frac{1}{2}$.

$\begin{vmatrix} 1 & -1 & -3 & \frac{1}{2} \\ 0 & 6 & 15 & -1 \\ 0 & -1 & 3 & 2 \end{vmatrix}$ Multiply row 1 by -4 and add to row 2. Multiply row 1 by -2 and add to row 3.

$\begin{vmatrix} 1 & -1 & -3 & \frac{1}{2} \\ 0 & 1 & \frac{5}{2} & -\frac{1}{6} \\ 0 & -1 & 3 & 2 \end{vmatrix}$ Multiply row 2 by $\frac{1}{6}$.

$\begin{vmatrix} 1 & -1 & -3 & \frac{1}{2} \\ 0 & 1 & \frac{5}{2} & -\frac{1}{6} \\ 0 & 0 & \frac{11}{2} & \frac{11}{6} \end{vmatrix}$ Add row 2 and row 3. Replace row 3 by the sum.

$\begin{vmatrix} 1 & -1 & -3 & \frac{1}{2} \\ 0 & 1 & \frac{5}{2} & -\frac{1}{6} \\ 0 & 0 & 1 & \frac{1}{3} \end{vmatrix}$ Multiply row 3 by $\frac{2}{11}$.

$x - y - 3z = \frac{1}{2}$
$y + \left(\frac{5}{2}\right)z = -\frac{1}{6}$
$z = \frac{1}{3}$

$y + \left(\frac{5}{2}\right)z = -\frac{1}{6}$
$y + \left(\frac{5}{2}\right)\left(\frac{1}{3}\right) = -\frac{1}{6}$
$y + \frac{5}{6} = -\frac{1}{6}$
$y = -1$

$x - y - 3z = \frac{1}{2}$
$x - (-1) - 3\left(\frac{1}{3}\right) = \frac{1}{2}$
$x + 1 - 1 = \frac{1}{2}$
$x = \frac{1}{2}$

The solution is $\left(\frac{1}{2}, -1, \frac{1}{3}\right)$.

16. $\begin{vmatrix} 3 & -2 & 5 \\ 4 & 6 & 3 \\ 1 & 2 & 1 \end{vmatrix} = 3\begin{vmatrix} 6 & 3 \\ 2 & 1 \end{vmatrix} + 2\begin{vmatrix} 4 & 3 \\ 1 & 1 \end{vmatrix} + 5\begin{vmatrix} 4 & 6 \\ 1 & 2 \end{vmatrix}$

$= 3(6-6) + 2(4-3) + 5(8-6)$
$= 3(0) + 2(1) + 5(2)$
$= 0 + 2 + 10 = 12$

17. $4x - 3y = 17$
$3x - 2y = 12$

$D = \begin{vmatrix} 4 & -3 \\ 3 & -2 \end{vmatrix} = 1$

$D_x = \begin{vmatrix} 17 & -3 \\ 12 & -2 \end{vmatrix} = 2$

$D_y = \begin{vmatrix} 4 & 17 \\ 3 & 12 \end{vmatrix} = -3$

$x = \frac{D_x}{D} = \frac{2}{1} = 2$

$y = \frac{D_y}{D} = \frac{-3}{1} = -3$

The solution is $(2, -3)$.

18. $3x + 2y - z = -1$
$x + 2y + 3z = -1$
$3x + 4y + 6z = 0$

$\begin{vmatrix} 3 & 2 & -1 & -1 \\ 1 & 2 & 3 & -1 \\ 3 & 4 & 6 & 0 \end{vmatrix}$

$\begin{vmatrix} 1 & 2 & 3 & -1 \\ 3 & 2 & -1 & -1 \\ 3 & 4 & 6 & 0 \end{vmatrix}$ Interchange rows 1 and 2.

$\begin{vmatrix} 1 & 2 & 3 & -1 \\ 0 & -4 & -10 & 2 \\ 0 & -2 & -3 & 3 \end{vmatrix}$ Multiply row 1 by -3 and add to row 2. Multiply row 1 by -3 and add to row 3.

$\begin{vmatrix} 1 & 2 & 3 & -1 \\ 0 & 1 & \frac{5}{2} & -\frac{1}{2} \\ 0 & -2 & -3 & 3 \end{vmatrix}$ Multiply row 2 by $-\frac{1}{4}$.

$\begin{vmatrix} 1 & 2 & 3 & -1 \\ 0 & 1 & \frac{5}{2} & -\frac{1}{2} \\ 0 & 0 & 2 & 2 \end{vmatrix}$ Multiply row 2 by 2 and add to row 3.

$\begin{vmatrix} 1 & 2 & 3 & -1 \\ 0 & 1 & \frac{5}{2} & -\frac{1}{2} \\ 0 & 0 & 1 & 1 \end{vmatrix}$ Multiply row 3 by $\frac{1}{2}$.

$x + 2y + 3z = -1$
$y + \frac{5}{2}z = -\frac{1}{2}$
$z = 1$

$y + \left(\frac{5}{2}\right)z = -\frac{1}{2}$
$y + \frac{5}{2}(1) = -\frac{1}{2}$
$y + \frac{5}{2} = -\frac{1}{2}$
$y = -3$

$x + 2y + 3z = -1$
$x + 2(-3) + 3(1) = -1$
$x - 6 + 3 = -1$
$x - 3 = -1$
$x = 2$

The solution is $(2, -3, 1)$.

19.

The solution is $(0, 3)$.

20.

The two equations represent the same line. The solutions are the ordered pairs $(x, 2x - 4)$.

21. Solve each inequality for y.

$x + 3y \le 6$ $2x - y \ge 4$
$3y \le -x + 6$ $-y \ge -2x + 4$
$y \le -\frac{1}{3}x + 2$ $y \le 2x - 4$

22. Solve each inequality for y.

$2x + 4y \ge 8$ $x + y \le 3$
$4y \ge -2x + 8$ $y \le -x + 3$
$y \ge -\frac{1}{2}x + 2$

23. Strategy
Rate of the cabin cruiser in calm water: x
Rate of the current: y

	Rate	Time	Distance
With current	$x+y$	3	$3(x+y)$
Against current	$x-y$	5	$5(x-y)$

• The distance traveled with the current is 60 mi. The distance traveled against the current is 60 mi.
$3(x+y) = 60$
$5(x-y) = 60$

Solution
$3(x+y) = 60$
$5(x-y) = 60$

$\frac{1}{3} \cdot 3(x+y) = \frac{1}{3}(60)$
$\frac{1}{5} \cdot 5(x-y) = \frac{1}{5}(60)$

$x + y = 20$
$x - y = 12$

$2x = 32$
$x = 16$

$x + y = 20$
$16 + y = 20$
$y = 4$

The rate of the cabin cruiser in calm water is 16 mph. The rate of the current is 4 mph.

24. Strategy
Rate of the plane in calm air: p
Rate of the wind: w

	Rate	Time	Distance
With wind	$p+w$	3	$3(p+w)$
Against wind	$p-w$	4	$4(p-w)$

• The distance traveled with the wind is 600 mi. The distance traveled against the wind is 600 mi.
$3(p+w) = 600$
$4(p-w) = 600$

Solution
$3(p+w) = 600$
$4(p-w) = 600$

$\frac{1}{3} \cdot 3(p+w) = \frac{1}{3}(600)$
$\frac{1}{4} \cdot 4(p-w) = \frac{1}{4}(600)$

$p + w = 200$
$p - w = 150$

$2p = 350$
$p = 175$

$p + w = 200$
$175 + w = 200$
$w = 25$

The rate of the plane in calm air is 175 mph. The rate of the wind is 25 mph.

116 *Chapter 4: Systems of Equations and Inequalities*

25. Strategy
Number of children's tickets sold Friday: x
Number of adult's tickets sold Friday: y
Friday:

	Number	Value	Total Value
Children	x	5	$5x$
Adults	y	8	$8y$

Saturday:

	Number	Value	Total Value
Children	$3x$	5	$5(3x)$
Adults	$\frac{1}{2}y$	8	$8\left(\frac{1}{2}y\right)$

• The total receipts for Friday were $2500.
The total receipts for Saturday were $2500.

$$5x + 8y = 2500$$
$$5(3x) + 8\left(\frac{1}{2}y\right) = 2500$$

Solution
$5x + 8y = 2500$
$15x + 4y = 2500$

$5x + 8y = 2500$
$-2(15x + 4y) = -2(2500)$

$5x + 8y = 2500$
$-30x - 8y = -5000$

$-25x = -2500$
$x = 100$

The number of children attending on Friday was 100.

26. Strategy
Amount of meat to cook: x
Amount of potatoes to cook: y
Amount of green beans to cook: z

Calories	Amount	Calories	Total Calories
Meat	x	50	$50x$
Potatoes	y	9	$9y$
Beans	z	12	$12z$

Protein	Amount	Protein	Total Protein
Meat	x	20	$20x$
Potatoes	y	1	$1y$
Beans	z	75	$75z$

Sodium	Amount	Sodium	Total Sodium
Meat	x	16	$16x$
Potatoes	y	3	$3y$
Beans	z	17	$17z$

The chef wants the meal to contain 243 cal, 365 g of protein and 131 mg of sodium.

$50x + 9y + 12z = 243$
$20x + 1y + 75z = 365$
$16x + 3y + 17z = 131$

Solution
$50x + 9y + 12z = 243$
$20x + 1y + 75z = 365$
$16x + 3y + 17z = 131$

$50x + 9y + 12z = 243$
$-9(20x + 1y + 75z) = -9(365)$

$50x + 9y + 12z = 243$
$-180x - 9y - 675z = -3285$

$-130x - 663z = -3042$

$50x + 9y + 12z = 243$
$-3(16x + 3y + 17z) = -3(131)$

$50x + 9y + 12z = 243$
$-48x - 9y - 51z = -393$

$2x - 39z = -150$

$-130x - 663z = -3042$
$65(2x - 39z) = 65(-150)$

$-130x - 663z = -3042$
$130x - 2535z = -9750$
$-3198z = -12{,}792$
$\dfrac{-3198z}{-3198} = \dfrac{-12{,}792}{-3198}$
$z = 4$

$2x - 39z = -150$
$2x - 39(4) = -150$
$2x - 156 = -150$
$2x = 6$
$x = 3$

$20x + 1y + 75z = 365$
$20(3) + y + 75(4) = 365$
$60 + y + 300 = 365$
$y = 5$

3 oz of meat, 5 oz of potatoes, and 4 oz of green beans should be prepared.

Chapter Test

1. (1) $3x + 2y = 4$
   (2) $\quad\quad x = 2y - 1$
   Substitute $2y - 1$ for x in equation (1).
   $3(2y - 1) + 2y = 4$
   $6y - 3 + 2y = 4$
   $8y = 7$
   $y = \dfrac{7}{8}$
   Substitute into equation (2).
   $x = 2y - 1$
   $x = 2\left(\dfrac{7}{8}\right) - 1$
   $x = \dfrac{7}{4} - 1 = \dfrac{3}{4}$
   The solution is $\left(\dfrac{3}{4}, \dfrac{7}{8}\right)$.

2. (1) $5x + 2y = -23$
   (2) $2x + y = -10$
   Solve equation (2) for y.
   $2x + y = -10$
   $y = -2x - 10$
   Substitute $-2x - 10$ for y in equation (1).
   $5x + 2y = -23$
   $5x + 2(-2x - 10) = -23$
   $5x - 4x - 20 = -23$
   $x - 20 = -23$
   $x = -3$
   Substitute into equation (2).
   $2x + y = -10$
   $2(-3) + y = -10$
   $-6 + y = -10$
   $y = -4$
   The solution is $(-3, -4)$.

3. (1) $y = 3x - 7$
   (2) $y = -2x + 3$
   Substitute equation (2) into equation (1).
   $-2x + 3 = 3x - 7$
   $-5x + 3 = -7$
   $-5x = -10$
   $x = 2$
   Substitute into equation (1).
   $y = 3x - 7$
   $y = 3(2) - 7 = 6 - 7 = -1$
   The solution is $(2, -1)$.

4. $3x + 4y = -2$
   $2x + 5y = 1$
   $\begin{vmatrix} 3 & 4 & -2 \\ 2 & 5 & 1 \end{vmatrix}$
   $\begin{vmatrix} 1 & -1 & -3 \\ 2 & 5 & 1 \end{vmatrix}$ Multiply row 2 by -1 and add to row 1.
   $\begin{vmatrix} 1 & -1 & -3 \\ 0 & 7 & 7 \end{vmatrix}$ Multiply row 1 by -1 and add to row 2.
   $\begin{vmatrix} 1 & -1 & -3 \\ 0 & 1 & 1 \end{vmatrix}$ Multiply row 2 by $\dfrac{1}{7}$.
   $x - y = -3$
   $y = 1$
   $x - 1 = -3$
   $x = -2$
   The solution is $(-2, 1)$.

5. (1) $4x - 6y = 5$
   (2) $6x - 9y = 4$
   Multiply equation (1) by -3. Multiply equation (2) by 2. Add the new equations.
   $-3(4x - 6y) = -3(5)$
   $2(6x - 9y) = 2(4)$

   $-12x + 18y = -15$
   $12x - 18y = 8$

   $0 = -7$
   This is not a true equation. The system of equations is inconsistent and therefore has no solution.

6. (1) $3x - y = 2x + y - 1$
   (2) $5x + 2y = y + 6$
   Write the equation in the form $Ax + By = C$.
   (3) $x - 2y = -1$
   (4) $5x + y = 6$
   Multiply equation (4) by 2 and add to equation (3).
   $x - 2y = -1$
   $2(5x + y) = 2(6)$

   $x - 2y = -1$
   $10x + 2y = 12$

   $11x = 11$
   $x = 1$
   Substitute into equation (4).
   $5x + y = 6$
   $5(1) + y = 6$
   $y = 1$
   The solution is $(1, 1)$.

7. (1) $2x + 4y - z = 3$
   (2) $x + 2y + z = 5$
   (3) $4x + 8y - 2z = 7$
   Eliminate z. Add equations (1) and (2).
   $2x + 4y - z = 3$
   $x + 2y + z = 5$

   (4) $3x + 6y = 8$
   Multiply equation (2) by 2 and add to equation (3).
   $2(x + 2y + z) = 2(5)$
   $4x + 8y - 2z = 7$

   $2x + 4y + 2z = 10$
   $4x + 8y - 2z = 7$

   (5) $6x + 12y = 17$
   Multiply equation (4) by -2 and add to equation (5).
   $-2(3x + 6y) = -2(8)$
   $6x + 12y = 17$

   $-6x - 12y = -16$
   $6x + 12y = 17$
   $0 = 1$
   This is not a true equation. The system of equations is inconsistent and therefore has no solution.

8. $x - y - z = 5$
   $2x + z = 2$
   $3y - 2z = 1$

   $\begin{vmatrix} 1 & -1 & -1 & 5 \\ 2 & 0 & 1 & 2 \\ 0 & 3 & -2 & 1 \end{vmatrix}$

   $\begin{vmatrix} 1 & -1 & -1 & 5 \\ 0 & 2 & 3 & -8 \\ 0 & 3 & -2 & 1 \end{vmatrix}$ Multiply row 1 by -2 and add to row 2.

   $\begin{vmatrix} 1 & -1 & -1 & 5 \\ 0 & 3 & -2 & 1 \\ 0 & 2 & 3 & -8 \end{vmatrix}$ Interchange rows 2 and 3.

   $\begin{vmatrix} 1 & -1 & -1 & 5 \\ 0 & 1 & -5 & 9 \\ 0 & 2 & 3 & -8 \end{vmatrix}$ Multiply row 3 by -1 and add to row 2.

   $\begin{vmatrix} 1 & -1 & -1 & 5 \\ 0 & 1 & -5 & 9 \\ 0 & 0 & 13 & -26 \end{vmatrix}$ Multiply row 2 by -2 and add to row 3.

   $\begin{vmatrix} 1 & -1 & -1 & 5 \\ 0 & 1 & -5 & 9 \\ 0 & 0 & 1 & -2 \end{vmatrix}$ Multiply row 3 by $\frac{1}{13}$.

   $x - y - z = 5$
   $y - 5z = 9$
   $z = -2$

   $y - 5z = 9$
   $y - 5(-2) = 9$
   $y + 10 = 9$
   $y = -1$

   $x - y - z = 5$
   $x - (-1) - (-2) = 5$
   $x + 1 + 2 = 5$
   $x + 3 = 5$
   $x = 2$
   The solution is $(2, -1, -2)$.

9. $\begin{vmatrix} 3 & -1 \\ -2 & 4 \end{vmatrix} = 3(4) - (-2)(-1) = 12 - 2 = 10$

10. $\begin{vmatrix} 1 & -2 & 3 \\ 3 & 1 & 1 \\ 2 & -1 & -2 \end{vmatrix} = 1\begin{vmatrix} 1 & 1 \\ -1 & -2 \end{vmatrix} - (-2)\begin{vmatrix} 3 & 1 \\ 2 & -2 \end{vmatrix} + 3\begin{vmatrix} 3 & 1 \\ 2 & -1 \end{vmatrix}$
    $= 1(-2 - (-1)) + 2(-6 - 2) + 3(-3 - 2)$
    $= 1(-2 + 1) + 2(-8) + 3(-5)$
    $= -1 - 16 - 15$
    $= -32$

11. $x - y = 3$
    $2x + y = -4$

$D = \begin{vmatrix} 1 & -1 \\ 2 & 1 \end{vmatrix} = 3$

$D_x = \begin{vmatrix} 3 & -1 \\ -4 & 1 \end{vmatrix} = -1$

$D_y = \begin{vmatrix} 1 & 3 \\ 2 & -4 \end{vmatrix} = -10$

$x = \dfrac{D_x}{D} = -\dfrac{1}{3}$

$y = \dfrac{D_y}{D} = -\dfrac{10}{3}$

The solution is $\left(-\dfrac{1}{3}, -\dfrac{10}{3}\right)$.

12. $x - y + z = 2$
    $2x - y - z = 1$
    $x + 2y - 3z = -4$

$D = \begin{vmatrix} 1 & -1 & 1 \\ 2 & -1 & -1 \\ 1 & 2 & -3 \end{vmatrix} = 5$

$D_x = \begin{vmatrix} 2 & -1 & 1 \\ 1 & -1 & -1 \\ -4 & 2 & -3 \end{vmatrix} = 1$

$D_y = \begin{vmatrix} 1 & 2 & 1 \\ 2 & 1 & -1 \\ 1 & -4 & -3 \end{vmatrix} = -6$

$D_z = \begin{vmatrix} 1 & -1 & 2 \\ 2 & -1 & 1 \\ 1 & 2 & -4 \end{vmatrix} = 3$

$x = \dfrac{D_x}{D} = \dfrac{1}{5}$

$y = \dfrac{D_y}{D} = -\dfrac{6}{5}$

$z = \dfrac{D_z}{D} = \dfrac{3}{5}$

The solution is $\left(\dfrac{1}{5}, -\dfrac{6}{5}, \dfrac{3}{5}\right)$.

13. $3x + 2y + 2z = 2$
    $x - 2y - z = 1$
    $2x - 3y - 3z = -3$

$D = \begin{vmatrix} 3 & 2 & 2 \\ 1 & -2 & -1 \\ 2 & -3 & -3 \end{vmatrix} = 13$

$D_x = \begin{vmatrix} 2 & 2 & 2 \\ 1 & -2 & -1 \\ -3 & -3 & -3 \end{vmatrix} = 0$

$D_y = \begin{vmatrix} 3 & 2 & 2 \\ 1 & 1 & -1 \\ 2 & -3 & 3 \end{vmatrix} = -26$

$D_z = \begin{vmatrix} 3 & 2 & 2 \\ 1 & -2 & 1 \\ 2 & -3 & -3 \end{vmatrix} = 39$

$\dfrac{D_x}{D} = \dfrac{0}{13} = 0$

$\dfrac{D_y}{D} = \dfrac{-26}{13} = -2$

$\dfrac{D_z}{D} = \dfrac{39}{13} = 3$

The solution is $(0, -2, 3)$.

14.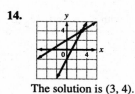

The solution is $(3, 4)$.

15.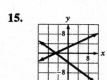

The solution is $(-5, 0)$.

16. Solve each inequality for y.

$2x - y < 3 \qquad\qquad 4x + 3y < 11$
$-y < -2x + 3 \qquad\quad 3y < -4x + 11$
$y > 2x - 3 \qquad\qquad y < -\dfrac{4}{3}x + \dfrac{11}{3}$

$y > 2x - 3$
$y < -\dfrac{4}{3}x + \dfrac{11}{3}$

17. Solve each inequality for y.

$x + y > 2$ $\quad$ $2x - y < -1$
$y > -x + 2$ $\quad$ $-y < -2x - 1$
$y > 1 + 2x$

18. $(-0.14, 2.43)$

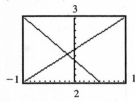

19. **Strategy**
 • Rate of the plane in calm air: x
 Rate of the wind: y

	Rate	Time	Distance
With wind	$x + y$	2	$2(x+y)$
Against wind	$x - y$	2.8	$2.8(x-y)$

 • The distance traveled with the wind is 350 mi.
 The distance traveled against the wind is 350 mi.
 $2(x+y) = 350$
 $2.8(x-y) = 350$

Solution
$2(x+y) = 350$
$2.8(x-y) = 350$

$\dfrac{1}{2} \cdot 2(x+y) = \dfrac{1}{2} \cdot 350$

$\dfrac{1}{2.8} \cdot 2.8(x-y) = \dfrac{1}{2.8} \cdot 350$

$x + y = 175$
$x - y = 125$

$2x = 300$
$x = 150$

$x + y = 175$
$150 + y = 175$
$y = 25$

The rate of the plane in calm air is 150 mph. The rate of the wind is 25 mph.

20. **Strategy**
Cost per yard of cotton: x
Cost per yard of wool: y
First purchase:

	Amount	Cost	Total Value
Cotton	60	x	$60x$
Wool	90	y	$90y$

Second purchase:

	Amount	Cost	Total Value
Cotton	80	x	$80x$
Wool	20	y	$20y$

 • The total cost of the first purchase was $1800.
 The total cost of the second purchase was $1000.
 $60x + 90y = 1800$
 $80x + 20y = 1000$

Solution
$-4(60x + 90y) = -4(1800)$
$3(80x + 20y) = 3(1000)$

$-240x - 360y = -7200$
$240x + 60y = 3000$

$-300y = -4200$
$y = 14$

$60x + 90(14) = 1800$
$60x + 1260 = 1800$
$60x = 540$
$x = 9$

The cost per yard of cotton is $9.00. The cost per yard of wool is $14.00.

Cumulative Review Exercises

1. $\dfrac{3}{2}x - \dfrac{3}{8} + \dfrac{1}{4}x = \dfrac{7}{12}x - \dfrac{5}{6}$

$24\left(\dfrac{3}{2}x - \dfrac{3}{8} + \dfrac{1}{4}x\right) = 24\left(\dfrac{7}{12}x - \dfrac{5}{6}\right)$

$36x - 9 + 6x = 14x - 20$
$42x - 9 = 14x - 20$
$28x - 9 = -20$
$28x = -11$
$x = -\dfrac{11}{28}$

The solution is $-\dfrac{11}{28}$.

2. $(x_1, y_1) = (2, -1)$, $(x_2, y_2) = (3, 4)$

$m = \dfrac{y_2 - y_1}{x_2 - x_1} = \dfrac{4 - (-1)}{3 - 2} = \dfrac{5}{1} = 5$

$y - y_1 = m(x - x_1)$
$y - (-1) = 5(x - 2)$
$y + 1 = 5x - 10$
$y = 5x - 11$

The equation of the line is $y = 5x - 11$.

3. $3[x-2(5-2x)-4x]+6 = 3(x-10+4x-4x)+6$
$= 3(x-10)+6$
$= 3x-30+6$
$= 3x-24$

4. $a = 4, \ b = 8, \ c = -2$
$a + bc \div 2 = 4 + 8(-2) \div 2$
$= 4 - 16 \div 2$
$= 4 - 8$
$= -4$

5. $2x - 3 < 9$ or $5x - 1 < 4$
Solve each inequality.
$2x - 3 < 9$ or $5x - 1 < 4$
$2x < 12$ $\qquad$ $5x < 5$
$x < 6$ $\qquad$ $x < 1$
is the same as $\{x | x < 6\} \cup \{x | x < 1\}$, which is $\{x | x < 6\}$.

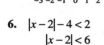

6. $|x - 2| - 4 < 2$
$|x - 2| < 6$

$-6 < x - 2 < 6$
$-6 + 2 < x - 2 + 2 < 6 + 2$
$-4 < x < 8$
$\{x | -4 < x < 8\}$

7. $|2x - 3| > 5$
Solve each inequality.
$2x - 3 < -5$ or $2x - 3 > 5$
$2x < -2$ $\qquad$ $2x > 8$
$x < -1$ $\qquad$ $x > 4$
This is the set $\{x | x < -1\} \cup \{x | x > 4\}$ or $\{x | x < -1 \text{ or } x > 4\}$.

8. $f(x) = 3x^3 - 2x^2 + 1$
$f(-3) = 3(-3)^3 - 2(-3)^2 + 1$
$f(-3) = 3(-27) - 2(9) + 1$
$f(-3) = -98$

9. The range is the set of numbers found by substituting in the set of numbers in the domain.
$f(-2) = 3(-2)^2 - 2(-2) = 16$
$f(-1) = 3(-1)^2 - 2(-1) = 5$
$f(0) = 3(0)^2 - 2(0) = 0$
$f(1) = 3(1)^2 - 2(1) = 1$
$f(2) = 3(2)^2 - 2(2) = 8$
The range is $\{0, 1, 5, 8, 16\}$.

10. $F(x) = x^2 - 3$
$F(2) = 2^2 - 3 = 1$

11. $f(x) = 3x - 4$
$f(2 + h) = 3(2 + h) - 4$
$= 6 + 3h - 4$
$= 2 + 3h$

$f(2) = 3(2) - 4 = 2$
$f(2 + h) - f(2) = 2 + 3h - 2 = 3h$

12. $\{x | x \leq 2\} \cap \{x | x > -3\}$

13. Slope $= -\dfrac{2}{3}$, Point $= (-2, 3)$
$y - 3 = -\dfrac{2}{3}[x - (-2)]$
$y - 3 = -\dfrac{2}{3}(x + 2)$
$y = -\dfrac{2}{3}x - \dfrac{4}{3} + 3$
$y = -\dfrac{2}{3}x + \dfrac{5}{3}$

14. The slope of the line $2x - 3y = 7$ is found by rearranging the equation as follows:
$-3y = 7 - 2x$
$y = -\dfrac{7}{3} + \dfrac{2}{3}x$
Slope $= \dfrac{2}{3}$

The perpendicular line has slope $= -\dfrac{3}{2}$.
The line is found using
$y - 2 = -\dfrac{3}{2}[x - (-1)]$
$y = -\dfrac{3}{2}(x + 1) + 2$
$y = -\dfrac{3}{2}x - \dfrac{3}{2} + 2$
$y = -\dfrac{3}{2}x + \dfrac{1}{2}$

15. The distance between points is
$\sqrt{(x_2 - x_1)^2 + (y_2 - y_1)^2}$.
$d = \sqrt{[2 - (-4)]^2 + (0 - 2)^2}$
$d = \sqrt{6^2 + (-2)^2}$
$d = \sqrt{36 + 4}$
$d = \sqrt{40}$
$d = 2\sqrt{10}$

16. The midpoint is found using $\left(\dfrac{x_1 + x_2}{2}, \dfrac{y_1 + y_2}{2}\right)$.
Midpoint $= \left(\dfrac{-4 + 3}{2}, \dfrac{3 + 5}{2}\right)$
$= \left(-\dfrac{1}{2}, 4\right)$

17. $2x - 5y = 10$
    $-5y = -2x + 10$
    $y = \dfrac{2}{5}x - 2$
    The y-intercept is -2.
    The slope is $\dfrac{2}{5}$.

18. $3x - 4y \geq 8$
    $-4y \geq -3x + 8$
    $y \leq \dfrac{3}{4}x - 2$
    The y-intercept is -2. The slope is $\dfrac{3}{4}$.

19. (1) $3x - 2y = 7$
    (2) $y = 2x - 1$
    Solve by the substitution method.
    $3x - 2(2x - 1) = 7$
    $3x - 4x + 2 = 7$
    $-x + 2 = 7$
    $-x = 5$
    $x = -5$
    Substitute -5 for x in equation (2).
    $y = 2x - 1$
    $y = 2(-5) - 1 = -10 - 1 = -11$
    The solution is $(-5, -11)$.

20. (1) $3x + 2z = 1$
    (2) $2y - z = 1$
    (3) $x + 2y = 1$
    Multiply equation (2) by -1 and add to equation (3).
    $-1(2y - z) = -1(1)$
    $\quad x + 2y = 1$

    $-2y + z = -1$
    $\ x + 2y = 1$

    (4) $x + z = 0$
    Multiply equation (4) by -2 and add to equation (1).
    $(-2)(x + z) = -2(0)$
    $3x + 2z = 1$

    $-2x - 2z = 0$
    $\ 3x + 2z = 1$

    $x = 1$
    Substitute 1 for x in equation (4).
    $x + z = 0$
    $1 + z = 0$
    $z = -1$
    Substitute 1 for x in equation (3).
    $x + 2y = 1$
    $1 + 2y = 1$
    $2y = 0$
    $y = 0$
    The solution is $(1, 0, -1)$.

21. $\begin{vmatrix} 2 & -5 & 1 \\ 3 & 1 & 2 \\ 6 & -1 & 4 \end{vmatrix} = 2\begin{vmatrix} 1 & 2 \\ -1 & 4 \end{vmatrix} - 3\begin{vmatrix} -5 & 1 \\ -1 & 4 \end{vmatrix} + 6\begin{vmatrix} -5 & 1 \\ 1 & 2 \end{vmatrix}$
    $= 2(4 + 2) - 3(-20 + 1) + 6(-10 - 1)$
    $= 2(6) - 3(-19) + 6(-11)$
    $= 12 + 57 - 66$
    $= 3$

22.

    The solution is $(2, 0)$.

23. $D = \begin{vmatrix} 4 & -3 \\ 3 & -2 \end{vmatrix} = 4(-2) - 3(-3) = 1$
    $D_x = \begin{vmatrix} 17 & -3 \\ 12 & -2 \end{vmatrix} = 17(-2) - 12(-3) = 2$
    $D_y = \begin{vmatrix} 4 & 17 \\ 3 & 12 \end{vmatrix} = 4(12) - 3(17) = -3$
    $x = \dfrac{D_x}{D} = \dfrac{2}{1} = 2$
    $y = \dfrac{D_y}{D} = \dfrac{-3}{1} = -3$
    The solution is $(2, -3)$.

24. Solve each inequality for y.
$3x - 2y \geq 4$ $x + y < 3$
$-2y \geq -3x + 4$ $y < -x + 3$
$y \leq \dfrac{3}{2}x - 2$

25. Strategy
• The unknown number of quarters: x
The unknown number of dimes: $3x$
The unknown number of nickels: $40 - (x + 3x)$

	Number	Value	Total Value
Quarters	x	25	$25x$
Dimes	$3x$	10	$10(3x)$
Nickels	$40 - 4x$	5	$5(40 - 4x)$

• The sum of the total values of the denominations is $4.10 (410 cents).
$25x + 10(3x) + 5(40 - 4x) = 410$

Solution
$25x + 10(3x) + 5(40 - 4x) = 410$
$25x + 30x + 200 - 20x = 410$
$35x + 200 = 410$
$35x = 210$
$x = 6$
$40 - 4x = 40 - 4(6) = 40 - 24 = 16$
There are 16 nickels in the purse.

26. Strategy
The unknown amount of pure water: x

	Amount	Percent	Quantity
Water	x	0	$0x$
4%	100	0.04	$100(0.04)$
2.5%	$100 + x$	0.025	$(100 + x)(0.025)$

The sum of the quantities before mixing equals the quantity after mixing.
$0 \cdot x + 100(0.04) = (100 + x)0.025$

Solution
$0 \cdot x + 100(0.04) = (100 + x)0.025$
$0 + 4 = 2.5 + 0.025x$
$1.5 = 0.025x$
$60 = x$
The amount of water that should be added is 60 ml.

27. Strategy
Rate of the plane in calm air: x
Rate of the wind: y

	Rate	Time	Distance
With wind	$x + y$	2	$2(x + y)$
Against wind	$x - y$	3	$3(x - y)$

• The distance traveled with the wind is 150 mi. The distance traveled against the wind is 150 mi.
$2(x + y) = 150$
$3(x - y) = 150$

Solution
$2(x + y) = 150$
$3(x - y) = 150$

$\dfrac{1}{2} \cdot 2(x + y) = \dfrac{1}{2} \cdot 150$
$\dfrac{1}{3} \cdot 3(x - y) = \dfrac{1}{3} \cdot 150$

$x + y = 75$
$x - y = 50$

$2x = 125$
$x = 62.5$

$x + y = 75$
$62.5 + y = 75$
$y = 12.5$
The rate of the wind is 12.5 mph.

28. Strategy
Cost per pound of hamburger: x
Cost per pound of steak: y
First purchase:

	Amount	Cost	Value
Hamburger	100	x	$100x$
Steak	50	y	$50y$

Second purchase:

	Amount	Cost	Value
Hamburger	150	x	$150x$
Steak	100	y	$100y$

• The total cost of the first purchase is $490.
The total cost of the second purchase is $860.
$100x + 50y = 490$
$150x + 100y = 860$

Solution
$100x + 50y = 490$
$150x + 100y = 860$

$3(100x + 50y) = 3(490)$
$-2(150x + 100y) = -2(860)$

$300x + 150y = 1470$
$-300x - 200y = -1720$

$-50y = -250$
$y = 5$

The cost per pound of steak is $5.

29. Strategy
Let M be the number of ohms, T the tolerance, and r the given amount of resistance. Find the tolerance and solve $|M - r| \leq T$ for M.

Solution
$T = 0.15 \cdot 12,000 = 1800$ ohms

$$|M - 12,000| \leq 1800$$
$$-1800 \leq M - 12,000 \leq 1800$$
$$-1800 + 12,000 \leq M - 12,000 + 12,000$$
$$\leq 1800 + 12,000$$
$$10,200 \leq M \leq 13,800$$

The lower and upper limits of the resistance are 10,200 ohms and 13,800 ohms.

30.
The slope of the line is
$$\frac{5000 - 1000}{100 - 0} = \frac{4000}{100} = 40$$
The slope represents the marginal income or the income generated per number of sales. The account executive earns $40 for each $1000 of sales.

Chapter 5: Polynomials and Exponents

Section 5.1

Concept Review 5.1

1. Never true
$$2^{-4} = \frac{1}{2^4} = \frac{1}{16}$$

3. Never true
$$(2+3)^{-1} = 5^{-1} = \frac{1}{5}$$

5. Never true
There is no rule to add two numbers with the same base.

Objective 1 Exercises

3. $(ab^3)(a^3b) = a^4b^4$

5. $(9xy^2)(-2x^2y^2) = -18x^3y^4$

7. $(x^4y^4)^4 = x^8y^{16}$

9. $(-3x^2y^3)^4 = (-3)^4x^8y^{12} = 81x^8y^{12}$

11. $(3^3a^5b^3)^2 = 3^6a^{10}b^6 = 729a^{10}b^6$

13. $(x^2y^2)(xy^3)^3 = (x^2y^2)(x^3y^9) = x^5y^{11}$

15. $[(3x)^3]^2 = (3x)^6 = 3^6x^6 = 729x^6$

17. $[(ab)^3]^6 = (ab)^{18} = a^{18}b^{18}$

19. $[(2xy)^3]^4 = (2xy)^{12} = 2^{12}x^{12}y^{12} = 4096x^{12}y^{12}$

21. $[(2a^4b^3)^3]^2 = (2a^4b^3)^6 = 2^6a^{24}b^{18} = 64a^{24}b^{18}$

23. $x^n \cdot x^{n+1} = x^{n+n+1} = x^{2n+1}$

25. $y^{3n} \cdot y^{3n-2} = y^{3n+3n-2} = y^{6n-2}$

27. $(a^{n-3})^{2n} = a^{(n-3)2n} = a^{2n^2-6n}$

29. $(x^{3n+2})^5 = x^{(3n+2)5} = x^{15n+10}$

31. $(2xy)(-3x^2yz)(x^2y^3z^3) = -6x^5y^5z^4$

33. $(3b^5)(2ab^2)(-2ab^2c^2) = -12a^2b^9c^2$

35. $(-2x^2y^3z)(3x^2yz^4) = -6x^4y^4z^5$

37. $(-3ab^3)^3(-2^2a^2b)^2 = [(-3)^3a^3b^9][(-2^2)^2a^4b^2]$
$= (-27a^3b^9)(16a^4b^2)$
$= -432a^7b^{11}$

39. $(-2ab^2)(-3a^4b^5)^3 = (-2ab^2)[(-3)^3a^{12}b^{15}]$
$= (-2ab^2)(-27a^{12}b^{15})$
$= 54a^{13}b^{17}$

Objective 2 Exercises

43. $\dfrac{a^8}{a^5} = a^{8-5} = a^3$

45. $\dfrac{a^7b}{a^2b^4} = a^{7-2}b^{1-4} = a^5b^{-3} = \dfrac{a^5}{b^3}$

47. $\dfrac{1}{3^{-5}} = 3^5 = 243$

49. $\dfrac{1}{y^{-3}} = y^3$

51. $\dfrac{a^3}{4b^{-2}} = \dfrac{a^3b^2}{4}$

53. $x^{-3} \cdot x^{-5} = x^{-8} = \dfrac{1}{x^8}$

55. $(5x^2)^{-3} = 5^{-3}x^{-6} = \dfrac{1}{5^3x^6} = \dfrac{1}{125x^6}$

57. $\dfrac{x^4}{x^{-5}} = x^9$

59. $a^{-5} \cdot a^7 = a^2$

61. $(x^3y^5)^{-2} = x^{-6}y^{-10} = \dfrac{1}{x^6y^{10}}$

63. $(3a)^{-3}(9a^{-1})^{-2} = (3a)^{-3}(3^2a^{-1})^{-2}$
$= (3^{-3}a^{-3})(3^{-4}a^2)$
$= 3^{-7}a^{-1}$
$= \dfrac{1}{3^7a}$
$= \dfrac{1}{2187a}$

65. $(x^{-1}y^2)^{-3}(x^2y^{-4})^{-3} = (x^3y^{-6})(x^{-6}y^{12})$
$= x^{-3}y^6$
$= \dfrac{y^6}{x^3}$

67. $\left(\dfrac{x^2y^{-1}}{xy}\right)^{-4} = \left(\dfrac{x}{y^2}\right)^{-4} = \dfrac{x^{-4}}{y^{-8}} = \dfrac{y^8}{x^4}$

69. $\dfrac{a^2b^3c^7}{a^6bc^5} = \dfrac{b^2c^2}{a^4}$

71. $\dfrac{(-3a^2b^3)^2}{(-2ab^4)^3} = \dfrac{(-3)^2a^4b^6}{(-2)^3a^3b^{12}} = \dfrac{9a^4b^6}{-8a^3b^{12}} = -\dfrac{9a}{8b^6}$

73. $\left(\dfrac{a^{-2}b}{a^3b^{-4}}\right)^2 = \left(\dfrac{b^5}{a^5}\right)^2 = \dfrac{b^{10}}{a^{10}}$

75. $\dfrac{y^{2n}}{-y^{8n}} = -\dfrac{1}{y^{8n-2n}} = -\dfrac{1}{y^{6n}}$

77. $\dfrac{x^{2n-1}y^{n-3}}{x^{n+4}y^{n+3}} = x^{2n-1-(n+4)}y^{n-3-(n+3)}$
$= x^{2n-1-n-4}y^{n-3-n-3}$
$= x^{n-5}y^{-6}$
$= \dfrac{x^{n-5}}{y^6}$

79. $\dfrac{(3x^{-2}y)^{-2}}{(4xy^{-2})^{-1}} = \dfrac{3^{-2}x^4 y^{-2}}{4^{-1}x^{-1}y^2} = \dfrac{4x^5}{9y^4}$

81. $\left(\dfrac{9ab^{-2}}{8a^{-2}b}\right)^{-2}\left(\dfrac{3a^{-2}b}{2a^2 b^{-2}}\right)^3 = \left(\dfrac{9a^3 b^{-3}}{8}\right)^{-2}\left(\dfrac{3a^{-4}b^3}{2}\right)^3$
$= \dfrac{9^{-2}a^{-6}b^6}{8^{-2}} \cdot \dfrac{3^3 a^{-12}b^9}{2^3}$
$= \dfrac{8^2 \cdot 3^3 a^{-18} b^{15}}{9^2 \cdot 2^3}$
$= \dfrac{64 \cdot 27 b^{15}}{81 \cdot 8 a^{18}}$
$= \dfrac{8b^{15}}{3a^{18}}$

83. $[(x^{-2}y^{-1})^2]^{-3} = (x^{-4}y^{-2})^{-3} = x^{12}y^6$

85. $\left[\left(\dfrac{a^2}{b}\right)^{-1}\right]^2 = \left(\dfrac{a^{-2}}{b^{-1}}\right)^2 = \left(\dfrac{b}{a^2}\right)^2 = \dfrac{b^2}{a^4}$

Objective 3 Exercises

87. $0.00000005 = 5 \times 10^{-8}$

89. $4,300,000 = 4.3 \times 10^6$

91. $9,800,000,000 = 9.8 \times 10^9$

93. $6.2 \times 10^{-12} = 0.0000000000062$

95. $6.34 \times 10^5 = 634,000$

97. $4.35 \times 10^9 = 4,350,000,000$

99. $(8.9 \times 10^{-5})(3.2 \times 10^{-6}) = (8.9)(3.2) \times 10^{-5+(-6)}$
$= 28.48 \times 10^{-11}$
$= 2.848 \times 10^{-10}$

101. $(480,000)(0.0000000096)$
$= (4.8 \times 10^5)(9.6 \times 10^{-9})$
$= (4.8)(9.6) \times 10^{5+(-9)}$
$= 46.08 \times 10^{-4}$
$= 4.608 \times 10^{-3}$

103. $\dfrac{2.7 \times 10^4}{3 \times 10^{-6}} = 0.9 \times 10^{4-(-6)}$
$= 0.9 \times 10^{10}$
$= 9 \times 10^9$

105. $\dfrac{4,800}{0.00000024} = \dfrac{4.8 \times 10^3}{2.4 \times 10^{-7}}$
$= 2 \times 10^{3-(-7)}$
$= 2 \times 10^{10}$

107. $\dfrac{0.000000346}{0.0000005} = \dfrac{3.46 \times 10^{-7}}{5 \times 10^{-7}}$
$= 0.692 \times 10^{-7-(-7)}$
$= 0.692 \times 10^0 = 6.92 \times 10^{-1}$

109. $\dfrac{(6.9 \times 10^{27})(8.2 \times 10^{-13})}{4.1 \times 10^{15}}$
$= \dfrac{(6.9)(8.2) \times 10^{27+(-13)-15}}{4.1}$
$= 13.8 \times 10^{-1}$
$= 1.38$

111. $\dfrac{(720)(0.0000000039)}{(26,000,000,000)(0.018)}$
$= \dfrac{7.2 \times 10^2 \times 3.9 \times 10^{-9}}{2.6 \times 10^{10} \times 1.8 \times 10^{-2}}$
$= \dfrac{(7.2)(3.9) \times 10^{2+(-9)-10-(-2)}}{(2.6)(1.8)}$
$= 6 \times 10^{-15}$

Objective 4 Exercises

113. Strategy
To find the number of arithmetic operations:
• Find the reciprocal of 2×10^{-9}, which is the number of operations performed in one second.
• Write the number of seconds in one minute (60) in scientific notation.
• Multiply the number of arithmetic operations per second by the number of seconds in one minute.

Solution
$\dfrac{1}{2 \times 10^{-9}} = \dfrac{1}{2} \times 10^9$
$60 = 6 \times 10$
$\left(\dfrac{1}{2} \times 10^9\right)(6 \times 10) = \dfrac{1}{2} \times 6 \times 10^{10}$
$\qquad\qquad\qquad\quad = 3 \times 10^{10}$

The computer can perform 3×10^{10} operations in one minute.

115. Strategy
To find the distance traveled:
Write the number of seconds in one day in scientific notation.
Use the equation $d = rt$, where r is the speed of light and t is the number of seconds in one day.

Solution
$r = 3 \times 10^8$
$24 \cdot 60 \cdot 60 = 86,400 = 8.64 \times 10^4$
$d = rt$
$d = (3 \times 10^8)(8.64 \times 10^4)$
$d = 3 \times 8.64 \times 10^{12}$
$d = 25.92 \times 10^{12}$
$d = 2.592 \times 10^{13}$
Light travels 2.592×10^{13} m in one day.

117. Strategy
To find the number of times heavier the proton is, divide the mass of the proton by the mass of the electron.

Solution
$\dfrac{1.673 \times 10^{-27}}{9.109 \times 10^{-31}} = 0.183664508 \times 10^4$
$= 1.83664508 \times 10^3$
The proton is 1.83664508×10^3 times heavier than the electron.

119. Strategy
To find the number of miles:
Find the number of seconds in one year.
Use the equation $d = rt$, where r is the speed of light and t is the time in seconds for one year.

Solution
$60 \cdot 60 \cdot 24 \cdot 365 = 31,536,000$
$= 3.1536 \times 10^7$
$d = rt$
$d = (1.86 \times 10^5)(3.1536 \times 10^7)$
$d = 1.86 \times 3.1536 \times 10^{12}$
$d = 5.865696 \times 10^{12}$
Light travels 5.865696×10^{12} mi in one year.

121. Strategy
To find the weight of one seed:
Write the number of seeds per ounce in scientific notation.
Find the reciprocal of the number of seeds per ounce, which is the number of ounces per seed.

Solution
$31,000,000 = 3.1 \times 10^7$
$\dfrac{1}{3.1 \times 10^7} = 0.32258065 \times 10^{-7}$
$= 3.2258065 \times 10^{-8}$
The weight of one orchid seed is 3.2258065×10^{-8} oz.

123. Strategy
To find the time, use the equation $d = rt$, where r is the speed of the satellite and d is the distance to Saturn.

Solution
$d = rt$
$8.86 \times 10^8 = (1 \times 10^5)t$
$\dfrac{8.86 \times 10^8}{1 \times 10^5} = t$
$8.86 \times 10^3 = t$
It will take the satellite 8.86×10^3 h to reach Saturn.

125. Strategy
To find the volume, use the formula $V = \dfrac{4}{3}\pi r^3$, where $r = 1.5 \times 10^{-4}$ mm.

Solution
$V = \dfrac{4}{3}\pi r^3$
$= \dfrac{4}{3}\pi(1.5 \times 10^{-4})^3$
$= \dfrac{4}{3}\pi(1.5)^3 \times 10^{-12}$
$= 14.1371669 \times 10^{-12}$
$= 1.41371669 \times 10^{-11}$
The volume of the cell is $1.41371669 \times 10^{-11}$ mm^3.

127. Strategy
To find the time:
Write the speed of the space ship in scientific notation.
Use the equation $d = rt$, where r is the speed of the space ship and d is the distance across the galaxy.

Solution
$25,000 = 2.5 \times 10^4$
$d = rt$
$6 \times 10^{17} = (2.5 \times 10^4)t$
$\dfrac{6 \times 10^{17}}{2.5 \times 10^4} = t$
$2.4 \times 10^{13} = t$
The space ship travels across the galaxy in 2.4×10^{13} h.

Applying Concepts 5.1

129. $\dfrac{3}{x} - 1$
No, the expression is not a polynomial because there is a variable in the denominator.

131. $\sqrt{5}x + 2$
Yes, the expression is a polynomial.

133. $x + \sqrt{3}$
Yes, the expression is a polynomial.

128   *Chapter 5: Polynomials and Exponents*

135. $(6x^3 + kx^2 - 2x - 1) - (4x^3 - 3x^2 + 1) = 2x^3 - x^2 - 2x - 2$
$(6x^3 + kx^2 - 2x - 1) + (-4x^3 + 3x^2 - 1) = 2x^3 - x^2 - 2x - 2$
$\qquad\qquad 2x^3 + (k+3)x^2 - 2x - 2 = 2x^3 - x^2 - 2x - 2$
$\qquad\qquad\qquad\qquad\qquad k + 3 = -1$
$\qquad\qquad\qquad\qquad\qquad\qquad k = -4$

137. Strategy
To find the perimeter, replace the variables a, b, and c in the equation $P = a + b + c$ by the given values and solve for P.

Solution
$P = a + b + c$
$P = 4x^n + 3x^n + 3x^n$
$P = 10x^n$
The perimeter is $10x^n$.

139. Strategy
To find the area, replace the variables b and h in the equation $A = \frac{1}{2}bh$ by the given values and solve for A.

Solution
$A = \frac{1}{2}bh$
$A = \frac{1}{2}(8xy)(5xy)$
$A = 4xy(5xy)$
$A = 20x^2y^2$
The area is $20x^2y^2$.

141. $\dfrac{5x^3}{y^{-6}} + \left(\dfrac{x^{-1}}{y^2}\right)^{-3} = 5x^3y^6 + (x^{-1}y^{-2})^{-3}$
$\qquad\qquad\qquad\qquad = 5x^3y^6 + x^3y^6$
$\qquad\qquad\qquad\qquad = 6x^3y^6$

143. $\left(\dfrac{2m^3n^{-2}}{4m^4n}\right)^{-2} \div \left(\dfrac{mn^5}{m^{-1}n^3}\right)^3$

$= \left(\dfrac{m^{-1}n^{-3}}{2}\right)^{-2} \div (m^2n^2)^3$

$= \dfrac{m^2n^6}{2^{-2}} \div m^6n^6$

$= 2^2 m^2 n^6 \div m^6 n^6$

$= \dfrac{4m^2n^6}{m^6n^6}$

$= \dfrac{4}{m^4}$

145. No, let $a = -1$, and $b = 1$. Then $a < b$, but $a^{-1} = \dfrac{1}{-1} = -1 < b^{-1} = \dfrac{1}{1} = 1$ because $-1 < 1$.

Section 5.2

Concept Review 5.2

1. Always true

3. Always true

5. Sometimes true

   The sum of $-x^2 + 2x - 3$ and $x^2 + 5x - 10$ is the binomial $7x - 13$.

Objective 1 Exercises

3. $P(3) = 3(3)^2 - 2(3) - 8$
   $P(3) = 13$

5. $R(2) = 2(2)^3 - 3(2)^2 + 4(2) - 2$
   $R(2) = 10$

7. $f(-1) = (-1)^4 - 2(-1)^2 - 10$
   $f(-1) = -11$

9. $L(s) = 0.641s^2$
   $L(6) = 0.641(6)^2$
   $L(6) = 23.076$
   The length of the wave is 23.1 m.

11. $T(n) = n^2 - n$
    $T(8) = (8)^2 - 8$
    $T(8) = 56$
    The league must schedule 56 games.

13. $M(r) = 6.14r^2 + 6.14r + 2.094$
    $M(6) = 6.14(6)^2 + 6.14(6) + 2.094$
    $M(6) = 259.974$
    260 in^3 of meringue are needed.

15. Polynomial: (a) –1   (b) 8   (c) 2

17. Not a polynomial.

19. Not a polynomial.

21. Polynomial: (a) 3   (b) π   (c) 5

23. Polynomial: (a) –5   (b) 2   (c) 3

25. Polynomial: (a) 14   (b) 14   (c) 0

27.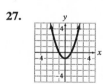

29.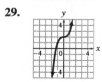

31.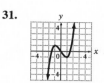

Objective 2 Exercises

35. $5x^2 + 2x - 7$
    $x^2 - 8x + 12$
    $\overline{6x^2 - 6x + 5}$

37. $x^2 - 3x + 8$
    $-2x^2 + 3x - 7$
    $\overline{-x^2 + 1}$

39. $(3y^2 - 7y) + (2y^2 - 8y + 2)$
    $= (3y^2 + 2y^2) + (-7y - 8y) + 2$
    $= 5y^2 - 15y + 2$

41. $(2a^2 - 3a - 7) - (-5a^2 - 2a - 9)$
    $= (2a^2 + 5a^2) + (-3a + 2a) + (-7 + 9)$
    $= 7a^2 - a + 2$

43. $P(x) + R(x) = (3x^3 - 4x^2 - x + 1) + (2x^3 + 5x - 8)$
    $ = 5x^3 - 4x^2 + 4x - 7$

45. $P(x) + R(x) = (x^{2n} + 7x^n - 3) + (-x^{2n} + 2x^n + 8)$
    $ = 9x^n + 5$

47. $S(x) = P(x) + R(x)$
    $ = (3x^4 - 3x^3 - x^2) + (3x^3 - 7x^2 + 2x)$
    $ = 3x^4 - 8x^2 + 2x$

49. $D(x) = P(x) - R(x)$
    $ = (x^2 + 2x + 1) - (2x^3 - 3x^2 + 2x - 7)$
    $ = (x^2 + 2x + 1) + (-2x^3 + 3x^2 - 2x + 7)$
    $ = -2x^3 + 4x^2 + 8$

Applying Concepts 5.2

51.

    $x = 1.5$

53.

    $x = 0.6, 3.4$

130 *Chapter 5: Polynomials and Exponents*

55.
$x = -1.9, 0.35, 1.5$

57.
$x = -0.6, 1.6$

59.
$$(2x^3 + 3x^2 + kx + 5) - (x^3 + x^2 - 5x + 2) = x^3 + 2x^2 + 3x + 7$$
$$(2x^3 - x^3) + (3x^2 - x^2) + (kx + 5x) + (5 - 2) = x^3 + 2x^2 + 3x + 7$$
$$x^3 + 2x^2 + (k + 5)x + 3 = x^3 + 2x^2 + 3x + 7$$
$$(k + 5)x = 3x$$
$$k + 5 = 3$$
$$k = -2$$

61. The degree of $P(x) + Q(x)$ will be 4.
Example:
$$P(x) = 2x^3 + 4x^2 - 3x + 9$$
$$Q(x) = x^4 - 5x + 1$$
$$P(x) + Q(x) = x^4 + 2x^3 + 4x^2 - 8x + 10,$$
a fourth-degree polynomial.

63. Strategy
Determine the midpoint of the beam, x, and substitute x into $D(x)$ to determine the deflection.

Solution
Length of beam = 10 ft
Midpoint of beam = 5
$$D(x) = 0.005x^4 - 0.1x^3 + 0.5x^2$$
$$D(5) = 0.005(5)^4 - 0.1(5)^3 + 0.5(5)^2$$
$$D(5) = 3.125$$
The maximum deflection of the beam is 3.125 in.

65. $P(x) = 4x^4 - 3x^2 + 6x + c$
$$-3 = 4(-1)^4 - 3(-1)^2 + 6(-1) + c$$
$$-3 = 4 - 3 - 6 + c$$
$$-3 = -5 + c$$
$$2 = c$$
The value of c is 2.

67.
The graph of k is the graph of f moved 2 units down.

Section 5.3

Concept Review 5.3

1. Always true
3. Always true
5. Always true

Objective 1 Exercises

3. $2x(x-3) = 2x^2 - 6x$

5. $3x^2(2x^2 - x) = 6x^4 - 3x^3$

7. $3xy(2x - 3y) = 6x^2y - 9xy^2$

9. $x^n(x+1) = x^{n+1} + x^n$

11. $x^n(x^n + y^n) = x^{2n} + x^n y^n$

13. $2b + 4b(2-b) = 2b + 8b - 4b^2 = -4b^2 + 10b$

15. $-2a^2(3a^2 - 2a + 3) = -6a^4 + 4a^3 - 6a^2$

17. $3b(3b^4 - 3b^2 + 8) = 9b^5 - 9b^3 + 24b$

19. $-5x^2(4 - 3x + 3x^2 + 4x^3)$
    $= -20x^2 + 15x^3 - 15x^4 - 20x^5$
    $= -20x^5 - 15x^4 + 15x^3 - 20x^2$

21. $-2x^2y(x^2 - 3xy + 2y^2) = -2x^4y + 6x^3y^2 - 4x^2y^3$

23. $x^n(x^{2n} + x^n + x) = x^{3n} + x^{2n} + x^{n+1}$

25. $a^{n+1}(a^n - 3a + 2) = a^{2n+1} - 3a^{n+2} + 2a^{n+1}$

27. $2y^2 - y[3 - 2(y-4) - y] = 2y^2 - y[3 - 2y + 8 - y]$
    $= 2y^2 - y[11 - 3y]$
    $= 2y^2 - 11y + 3y^2$
    $= 5y^2 - 11y$

29. $2y - 3[y - 2y(y-3) + 4y]$
    $= 2y - 3[y - 2y^2 + 6y + 4y]$
    $= 2y - 3[11y - 2y^2]$
    $= 2y - 33y + 6y^2$
    $= 6y^2 - 31y$

Objective 2 Exercises

31. $(5x - 7)(3x - 8) = 15x^2 - 40x - 21x + 56$
    $= 15x^2 - 61x + 56$

33. $(7x - 3y)(2x - 9y) = 14x^2 - 63xy - 6xy + 27y^2$
    $= 14x^2 - 69xy + 27y^2$

35. $(3a - 5b)(a + 7b) = 3a^2 + 21ab - 5ab - 35b^2$
    $= 3a^2 + 16ab - 35b^2$

37. $(5x + 9y)(3x + 2y) = 15x^2 + 10xy + 27xy + 18y^2$
    $= 15x^2 + 37xy + 18y^2$

39. $(5x - 9y)(6x - 5y) = 30x^2 - 25xy - 54xy + 45y^2$
    $= 30x^2 - 79xy + 45y^2$

41. $(xy - 5)(2xy + 7) = 2x^2y^2 + 7xy - 10xy - 35$
    $= 2x^2y^2 - 3xy - 35$

43. $(x^2 - 4)(x^2 - 6) = x^4 - 6x^2 - 4x^2 + 24$
    $= x^4 - 10x^2 + 24$

45. $(x^2 - 2y^2)(x^2 + 4y^2)$
    $= x^4 + 4x^2y^2 - 2x^2y^2 - 8y^4$
    $= x^4 + 2x^2y^2 - 8y^4$

47. $(x^n - 4)(x^n - 5) = x^{2n} - 5x^n - 4x^n + 20$
    $= x^{2n} - 9x^n + 20$

49. $(5b^n - 1)(2b^n + 4) = 10b^{2n} + 20b^n - 2b^n - 4$
    $= 10b^{2n} + 18b^n - 4$

51. $(3x^n + b^n)(x^n + 2b^n)$
    $= 3x^{2n} + 6x^n b^n + x^n b^n + 2b^{2n}$
    $= 3x^{2n} + 7x^n b^n + 2b^{2n}$

53. $ x^2 + 5x - 8$
    $\underline{\times x + 3}$
    $ 3x^2 + 15x - 24$
    $\underline{x^3 + 5x^2 - 8x }$
    $x^3 + 8x^2 + 7x - 24$

55. $ a^3 - 3a^2 + 7$
    $\underline{\times a + 2}$
    $ 2a^3 - 6a^2 + 14$
    $\underline{a^4 - 3a^3 + 7a}$
    $a^4 - a^3 - 6a^2 + 7a + 14$

57. $ 2a^2 - 5ab - 3b^2$
    $\underline{\times 3a + b}$
    $ 2a^2b - 5ab^2 - 3b^3$
    $\underline{6a^3 - 15a^2b - 9ab^2 }$
    $6a^3 - 13a^2b - 14ab^2 - 3b^3$

59. $ 3b^2 - 3b + 6$
    $\underline{\times 2b^2 - 3}$
    $ -9b^2 + 9b - 18$
    $\underline{6b^4 - 6b^3 + 12b^2 }$
    $6b^4 - 6b^3 + 3b^2 + 9b - 18$

Chapter 5: Polynomials and Exponents

61.
$$\begin{array}{r} 3a^4 \quad -3a^2+2a-5 \\ \times \quad\quad\quad\quad\quad 2a-5 \\ \hline -15a^4 \quad +15a^2-10a+25 \\ 6a^5 \quad\quad -6a^3+4a^2-10a \\ \hline 6a^5-15a^4-6a^3+19a^2-20a+25 \end{array}$$

63.
$$\begin{array}{r} x^2-3x+1 \\ \times \quad x^2-2x+7 \\ \hline 7x^2-21x+7 \\ -2x^3+6x^2-2x \\ x^4-3x^3+\quad x^2 \\ \hline x^4-5x^3+14x^2-23x+7 \end{array}$$

65. $(b-3)(3b-2)(b-1)$
$= (3b^2 - 2b - 9b + 6)(b-1)$
$= (3b^2 - 11b + 6)(b-1)$
$$\begin{array}{r} = \quad 3b^2-11b+6 \\ \times \quad\quad\quad b-1 \\ \hline -3b^2+11b-6 \\ 3b^3-11b^2+\quad 6b \\ \hline 3b^3-14b^2+17b-6 \end{array}$$

67.
$$\begin{array}{r} x^{2n}-3x^n y^n - y^{2n} \\ \times \quad\quad\quad x^n - y^n \\ \hline -x^{2n}y^n+3x^n y^{2n}+y^{3n} \\ x^{3n}-3x^{2n}y^n-\quad x^n y^{2n} \\ \hline x^{3n}-4x^{2n}y^n+2x^n y^{2n}+y^{3n} \end{array}$$

Objective 3 Exercises

69. $(b-7)(b+7) = b^2 - 49$

71. $(b-11)(b+11) = b^2 - 121$

73. $(5x-4y)^2 = 25x^2 - 40xy + 16y^2$

75. $(x^2+y^2)^2 = x^4 + 2x^2 y^2 + y^4$

77. $(2a-3b)(2a+3b) = 4a^2 - 9b^2$

79. $(x^2+1)(x^2-1) = x^4 - 1$

81. $(2x^n + y^n)^2 = 4x^{2n} + 4x^n y^n + y^{2n}$

83. $(5a-9b)(5a+9b) = 25a^2 - 81b^2$

85. $(2x^n - 5)(2x^n + 5) = 4x^{2n} - 25$

87. $(6-x)(6+x) = 36 - x^2$

89. $(3a-4b)^2 = 9a^2 - 24ab + 16b^2$

91. $(3x^n + 2)^2 = 9x^{2n} + 12x^n + 4$

93. $(x^n + 3)(x^n - 3) = x^{2n} - 9$

95. $(x^n - 1)^2 = x^{2n} - 2x^n + 1$

97. $(2x^n + 5y^n)^2 = 4x^{2n} + 20x^n y^n + 25y^{2n}$

Objective 4 Exercises

99. **Strategy**

To find the area, replace the variables b and h in the equation $A = \frac{1}{2}bh$ by the given values and solve for A.

Solution

$A = \frac{1}{2}bh$

$A = \frac{1}{2}(x+2)(2x-3)$

$A = \left(\frac{1}{2}x+1\right)(2x-3)$

$A = x^2 - \frac{3}{2}x + 2x - 3$

$A = x^2 + \frac{x}{2} - 3$

The area is $\left(x^2 + \frac{x}{2} - 3\right)$ ft^2.

101. Strategy
To find the area, subtract the four small rectangles from the large rectangle.
Large rectangle:
Length = $L_1 = x+8$
Width = $W_1 = x+4$
Small rectangles:
Length = $L_2 = 2$
Width = $W_2 = 2$

Solution
A = Area of the large rectangle $-$ 4(area of small rectangle)
$A = (L_1 \cdot W_1) - 4(L_2 \cdot W_2)$
$A = (x+8)(x+4) - 4(2)(2)$
$A = x^2 + 4x + 8x + 32 - 16$
$A = x^2 + 12x + 16$
The area is $(x^2 + 12x + 16)$ ft^2.

103. Strategy
Length of the box: $18 - 2x$
Width of the box: $18 - 2x$
Height of the box: x
To find the volume, replace the variables L, W, and H in the equation $V = LWH$ and solve for V.

Solution
$V = LWH$
$V = (18 - 2x)(18 - 2x)x$
$V = (324 - 36x - 36x + 4x^2)x$
$V = (324 - 72x + 4x^2)x$
$V = 324x - 72x^2 + 4x^3$
The volume is $(4x^3 - 72x^2 + 324x)$ in^3.

105. Strategy
To find the volume, replace the variables, L, W, and H in the equation $V = L \cdot W \cdot H$ by the given values and solve for V.

Solution
$V = L \cdot W \cdot H$
$V = (2x+3)(x-5)(x)$
$V = (2x^2 - 7x - 15)(x)$
$V = 2x^3 - 7x^2 - 15x$
The volume is $(2x^3 - 7x^2 - 15x)$ cm^3.

107. Strategy
To find the volume, add the volume of the small rectangular solid to the volume of the large rectangular solid.
Large rectangular solid:
Length = $L_1 = 3x+4$
Width = $W_1 = x+6$
Height = $H_1 = x$
Small rectangular solid:
Length = $L_2 = x+4$
Width = $W_2 = x+6$
Height = $H_2 = x$

Solution
$V = (L_1 \cdot W_1 \cdot H_1) + (L_2 \cdot W_2 \cdot H_2)$
$V = (3x+4)(x+6)(x) + (x+4)(x+6)(x)$
$V = (3x^2 + 22x + 24)(x) + (x^2 + 10x + 24)(x)$
$V = 3x^3 + 22x^2 + 24x + x^3 + 10x^2 + 24x$
$V = 4x^3 + 32x^2 + 48x$
The volume is $(4x^3 + 32x^2 + 48x)$ cm^3.

Applying Concepts 5.3

109. $\dfrac{(3x-5)^6}{(3x-5)^4} = (3x-5)^{6-4}$
$= (3x-5)^2$
$= (3x-5)(3x-5)$
$= 9x^2 - 30x + 25$

111. $(x+2y)^2 + (x+2y)(x-2y) = (x+2y)(x+2y) + (x+2y)(x-2y)$
$= (x^2 + 4xy + 4y^2) + (x^2 - 4y^2)$
$= 2x^2 + 4xy$

113. $2x^2(3x^3 + 4x - 1) - 5x^2(x^2 - 3) = (6x^5 + 8x^3 - 2x^2) - (5x^4 - 15x^2)$
$= (6x^5 + 8x^3 - 2x^2) + (-5x^4 + 15x^2)$
$= 6x^5 - 5x^4 + 8x^3 + 13x^2$

115. $(3x-2y)^2 - (2x-3y)^2 = (3x-2y)(3x-2y) - (2x-3y)(2x-3y)$
$= (9x^2 - 12xy + 4y^2) - (4x^2 - 12xy + 9y^2)$
$= (9x^2 - 12xy + 4y^2) + (-4x^2 + 12xy - 9y^2)$
$= 5x^2 - 5y^2$

117. $[x^2(2y-1)]^2 = x^{2 \cdot 2}(2y-1)^2$
$= x^4(4y^2 - 4y + 1)$
$= 4x^4 y^2 - 4x^4 y + x^4$

119. $(kx - 7)(kx + 2) = k^2 x^2 + 5x - 14$
$k^2 x^2 + 2kx - 7kx - 14 = k^2 x^2 + 5x - 14$
$k^2 x^2 - 5kx - 14 = k^2 x^2 + 5x - 14$
$-5kx = 5x$
$-5k = 5$
$k = -1$

121. $\dfrac{a^m}{a^n}$, $m = n + 2$
$\dfrac{a^{n+2}}{a^n} = a^{(n+2)-n} = a^2$

123. $(x+7)(2x-3) = 2x^2 - 3x + 14x - 21$
$= 2x^2 + 11x - 21$

125. $(6x^2 + 12xy - 2y^2) - (5x - y)(x + 3y)$
$= (6x^2 + 12xy - 2y^2) - (5x^2 + 15xy - xy - 3y^2)$
$= (6x^2 + 12xy - 2y^2) - (5x^2 + 14xy - 3y^2)$
$= (6x^2 + 12xy - 2y^2) + (-5x^2 - 14xy + 3y^2)$
$= x^2 - 2xy + y^2$

127. Find the value of n.
$3(2n - 1) = 5(n - 1)$
$6n - 3 = 5n - 5$
$n - 3 = -5$
$n = -2$
Substitute the value of n into the expression.
$(-2n^3)^2 = [-2(-2)^3]^2 = [-2(-8)]^2 = 16^2 = 256$

Section 5.4

Concept Review 5.4

1. Sometimes true

   For example, 1 and x are monomials but $\dfrac{1}{x}$ is not a monomial.

3. Always true

Objective 1 Exercises

3.
$$\begin{array}{r} x+8 \\ x-5 \overline{\smash{)}x^2+3x-40} \\ \underline{x^2-5x} \\ 8x-40 \\ \underline{8x-40} \\ 0 \end{array}$$

$(x^2+3x-40) \div (x-5) = x+8$

5.
$$\begin{array}{r} x^2 \\ x-3 \overline{\smash{)}x^3-3x^2+0x+2} \\ \underline{x^3-3x^2} \\ 2 \end{array}$$

$(x^3-3x^2+2) \div (x-3) = x^2 + \dfrac{2}{x-3}$

7.
$$\begin{array}{r} 3x+5 \\ 2x+1 \overline{\smash{)}6x^2+13x+8} \\ \underline{6x^2+3x} \\ 10x+8 \\ \underline{10x+5} \\ 3 \end{array}$$

$(6x^2+13x+8) \div (2x+1) = 3x+5 + \dfrac{3}{2x+1}$

9.
$$\begin{array}{r} 5x+7 \\ 2x-1 \overline{\smash{)}10x^2+9x-5} \\ \underline{10x^2-5x} \\ 14x-5 \\ \underline{14x-7} \\ 2 \end{array}$$

$(10x^2+9x-5) \div (2x-1) = 5x+7 + \dfrac{2}{2x-1}$

11.
$$\begin{array}{r} 4x^2+6x+9 \\ 2x-3 \overline{\smash{)}8x^3+0x^2+0x-9} \\ \underline{8x^3-12x^2} \\ 12x^2+0x \\ \underline{12x^2-18x} \\ 18x-9 \\ \underline{18x-27} \\ 18 \end{array}$$

$(8x^3-9) \div (2x-3) = 4x^2+6x+9 + \dfrac{18}{2x-3}$

13.
$$\begin{array}{r} 3x^2+1 \\ 2x^2-5 \overline{\smash{)}6x^4+0x^3-13x^2+0x-4} \\ \underline{6x^4-15x^2} \\ 2x^2+0x-4 \\ \underline{2x^2-5} \\ 1 \end{array}$$

$(6x^4-13x^2-4) \div (2x^2-5) = 3x^2+1 + \dfrac{1}{2x^2-5}$

15.
$$\begin{array}{r} x^2-3x-10 \\ 3x+1 \overline{\smash{)}3x^3-8x^2-33x-10} \\ \underline{3x^3+x^2} \\ -9x^2-33x \\ \underline{-9x^2-3x} \\ -30x-10 \\ \underline{-30x-10} \\ 0 \end{array}$$

$\dfrac{3x^3-8x^2-33x-10}{3x+1} = x^2-3x-10$

17.
$$\begin{array}{r} -x^2+2x-1 \\ x-3 \overline{\smash{)}-x^3+5x^2-7x+4} \\ \underline{-x^3+3x^2} \\ 2x^2-7x \\ \underline{2x^2-6x} \\ -x+4 \\ \underline{-x+3} \\ 1 \end{array}$$

$\dfrac{4-7x+5x^2-x^3}{x-3} = -x^2+2x-1 + \dfrac{1}{x-3}$

19.
$$\begin{array}{r} 2x^3-3x^2+x-4 \\ x-5 \overline{\smash{)}2x^4-13x^3+16x^2-9x+20} \\ \underline{2x^4-10x^3} \\ -3x^3+16x^2 \\ \underline{-3x^3+15x^2} \\ x^2-9x \\ \underline{x^2-5x} \\ -4x+20 \\ \underline{-4x+20} \\ 0 \end{array}$$

$\dfrac{16x^2-13x^3+2x^4-9x+20}{x-5} = 2x^3-3x^2+x-4$

21.
$$\begin{array}{r} x-4 \\ x^2+1 \overline{\smash{)}x^3-4x^2+2x-1} \\ \underline{x^3+x} \\ -4x^2+x-1 \\ \underline{-4x^2-4} \\ x+3 \end{array}$$

$\dfrac{x^3-4x^2+2x-1}{x^2+1} = x-4 + \dfrac{x+3}{x^2+1}$

23. $x^2-1 \overline{)2x^3-3x^2-x+4}$ with quotient $2x-3$

$\underline{2x^3-2x}$
$-3x^2+x+4$
$\underline{-3x^2+3}$
$x+1$

$$\frac{2x^3-x+4-3x^2}{x^2-1} = 2x-3+\frac{x+1}{x^2-1}$$
$$= 2x-3+\frac{1}{x-1}$$

25. $2x^2-3 \overline{)6x^3+2x^2+x+4}$ with quotient $3x+1$

$\underline{6x^3-9x}$
$2x^2+10x+4$
$\underline{2x^2-3}$
$10x+7$

$$\frac{6x^3+2x^2+x+4}{2x^2-3} = 3x+1+\frac{10x+7}{2x^2-3}$$

Objective 2 Exercises

27. $-1\ |\ \begin{array}{rrr} 2 & -6 & -8 \\ & -2 & 8 \\ \hline 2 & -8 & 0 \end{array}$

$(2x^2-6x-8) \div (x+1) = 2x-8$

29. $1\ |\ \begin{array}{rrr} 3 & 0 & -4 \\ & 3 & 3 \\ \hline 3 & 3 & -1 \end{array}$

$(3x^2-4) \div (x-1) = 3x+3-\dfrac{1}{x-1}$

31. $-4\ |\ \begin{array}{rrr} 1 & 0 & -9 \\ & -4 & 16 \\ \hline 1 & -4 & 7 \end{array}$

$(x^2-9) \div (x+4) = x-4+\dfrac{7}{x+4}$

33. $-2\ |\ \begin{array}{rrr} 1 & 0 & 12 \\ & -2 & 4 \\ \hline 1 & -2 & 16 \end{array}$

$(2x^2+24) \div (2x+4) = x-2+\dfrac{16}{x+2}$

35. $-1\ |\ \begin{array}{rrrr} 2 & -1 & 6 & 9 \\ & -2 & 3 & -9 \\ \hline 2 & -3 & 9 & 0 \end{array}$

$(2x^3-x^2+6x+9) \div (x+1) = 2x^2-3x+9$

37. $3\ |\ \begin{array}{rrr} 1 & -6 & 11 & -6 \\ & 3 & -9 & 6 \\ \hline 1 & -3 & 2 & 0 \end{array}$

$(x^3-6x^2+11x-6) \div (x-3) = x^2-3x+2$

39. $-2\ |\ \begin{array}{rrrr} 1 & -3 & 6 & -9 \\ & -2 & 10 & -32 \\ \hline 1 & -5 & 16 & -41 \end{array}$

$(6x-3x^2+x^3-9) \div (x+2)$
$= x^2-5x+16-\dfrac{41}{x+2}$

41. $-1\ |\ \begin{array}{rrrr} 1 & 0 & 1 & -2 \\ & -1 & 1 & -2 \\ \hline 1 & -1 & 2 & -4 \end{array}$

$(x^3+x-2) \div (x+1) = x^2-x+2-\dfrac{4}{x+1}$

43. $2\ |\ \begin{array}{rrrr} 4 & 0 & -1 & -18 \\ & 8 & 16 & 30 \\ \hline 4 & 8 & 15 & 12 \end{array}$

$(18+x-4x^3) \div (2-x) = 4x^2+8x+15+\dfrac{12}{x-2}$

45. $5\ |\ \begin{array}{rrrrr} 2 & -13 & 16 & -9 & 20 \\ & 10 & -15 & 5 & -20 \\ \hline 2 & -3 & 1 & -4 & 0 \end{array}$

$\dfrac{16x^2-13x^3+2x^4-9x+20}{x-5} = 2x^3-3x^2+x-4$

47. $2\ |\ \begin{array}{rrrrr} 3 & -4 & 8 & -5 & -5 \\ & 6 & 4 & 24 & 38 \\ \hline 3 & 2 & 12 & 19 & 33 \end{array}$

$\dfrac{5+5x-8x^2+4x^3-3x^4}{2-x}$
$= 3x^3+2x^2+12x+19+\dfrac{33}{x-2}$

49. $-1\ |\ \begin{array}{rrrrr} 3 & 3 & -1 & 3 & 2 \\ & -3 & 0 & 1 & -4 \\ \hline 3 & 0 & -1 & 4 & -2 \end{array}$

$\dfrac{3x^4+3x^3-x^2+3x+2}{x+1}$
$= 3x^3-x+4-\dfrac{2}{x+1}$

51.

$$\begin{array}{r|rrrrr} 3 & 2 & 0 & -1 & 0 & 2 \\ & & 6 & 18 & 51 & 153 \\ \hline & 2 & 6 & 17 & 51 & 155 \end{array}$$

$$\frac{2x^4 - x^2 + 2}{x - 3} = 2x^3 + 6x^2 + 17x + 51 + \frac{155}{x - 3}$$

53.

$$\begin{array}{r|rrrr} -5 & 1 & 0 & 0 & 125 \\ & & -5 & 25 & -125 \\ \hline & 1 & -5 & 25 & 0 \end{array}$$

$$\frac{x^3 + 125}{x + 5} = x^2 - 5x + 25$$

Objective 3 Exercises

55.

$$\begin{array}{r|rrr} 3 & 2 & -3 & -1 \\ & & 6 & 9 \\ \hline & 2 & 3 & 8 \end{array}$$

$P(3) = 8$

57.

$$\begin{array}{r|rrrr} 4 & 1 & -2 & 3 & -1 \\ & & 4 & 8 & 44 \\ \hline & 1 & 2 & 11 & 43 \end{array}$$

$R(4) = 43$

59.

$$\begin{array}{r|rrrr} -2 & 2 & -4 & 3 & -1 \\ & & -4 & 16 & -38 \\ \hline & 2 & -8 & 19 & -39 \end{array}$$

$P(-2) = -39$

61.

$$\begin{array}{r|rrrrr} 2 & 1 & 3 & -2 & 4 & -9 \\ & & 2 & 10 & 16 & 40 \\ \hline & 1 & 5 & 8 & 20 & 31 \end{array}$$

$Q(2) = 31$

63.

$$\begin{array}{r|rrrrr} -3 & 2 & -1 & 0 & 2 & -5 \\ & & -6 & 21 & -63 & 183 \\ \hline & 2 & -7 & 21 & -61 & 178 \end{array}$$

$F(-3) = 178$

65.

$$\begin{array}{r|rrrr} 5 & 1 & 0 & 0 & -3 \\ & & 5 & 25 & 125 \\ \hline & 1 & 5 & 25 & 122 \end{array}$$

$P(5) = 122$

67.

$$\begin{array}{r|rrrrr} -3 & 4 & 0 & -3 & 0 & 5 \\ & & -12 & 36 & -99 & 297 \\ \hline & 4 & -12 & 33 & -99 & 302 \end{array}$$

$R(-3) = 302$

69.

$$\begin{array}{r|rrrrrr} 2 & 1 & 0 & -4 & -2 & 5 & -2 \\ & & 2 & 4 & 0 & -4 & 2 \\ \hline & 1 & 2 & 0 & -2 & 1 & 0 \end{array}$$

$Q(2) = 0$

Applying Concepts 5.4

71.

$$\begin{array}{r} x - y \\ 3x + 2y \overline{\smash{\big)}\, 3x^2 - xy - 2y^2} \\ \underline{3x^2 + 2xy } \\ -3xy - 2y^2 \\ \underline{-3xy - 2y^2} \\ 0 \end{array}$$

$$\frac{3x^2 - xy - 2y^2}{3x + 2y} = x - y$$

73.

$$\begin{array}{r} a^2 + ab + b^2 \\ a - b \overline{\smash{\big)}\, a^3 + 0a^2b + 0ab^2 - b^3} \\ \underline{a^3 - a^2b } \\ a^2b + 0ab^2 \\ \underline{a^2b - ab^2 } \\ ab^2 - b^3 \\ \underline{ab^2 - b^3} \\ 0 \end{array}$$

$$\frac{a^3 - b^3}{a - b} = a^2 + ab + b^2$$

75.

$$\begin{array}{r} x^4 - x^3y + x^2y^2 - xy^3 + y^4 \\ x+y \overline{\smash{\big)}\, x^5 + 0x^4y + 0x^3y^2 + 0x^2y^3 + 0xy^4 + y^5} \\ \underline{x^5 + \ x^4y} \\ -x^4y + 0x^3y^2 \\ \underline{-x^4y - \ x^3y^2} \\ x^3y^2 + 0x^2y^3 \\ \underline{x^3y^2 + \ x^2y^3} \\ -x^2y^3 + 0xy^4 \\ \underline{-x^2y^3 - \ xy^4} \\ xy^4 + y^5 \\ \underline{xy^4 + y^5} \\ 0 \end{array}$$

$$\frac{x^5 + y^5}{x+y} = x^4 - x^3y + x^2y^2 - xy^3 + y^4$$

77.

$$\begin{array}{r|rrrr} 3 & 1 & -3 & -1 & k \\ & & 3 & 0 & -3 \\ \hline & 1 & 0 & -1 & k-3 \end{array}$$

$k - 3 = 0$
$k = 3$
The remainder is zero when k equals 3.

79.

$$\begin{array}{r|rrr} 3 & 1 & k & -6 \\ & & 3 & 9+3k \\ \hline & 1 & 3+k & 3+3k \end{array}$$

$3 + 3k = 0$
$3k = -3$
$k = -1$
The remainder is zero when k equals -1.

81. Note: (Quotient)(Divisor) + Remainder = Dividend
(Quotient)(Divisor) = Dividend − Remainder
$$\text{Divisor} = \frac{\text{Dividend} - \text{Remainder}}{\text{Quotient}}$$

Therefore, $\text{Divisor} = \dfrac{(x^2 + x + 2) - 14}{x + 4}$

$= \dfrac{x^2 + x - 12}{x + 4}$

$$\begin{array}{r} x - 3 \\ x+4 \overline{\smash{\big)}\, x^2 + \ x - 12} \\ \underline{x^2 + 4x} \\ -3x - 12 \\ \underline{-3x - 12} \\ 0 \end{array}$$

The polynomial is $x - 3$.

Check: $(x + 4)(x - 3) + 14 = x^2 + x - 12 + 14$
$= x^2 + x + 2$

Section 5.5

Concept Review 5.5

1. Always true

3. Sometimes true
   There are some trinomials that are nonfactorable over the integers.

Objective 1 Exercises

3. The GCF of $6a^2$ and $15a$ is $3a$.
   $6a^2 - 15a = 3a(2a - 5)$

5. The GCF of $4x^3$ and $3x^2$ is x^2.
   $4x^3 - 3x^2 = x^2(4x - 3)$

7. There is no common factor.
   $3a^2 - 10b^3$ is nonfactorable over the integers.

9. The GCF of x^5, x^3, and x is x.
   $x^5 - x^3 - x = x(x^4 - x^2 - 1)$

11. The GCF of $16x^2$, $12x$ and 24 is 4.
    $16x^2 - 12x + 24 = 4(4x^2 - 3x + 6)$

13. The GCF of $5b^2$, $10b^3$, and $25b^4$ is $5b^2$.
    $5b^2 - 10b^3 + 25b^4 = 5b^2(1 - 2b + 5b^2)$

15. The GCF of x^{2n} and x^n is x^n.
    $x^{2n} - x^n = x^n(x^n - 1)$

17. The GCF of x^{3n} and x^{2n} is x^{2n}.
    $x^{3n} - x^{2n} = x^{2n}(x^n - 1)$

19. The GCF of a^{2n+2} and a^2 is a^2.
    $a^{2n+2} + a^2 = a^2(a^{2n} + 1)$

21. The GCF of $12x^2y^2$, $18x^3y$, and $24x^2y$ is $6x^2y$.
    $12x^2y^2 - 18x^3y + 24x^2y = 6x^2y(2y - 3x + 4)$

23. The GCF of $16a^2b^4$, $4a^2b^2$, and $24a^3b^2$ is $4a^2b^2$.
    $-16a^2b^4 - 4a^2b^2 + 24a^3b^2$
    $= 4a^2b^2(-4b^2 - 1 + 6a)$

25. The GCF of y^{2n+2}, y^{n+2}, and y^2 is y^2.
    $y^{2n+2} + y^{n+2} - y^2$
    $= y^2(y^{2n} + y^n - 1)$

Objective 2 Exercises

27. $x(a+2) - 2(a+2) = (a+2)(x-2)$

29. $a(x-2) - b(2-x) = a(x-2) + b(x-2)$
    $= (x-2)(a+b)$

31. $x^2 + 3x + 2x + 6 = x(x+3) + 2(x+3)$
    $= (x+3)(x+2)$

33. $xy + 4y - 2x - 8 = y(x+4) - 2(x+4)$
    $= (x+4)(y-2)$

35. $ax + bx - ay - by = x(a+b) - y(a+b)$
    $= (a+b)(x-y)$

37. $x^2y - 3x^2 - 2y + 6 = x^2(y-3) - 2(y-3)$
    $= (y-3)(x^2 - 2)$

39. $6 + 2y + 3x^2 + x^2y = 2(3+y) + x^2(3+y)$
    $= (3+y)(2+x^2)$

41. $2ax^2 + bx^2 - 4ay - 2by = x^2(2a+b) - 2y(2a+b)$
    $= (2a+b)(x^2 - 2y)$

43. $x^n y - 5x^n + y - 5 = x^n(y-5) + (y-5)$
    $= (y-5)(x^n + 1)$

45. $x^3 + x^2 + 2x + 2 = x^2(x+1) + 2(x+1)$
    $= (x+1)(x^2 + 2)$

47. $2x^3 - x^2 + 4x - 2 = x^2(2x-1) + 2(2x-1)$
    $= (2x-1)(x^2 + 2)$

Objective 3 Exercises

51. $x^2 - 8x + 15 = (x-5)(x-3)$

53. $a^2 + 12a + 11 = (a+11)(a+1)$

55. $b^2 + 2b - 35 = (b+7)(b-5)$

57. $y^2 - 16y + 39 = (y-3)(y-13)$

59. $b^2 + 4b - 32 = (b+8)(b-4)$

61. $a^2 - 15a + 56 = (a-7)(a-8)$

63. $y^2 + 13y + 12 = (y+12)(y+1)$

65. $x^2 + 4x - 5 = (x+5)(x-1)$

67. $a^2 + 11ab + 30b^2 = (a+6b)(a+5b)$

69. $x^2 - 14xy + 24y^2 = (x-12y)(x-2y)$

71. $y^2 + 2xy - 63x^2 = (y+9x)(y-7x)$

73. $x^2 - 35x - 36 = (x-36)(x+1)$

75. $a^2 + 13a + 36 = (a+9)(a+4)$

77. There are no binomial factors whose product $x^2 - 7x - 12$. The trinomial is nonfactorable over the integers.

Objective 4 Exercises

79. $2x^2 - 11x - 40 = (2x+5)(x-8)$

81. $4y^2 - 15y + 9 = (4y-3)(y-3)$

83. $2a^2 + 13a + 6 = (2a+1)(a+6)$

85. There are no binomial factors whose product is $12y^2 - 13y - 72$. The trinomial is nonfactorable over the integers.

87. $5x^2 + 26x + 5 = (5x+1)(x+5)$

89. $11x^2 - 122x + 11 = (11x-1)(x-11)$

91. $12x^2 - 17x + 5 = (12x-5)(x-1)$

93. $8y^2 - 18y + 9 = (4y-3)(2y-3)$

95. There are no binomial factors whose product is $6a^2 - 5a - 2$. The trinomial is nonfactorable over the integers.

97. There are no binomial factors whose product is $2x^2 + 5x + 12$. The trinomial is nonfactorable over the integers.

99. $6x^2 + 5xy - 21y^2 = (2x-3y)(3x+7y)$

101. $4a^2 + 43ab + 63b^2 = (4a+7b)(a+9b)$

103. $10x^2 - 23xy + 12y^2 = (5x-4y)(2x-3y)$

105. $24 + 13x - 2x^2 = (8-x)(3+2x)$

107. There are no binomial factors whose product is $8 - 13x + 6x^2$. The trinomial is nonfactorable over the integers.

109. $15 - 14a - 8a^2 = (3-4a)(5+2a)$

111. The GCF of $5y^4$, $29y^3$, and $20y^2$ is y^2.
$5y^4 - 29y^3 + 20y^2 = y^2(5y^2 - 29y + 20)$
$= y^2(5y-4)(y-5)$

113. The GCF of $4x^3$, $10x^2y$ and $24xy^2$ is $2x$.
$4x^3 + 10x^2y - 24xy^2 = 2x(2x^2 + 5xy - 12y^2)$
$= 2x(2x-3y)(x+4y)$

115. The GCF of 100, $5x$, and $5x^2$ is 5.
$100 - 5x - 5x^2 = 5(20 - x - x^2)$
$= 5(5+x)(4-x)$

117. The GCF of $320x$, $8x^2$, and $4x^3$ is $4x$.
$320x - 8x^2 - 4x^3 = 4x(80 - 2x - x^2)$
$= 4x(10+x)(8-x)$

119. The GCF of $20x^2$, $38x^3$, and $30x^4$ is $2x^2$.
$20x^2 - 38x^3 - 30x^4 = 2x^2(10 - 19x - 15x^2)$
$= 2x^2(5+3x)(2-5x)$

121. The GCF of a^4b^4, $3a^3b^3$, and $10a^2b^2$ is a^2b^2.
$a^4b^4 - 3a^3b^3 - 10a^2b^2 = a^2b^2(a^2b^2 - 3ab - 10)$
$= a^2b^2(ab-5)(ab+2)$

123. The GCF of $90a^2b^2$, $45ab$, and 10 is 5.
$90a^2b^2 + 45ab + 10 = 5(18a^2b^2 + 9ab + 2)$

125. There is no common factor.
$4x^4 - 45x^2 + 80$ is nonfactorable over the integers.

127. The GCF of $2a^5$, $14a^3$, and $20a$ is $2a$.
$2a^5 + 14a^3 + 20a = 2a(a^4 + 7a^2 + 10)$
$= 2a(a^2 + 5)(a^2 + 2)$

129. The GCF of $3x^4y^2$, $39x^2y^2$, and $120y^2$ is $3y^2$.
$3x^4y^2 - 39x^2y^2 + 120y^2 = 3y^2(x^4 - 13x^2 + 40)$
$= 3y^2(x^2 - 5)(x^2 - 8)$

131. The GCF of $45a^2b^2$, $6ab^2$, and $72b^2$ is $3b^2$.
$45a^2b^2 + 6ab^2 - 72b^2 = 3b^2(15a^2 + 2a - 24)$
$= 3b^2(3a+4)(5a-6)$

133. The GCF of $36x^3y$, $24x^2y^2$, and $45xy^3$ is $3xy$.
$36x^3y + 24x^2y^2 - 45xy^3$
$= 3xy(12x^2 + 8xy - 15y^2)$
$= 3xy(6x-5y)(2x+3y)$

135. The GCF of $48a^2b^2$, $36ab^3$, and $54b^4$ is $6b^2$.
$48a^2b^2 - 36ab^3 - 54b^4$
$= 6b^2(8a^2 - 6ab - 9b^2)$
$= 6b^2(2a-3b)(4a+3b)$

137. The GCF of $10x^{2n}$, $25x^n$, and 60 is 5.
$10x^{2n} + 25x^n - 60 = 5(2x^{2n} + 5x^n - 12)$
$= 5(2x^n - 3)(x^n + 4)$

Applying Concepts 5.5

139. $3x^3y - xy^3 - 2x^2y^2 = 3x^3y - 2x^2y^2 - xy^3$
$= xy(3x^2 - 2xy - y^2)$
$= xy(3x+y)(x-y)$

141. $9b^3 + 3b^5 - 30b = 3b^5 + 9b^3 - 30b$
$= 3b(b^4 + 3b^2 - 10)$
$= 3b(b^2 + 5)(b^2 - 2)$

143. $x^2 + kx - 6$

Factors of –6	Sum
1, –6	–5
–1, 6	5
2, –3	–1
–2, 3	1

The possible integer values of k are 5, –5, 1, and –1.

145. $2x^2 - kx - 5$

$2x^2 + (-k)x - 5$

Factors of 2	Factors of –5
1, 2	1, –5
	–1, 5

Trial Factors	Middle Term
$(x + 1)(2x - 5)$	$-5x + 2x = -3x$
$(x - 5)(2x + 1)$	$x - 10x = -9x$
$(x - 1)(2x + 5)$	$5x - 2x = 3x$
$(x + 5)(2x - 1)$	$-x + 10x = 9x$

The possible integer values of $-k$ are 3, –3, 9, and –9.

The possible integer values of k are 3, 9, –3, and –9.

147. $2x^2 + kx - 3$

Factors of 2	Factors of –3
1, 2	1, –3
	–1, 3

Trial Factors	Middle Term
$(x + 1)(2x - 3)$	$-3x + 2x = -x$
$(x - 3)(2x + 1)$	$x - 6x = -5x$
$(x - 1)(2x + 3)$	$3x - 2x = x$
$(x + 3)(2x - 1)$	$-x + 6x = 5x$

The possible integer values of k are –1, –5, 1, and 5.

Section 5.6

Concept Review 5.6

1. Never true
   $a^2b^2c^2$ is a monomial.

3. Never true
   The difference of two perfect cubes is always factorable. For example,
   $8^2 - x^3 = 2^3 - x^3$
   $= (2 - x)(4 + 2x + x^2)$

5. Never true
   $b^3 + 1$ is the sum of two perfect cubes.
   $b^3 + 1^3 = (b + 1)(b^2 - b + 1)$

Objective 1 Exercises

3. $x^2 - 16 = x^2 - 4^2$
$= (x + 4)(x - 4)$

5. $4x^2 - 1 = (2x)^2 - 1^2$
$= (2x + 1)(2x - 1)$

7. $b^2 - 2b + 1 = (b - 1)^2$

9. $16x^2 - 40x + 25 = (4x - 5)^2$

11. $x^2y^2 - 100 = (xy)^2 - 10^2$
$= (xy + 10)(xy - 10)$

13. $x^2 + 4$ is nonfactorable over the integers.

15. $x^2 + 6xy + 9y^2 = (x + 3y)^2$

17. $4x^2 - y^2 = (2x)^2 - 4^2 = (2x + y)(2x - y)$

19. $a^{2n} - 1 = (a^n)^2 - 1^2$
$= (a^n + 1)(a^n - 1)$

21. $a^2 + 4a + 4 = (a + 2)^2$

23. $x^2 - 12x + 36 = (x - 6)^2$

25. $16x^2 - 121 = (4x)^2 - 11^2$
$= (4x + 11)(4x - 11)$

27. $1 - 9a^2 = 1^2 - (3a)^2$
$= (1 + 3a)(1 - 3a)$

29. $4a^2 + 4a - 1$ is nonfactorable over the integers.

31. $b^2 + 7b + 14$ is nonfactorable over the integers.

33. $25 - a^2b^2 = 5^2 - (ab)^2$
$= (5 + ab)(5 - ab)$

35. $25a^2 - 40ab + 16b^2 = (5a - 4b)^2$

37. $x^{2n} + 6x^n + 9 = (x^n + 3)^2$

Objective 2 Exercises

39. $x^3 - 27 = x^3 - 3^3$
$= (x - 3)(x^2 + 3x + 9)$

41. $8x^3 - 1 = (2x)^3 - 1^3$
$= (2x - 1)(4x^2 + 2x + 1)$

43. $x^3 - y^3 = (x - y)(x^2 + xy + y^2)$

45. $m^3 + n^3 = (m + n)(m^2 - mn + n^2)$

47. $64x^3 + 1 = (4x)^3 + 1^3$
$= (4x + 1)(16x^2 - 4x + 1)$

49. $27x^3 - 8y^3 = (3x)^3 - (2y)^3$
$= (3x - 2y)(9x^2 + 6xy + 4y^2)$

51. $x^3y^3 + 64 = (xy)^3 + 4^3$
   $= (xy+4)(x^2y^2 - 4xy + 16)$

53. $16x^3 - y^3$ is nonfactorable over the integers.

55. $8x^3 - 9y^3$ is nonfactorable over the integers.

57. $(a-b)^3 - b^3$
   $= [(a-b)-b][(a-b)^2 + b(a-b) + b^2]$
   $= (a-2b)(a^2 - 2ab + b^2 + ab - b^2 + b^2)$
   $= (a-2b)(a^2 - ab + b^2)$

59. $x^{6n} + y^{3n} = (x^{2n})^3 + (y^n)^3$
   $= (x^{2n} + y^n)(x^{4n} - x^{2n}y^n + y^{2n})$

Objective 3 Exercises

63. Let $u = xy$.
   $x^2y^2 - 8xy + 15 = u^2 - 8u + 15$
   $= (u-3)(u-5)$
   $= (xy-3)(xy-5)$

65. Let $u = xy$.
   $x^2y^2 - 17xy + 60 = u^2 - 17u + 60$
   $= (u-5)(u-12)$
   $= (xy-5)(xy-12)$

67. Let $u = x^2$.
   $x^4 - 9x^2 + 18 = u^2 - 9u + 18$
   $= (u-6)(u-3)$
   $= (x^2-6)(x^2-3)$

69. Let $u = b^2$.
   $b^4 - 13b^2 - 90 = u^2 - 13u - 90$
   $= (u-18)(u+5)$
   $= (b^2-18)(b^2+5)$

71. Let $u = x^2y^2$.
   $x^4y^4 - 8x^2y^2 + 12 = u^2 - 8u + 12$
   $= (u-6)(u-2)$
   $= (x^2y^2-6)(x^2y^2-2)$

73. Let $u = x^n$.
   $x^{2n} + 3x^n + 2 = u^2 + 3u + 2$
   $= (u+2)(u+1)$
   $= (x^n+2)(x^n+1)$

75. Let $u = xy$.
   $3x^2y^2 - 14xy + 15 = 3u^2 - 14u + 15$
   $= (3u-5)(u-3)$
   $= (3xy-5)(xy-3)$

77. Let $u = ab$.
   $6a^2b^2 - 23ab + 21 = 6u^2 - 23u + 21$
   $= (2u-3)(3u-7)$
   $= (2ab-3)(3ab-7)$

79. Let $u = x^2$.
   $2x^4 - 13x^2 - 15 = 2u^2 - 13u - 15$
   $= (2u-15)(u+1)$
   $= (2x^2-15)(x^2+1)$

81. Let $u = x^n$.
   $2x^{2n} - 7x^n + 3 = 2u^2 - 7u + 3$
   $= (2u-1)(u-3)$
   $= (2x^n-1)(x^n-3)$

83. Let $u = a^n$.
   $6a^{2n} + 19a^n + 10 = 6u^2 + 19u + 10$
   $= (2u+5)(3u+2)$
   $= (2a^n+5)(3a^n+2)$

Objective 4 Exercises

85. $12x^2 - 36x + 27 = 3(4x^2 - 12x + 9)$
   $= 3(2x-3)^2$

87. $27a^4 - a = a(27a^3 - 1)$
   $= a(3a-1)(9a^2 + 3a + 1)$

89. $20x^2 - 5 = 5(4x^2 - 1)$
   $= 5(2x+1)(2x-1)$

91. $y^5 + 6y^4 - 55y^3 = y^3(y^2 + 6y - 55)$
   $= y^3(y+11)(y-5)$

93. $16x^4 - 81 = (4x^2+9)(4x^2-9)$
   $= (4x^2+9)(2x+3)(2x-3)$

95. $16a - 2a^4 = 2a(8 - a^3)$
   $= 2a(2-a)(4 + 2a + a^2)$

97. $x^3 + 2x^2 - x - 2 = x^2(x+2) - 1(x+2)$
   $= (x+2)(x^2-1)$
   $= (x+2)(x+1)(x-1)$

99. $2x^3 + 4x^2 - 3x - 6 = 2x^2(x+2) - 3(x+2)$
   $= (x+2)(2x^2-3)$

101. $x^3 + x^2 - 16x - 16 = x^2(x+1) - 16(x+1)$
   $= (x+1)(x^2-16)$
   $= (x+1)(x+4)(x-4)$

103. $a^3b^6 - b^3 = b^3(a^3b^3 - 1)$
   $= b^3(ab-1)(a^2b^2 + ab + 1)$

105. $x^4 - 2x^3 - 35x^2 = x^2(x^2 - 2x - 35)$
   $= x^2(x-7)(x+5)$

107. $4x^2 + 4x - 1$ is nonfactorable over the integers.

109. $6x^5 + 74x^4 + 24x^3 = 2x^3(3x^2 + 37x + 12)$
   $= 2x^3(3x+1)(x+12)$

111. $16a^4 - b^4 = (4a^2 + b^2)(4a^2 - b^2)$
$= (4a^2 + b^2)(2a+b)(2a-b)$

113. $x^4 - 5x^2 - 4$ is nonfactorable over the integers.

115. $3b^5 - 24b^2 = 3b^2(b^3 - 8)$
$= 3b^2(b-2)(b^2 + 2b + 4)$

117. $x^4y^2 - 5x^3y^3 + 6x^2y^4 = x^2y^2(x^2 - 5xy + 6y^2)$
$= x^2y^2(x - 3y)(x - 2y)$

119. $16x^3y + 4x^2y^2 - 42xy^3$
$= 2xy(8x^2 + 2xy - 21y^2)$
$= 2xy(4x + 7y)(2x - 3y)$

121. $x^3 - 2x^2 - x + 2 = x^2(x - 2) - (x - 2)$
$= (x - 2)(x^2 - 1)$
$= (x - 2)(x + 1)(x - 1)$

123. $8xb - 8x - 4b + 4 = 4(2xb - 2x - b + 1)$
$= 4[2x(b - 1) - (b - 1)]$
$= 4(b - 1)(2x - 1)$

125. $4x^2y^2 - 4x^2 - 9y^2 + 9$
$= 4x^2(y^2 - 1) - 9(y^2 - 1)$
$= (y^2 - 1)(4x^2 - 9)$
$= (y + 1)(y - 1)(2x + 3)(2x - 3)$

127. $x^5 - 4x^3 - 8x^2 + 32$
$= x^3(x^2 - 4) - 8(x^2 - 4)$
$= (x^2 - 4)(x^3 - 8)$
$= (x + 2)(x - 2)(x - 2)(x^2 + 2x + 4)$
$= (x + 2)(x - 2)^2(x^2 + 2x + 4)$

129. $a^{2n+2} - 6a^{n+2} + 9a^2 = a^2(a^{2n} - 6a^n + 9)$
$= a^2(a^n - 3)^2$

131. $2x^{n+2} - 7x^{n+1} + 3x^n = x^n(2x^2 - 7x + 3)$
$= x^n(2x - 1)(x - 3)$

Applying Concepts 5.6

133. $4x^2 - kx + 25 = (2x + 5)^2$ or $(2x - 5)^2$
$(2x + 5)(2x + 5) = 4x^2 + 20x + 25$
$(2x - 5)(2x - 5) = 4x^2 - 20x + 25$
The possible values of $-k$ are 20 and -20.
The possible values of k are -20 and 20.

135. $16x^2 + kxy + y^2 = (4x + y)^2$ or $(4x - y)^2$
$(4x + y)(4x + y) = 16x^2 + 8xy + y^2$
$(4x - y)(4x - y) = 16x^2 - 8xy + y^2$
The possible values of k are 8 and -8.

137. $ax^3 + b - bx^3 - a = (ax^3 - bx^3) + (b - a)$
$= x^3(a - b) + [-(a - b)]$
$= x^3(a - b) - (a - b)$
$= (a - b)(x^3 - 1)$
$= (a - b)(x^3 - 1^3)$
$= (a - b)(x - 1)(x^2 + x + 1)$

139. $y^{8n} - 2y^{4n} + 1 = (y^{4n} - 1)^2$
$= [(y^{2n})^2 - 1][(y^{2n})^2 - 1]$
$= (y^{2n} + 1)(y^{2n} - 1)(y^{2n} + 1)(y^{2n} - 1)$
$= (y^{2n} + 1)^2(y^{2n} - 1)^2$
$= (y^{2n} + 1)^2[(y^n)^2 - 1]^2$
$= (y^{2n} + 1)^2[(y^n + 1)(y^n - 1)]^2$
$= (y^{2n} + 1)^2(y^n + 1)^2(y^n - 1)^2$

141. One number is a perfect square less than 63: 1, 4, 9, 16, 25, 36, 49
One number is a prime number less than 63: 2, 3, 5, 7, 11, 13, 17, 19, 23, 29, 31, 37, 41, 43, 47, 53, 59, 61
The product of the numbers is 63: $9 \cdot 7 = 63$
$9 + 7 = 16$
The sum of the numbers is 16.

143. Perfect squares less than 500

1	144
4	169
9	196
16	225
25	256
36	289
49	324
64	361
81	400
100	441
121	484

The palindromic perfect squares less than 500 are 1, 4, 9, 121, and 484.

145. $x^4 + 64$
$= (x^4 + 16x^2 + 64) - 16x^2$
$= (x^2 + 8)(x^2 + 8) - 16x^2$
$= (x^2 + 8)^2 - 16x^2$
$= (x^2 + 8 - 4x)(x^2 + 8 + 4x)$
$= (x^2 - 4x + 8)(x^2 + 4x + 8)$

Section 5.7

Concept Review 5.7

1. Always true

3. Sometimes true
The equation $(x - 2)(x + 3)(x - 1) = 0$ has three solutions.

5. Sometimes true
If $n = 1$, then $n + 1 = 2$, and $n + 3 = 4$, which are even integers.

Objective 1 Exercises

3. $(y + 4)(y + 6) = 0$
$y + 4 = 0 \quad y + 6 = 0$
$y = -4 \quad y = -6$
The solutions are −4 and −6.

5. $x(x - 7) = 0$
$x = 0 \quad x - 7 = 0$
$\quad\quad\quad\quad x = 7$
The solutions are 0 and 7.

7. $3z(2z + 5) = 0$
$3z = 0 \quad 2z + 5 = 0$
$z = 0 \quad 2z = -5$
$\quad\quad\quad\quad z = -\dfrac{5}{2}$
The solutions are 0 and $-\dfrac{5}{2}$.

9. $(2x + 3)(x - 7) = 0$
$2x + 3 = 0 \quad x - 7 = 0$
$2x = -3 \quad x = 7$
$x = -\dfrac{3}{2}$
The solutions are $-\dfrac{3}{2}$ and 7.

11. $b^2 - 49 = 0$
$b^2 - 7^2 = 0$
$(b + 7)(b - 7) = 0$
$b + 7 = 0 \quad b - 7 = 0$
$b = -7 \quad b = 7$
The solutions are 7 and −7.

13. $9t^2 - 16 = 0$
$(3t)^2 - 4^2 = 0$
$(3t + 4)(3t - 4) = 0$
$3t + 4 = 0 \quad 3t - 4 = 0$
$3t = -4 \quad 3t = 4$
$t = -\dfrac{4}{3} \quad t = \dfrac{4}{3}$
The solutions are $\dfrac{4}{3}$ and $-\dfrac{4}{3}$.

15. $y^2 + 4y - 5 = 0$
$(y + 5)(y - 1) = 0$
$y + 5 = 0 \quad y - 1 = 0$
$y = -5 \quad y = 1$
The solutions are −5 and 1.

17. $2b^2 - 5b - 12 = 0$
$(2b + 3)(b - 4) = 0$
$2b + 3 = 0 \quad b - 4 = 0$
$2b = -3 \quad b = 4$
$b = -\dfrac{3}{2}$
The solutions are $-\dfrac{3}{2}$ and 4.

19. $x^2 - 9x = 0$
$x(x - 9) = 0$
$x = 0 \quad x - 9 = 0$
$\quad\quad\quad\quad x = 9$
The solutions are 0 and 9.

21. $3a^2 - 12a = 0$
$3a(a - 4) = 0$
$3a = 0 \quad a - 4 = 0$
$a = 0 \quad a = 4$
The solutions are 0 and 4.

23. $z^2 - 3z = 28$
$z^2 - 3z - 28 = 0$
$(z-7)(z+4) = 0$
$z - 7 = 0 \quad z + 4 = 0$
$z = 7 \quad z = -4$
The solutions are 7 and –4.

25. $3t^2 + 13t = 10$
$3t^2 + 13t - 10 = 0$
$(3t - 2)(t + 5) = 0$
$3t - 2 = 0 \quad t + 5 = 0$
$3t = 2 \quad t = -5$
$t = \frac{2}{3}$
The solutions are $\frac{2}{3}$ and –5.

27. $5b^2 - 17b = -6$
$5b^2 - 17b + 6 = 0$
$(5b - 2)(b - 3) = 0$
$5b - 2 = 0 \quad b - 3 = 0$
$5b = 2 \quad b = 3$
$b = \frac{2}{5}$
The solutions are $\frac{2}{5}$ and 3.

29. $8x^2 - 10x = 3$
$8x^2 - 10x - 3 = 0$
$(2x - 3)(4x + 1) = 0$
$2x - 3 = 0 \quad 4x + 1 = 0$
$2x = 3 \quad 4x = -1$
$x = \frac{3}{2} \quad x = -\frac{1}{4}$
The solutions are $\frac{3}{2}$ and $-\frac{1}{4}$.

31. $y(y - 2) = 35$
$y^2 - 2y = 35$
$y^2 - 2y - 35 = 0$
$(y - 7)(y + 5) = 0$
$y - 7 = 0 \quad y + 5 = 0$
$y = 7 \quad y = -5$
The solutions are 7 and –5.

33. $x(x - 12) = -27$
$x^2 - 12x = -27$
$x^2 - 12x + 27 = 0$
$(x - 9)(x - 3) = 0$
$x - 9 = 0 \quad x - 3 = 0$
$x = 9 \quad x = 3$
The solutions are 9 and 3.

35. $y(3y - 2) = 8$
$3y^2 - 2y = 8$
$3y^2 - 2y - 8 = 0$
$(3y + 4)(y - 2) = 0$
$3y + 4 = 0 \quad y - 2 = 0$
$3y = -4 \quad y = 2$
$y = -\frac{4}{3}$
The solutions are $-\frac{4}{3}$ and 2.

37. $3a^2 - 4a = 20 - 15a$
$3a^2 + 11a = 20$
$3a^2 + 11a - 20 = 0$
$(3a - 4)(a + 5) = 0$
$3a - 4 = 0 \quad a + 5 = 0$
$3a = 4 \quad a = -5$
$a = \frac{4}{3}$
The solutions are $\frac{4}{3}$ and –5.

39. $(y + 5)(y - 7) = -20$
$y^2 - 2y - 35 = -20$
$y^2 - 2y - 15 = 0$
$(y - 5)(y + 3) = 0$
$y - 5 = 0 \quad y + 3 = 0$
$y = 5 \quad y = -3$
The solutions are 5 and –3.

41. $(b + 5)(b + 10) = 6$
$b^2 + 15b + 50 = 6$
$b^2 + 15b + 44 = 0$
$(b + 11)(b + 4) = 0$
$b + 11 = 0 \quad b + 4 = 0$
$b = -11 \quad b = -4$
The solutions are –11 and –4.

43. $(t - 3)^2 = 1$
$t^2 - 6t + 9 = 1$
$t^2 - 6t + 8 = 0$
$(t - 2)(t - 4) = 0$
$t - 2 = 0 \quad t - 4 = 0$
$t = 2 \quad t = 4$
The solutions are 2 and 4.

45. $(3 - x)^2 + x^2 = 5$
$9 - 6x + x^2 + x^2 = 5$
$2x^2 - 6x + 9 = 5$
$2x^2 - 6x + 4 = 0$
$2(x^2 - 3x + 2) = 0$
$2(x - 1)(x - 2) = 0$
$x - 1 = 0 \quad x - 2 = 0$
$x = 1 \quad x = 2$
The solutions are 1 and 2.

47. $(a-1)^2 = 3a - 5$
$a^2 - 2a + 1 = 3a - 5$
$a^2 - 5a + 1 = -5$
$a^2 - 5a + 6 = 0$
$(a-2)(a-3) = 0$
$a - 2 = 0 \quad a - 3 = 0$
$a = 2 \quad\quad a = 3$
The solutions are 2 and 3.

49. $x^3 + 4x^2 - x - 4 = 0$
$x^2(x+4) - 1(x+4) = 0$
$(x+4)(x^2 - 1) = 0$
$(x+4)(x+1)(x-1) = 0$
$x + 4 = 0 \quad x + 1 = 0 \quad x - 1 = 0$
$x = -4 \quad\quad x = -1 \quad\quad x = 1$
The solutions are $-4, -1,$ and 1.

51. $f(c) = 1$
$c^2 - 3c + 3 = 1$
$c^2 - 3c + 2 = 0$
$(c-2)(c-1) = 0$
$c - 2 = 0 \quad c - 1 = 0$
$c = 2 \quad\quad c = 1$
The values of c are 1 and 2.

53. $f(c) = -4$
$2c^2 - c - 5 = -4$
$2c^2 - c - 1 = 0$
$(2c+1)(c-1) = 0$
$2c + 1 = 0 \quad c - 1 = 0$
$2c = -1 \quad\quad c = 1$
$c = -\dfrac{1}{2}$
The values of c are $-\dfrac{1}{2}$ and 1.

55. $f(c) = 2$
$4c^2 - 4c + 3 = 2$
$4c^2 - 4c + 1 = 0$
$(2c-1)(2c-1) = 0$
$2c - 1 = 0 \quad 2c - 1 = 0$
$2c = 1 \quad\quad 2c = 1$
$c = \dfrac{1}{2} \quad\quad c = \dfrac{1}{2}$
The value of c is $\dfrac{1}{2}$.

57. $f(c) = -5$
$c^3 + 9c^2 - c - 14 = -5$
$c^3 + 9c^2 - c - 9 = 0$
$c^2(c+9) - (c+9) = 0$
$(c+9)(c^2 - 1) = 0$
$(c+9)(c+1)(c-1) = 0$
$c + 9 = 0 \quad c + 1 = 0 \quad c - 1 = 0$
$c = -9 \quad\quad c = -1 \quad\quad c = 1$
The values of c are $-9, -1,$ and 1.

Objective 2 Exercises

59. Strategy
This is an integer problem.
The unknown integer is n.
The sum of the integer and its square is 90.
$n + n^2 = 90$

Solution
$n + n^2 = 90$
$n^2 + n - 90 = 0$
$(n+10)(n-9) = 0$
$n + 10 = 0 \quad n - 9 = 0$
$n = -10 \quad\quad n = 9$
The integer is -10 or 9.

61. Strategy
This is an integer problem.
The first positive integer is n.
The next consecutive positive integer is $n + 1$.
The sum of the squares of the two consecutive positive integers is equal to 145.
$n^2 + (n+1)^2 = 145$

Solution
$n^2 + (n+1)^2 = 145$
$n^2 + n^2 + 2n + 1 = 145$
$2n^2 + 2n - 144 = 0$
$2(n^2 + n - 72) = 0$
$2(n+9)(n-8) = 0$
$n + 9 = 0 \quad n - 8 = 0$
$n = -9 \quad\quad n = 8$
The solutions -9 and -8 are not possible because they are not positive. The integers are 8 and 9.

63. Strategy
This is an integer problem.
The unknown integer is n.
The sum of the cube of the integer and the product of the integer and twelve is equal to seven times the square of the integer.
$x^3 + 12x = 7x^2$

Solution
$$x^3 + 12x = 7x^2$$
$$x^3 - 7x^2 + 12x = 0$$
$$x(x^2 - 7x + 12) = 0$$
$$x(x-3)(x-4) = 0$$
$$x = 0 \quad x - 3 = 0 \quad x - 4 = 0$$
$$\quad x = 3 \quad\quad x = 4$$
The integer is 0, 3, or 4.

65. Strategy
This is a geometry problem.
The width of the rectangle: x
The length of the rectangle: $2x + 5$

The area of the rectangle is 168 in^2. Use the equation for the area of a rectangle ($A = L \cdot W$).

Solution
$$A = L \cdot W$$
$$168 = (2x + 5)x$$
$$168 = 2x^2 + 5x$$
$$0 = 2x^2 + 5x - 168$$
$$0 = (2x + 21)(x - 8)$$
$$2x + 21 = 0 \quad\quad x - 8 = 0$$
$$2x = -21 \quad\quad x = 8$$
$$x = -\frac{21}{2}$$
Since the width cannot be negative, $-\frac{21}{2}$ cannot be a solution.
$2x + 5 = 2(8) + 5 = 21$
The width is 8 in. The length is 21 in.

67. Strategy
Length of the trough: 6
Width of the trough: $10 - 2x$
Height of the trough: x

The volume of the trough is 72 m^3.
Use the equation for the volume of a rectangular solid ($V = LWH$).

Solution
$$V = LWH$$
$$72 = 6(10 - 2x)(x)$$
$$72 = (60 - 12x)x$$
$$72 = 60x - 12x^2$$
$$12x^2 - 60x + 72 = 0$$
$$12(x^2 - 5x + 6) = 0$$
$$12(x - 3)(x - 2) = 0$$
$$x - 3 = 0 \quad x - 2 = 0$$
$$x = 3 \quad\quad x = 2$$
The value of x should be 2 m or 3 m.

69. Strategy
To find the velocity of the rocket, substitute 500 for s in the equation and solve for v.

Solution
$$v^2 = 20s$$
$$v^2 = 20(500)$$
$$v^2 = 10,000$$
$$v^2 - 10,000 = 0$$
$$(v - 100)(v + 100) = 0$$
$$v - 100 = 0 \quad\quad v + 100 = 0$$
$$v = 100 \quad\quad v = -100$$
Since the rocket is traveling in the direction it was launched, the velocity is not negative, so –100 is not a solution. The velocity of the rocket is 100 m/s.

71. Strategy
This is a geometry problem.
The height of the triangle: x
The base of the triangle: $3x$

The area of the triangle is 24 cm^2. Use the equation for the area of a triangle $\left(A = \frac{1}{2}bh\right)$.

Solution
$$A = \frac{1}{2}bh$$
$$24 = \frac{1}{2}(3x)x$$
$$24 = \frac{1}{2}(3x^2)$$
$$48 = 3x^2$$
$$0 = 3x^2 - 48$$
$$0 = 3(x^2 - 16)$$
$$0 = 3(x + 4)(x - 4)$$
$$x + 4 = 0 \quad\quad x - 4 = 0$$
$$x = -4 \quad\quad x = 4$$
Since the height cannot be negative, –4 cannot be a solution.
$3x = 3(4) = 12$
The height is 4 cm. The base is 12 cm.

73. Strategy
To find the time for the object to reach the ground, replace the variables d and v in the equation by their given values and solve for t.

Solution
$$d = vt + 16t^2$$
$$480 = 16t + 16t^2$$
$$0 = 16t^2 + 16t - 480$$
$$0 = 16(t^2 + t - 30)$$
$$0 = t^2 + t - 30$$
$$0 = (t+6)(t-5)$$
$$t+6 = 0 \quad t-5 = 0$$
$$t = -6 \quad t = 5$$
Since the time cannot be a negative number, −6 is not a solution. The time is 5 s.

75. Strategy
Increase in length and width: x
Length of the larger rectangle: $6 + x$
Width of the larger rectangle: $3 + x$
Use the equation $A = L \cdot W$.

Solution
Smaller rectangle:
$A = L \cdot W$
$A = 6 \cdot 3 = 18$
Larger rectangle:
$A = L \cdot W$
$$18 + 70 = (6+x)(3+x)$$
$$88 = 18 + 9x + x^2$$
$$0 = x^2 + 9x - 70$$
$$0 = (x+14)(x-5)$$
$$x+14 = 0 \quad x-5 = 0$$
$$x = -14 \quad x = 5$$
Since an increase in length and width cannot be a negative number, −14 is not a solution.
Length = $6 + x = 6 + 5 = 11$
Width = $3 + x = 3 + 5 = 8$
The length of the larger rectangle is 11 cm. The width is 8 cm.

Applying Concepts 5.7

77. $x^2 + 3ax - 10a^2 = 0$
$(x - 2a)(x + 5a) = 0$
$x - 2a = 0 \quad x + 5a = 0$
$x = 2a \quad x = -5a$
The solutions are $2a$ and $-5a$.

79. $x^2 - 9a^2 = 0$
$x^2 - (3a)^2 = 0$
$(x - 3a)(x + 3a) = 0$
$x - 3a = 0 \quad x + 3a = 0$
$x = 3a \quad x = -3a$
The solutions are $3a$ and $-3a$.

81. $x^2 - 5ax = 15a^2 - 3ax$
$x^2 - 2ax = 15a^2$
$x^2 - 2ax - 15a^2 = 0$
$(x - 5a)(x + 3a) = 0$
$x - 5a = 0 \quad x + 3a = 0$
$x = 5a \quad x = -3a$
The solutions are $5a$ and $-3a$.

83. Strategy
Solve the equation $n(n + 6) = 16$ for n, and then substitute the values of n into the expression $3n^2 + 2n - 1$ and evaluate.

Solution
$$n(n+6) = 16$$
$$n^2 + 6n = 16$$
$$n^2 + 6n - 16 = 0$$
$$(n+8)(n-2) = 0$$
$$n+8 = 0 \quad n-2 = 0$$
$$n = -8 \quad n = 2$$

$3n^2 + 2n - 1$ $\quad$ $3n^2 + 2n - 1$
$= 3(-8)^2 + 2(-8) - 1$ $\quad$ $= 3(2)^2 + 2(2) - 1$
$= 3(64) + 2(-8) - 1$ $\quad$ $= 3(4) + 2(2) - 1$
$= 192 - 16 - 1$ $\quad$ $= 12 + 4 - 1$
$= 175$ $\quad$ $= 15$

$3n^2 + 2n - 1$ is equal to 175 or 15.

85. Strategy
Width of rectangular piece of cardboard: x
Length of rectangular piece of cardboard: $x + 10$
Width of cardboard box: $x - 4$
Length of cardboard box: $(x + 10) - 4 = x + 6$
Height of cardboard box: 2
The formula for the volume of a rectangular solid is $V = LWH$.

Solution
$$V = LWH$$
$$112 = (x+6)(x-4)(2)$$
$$56 = (x+6)(x-4)$$
$$56 = x^2 + 2x - 24$$
$$0 = x^2 + 2x - 80$$
$$0 = (x+10)(x-8)$$
$$x+10 = 0 \quad x-8 = 0$$
$$x = -10 \quad x = 8$$
Since width cannot be a negative number, −10 is not a solution.
Length = $x + 10 = 8 + 10 = 18$
The length is 18 in. The width is 8 in.

87.
$$(x-3)(x-2)(x+1) = 0$$
$$(x^2 - 5x + 6)(x+1) = 0$$
$$(x^3 - 5x^2 + 6x) + (x^2 - 5x + 6) = 0$$
$$x^3 - 4x^2 + x + 6 = 0$$

89. a. $P(x) = 0.3199x^2 - 58.4992x + 2671.3086$
To find the profit for 1997, use $x = 97$.
$P(97) = 0.3199(97)^2 - 58.4992(97) + 2671.3086$
$P(97) = 6.8253$
The U.S. airline industry had a profit of $6.8253 billion in 1997.

b. % error
$= \left| \dfrac{\text{actual value} - \text{predicted value}}{\text{actual value}} \right| \times 100$
% error
$= \left| \dfrac{7.7 - 6.8253}{7.7} \right| \times 100 \approx 11.4$
There is an 11.4% error between the actual value and the predicted value for 1997.

Focus on Problem Solving

1. False
   0 is a real number, and $0^2 = 0$ is not positive.
2. True
3. True
4. False
   Let $x = -4$ and $y = 2$. Then the expression $(-4)^2 < 2^2$ is not true.
5. False
   It is impossible to construct a triangle with $a = 2$, $b = 3$, and $c = 10$. In a triangle, the sum of two sides must be greater than the third side.
   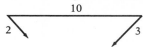
6. False
   The product of $\sqrt{3} \cdot \sqrt{3} = 3$.
7. False
   For $n = 4$, $1 \cdot 2 \cdot 3 \cdot 4 + 1 = 25$. The number is not a prime number.
8. False
   Part of AB is outside the polygon.
   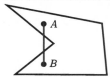
9. True
10. False
    If the points are selected so that $AC + CD < DB$, a triangle cannot be formed.

Projects and Group Activities

Reverse Polish Notation

1. 12 r 6 f
2. 7 r 5 _ 6 +
3. 9 r 3 + 4 f
4. 5 r 7 + 2 r 4 + f
5. 6 r 7 _ 10 _
6. 1 r 4 r 5 _ + 7 f
7. $18 \div 2 = 9$
8. $6 + 5 \times 3 = 21$
9. $3 \times 5 + 4 = 19$
10. $[7 + (4 \times 5)] \div 3 = 9$
11. $228 \div [6 + (9 \times 12)] = 2$
12. $1 + [1 \div (1 + 3)] = 1.25$

Pythagorean Triples

1. **a.** $5^2 + 7^2 = 25 + 49 = 74 \neq 81 = 9^2$
   5, 7, and 9 are not a Pythagorean triple.

   b. $8^2 + 15^2 = 64 + 225 = 289 = 17^2$
   8, 15, and 17 are a Pythagorean triple.

   c. $11^2 + 60^2 = 121 + 3600 = 3721 = 61^2$
   11, 60, and 61 are a Pythagorean triple.

   d. $28^2 + 45^2 = 784 + 2025 = 2809 = 53^2$
   28, 45, and 53 are a Pythagorean triple.

2. $a = m^2 - n^2$
   $b = 2mn$
   $c = m^2 + n^2$

   a. $m = 3, n = 1$
   $a = 3^2 - 1^2 = 8$
   $b = 2(3)(1) = 6$
   $c = 3^2 + 1^2 = 10$
   The Pythagorean triple is 6, 8, and 10.

   b. $m = 5, n = 2$
   $a = 5^2 - 2^2 = 21$
   $b = 2(5)(2) = 20$
   $c = 5^2 + 2^2 = 29$
   The Pythagorean triple is 20, 21, and 29.

   c. $m = 4, n = 2$
   $a = 4^2 - 2^2 = 12$
   $b = 2(4)(2) = 16$
   $c = 4^2 + 2^2 = 20$
   The Pythagorean triple is 12, 16, and 20.

   d. $m = 6, n = 1$
   $a = 6^2 - 1^2 = 35$
   $b = 2(6)(1) = 12$
   $c = 6^2 + 1^2 = 37$
   The Pythagorean triple is 12, 35, and 37.

3. $a = m^2 - n^2 = 11$
   $b = 2mn = 60$
   $c = m^2 + n^2 = 61$

   $m^2 - n^2 = 11$
   $m^2 + n^2 = 61$
   $2m^2 = 72$
   $m^2 = 36$
   $m = 6$

   -6 is not a solution since m and n must be positive.

   $2mn = 60$
   $2(6)n = 60$
   $12n = 60$
   $n = 5$

   So $m = 6$ and $n = 5$ generate the Pythagorean triple 11, 60, and 61.

4. $a^2 = (m^2 - n^2)^2$
   $= (m^2 - n^2)(m^2 - n^2)$
   $= m^4 - 2m^2n^2 + n^4$

   $b^2 = (2mn)^2 = 4m^2n^2$

   $c^2 = (m^2 + n^2)^2$
   $= (m^2 + n^2)(m^2 + n^2)$
   $= m^4 + 2m^2n^2 + n^4$

   $a^2 + b^2 = m^4 - 2m^2n^2 + n^4 + 4m^2n^2$
   $= m^4 + 2m^2n^2 + n^4$
   $= c^2$

5. Using the Pythagorean triple 3, 4, and 5, the rope could have been held as shown in the diagram below.

6. It is impossible to find odd numbers a, b, and c so that $a^2 + b^2 = c^2$. If it were possible, then a^2, b^2, and c^2 would also be odd (since an odd number times an odd number is odd). Then $a^2 + b^2$ would be even, so $a^2 + b^2$ could not equal c^2.

Chapter Review Exercises

1. $(3x^2 - 2x - 6) + (-x^2 - 3x + 4) = 2x^2 - 5x - 2$

2. $(5x^2 - 8xy + 2y^2) - (x^2 - 3y^2)$
   $= (5x^2 - 8xy + 2y^2) + (-x^2 + 3y^2)$
   $= 4x^2 - 8xy + 5y^2$

3. $(5x^2yz^4)(2xy^3z^{-1})(7x^{-2}y^{-2}z^3) = 70xy^2z^6$

4. $(2x^{-1}y^2z^5)^4(-3x^3yz^{-3})$
   $= (16x^{-4}y^8z^{20})(-3x^3yz^{-3})$
   $= -48x^{-1}y^9z^{17}$
   $= -\dfrac{48y^9z^{17}}{x}$

5. $\dfrac{3x^4yz^{-1}}{-12xy^3z^2} = -\dfrac{x^3}{4y^2z^3}$

6. $\dfrac{(2a^4b^{-3}c^2)^3}{(2a^3b^2c^{-1})^4} = \dfrac{8a^{12}b^{-9}c^6}{16a^{12}b^8c^{-4}}$
   $= \dfrac{1}{2}a^{12-12}b^{-9-8}c^{6-(-4)}$
   $= \dfrac{1}{2}b^{-17}c^{10}$
   $= \dfrac{c^{10}}{2b^{17}}$

7. $93{,}000{,}000 = 9.3 \times 10^7$

8. $2.54 \times 10^{-3} = 0.00254$

9. $\dfrac{3 \times 10^{-3}}{15 \times 10^2} = \dfrac{3}{15} \times \dfrac{10^{-3}}{10^2} = 0.2 \times 10^{-5} = 2 \times 10^{-6}$

10. $P(-2) = 2(-2)^3 - (-2) + 7$
    $P(-2) = -16 + 2 + 7$
    $P(-2) = -7$

11. $y = x^2 + 1$

12. a. 3
    b. 8
    c. 5

13. $\begin{array}{r|rrrr} -3 & -2 & 2 & 0 & -4 \\ & & 6 & -24 & 72 \\ \hline & -2 & 8 & -24 & 68 \end{array}$
    $P(-3) = 68$

14. $\begin{array}{r} 5x + 4 \\ 3x-2\overline{)15x^2 + 2x - 2} \\ \underline{15x^2 - 10x} \\ 12x - 2 \\ \underline{12x - 8} \\ 6 \end{array}$

    $\dfrac{15x^2 + 2x - 2}{3x - 2} = 5x + 4 + \dfrac{6}{3x - 2}$

15.
$$
\begin{array}{r}
2x-3 \\
6x+1 \overline{\smash{)}12x^2-16x-7} \\
\underline{12x^2+\ 2x} \\
-18x-7 \\
\underline{-18x-3} \\
-4
\end{array}
$$

$$\frac{12x^2-16x-7}{6x+1} = 2x-3 - \frac{4}{6x+1}$$

16.
$$
\begin{array}{r|rrrr}
-6 & 4 & 27 & 10 & 2 \\
   &   & -24 & -18 & 48 \\ \hline
   & 4 & 3 & -8 & 50
\end{array}
$$

$$\frac{4x^3+27x^2+10x+2}{x+6} = 4x^2+3x-8+\frac{50}{x+6}$$

17.
$$
\begin{array}{r|rrrrr}
4 & 1 & 0 & 0 & 0 & -4 \\
  &   & 4 & 16 & 64 & 256 \\ \hline
  & 1 & 4 & 16 & 64 & 252
\end{array}
$$

$$\frac{x^4-4}{x-4} = x^3+4x^2+16x+64+\frac{252}{x-4}$$

18. $4x^2y(3x^3y^2+2xy-7y^3)$
$= 12x^5y^3 + 8x^3y^2 - 28x^2y^4$

19. $a^{2n+3}(a^n-5a+2) = a^{3n+3} - 5a^{2n+4} + 2a^{2n+3}$

20. $5x^2 - 4x[x-(3x+2)+x]$
$= 5x^2 - 4x(x-3x-2+x)$
$= 5x^2 - 4x(-x-2)$
$= 5x^2 + 4x^2 + 8x$
$= 9x^2 + 8x$

21. $(x^{2n}-x)(x^{n+1}-3) = x^{3n+1} - 3x^{2n} - x^{n+2} + 3x$

22.
$$
\begin{array}{r}
x^3 - 3x^2 - 5x + 1 \\
\times \quad x + 6 \\ \hline
6x^3 - 18x^2 - 30x + 6 \\
x^4 - 3x^3 - 5x^2 + x \\ \hline
x^4 + 3x^3 - 23x^2 - 29x + 6
\end{array}
$$

$(x+6)(x^3-3x^2-5x+1)$
$= x^4+3x^3-23x^2-29x+6$

23. $(x-4)(3x+2)(2x-3)$
$= (x-4)(6x^2-5x-6)$
$= 6x^3 - 5x^2 - 6x - 24x^2 + 20x + 24$
$= 6x^3 - 29x^2 + 14x + 24$

24. $(5a+2b)(5a-2b) = 25a^2 - 4b^2$

25. $(4x-3y)^2 = 16x^2 - 24xy + 9y^2$

26. The GCF of $18a^5b^2$, $12a^3b^3$, and $30a^2b$ is $6a^2b$.
$18a^5b^2 - 12a^3b^3 + 30a^2b$
$= 6a^2b(3a^3b - 2ab^2 + 5)$

27. The GCF of $5x^{n+5}$, x^{n+3}, and $4x^2$ is x^2.
$5x^{n+5} + x^{n+3} + 4x^2 = x^2(5x^{n+3} + x^{n+1} + 4)$

28. $x(y-3) + 4(3-y) = x(y-3) + 4[(-1)(y-3)]$
$= x(y-3) - 4(y-3)$
$= (y-3)(x-4)$

29. $2ax + 4bx - 3ay - 6by = 2x(a+2b) - 3y(a+2b)$
$= (a+2b)(2x-3y)$

30. $x^2 + 12x + 35 = (x+5)(x+7)$

31. $12 + x - x^2 = (3+x)(4-x)$

32. $x^2 - 16x + 63 = (x-7)(x-9)$

33. $6x^2 - 31x + 18 = (3x-2)(2x-9)$

34. $24x^2 + 61x - 8 = (8x-1)(3x+8)$

35. Let $u = xy$.
$x^2y^2 - 9 = u^2 - 9$
$= (u+3)(u-3)$
$= (xy+3)(xy-3)$

36. $4x^2 + 12xy + 9y^2 = (2x+3y)^2$

37. Let $u = x^n$.
$x^{2n} - 12x^n + 36 = u^2 - 12u + 36$
$= (u-6)^2$
$= (x^n-6)^2$

38. $36 - a^{2n} = 6^2 - (a^n)^2 = (6+a^n)(6-a^n)$

39. $64a^3 - 27b^3 = (4a)^3 - (3b)^3$
$= (4a-3b)(16a^2+12ab+9b^2)$

40. $8 - y^{3n} = (2)^3 - (y^n)^3 = (2-y^n)(4+2y^n+y^{2n})$

41. Let $u = x^2$.
$15x^4 + x^2 - 6 = 15u^2 + u - 6$
$= (3u+2)(5u-3)$
$= (3x^2+2)(5x^2-3)$

42. Let $u = x^4$.
$36x^8 - 36x^4 + 5 = 36u^2 - 36u + 5$
$= (6u-5)(6u-1)$
$= (6x^4-5)(6x^4-1)$

43. Let $u = x^2 y^2$.
$21x^4 y^4 + 23x^2 y^2 + 6 = 21u^2 + 23u + 6$
$= (7u+3)(3u+2)$
$= (7x^2 y^2 + 3)(3x^2 y^2 + 2)$

44. $3a^6 - 15a^4 - 18a^2 = 3a^2(a^4 - 5a^2 - 6)$
$= 3a^2(a^2 - 6)(a^2 + 1)$

45. $x^{4n} - 8x^{2n} + 16 = (x^{2n} - 4)^2$
$= [(x^n + 2)(x^n - 2)]^2$
$= (x^n + 2)^2 (x^n - 2)^2$

46. $3a^4 b - 3ab^4 = 3ab(a^3 - b^3)$
$= 3ab(a-b)(a^2 + ab + b^2)$

47. $x^3 - x^2 - 6x = 0$
$x(x^2 - x - 6) = 0$
$x(x-3)(x+2) = 0$
$x = 0 \quad x - 3 = 0 \quad x + 2 = 0$
$\quad\quad\quad x = 3 \quad\quad x = -2$
The solutions are 0, 3, and –2.

48. $\quad\quad 6x^2 + 60 = 39x$
$\quad\quad 6x^2 - 39x + 60 = 0$
$\quad\quad 3(2x^2 - 13x + 20) = 0$
$\quad\quad 3(2x - 5)(x - 4) = 0$
$2x - 5 = 0 \quad x - 4 = 0$
$\quad 2x = 5 \quad\quad x = 4$
$\quad\quad x = \dfrac{5}{2}$
The solutions are $\dfrac{5}{2}$ and 4.

49. $\quad x^3 - 16x = 0$
$\quad x(x^2 - 16) = 0$
$x(x+4)(x-4) = 0$
$x = 0 \quad x + 4 = 0 \quad x - 4 = 0$
$\quad\quad\quad x = -4 \quad\quad x = 4$
The solutions are 0, –4, and 4.

50. $y^3 + y^2 - 36y - 36 = 0$
$y^2(y+1) - 36(y+1) = 0$
$\quad (y+1)(y^2 - 36) = 0$
$(y+1)(y+6)(y-6) = 0$
$y + 1 = 0 \quad y + 6 = 0 \quad y - 6 = 0$
$\quad y = -1 \quad\quad y = -6 \quad\quad y = 6$
The solutions are –1, –6, and 6.

51. **Strategy**
To find how far Earth is from the Great Galaxy of Andromeda, use the equation $d = rt$, where $r = 5.9 \times 10^{12}$ mph and $t = 2.2 \times 10^6$ years.
$2.2 \times 10^6 \times 24 \times 365 = 1.9272 \times 10^{10}$

Solution
$d = r \cdot t$
$\quad = (5.9 \times 10^{12})(1.9272 \times 10^{10})$
$\quad = 5.9 \times 1.9272 \times 10^{22}$
$\quad = 11.37048 \times 10^{22}$
$\quad = 1.137048 \times 10^{23}$
The distance from Earth to the Great Galaxy of Andromeda is 1.137048×10^{23} mi.

52. **Strategy**
To find how much power is generated by the sun, divide the amount of horsepower that Earth receives by the proportion that this is of the power generated by the sun.

Solution
$\dfrac{2.4 \times 10^{14}}{2.2 \times 10^{-7}} = \dfrac{2.4}{2.2} \times \dfrac{10^{14}}{10^{-7}} = 1.09 \times 10^{21}$
The sun generates 1.09×10^{21} horsepower.

53. **Strategy**
To find the area, replace the variables L and W in the equation $A = LW$ by the given values and solve for A.

Solution
$A = L \cdot W$
$A = (5x+3)(2x-7) = 10x^2 - 29x - 21$
The area is $(10x^2 - 29x - 21)$ cm^2.

54. **Strategy**
This is a geometry problem.
To find the volume, replace the variable s in the equation $V = s^3$ by the given value and solve for V.

Solution
$V = s^3$
$V = (3x-1)^3$
$V = (3x-1)(9x^2 - 6x + 1)$
$V = 27x^3 - 18x^2 + 3x - 9x^2 + 6x - 1$
$V = 27x^3 - 27x^2 + 9x - 1$
The volume is $(27x^3 - 27x^2 + 9x - 1)$ ft^3.

55. Strategy
To find the area, subtract the area of the small square from the area of the large rectangle.
Large rectangle:
$L_1 = 3x - 2$
$W_1 = (x+4) + x$
Small square:
side $= x$

Solution
A = Area of large rectangle − Area of square
$A = L_1 W_1 - s^2$
$A = (3x-2)[(x+4)+x] - x^2$
$A = (3x-2)(2x+4) - x^2$
$A = 6x^2 + 8x - 8 - x^2$
$A = 5x^2 + 8x - 8$
The area is $(5x^2 + 8x - 8)$ in^2.

56. Strategy
This is an integer problem.
The first even integer: n
The next consecutive even integer: $n + 2$
The sum of the squares of the 2 consecutive even integers is 52.
$n^2 + (n+2)^2 = 52$

Solution
$n^2 + (n+2)^2 = 52$
$n^2 + n^2 + 4n + 4 = 52$
$2n^2 + 4n - 48 = 0$
$2(n^2 + 2n - 24) = 0$
$2(n+6)(n-4) = 0$
$n + 6 = 0 \qquad n - 4 = 0$
$n = -6 \qquad n = 4$
$n + 2 = -6 + 2 = -4 \quad n + 2 = 4 + 2 = 6$
The two integers are −6 and −4, or 4 and 6.

57. Strategy
This is an integer problem.
The unknown integer is n.
The sum of this number and its square is 56.

Solution
$n^2 + n = 56$
$n^2 + n - 56 = 0$
$(n+8)(n-7) = 0$
$n + 8 = 0 \qquad n - 7 = 0$
$n = -8 \qquad n = 7$
The integer is −8 or 7.

58. Strategy
This is a geometry problem.
Width of the rectangle: x
Length of the rectangle: $2x + 2$

The area of the rectangle is 60 m^2.
Use the equation for the area of a rectangle $(A = LW)$.

Solution
$A = LW$
$60 = (2x+2)(x)$
$60 = 2x^2 + 2x$
$0 = 2x^2 + 2x - 60$
$0 = 2(x^2 + x - 30)$
$0 = 2(x+6)(x-5)$
$x + 6 = 0 \qquad x - 5 = 0$
$x = -6 \qquad x = 5$
Since the width of a rectangle cannot be negative, −6 cannot be a solution.
$2x + 2 = 2(5) + 2 = 10 + 2 = 12$
The length of the rectangle is 12 m.

Chapter Test

1. $(6x^3 - 7x^2 + 6x - 7) - (4x^3 - 3x^2 + 7)$
$= (6x^3 - 7x^2 + 6x - 7) + (-4x^3 + 3x^2 - 7)$
$= 2x^3 - 4x^2 + 6x - 14$

2. $(-4a^2 b)^3(-ab^4) = (-64a^6 b^3)(-ab^4) = 64a^7 b^7$

3. $\dfrac{(2a^{-4} b^2)^3}{4a^{-2} b^{-1}} = \dfrac{8a^{-12} b^6}{4a^{-2} b^{-1}}$
$= 2a^{-12-(-2)} b^{6-(-1)}$
$= 2a^{-10} b^7$
$= \dfrac{2b^7}{a^{10}}$

4. $0.000000501 = 5.01 \times 10^{-7}$

5. Strategy
To find the number of seconds in one week in scientific notation:
Multiply the number of seconds in a minute (60) by the minutes in an hour (60) by the hours in a day (24) by the days in a week (7).
Convert that product to scientific notation.

Solution
$60 \times 60 \times 24 \times 7 = 604{,}800 = 6.048 \times 10^5$
The number of seconds in a week is 6.048×10^5.

6. $(2x^{-3} y)^{-4} = 2^{-4} x^{12} y^{-4} = \dfrac{x^{12}}{2^4 y^4} = \dfrac{x^{12}}{16y^4}$

7. $-5x[3 - 2(2x-4) - 3x] = -5x[3 - 4x + 8 - 3x]$
$= -5x[-7x + 11]$
$= 35x^2 - 55x$

8. $(3a + 4b)(2a - 7b) = 6a^2 - 13ab - 28b^2$

9. $3t^3 - 4t^2 + 1$
     $\underline{\times 2t^2 - 5}$
     $ -15t^3 + 20t^2 - 5$
     $\underline{6t^5 - 8t^4 + 2t^2 }$
     $6t^5 - 8t^4 - 15t^3 + 22t^2 - 5$

10. $(3z - 5)^2 = 9z^2 - 30z + 25$

11. $ 2x^2 + 3x + 5$
    $2x - 3 \overline{)4x^3 + 0x^2 + x - 15}$
    $\underline{4x^3 - 6x^2}$
    $ 6x^2 + x$
    $ \underline{6x^2 - 9x}$
    $ 10x - 15$
    $ \underline{10x - 15}$
    $ 0$
    $(4x^3 + x - 15) \div (2x - 3) = 2x^2 + 3x + 5$

12.
3	1	−5	5	5
		3	−6	−3
	1	−2	−1	2

    $(x^3 - 5x^2 + 5x + 5) \div (x - 3) = x^2 - 2x - 1 + \dfrac{2}{x-3}$

13. $P(2) = 3(2)^2 - 8(2) + 1$
    $P(2) = 12 - 16 + 1$
    $P(2) = -3$

14.
−2	−1	0	4	−8
		2	−4	0
	−1	2	0	−8

    $P(-2) = -8$

15. Let $u = a^2$.
    $6a^4 - 13a^2 - 5 = 6u^2 - 13u - 5$
    $ = (2u - 5)(3u + 1)$
    $ = (2a^2 - 5)(3a^2 + 1)$

16. $12x^3 + 12x^2 - 45x = 3x(4x^2 + 4x - 15)$
    $ = 3x(2x - 3)(2x + 5)$

17. $16x^2 - 25 = (4x - 5)(4x + 5)$

18. $16t^2 + 24t + 9 = (4t + 3)^2$

19. $27x^3 - 8 = (3x)^3 - (2)^3 = (3x - 2)(9x^2 + 6x + 4)$

20. $6x^2 - 4x - 3xa + 2a = 2x(3x - 2) - a(3x - 2)$
    $ = (3x - 2)(2x - a)$

21. $ 6x^2 = x + 1$
    $6x^2 - x - 1 = 0$
    $(2x - 1)(3x + 1) = 0$
    $2x - 1 = 0 3x + 1 = 0$
    $ 2x = 1 3x = -1$
    $ x = \dfrac{1}{2} x = -\dfrac{1}{3}$

    The solutions are $\dfrac{1}{2}$ and $-\dfrac{1}{3}$.

22. $6x^3 + x^2 - 6x - 1 = 0$
    $x^2(6x + 1) - 1(6x + 1) = 0$
    $(x^2 - 1)(6x + 1) = 0$
    $(x + 1)(x - 1)(6x + 1) = 0$
    $x + 1 = 0 x - 1 = 0 6x + 1 = 0$
    $ x = -1 x = 1 6x = -1$
    $ x = -\dfrac{1}{6}$

    The solutions are -1, 1, and $-\dfrac{1}{6}$.

23. **Strategy**
    This is a geometry problem.
    To find the area of the rectangle, replace the variables L and W in the equation $A = LW$ by the given values and solve for A.

    Solution
    $A = LW$
    $A = (5x + 1)(2x - 1)$
    $A = 10x^2 - 3x - 1$
    The area of the rectangle is $(10x^2 - 3x - 1)$ ft^2.

24. **Strategy**
    To find the time:
    Use the equation $d = rt$, where d is the distance from Earth to the moon and r is the average velocity.

    Solution
    $ d = rt$
    $2.4 \times 10^5 = (2 \times 10^4)t$
    $\dfrac{2.4 \times 10^5}{2 \times 10^4} = t$
    $ 1.2 \times 10 = t$
    $ 12 = t$
    It takes 12 h for the space vehicle to reach the moon.

Cumulative Review Exercises

1. $8 - 2[-3 - (-1)]^2 + 4 = 8 - 2[-3 + 1]^2 + 4$
   $ = 8 - 2[-2]^2 + 4$
   $ = 8 - 2(4) + 4$
   $ = 8 - 8 + 4$
   $ = 0 + 4$
   $ = 4$

2. $\dfrac{2(4) - (-2)}{-2 - 6} = \dfrac{8 + 2}{-8} = \dfrac{10}{-8} = -\dfrac{5}{4}$

3. The Inverse Property of Addition

4. $2x - 4[x - 2(3 - 2x) + 4] = 2x - 4[x - 6 + 4x + 4]$
   $= 2x - 4(5x - 2)$
   $= 2x - 20x + 8$
   $= -18x + 8$

5. $\dfrac{2}{3} - y = \dfrac{5}{6}$
   $\dfrac{2}{3} - \dfrac{2}{3} - y = \dfrac{5}{6} - \dfrac{2}{3}$
   $-y = \dfrac{1}{6}$
   $(-1)(-y) = (-1)\dfrac{1}{6}$
   $y = -\dfrac{1}{6}$

6. $8x - 3 - x = -6 + 3x - 8$
   $7x - 3 = 3x - 14$
   $7x - 3 - 3x = 3x - 14 - 3x$
   $4x - 3 = -14$
   $4x - 3 + 3 = -14 + 3$
   $4x = -11$
   $\dfrac{4x}{4} = -\dfrac{11}{4}$
   $x = -\dfrac{11}{4}$

7. $\begin{array}{r|rrrr} 3 & 1 & 0 & 0 & -3 \\ & & 3 & 9 & 27 \\ \hline & 1 & 3 & 9 & 24 \end{array}$

   $\dfrac{x^3 - 3}{x - 3} = x^2 + 3x + 9 + \dfrac{24}{x - 3}$

8. $3 - |2 - 3x| = -2$
   $-|2 - 3x| = -5$
   $|2 - 3x| = 5$
   $2 - 3x = 5 \quad 2 - 3x = -5$
   $-3x = 3 \quad -3x = -7$
   $x = -1 \quad x = \dfrac{7}{3}$

   The solutions are -1 and $\dfrac{7}{3}$.

9. $P(-2) = 3(-2)^2 - 2(-2) + 2$
   $P(-2) = 12 + 4 + 2$
   $P(-2) = 18$

10. $x = -2$

11. $f(x) = 3x^2 - 4$
    $f(-2) = 3(-2)^2 - 4 = 8$
    $f(-1) = 3(-1)^2 - 4 = -1$
    $f(0) = 3(0)^2 - 4 = -4$
    $f(1) = 3(1)^2 - 4 = -1$
    $f(2) = 3(2)^2 - 4 = 8$
    Range = $\{-4, -1, 8\}$

12. $m = \dfrac{y_2 - y_1}{x_2 - x_1} = \dfrac{2 - 3}{4 - (-2)} = -\dfrac{1}{6}$

13. Use the point-slope form.
    $y - y_1 = m(x - x_1)$
    $y - 2 = -\dfrac{3}{2}[x - (-1)]$
    $y - 2 = -\dfrac{3}{2}(x + 1)$
    $y - 2 = -\dfrac{3}{2}x - \dfrac{3}{2}$
    $y = -\dfrac{3}{2}x + \dfrac{1}{2}$

14. $3x + 2y = 4$
    $2y = -3x + 4$
    $y = -\dfrac{3}{2}x + 2$
    $m_1 = -\dfrac{3}{2}$
    $m_1 \cdot m_2 = -1$
    $-\dfrac{3}{2} \cdot m_2 = -1$
    $m_2 = \dfrac{2}{3}$

    Now use the point-slope formula to find the equation of the line.
    $y - y_1 = m(x - x_1)$
    $y - 4 = \dfrac{2}{3}[x - (-2)]$
    $y - 4 = \dfrac{2}{3}(x + 2)$
    $y - 4 = \dfrac{2}{3}x + \dfrac{4}{3}$
    $y = \dfrac{2}{3}x + \dfrac{16}{3}$

    The equation of the perpendicular line is $y = \dfrac{2}{3}x + \dfrac{16}{3}$.

15. $2x - 3y = 2$
    $x + y = -3$
    $D = \begin{vmatrix} 2 & -3 \\ 1 & 1 \end{vmatrix} = 2 - (-3) = 5$
    $D_x = \begin{vmatrix} 2 & -3 \\ -3 & 1 \end{vmatrix} = 2 - 9 = -7$
    $D_y = \begin{vmatrix} 2 & 2 \\ 1 & -3 \end{vmatrix} = -6 - 2 = -8$
    $x = \dfrac{D_x}{D} = -\dfrac{7}{5}$
    $y = \dfrac{D_y}{D} = -\dfrac{8}{5}$

    The solution is $\left(-\dfrac{7}{5}, -\dfrac{8}{5}\right)$.

16. (1) $x - y + z = 0$
(2) $2x + y - 3z = -7$
(3) $-x + 2y + 2z = 5$
Add equations (1) and (3) to eliminate x.
$$\begin{aligned} x - y + z &= 0 \\ -x + 2y + 2z &= 5 \\ \hline y + 3z &= 5 \end{aligned}$$
Add -2 times equation (1) and equation (2) to eliminate x.
$$\begin{aligned} -2x + 2y - 2z &= 0 \\ 2x + y - 3z &= -7 \\ \hline 3y - 5z &= -7 \end{aligned}$$
Now solve the system in two variables.
$y + 3z = 5$
$3y - 5z = -7$
Add -3 times the first of these equations to the second.
$$\begin{aligned} -3y - 9z &= -15 \\ 3y - 5z &= -7 \\ \hline -14z &= -22 \end{aligned}$$
$z = \dfrac{11}{7}$
Next find y.
$y + 3z = 5$
$y + 3\left(\dfrac{11}{7}\right) = 5$
$y + \dfrac{33}{7} = 5$
$y = \dfrac{2}{7}$
Replace y and z in equation (1) and solve for x.
$x - \dfrac{2}{7} + \dfrac{11}{7} = 0$
$x = -\dfrac{9}{7}$
The solution is $\left(-\dfrac{9}{7}, \dfrac{2}{7}, \dfrac{11}{7}\right)$.

17. x-intercept: y-intercept:
$3x - 4y = 12$ $3x - 4y = 12$
$3x - 4(0) = 12$ $3(0) - 4y = 12$
$3x = 12$ $-4y = 12$
$x = 4$ $y = -3$
The x-intercept is $(4, 0)$.
The y-intercept is $(0, -3)$.

18. $-3x + 2y < 6$
$2y < 3x + 6$
$y < \dfrac{3}{2}x + 3$

19.

The lines intersect at $(1, -1)$.

20. Solve each inequality for y.
$2x + y < 3$ $-6x + 3y \geq 4$
$y < 3 - 2x$ $3y \geq 4 + 6x$
                    $y \geq \dfrac{4}{3} + 2x$

21. $(4a^{-2}b^3)(2a^3b^{-1})^{-2} = (4a^{-2}b^3)(2^{-2}a^{-6}b^2)$
$= 4 \cdot 2^{-2} a^{-2-6} b^{3+2}$
$= 4 \cdot \dfrac{1}{4} a^{-8} b^5$
$= \dfrac{b^5}{a^8}$

22. $\dfrac{(5x^3y^{-3}z)^{-2}}{y^4z^{-2}} = \dfrac{5^{-2}x^{-6}y^6z^{-2}}{y^4z^{-2}}$
$= 5^{-2} x^{-6} y^{6-4} z^{-2-(-2)}$
$= \dfrac{1}{25} x^{-6} y^2$
$= \dfrac{y^2}{25x^6}$

23. $3 - (3 - 3^{-1})^{-1} = 3 - \left(3 - \dfrac{1}{3}\right)^{-1}$
$= 3 - \left(\dfrac{8}{3}\right)^{-1}$
$= 3 - \dfrac{3}{8}$
$= \dfrac{21}{8}$

24. $2x^2 - 3x + 1$
$\underline{\times 2x + 3}$
$6x^2 - 9x + 3$
$\underline{4x^3 - 6x^2 + 2x}$
$4x^3 -7x + 3$

25. $-4x^3 + 14x^2 - 12x = -2x(2x^2 - 7x + 6)$
    $= -2x(2x-3)(x-2)$

26. $a(x-y) - b(y-x) = a(x-y) + b(x-y)$
    $= (x-y)(a+b)$

27. $x^4 - 16 = (x^2+4)(x^2-4)$
    $= (x^2+4)(x+2)(x-2)$

28. $2x^3 - 16 = 2(x^3 - 8)$
    $= 2(x-2)(x^2+2x+4)$

29. **Strategy**
    Smaller integer: x
    Larger integer: $24 - x$
    The difference between four times the smaller and nine is 3 less than twice the larger.
    $4x - 9 = 2(24 - x) - 3$

    Solution
    $4x - 9 = 2(24 - x) - 3$
    $4x - 9 = 48 - 2x - 3$
    $4x - 9 = 45 - 2x$
    $6x - 9 = 45$
    $6x = 54$
    $x = 9$
    $24 - x = 15$
    The integers are 9 and 15.

30. **Strategy**
    The number of ounces of pure gold: x

	Amount	Cost	Value
Pure gold	x	360	$360x$
Alloy	80	120	80(120)
Mixture	$x + 80$	200	$200(x+80)$

    The sum of the values before mixing equals the value after mixing.
    $360x + 80(120) = 200(x+80)$

    Solution
    $360x + 80(120) = 200(x+80)$
    $360x + 9600 = 200x + 16{,}000$
    $160x + 9600 = 16{,}000$
    $160x = 6400$
    $x = 40$
    40 oz of pure gold must be mixed with the alloy.

31. **Strategy**
    Faster cyclist: x
    Slower cyclist: $\frac{2}{3}x$

	Rate	Time	Distance
Faster cyclist	x	2	$2x$
Slower cyclist	$\frac{2}{3}x$	2	$2\left(\frac{2}{3}x\right)$

    The sum of the distances is 25 mi.

    Solution
    $$2x + 2\left(\frac{2}{3}x\right) = 25$$
    $$2x + \frac{4}{3}x = 25$$
    $$\frac{10}{3}x = 25$$
    $$x = 7.5$$
    $$\frac{2}{3}x = 5$$
    The slower cyclist travels at 5 mph, the faster cyclist at 7.5 mph.

32. **Strategy**
    Amount invested at 10%: x

	Principal	Rate	Interest
Amount at 7.5%	3000	0.075	0.075(3000)
Amount at 10%	x	0.10	$0.10x$
Amount at 9%	$3000 + x$	0.09	$0.09(3000+x)$

    The amount of interest earned at 9% equals the total amount of the interest earned at 7.5% and 10%.
    $0.075(3000) + 0.10x = 0.09(3000 + x)$

    Solution
    $0.075(3000) + 0.10x = 0.09(3000 + x)$
    $225 + 0.10x = 270 + 0.09x$
    $0.01x + 225 = 270$
    $0.01x = 45$
    $x = 4500$
    The additional investment is $4500.

33. $m = \dfrac{y_2 - y_1}{x_2 - x_1} = \dfrac{300 - 100}{6 - 2} = \dfrac{200}{4}$
    $m = 50$
    The slope represents the average speed of travel in miles per hour. The average speed was 50 mph.

Chapter 6: Rational Expressions

Section 6.1

Concept Review 6.1

1. Never true
   A rational expression is the quotient of polynomials. $x^{1/2} - 2x + 4$ is not a polynomial.

3. Always true

5. Never true
   The quotient $\dfrac{a+4}{a+4} = 1$. The correct solution is $\dfrac{1}{a+2}$.

Objective 1 Exercises

3. $f(x) = \dfrac{2}{x-3}$
   $f(4) = \dfrac{2}{4-3} = \dfrac{2}{1}$
   $f(4) = 2$

5. $f(x) = \dfrac{x-2}{x+4}$
   $f(-2) = \dfrac{-2-2}{-2+4} = \dfrac{-4}{2}$
   $f(-2) = -2$

7. $f(x) = \dfrac{1}{x^2 - 2x + 1}$
   $f(-2) = \dfrac{1}{(-2)^2 - 2(-2) + 1} = \dfrac{1}{9}$
   $f(-2) = \dfrac{1}{9}$

9. $f(x) = \dfrac{x-2}{2x^2 + 3x + 8}$
   $f(3) = \dfrac{3-2}{2(3)^2 + 3(3) + 8} = \dfrac{1}{35}$
   $f(3) = \dfrac{1}{35}$

11. $f(x) = \dfrac{x^2 - 2x}{x^3 - x + 4}$
    $f(-1) = \dfrac{(-1)^2 - 2(-1)}{(-1)^3 - (-1) + 4} = \dfrac{3}{4}$
    $f(-1) = \dfrac{3}{4}$

13. $x - 3 = 0$
    $x = 3$
    The domain of $H(x)$ is $\{x | x \neq 3\}$.

15. $x + 4 = 0$
    $x = -4$
    The domain $f(x)$ is $\{x | x \neq -4\}$.

17. $3x + 9 = 0$
    $3x = -9$
    $x = -3$
    The domain of $R(x)$ is $\{x | x \neq -3\}$.

19. $(x-4)(x+2) = 0$
    $x - 4 = 0 \qquad x + 2 = 0$
    $x = 4 \qquad x = -2$
    The domain of $q(x)$ is $\{x | x \neq -2, x \neq 4\}$.

21. $(2x+5)(3x-6) = 0$
    $2x + 5 = 0 \qquad 3x - 6 = 0$
    $2x = -5 \qquad 3x = 6$
    $x = -\dfrac{5}{2} \qquad x = 2$
    The domain of $V(x)$ is $\left\{x \middle| x \neq -\dfrac{5}{2}, x \neq 2\right\}$.

23. $x = 0$
    The domain of $f(x)$ is $\{x | x \neq 0\}$.

25. The domain must exclude values of x for which $x^2 + 1 = 0$. This is not possible, because $x^2 \geq 0$, and a positive number added to a number equal to or greater than zero cannot equal zero. Therefore, there are no real numbers that must be excluded from the domain of k.
    The domain of $k(x)$ is $\{x | x \text{ is a real number}\}$.

27. $x^2 + x - 6 = 0$
    $(x+3)(x-2) = 0$
    $x + 3 = 0 \qquad x - 2 = 0$
    $x = -3 \qquad x = 2$
    The domain of $f(x)$ is $\{x | x \neq -3, x \neq 2\}$.

29. $x^2 + 2x - 24 = 0$
    $(x+6)(x-4) = 0$
    $x + 6 = 0 \qquad x - 4 = 0$
    $x = -6 \qquad x = 4$
    The domain of $A(x)$ is $\{x | x \neq -6, x \neq 4\}$.

31. The domain must exclude values of x for which $3x^2 + 12 = 0$. This is not possible, because $3x^2 \geq 0$, and a positive number added to a number equal to or greater than zero cannot equal zero. Therefore, there are no real numbers that must be excluded from the domain of f.
    The domain of $f(x)$ is $\{x | x \text{ is a real number}\}$.

33. $6x^2 - 13x + 6 = 0$
    $(2x-3)(3x-2) = 0$
    $2x - 3 = 0 \qquad 3x - 2 = 0$
    $x = \dfrac{3}{2} \qquad\qquad x = \dfrac{2}{3}$
    The domain of $G(x)$ is $\left\{x \mid x \neq \dfrac{3}{2}, x \neq \dfrac{2}{3}\right\}$.

35. $2x^3 + 9x^2 - 5x = 0$
    $x(2x^2 + 9x - 5) = 0$
    $x(2x - 1)(x + 5) = 0$
    $x = 0 \qquad 2x - 1 = 0 \qquad x + 5 = 0$
    $\qquad\qquad x = \dfrac{1}{2} \qquad\qquad x = -5$
    The domain of $f(x)$ is $\left\{x \mid x \neq 0, x \neq \dfrac{1}{2}, x \neq -5\right\}$.

Objective 2 Exercises

39. $\dfrac{4 - 8x}{4} = \dfrac{4(1 - 2x)}{4} = 1 - 2x$

41. $\dfrac{6x^2 - 2x}{2x} = \dfrac{2x(3x - 1)}{2x} = 3x - 1$

43. $\dfrac{8x^2(x - 3)}{4x(x - 3)} = \dfrac{8x^2}{4x} = 2x$

45. $\dfrac{2x - 6}{3x - x^2} = \dfrac{2(x - 3)}{x(3 - x)} = \dfrac{2}{-x} = -\dfrac{2}{x}$

47. $\dfrac{6x^3 - 15x^2}{12x^2 - 30x} = \dfrac{3x^2(2x - 5)}{6x(2x - 5)} = \dfrac{3x^2}{6x} = \dfrac{x}{2}$

49. $\dfrac{a^2 + 4a}{4a - 16} = \dfrac{a(a + 4)}{4(a - 4)}$
    The expression is in simplest form.

51. $\dfrac{16x^3 - 8x^2 + 12x}{4x} = \dfrac{4x(4x^2 - 2x + 3)}{4x}$
    $= 4x^2 - 2x + 3$

53. $\dfrac{-10a^4 - 20a^3 + 30a^2}{-10a^2} = \dfrac{-10a^2(a^2 + 2a - 3)}{-10a^2}$
    $= a^2 + 2a - 3$

55. $\dfrac{3x^{3n} - 9x^{2n}}{12x^{2n}} = \dfrac{3x^{2n}(x^n - 3)}{12x^{2n}} = \dfrac{x^n - 3}{4}$

57. $\dfrac{x^2 - 7x + 12}{x^2 - 9x + 20} = \dfrac{(x - 3)(x - 4)}{(x - 4)(x - 5)} = \dfrac{x - 3}{x - 5}$

59. $\dfrac{x^2 - xy - 2y^2}{x^2 - 3xy + 2y^2} = \dfrac{(x + y)(x - 2y)}{(x - y)(x - 2y)} = \dfrac{x + y}{x - y}$

61. $\dfrac{6 - x - x^2}{3x^2 - 10x + 8} = \dfrac{(3 + x)(2 - x)}{(3x - 4)(x - 2)}$
    $= \dfrac{-1(x + 3)}{1(3x - 4)}$
    $= -\dfrac{x + 3}{3x - 4}$

63. $\dfrac{14 - 19x - 3x^2}{3x^2 - 23x + 14} = \dfrac{(7 + x)(2 - 3x)}{(3x - 2)(x - 7)}$
    $= \dfrac{-1(7 + x)}{1(x - 7)}$
    $= -\dfrac{x + 7}{x - 7}$

65. $\dfrac{a^2 - 7a + 10}{a^2 + 9a + 14} = \dfrac{(a - 5)(a - 2)}{(a + 7)(a + 2)}$
    The expression is in simplest form.

67. $\dfrac{a^2 - b^2}{a^3 + b^3} = \dfrac{(a + b)(a - b)}{(a + b)(a^2 - ab + b^2)} = \dfrac{a - b}{a^2 - ab + b^2}$

69. $\dfrac{x^3 + y^3}{3x^3 - 3x^2y + 3xy^2} = \dfrac{(x + y)(x^2 - xy + y^2)}{3x(x^2 - xy + y^2)}$
    $= \dfrac{x + y}{3x}$

71. $\dfrac{x^3 - 4xy^2}{3x^3 - 2x^2y - 8xy^2} = \dfrac{x(x^2 - 4y^2)}{x(3x^2 - 2xy - 8y^2)}$
    $= \dfrac{x(x + 2y)(x - 2y)}{x(3x + 4y)(x - 2y)}$
    $= \dfrac{x + 2y}{3x + 4y}$

73. $\dfrac{4x^3 - 14x^2 + 12x}{24x + 4x^2 - 8x^3} = \dfrac{2x(2x^2 - 7x + 6)}{4x(6 + x - 2x^2)}$
    $= \dfrac{2x(2x - 3)(x - 2)}{4x(3 + 2x)(2 - x)}$
    $= \dfrac{2x(2x - 3)(-1)}{4x(3 + 2x)(1)}$
    $= -\dfrac{2x - 3}{2(2x + 3)}$

75. $\dfrac{x^2 - 4}{a(x + 2) - b(x + 2)} = \dfrac{(x + 2)(x - 2)}{(x + 2)(a - b)}$
    $= \dfrac{(1)(x - 2)}{(1)(a - b)}$
    $= \dfrac{x - 2}{a - b}$

77. $\dfrac{x^4+3x^2+2}{x^4-1} = \dfrac{(x^2+1)(x^2+2)}{(x^2+1)(x^2-1)}$
$= \dfrac{(x^2+1)(x^2+2)}{(x^2+1)(x+1)(x-1)}$
$= \dfrac{(1)(x^2+2)}{(1)(x+1)(x-1)}$
$= \dfrac{x^2+2}{(x+1)(x-1)}$

79. $\dfrac{x^2y^2+4xy-21}{x^2y^2-10xy+21} = \dfrac{(xy+7)(xy-3)}{(xy-3)(xy-7)}$
$= \dfrac{(xy+7)(1)}{(1)(xy-7)}$
$= \dfrac{xy+7}{xy-7}$

81. $\dfrac{a^{2n}-a^n-2}{a^{2n}+3a^n+2} = \dfrac{(a^n+1)(a^n-2)}{(a^n+1)(a^n+2)}$
$= \dfrac{(1)(a^n-2)}{(1)(a^n+2)}$
$= \dfrac{a^n-2}{a^n+2}$

Applying Concepts 6.1

83. $h(x) = \dfrac{x+2}{x-3}$
$h(2.9) = -49$
$h(2.99) = -499$
$h(2.999) = -4999$
$h(2.9999) = -49{,}999$
As x becomes closer to 3, the values of $h(x)$ decrease.

85. **a.**

   (graph showing $f(x)$ vs. Distance between lens and object (in meters); Distance between lens and film (in millimeters) on vertical axis with values 20, 40, 60, 80, 100; horizontal axis 0, 2000, 4000)

b. The ordered pair (2000, 51) means that when the distance between the object and the lens is 2000 m, the distance between the lens and the film is 51 mm.

c. For $x = 50$, the expression $\dfrac{50x}{x-50}$ is undefined. For $0 \le x < 50$, $f(x)$ is negative, and distance cannot be negative. Therefore, the domain is $x > 50$.

d. For $x > 1000$, $f(x)$ changes very little for large changes in x.

Section 6.2

Concept Review 6.2

1. Always true

3. Never true
To add fractions with the same denominator, add the numerators.
$\dfrac{3}{x+y} + \dfrac{2}{x+y} = \dfrac{3+2}{x+y} = \dfrac{5}{x+y}$

Objective 1 Exercises

3. $\dfrac{27a^2b^5}{16xy^2} \cdot \dfrac{20x^2y^3}{9a^2b} = \dfrac{27a^2b^5 \cdot 20x^2y^3}{16xy^2 \cdot 9a^2b}$
$= \dfrac{15b^4xy}{4}$

5. $\dfrac{3x-15}{4x^2-2x} \cdot \dfrac{20x^2-10x}{15x-75} = \dfrac{3(x-5)}{2x(2x-1)} \cdot \dfrac{10x(2x-1)}{15(x-5)}$
$= \dfrac{3(x-5) \cdot 10x(2x-1)}{2x(2x-1) \cdot 15(x-5)}$
$= \dfrac{3(1) \cdot 10x(1)}{2x(1) \cdot 15(1)}$
$= 1$

7. $\dfrac{x^2y^3}{x^2-4x-5} \cdot \dfrac{2x^2-13x+15}{x^4y^3}$
$= \dfrac{x^2y^3}{(x+1)(x-5)} \cdot \dfrac{(2x-3)(x-5)}{x^4y^3}$
$= \dfrac{x^2y^3 \cdot (2x-3)(x-5)}{(x+1)(x-5) \cdot x^4y^3}$
$= \dfrac{x^2y^3 \cdot (2x-3)(1)}{(x+1)(1) \cdot x^4y^3}$
$= \dfrac{2x-3}{x^2(x+1)}$

9. $\dfrac{x^2-3x+2}{x^2-8x+15} \cdot \dfrac{x^2+x-12}{8-2x-x^2}$
$= \dfrac{(x-1)(x-2)}{(x-3)(x-5)} \cdot \dfrac{(x+4)(x-3)}{(4+x)(2-x)}$
$= \dfrac{(x-1)(x-2)(x+4)(x-3)}{(x-3)(x-5)(4+x)(2-x)}$
$= \dfrac{(x-1)(-1)(1)(1)}{(1)(x-5)(1)(1)}$
$= -\dfrac{x-1}{x-5}$

11. $\dfrac{x^{n+1}+2x^n}{4x^2-6x} \cdot \dfrac{8x^2-12x}{x^{n+1}-x^n} = \dfrac{x^n(x+2)}{2x(2x-3)} \cdot \dfrac{4x(2x-3)}{x^n(x-1)}$
$= \dfrac{x^n(x+2) \cdot 4x(2x-3)}{2x(2x-3) \cdot x^n(x-1)}$
$= \dfrac{x^n(x+2) \cdot 4x(1)}{2x(1) \cdot x^n(x-1)}$
$= \dfrac{2(x+2)}{x-1}$

13. $\dfrac{12+x-6x^2}{6x^2+29x+28} \cdot \dfrac{2x^2+x-21}{4x^2-9}$

$= \dfrac{(4+3x)(3-2x)}{(2x+7)(3x+4)} \cdot \dfrac{(2x+7)(x-3)}{(2x+3)(2x-3)}$

$= \dfrac{(4+3x)(3-2x)(2x+7)(x-3)}{(2x+7)(3x+4)(2x+3)(2x-3)}$

$= \dfrac{(1)(-1)(1)(x-3)}{(1)(1)(2x+3)(1)}$

$= -\dfrac{x-3}{2x+3}$

15. $\dfrac{x^{2n}-x^n-6}{x^{2n}+x^n-2} \cdot \dfrac{x^{2n}-5x^n-6}{x^{2n}-2x^n-3}$

$= \dfrac{(x^n+2)(x^n-3)}{(x^n+2)(x^n-1)} \cdot \dfrac{(x^n+1)(x^n-6)}{(x^n+1)(x^n-3)}$

$= \dfrac{(x^n+2)(x^n-3)(x^n+1)(x^n-6)}{(x^n+2)(x^n-1)(x^n+1)(x^n-3)}$

$= \dfrac{(1)(1)(1)(x^n-6)}{(1)(x^n-1)(1)(1)}$

$= \dfrac{x^n-6}{x^n-1}$

17. $\dfrac{x^3-y^3}{2x^2+xy-3y^2} \cdot \dfrac{2x^2+5xy+3y^2}{x^2+xy+y^2}$

$= \dfrac{(x-y)(x^2+xy+y^2)}{(2x+3y)(x-y)} \cdot \dfrac{(2x+3y)(x+y)}{x^2+xy+y^2}$

$= \dfrac{(x-y)(x^2+xy+y^2)(2x+3y)(x+y)}{(2x+3y)(x-y)(x^2+xy+y^2)}$

$= \dfrac{(1)(1)(1)(x+y)}{(1)(1)(1)}$

$= x+y$

19. $\dfrac{6x^2y^4}{35a^2b^5} \div \dfrac{12x^3y^3}{7a^4b^5} = \dfrac{6x^2y^4}{35a^2b^5} \cdot \dfrac{7a^4b^5}{12x^3y^3}$

$= \dfrac{6x^2y^4 \cdot 7a^4b^5}{35a^2b^5 \cdot 12x^3y^3}$

$= \dfrac{a^2y}{10x}$

21. $\dfrac{2x-6}{6x^2-15x} \div \dfrac{4x^2-12x}{18x^3-45x^2}$

$= \dfrac{2x-6}{6x^2-15x} \cdot \dfrac{18x^3-45x^2}{4x^2-12x}$

$= \dfrac{2(x-3)}{3x(2x-5)} \cdot \dfrac{9x^2(2x-5)}{4x(x-3)}$

$= \dfrac{2(x-3) \cdot 9x^2(2x-5)}{3x(2x-5) \cdot 4x(x-3)}$

$= \dfrac{2(1) \cdot 9x^2(1)}{3x(1) \cdot 4x(1)}$

$= \dfrac{3}{2}$

23. $\dfrac{2x^2-2y^2}{14x^2y^4} \div \dfrac{x^2+2xy+y^2}{35xy^3}$

$= \dfrac{2x^2-2y^2}{14x^2y^4} \cdot \dfrac{35xy^3}{x^2+2xy+y^2}$

$= \dfrac{2(x^2-y^2)}{14x^2y^4} \cdot \dfrac{35xy^3}{(x+y)(x+y)}$

$= \dfrac{2(x+y)(x-y)}{14x^2y^4} \cdot \dfrac{35xy^3}{(x+y)(x+y)}$

$= \dfrac{2(x+y)(x-y) \cdot 35xy^3}{14x^2y^4(x+y)(x+y)}$

$= \dfrac{2(1)(x-y) \cdot 35xy^3}{14x^2y^4(1)(x+y)}$

$= \dfrac{5(x-y)}{xy(x+y)}$

25. $\dfrac{2x^2-5x-3}{2x^2+7x+3} \div \dfrac{2x^2-3x-20}{2x^2-x-15}$

$= \dfrac{2x^2-5x-3}{2x^2+7x+3} \cdot \dfrac{2x^2-x-15}{2x^2-3x-20}$

$= \dfrac{(2x+1)(x-3)}{(2x+1)(x+3)} \cdot \dfrac{(2x+5)(x-3)}{(2x+5)(x-4)}$

$= \dfrac{(1)(x-3)(1)(x-3)}{(1)(x+3)(1)(x-4)}$

$= \dfrac{(x-3)^2}{(x+3)(x-4)}$

27. $\dfrac{x^2-8x+15}{x^2+2x-35} \div \dfrac{15-2x-x^2}{x^2+9x+14}$

$= \dfrac{x^2-8x+15}{x^2+2x-35} \cdot \dfrac{x^2+9x+14}{15-2x-x^2}$

$= \dfrac{(x-3)(x-5)}{(x+7)(x-5)} \cdot \dfrac{(x+2)(x+7)}{(5+x)(3-x)}$

$= \dfrac{(x-3)(x-5)(x+2)(x+7)}{(x+7)(x-5)(5+x)(3-x)}$

$= \dfrac{(-1)(1)(x+2)(1)}{(1)(1)(5+x)(1)}$

$= -\dfrac{x+2}{x+5}$

29. $\dfrac{x^{2n}+x^n}{2x-2} \div \dfrac{4x^n+4}{x^{n+1}-x^n} = \dfrac{x^{2n}+x^n}{2x-2} \cdot \dfrac{x^{n+1}-x^n}{4x^n+4}$

$= \dfrac{x^n(x^n+1)}{2(x-1)} \cdot \dfrac{x^n(x-1)}{4(x^n+1)}$

$= \dfrac{x^n(x^n+1) \cdot x^n(x-1)}{2(x-1) \cdot 4(x^n+1)}$

$= \dfrac{x^n(1) \cdot x^n(1)}{2(1) \cdot 4(1)}$

$= \dfrac{x^{2n}}{8}$

162 Chapter 6: Rational Expressions

31. $\dfrac{2x^2-13x+21}{2x^2+11x+15} \div \dfrac{2x^2+x-28}{3x^2+4x-15}$

$= \dfrac{2x^2-13x+21}{2x^2+11x+15} \cdot \dfrac{3x^2+4x-15}{2x^2+x-28}$

$= \dfrac{(2x-7)(x-3)}{(2x+5)(x+3)} \cdot \dfrac{(3x-5)(x+3)}{(2x-7)(x+4)}$

$= \dfrac{(2x-7)(x-3)(3x-5)(x+3)}{(2x+5)(x+3)(2x-7)(x+4)}$

$= \dfrac{(1)(x-3)(3x-5)(1)}{(2x+5)(1)(1)(x+4)}$

$= \dfrac{(x-3)(3x-5)}{(2x+5)(x+4)}$

33. $\dfrac{14+17x-6x^2}{3x^2+14x+8} \div \dfrac{4x^2-49}{2x^2+15x+28}$

$= \dfrac{14+17x-6x^2}{3x^2+14x+8} \cdot \dfrac{2x^2+15x+28}{4x^2-49}$

$= \dfrac{(2+3x)(7-2x)}{(3x+2)(x+4)} \cdot \dfrac{(2x+7)(x+4)}{(2x+7)(2x-7)}$

$= \dfrac{(2+3x)(7-2x)(2x+7)(x+4)}{(3x+2)(x+4)(2x+7)(2x-7)}$

$= \dfrac{(1)(-1)(1)(1)}{(1)(1)(1)(1)}$

$= -1$

35. $\dfrac{2x^{2n}-x^n-6}{x^{2n}-x^n-2} \div \dfrac{2x^{2n}+x^n-3}{x^{2n}-1}$

$= \dfrac{2x^{2n}-x^n-6}{x^{2n}-x^n-2} \cdot \dfrac{x^{2n}-1}{2x^{2n}+x^n-3}$

$= \dfrac{(2x^n+3)(x^n-2)}{(x^n+1)(x^n-2)} \cdot \dfrac{(x^n+1)(x^n-1)}{(2x^n+3)(x^n-1)}$

$= \dfrac{(2x^n+3)(x^n-2)(x^n+1)(x^n-1)}{(x^n+1)(x^n-2)(2x^n+3)(x^n-1)}$

$= \dfrac{(1)(1)(1)(1)}{(1)(1)(1)(1)}$

$= 1$

37. $\dfrac{6x^2+6x}{3x+6x^2+3x^3} \div \dfrac{x^2-1}{1-x^3}$

$= \dfrac{6x^2+6x}{3x+6x^2+3x^3} \cdot \dfrac{1-x^3}{x^2-1}$

$= \dfrac{6x(x+1)}{3x(1+2x+x^2)} \cdot \dfrac{(1-x)(1+x+x^2)}{(x+1)(x-1)}$

$= \dfrac{6x(x+1)}{3x(1+x)(1+x)} \cdot \dfrac{(1-x)(1+x+x^2)}{(x+1)(x-1)}$

$= \dfrac{6x(x+1)(1-x)(1+x+x^2)}{3x(1+x)(1+x)(x+1)(x-1)}$

$= \dfrac{6x(1)(-1)(1+x+x^2)}{3x(1)(1+x)(x+1)(1)}$

$= -\dfrac{2(1+x+x^2)}{(x+1)^2}$

Objective 2 Exercises

39. The LCM is $2xy$.

$\dfrac{3}{2xy} - \dfrac{7}{2xy} - \dfrac{9}{2xy} = \dfrac{3-7-9}{2xy} = -\dfrac{13}{2xy}$

41. The LCM is x^2-3x+2.

$\dfrac{x}{x^2-3x+2} - \dfrac{2}{x^2-3x+2} = \dfrac{x-2}{x^2-3x+2}$

$= \dfrac{(1)}{(1)(x-1)}$

$= \dfrac{1}{x-1}$

43. The LCM is $10x^2y$.

$\dfrac{3}{2x^2y} - \dfrac{8}{5x} - \dfrac{9}{10xy}$

$= \dfrac{3}{2x^2y} \cdot \dfrac{5}{5} - \dfrac{8}{5x} \cdot \dfrac{2xy}{2xy} - \dfrac{9}{10xy} \cdot \dfrac{x}{x}$

$= \dfrac{15-16xy-9x}{10x^2y}$

45. The LCM is $30xy$.

$\dfrac{2}{3x} - \dfrac{3}{2xy} + \dfrac{4}{5xy} - \dfrac{5}{6x}$

$= \dfrac{2}{3x} \cdot \dfrac{10y}{10y} - \dfrac{3}{2xy} \cdot \dfrac{15}{15} + \dfrac{4}{5xy} \cdot \dfrac{6}{6} - \dfrac{5}{6x} \cdot \dfrac{5y}{5y}$

$= \dfrac{20y-45+24-25y}{30xy}$

$= \dfrac{-5y-21}{30xy}$

$= -\dfrac{5y+21}{30xy}$

47. The LCM is $36x$.

$\dfrac{2x-1}{12x} - \dfrac{3x+4}{9x} = \dfrac{2x-1}{12x} \cdot \dfrac{3}{3} - \dfrac{3x+4}{9x} \cdot \dfrac{4}{4}$

$= \dfrac{(6x-3)-(12x+16)}{36x}$

$= \dfrac{6x-3-12x-16}{36x}$

$= \dfrac{-6x-19}{36x}$

$= -\dfrac{6x+19}{36x}$

49. The LCM is $12x^2y^2$.

$\dfrac{3x+2}{4x^2y} - \dfrac{y-5}{6xy^2} = \dfrac{3x+2}{4x^2y} \cdot \dfrac{3y}{3y} - \dfrac{y-5}{6xy^2} \cdot \dfrac{2x}{2x}$

$= \dfrac{(9xy+6y)-(2xy-10x)}{12x^2y^2}$

$= \dfrac{9xy+6y-2xy+10x}{12x^2y^2}$

$= \dfrac{10x+7xy+6y}{12x^2y^2}$

51. The LCM is $(x-3)(x-5)$.

$$\frac{2x}{x-3} - \frac{3x}{x-5} = \frac{2x}{x-3} \cdot \frac{x-5}{x-5} - \frac{3x}{x-5} \cdot \frac{x-3}{x-3}$$
$$= \frac{(2x^2-10x)-(3x^2-9x)}{(x-3)(x-5)}$$
$$= \frac{2x^2-10x-3x^2+9x}{(x-3)(x-5)}$$
$$= \frac{-x^2-x}{(x-3)(x-5)}$$
$$= -\frac{x^2+x}{(x-3)(x-5)}$$

53. $3-2a = -(2a-3)$
The LCM is $2a-3$.

$$\frac{3}{2a-3} + \frac{2a}{3-2a} = \frac{3}{2a-3} + \frac{2a}{3-2a} \cdot \frac{-1}{-1}$$
$$= \frac{3-2a}{2a-3}$$
$$= \frac{(-1)}{(1)}$$
$$= -1$$

55. $x^2-25 = (x+5)(x-5)$
The LCM is $(x+5)(x-5)$.

$$\frac{3}{x+5} + \frac{2x+7}{x^2-25} = \frac{3}{x+5} \cdot \frac{x-5}{x-5} + \frac{2x+7}{(x-5)(x+5)}$$
$$= \frac{3x-15}{(x+5)(x-5)} + \frac{2x+7}{(x-5)(x+5)}$$
$$= \frac{(3x-15)+(2x+7)}{(x-5)(x+5)}$$
$$= \frac{3x-15+2x+7}{(x-5)(x+5)}$$
$$= \frac{5x-8}{(x-5)(x+5)}$$

57. The LCM is $x(x-4)$.

$$\frac{2}{x} - 3 - \frac{10}{x-4} = \frac{2}{x} \cdot \frac{x-4}{x-4} - 3 \cdot \frac{x(x-4)}{x(x-4)} - \frac{10}{x-4} \cdot \frac{x}{x}$$
$$= \frac{(2x-8)-3x(x-4)-10x}{x(x-4)}$$
$$= \frac{2x-8-3x^2+12x-10x}{x(x-4)}$$
$$= \frac{-3x^2+4x-8}{x(x-4)}$$
$$= -\frac{3x^2-4x+8}{x(x-4)}$$

59. The LCM is $2x(2x-3)$.

$$\frac{1}{2x-3} - \frac{5}{2x} + 1$$
$$= \frac{1}{2x-3} \cdot \frac{2x}{2x} - \frac{5}{2x} \cdot \frac{2x-3}{2x-3} + \frac{2x(2x-3)}{2x(2x-3)}$$
$$= \frac{2x-5(2x-3)+4x^2-6x}{2x(2x-3)}$$
$$= \frac{2x-10x+15+4x^2-6x}{2x(2x-3)}$$
$$= \frac{4x^2-14x+15}{2x(2x-3)}$$

61. $x^2-1 = (x+1)(x-1)$
$x^2+2x+1 = (x+1)^2$
The LCM is $(x-1)(x+1)^2$.

$$\frac{3}{x^2-1} + \frac{2x}{x^2+2x+1}$$
$$= \frac{3}{(x+1)(x-1)} \cdot \frac{x+1}{x+1} + \frac{2x}{(x+1)^2} \cdot \frac{x-1}{x-1}$$
$$= \frac{3x+3+2x^2-2x}{(x+1)^2(x-1)}$$
$$= \frac{2x^2+x+3}{(x-1)(x+1)^2}$$

63. $x^2-9 = (x+3)(x-3)$
The LCM is $(x+3)(x-3)$.

$$\frac{x}{x+3} - \frac{3-x}{x^2-9} = \frac{x}{x+3} \cdot \frac{x-3}{x-3} - \frac{3-x}{(x+3)(x-3)}$$
$$= \frac{x^2-3x-(3-x)}{(x+3)(x-3)}$$
$$= \frac{x^2-3x-3+x}{(x+3)(x-3)}$$
$$= \frac{x^2-2x-3}{(x+3)(x-3)}$$
$$= \frac{(x-3)(x+1)}{(x+3)(x-3)}$$
$$= \frac{(1)(x+1)}{(x+3)(1)}$$
$$= \frac{x+1}{x+3}$$

65. $x^2 + 8x + 15 = (x+5)(x+3)$
The LCM is $(x+5)(x+3)$.

$$\frac{2x-3}{x+5} - \frac{x^2-4x-19}{x^2+8x+15}$$
$$= \frac{2x-3}{x+5} \cdot \frac{x+3}{x+3} - \frac{x^2-4x-19}{(x+5)(x+3)}$$
$$= \frac{2x^2+3x-9-(x^2-4x-19)}{(x+3)(x+5)}$$
$$= \frac{2x^2+3x-9-x^2+4x+19}{(x+3)(x+5)}$$
$$= \frac{x^2+7x+10}{(x+3)(x+5)}$$
$$= \frac{(x+2)(x+5)}{(x+3)(x+5)}$$
$$= \frac{(x+2)(1)}{(x+3)(1)}$$
$$= \frac{x+2}{x+3}$$

67. $x^{2n} - 1 = (x^n+1)(x^n-1)$
The LCM is $(x^n+1)(x^n-1)$.

$$\frac{x^n}{x^{2n}-1} - \frac{2}{x^n+1}$$
$$= \frac{x^n}{(x^n+1)(x^n-1)} - \frac{2}{x^n+1} \cdot \frac{x^n-1}{x^n-1}$$
$$= \frac{x^n - 2(x^n-1)}{(x^n+1)(x^n-1)}$$
$$= \frac{x^n - 2x^n + 2}{(x^n+1)(x^n-1)}$$
$$= \frac{-x^n + 2}{(x^n+1)(x^n-1)}$$
$$= \frac{-(x^n-2)}{(x^n+1)(x^n-1)}$$
$$= -\frac{x^n-2}{(x^n+1)(x^n-1)}$$

69. $4x^2 - 9 = (2x+3)(2x-3)$
$3 - 2x = -(2x-3)$
The LCM is $(2x+3)(2x-3)$.

$$\frac{2x-2}{4x^2-9} - \frac{5}{3-2x}$$
$$= \frac{2x-2}{(2x+3)(2x-3)} - \frac{(-5)}{2x-3} \cdot \frac{2x+3}{2x+3}$$
$$= \frac{2x-2+5(2x+3)}{(2x-3)(2x+3)}$$
$$= \frac{2x-2+10x+15}{(2x+3)(2x-3)}$$
$$= \frac{12x+13}{(2x+3)(2x-3)}$$

71. $2x^2 - x - 3 = (2x-3)(x+1)$
The LCM is $(2x-3)(x+1)$.

$$\frac{x-2}{x+1} - \frac{3-12x}{2x^2-x-3}$$
$$= \frac{x-2}{x+1} \cdot \frac{2x-3}{2x-3} - \frac{3-12x}{(2x-3)(x+1)}$$
$$= \frac{(x-2)(2x-3) - (3-12x)}{(2x-3)(x+1)}$$
$$= \frac{2x^2 - 7x + 6 - 3 + 12x}{(2x-3)(x+1)}$$
$$= \frac{2x^2 + 5x + 3}{(2x-3)(x+1)}$$
$$= \frac{(x+1)(2x+3)}{(2x-3)(x+1)}$$
$$= \frac{2x+3}{2x-3}$$

73. $x^2 + x - 6 = (x+3)(x-2)$
$x^2 + 4x + 3 = (x+3)(x+1)$
The LCM is $(x+3)(x-2)(x+1)$.

$$\frac{x+1}{x^2+x-6} - \frac{x+2}{x^2+4x+3} = \frac{x+1}{(x+3)(x-2)} \cdot \frac{x+1}{x+1} - \frac{x+2}{(x+3)(x+1)} \cdot \frac{x-2}{x-2}$$
$$= \frac{x^2+2x+1-(x^2-4)}{(x+3)(x-2)(x+1)}$$
$$= \frac{x^2+2x+1-x^2+4}{(x+3)(x-2)(x+1)}$$
$$= \frac{2x+5}{(x+3)(x-2)(x+1)}$$

75. $2x^2 + 11x + 12 = (2x+3)(x+4)$
$2x^2 - 3x - 9 = (2x+3)(x-3)$
The LCM is $(x+4)(2x+3)(x-3)$.
$$\frac{x-1}{2x^2+11x+12} + \frac{2x}{2x^2-3x-9} = \frac{x-1}{(2x+3)(x+4)} \cdot \frac{x-3}{x-3} + \frac{2x}{(2x+3)(x-3)} \cdot \frac{x+4}{x+4}$$
$$= \frac{x^2 - 4x + 3 + 2x^2 + 8x}{(x+4)(2x+3)(x-3)}$$
$$= \frac{3x^2 + 4x + 3}{(x+4)(2x+3)(x-3)}$$

77. $x^2 + x - 12 = (x+4)(x-3)$
The LCM is $(x+4)(x-3)$.
$$\frac{x}{x-3} - \frac{2}{x+4} - \frac{14}{x^2+x-12}$$
$$= \frac{x}{x-3} \cdot \frac{x+4}{x+4} - \frac{2}{x+4} \cdot \frac{x-3}{x-3} - \frac{14}{(x+4)(x-3)}$$
$$= \frac{x^2 + 4x - 2x + 6 - 14}{(x+4)(x-3)}$$
$$= \frac{x^2 + 2x - 8}{(x+4)(x-3)}$$
$$= \frac{(x+4)(x-2)}{(x+4)(x-3)}$$
$$= \frac{x-2}{x-3}$$

79. $x^2 + 3x - 18 = (x+6)(x-3)$
$3 - x = -(x-3)$
The LCM is $(x+6)(x-3)$.
$$\frac{x^2+6x}{x^2+3x-18} - \frac{2x-1}{x+6} + \frac{x-2}{3-x}$$
$$= \frac{x^2+6x}{(x+6)(x-3)} - \frac{2x-1}{x+6} \cdot \frac{x-3}{x-3} + \frac{-(x-2)}{x-3} \cdot \frac{x+6}{x+6}$$
$$= \frac{x^2 + 6x - (2x-1)(x-3) - (x-2)(x+6)}{(x+6)(x-3)}$$
$$= \frac{x^2 + 6x - 2x^2 + 7x - 3 - x^2 - 4x + 12}{(x+6)(x-3)}$$
$$= \frac{-2x^2 + 9x + 9}{(x+6)(x-3)}$$
$$= \frac{-(2x^2 - 9x - 9)}{(x+6)(x-3)}$$
$$= -\frac{2x^2 - 9x - 9}{(x+6)(x-3)}$$

81. $6x^2 + 11x - 10 = (3x-2)(2x+5)$
$2 - 3x = -(3x-2)$
The LCM is $(3x-2)(2x+5)$.
$$\frac{4-20x}{6x^2+11x-10} - \frac{4}{2-3x} + \frac{x}{2x+5} = \frac{4-20x}{(3x-2)(2x+5)} - \frac{(-4)}{3x-2} \cdot \frac{2x+5}{2x+5} + \frac{x}{2x+5} \cdot \frac{3x-2}{3x-2}$$
$$= \frac{4 - 20x + 8x + 20 + 3x^2 - 2x}{(3x-2)(2x+5)}$$
$$= \frac{3x^2 - 14x + 24}{(3x-2)(2x+5)}$$

83. $x^4 - 1 = (x+1)(x-1)(x^2+1)$
$x^2 - 1 = (x+1)(x-1)$
The LCM is $(x+1)(x-1)(x^2+1)$.

$$\frac{2x^2}{x^4-1} - \frac{1}{x^2-1} + \frac{1}{x^2+1} = \frac{2x^2}{(x+1)(x-1)(x^2+1)} - \frac{1}{(x+1)(x-1)} \cdot \frac{x^2+1}{x^2+1} + \frac{1}{x^2+1} \cdot \frac{(x+1)(x-1)}{(x+1)(x-1)}$$

$$= \frac{2x^2 - (x^2+1) + x^2 - 1}{(x+1)(x-1)(x^2+1)}$$

$$= \frac{2x^2 - x^2 - 1 + x^2 - 1}{(x+1)(x-1)(x^2+1)}$$

$$= \frac{2x^2 - 2}{(x+1)(x-1)(x^2+1)}$$

$$= \frac{2(x^2-1)}{(x+1)(x-1)(x^2+1)}$$

$$= \frac{2(x+1)(x-1)}{(x+1)(x-1)(x^2+1)}$$

$$= \frac{2}{x^2+1}$$

Applying Concepts 6.2

85. $\dfrac{(x+1)^2}{1-2x} \cdot \dfrac{2x-1}{x+1} = \dfrac{(x+1)(x+1)(2x-1)}{(1-2x)(x+1)}$
$\phantom{\dfrac{(x+1)^2}{1-2x} \cdot \dfrac{2x-1}{x+1}} = -(x+1)$
$\phantom{\dfrac{(x+1)^2}{1-2x} \cdot \dfrac{2x-1}{x+1}} = -x - 1$

87. $\left(\dfrac{y-2}{x^2}\right)^3 \cdot \left(\dfrac{x}{2-y}\right)^2 = \dfrac{(y-2)^3}{x^6} \cdot \dfrac{x^2}{(2-y)^2}$
$\phantom{\left(\dfrac{y-2}{x^2}\right)^3 \cdot \left(\dfrac{x}{2-y}\right)^2} = \dfrac{(y-2)(y-2)(y-2)x^2}{x^6(2-y)(2-y)}$
$\phantom{\left(\dfrac{y-2}{x^2}\right)^3 \cdot \left(\dfrac{x}{2-y}\right)^2} = \dfrac{y-2}{x^4}$

89. $\left(\dfrac{y+1}{y-1}\right)^2 - 1 = \left(\dfrac{y+1}{y-1}+1\right)\left(\dfrac{y+1}{y-1}-1\right)$
$\phantom{\left(\dfrac{y+1}{y-1}\right)^2 - 1} = \left(\dfrac{y+1}{y-1}+\dfrac{y-1}{y-1}\right)\left(\dfrac{y+1}{y-1}-\dfrac{y-1}{y-1}\right)$
$\phantom{\left(\dfrac{y+1}{y-1}\right)^2 - 1} = \left(\dfrac{y+1+y-1}{y-1}\right)\left(\dfrac{y+1-y+1}{y-1}\right)$
$\phantom{\left(\dfrac{y+1}{y-1}\right)^2 - 1} = \left(\dfrac{2y}{y-1}\right)\left(\dfrac{2}{y-1}\right)$
$\phantom{\left(\dfrac{y+1}{y-1}\right)^2 - 1} = \dfrac{4y}{(y-1)^2}$

91. $\dfrac{3x^2+6x}{4x^2-16} \cdot \dfrac{2x+8}{x^2+2x} \div \dfrac{3x-9}{5x-20} = \dfrac{3x^2+6x}{4x^2-16} \cdot \dfrac{2x+8}{x^2+2x} \cdot \dfrac{5x-20}{3x-9}$
$ = \dfrac{3x(x+2) \cdot 2(x+4) \cdot 5(x-4)}{4(x+2)(x-2) \cdot x(x+2) \cdot 3(x-3)}$
$ = \dfrac{5(x+4)(x-4)}{2(x+2)(x-2)(x-3)}$

93. $\dfrac{a^2+a-6}{4+11a-3a^2} \cdot \dfrac{15a^2-a-2}{4a^2+7a-2} \div \dfrac{6a^2-7a-3}{4-17a+4a^2} = \dfrac{a^2+a-6}{4+11a-3a^2} \cdot \dfrac{15a^2-a-2}{4a^2+7a-2} \cdot \dfrac{4-17a+4a^2}{6a^2-7a-3}$

$= \dfrac{(a+3)(a-2)(5a-2)(3a+1)(4-a)(1-4a)}{(4-a)(1+3a)(4a-1)(a+2)(3a+1)(2a-3)}$

$= -\dfrac{(a+3)(a-2)(5a-2)}{(a+2)(3a+1)(2a-3)}$

95. $\left(\dfrac{x+1}{2x-1} - \dfrac{x-1}{2x+1}\right) \cdot \left(\dfrac{2x-1}{x} - \dfrac{2x-1}{x^2}\right)$

$= \left[\dfrac{(x+1)(2x+1)}{(2x-1)(2x+1)} - \dfrac{(x-1)(2x-1)}{(2x+1)(2x-1)}\right] \cdot \left[\dfrac{x(2x-1)}{x^2} - \dfrac{2x-1}{x^2}\right]$

$= \dfrac{2x^2+3x+1-(2x^2-3x+1)}{(2x-1)(2x+1)} \cdot \dfrac{2x^2-x-(2x-1)}{x^2}$

$= \dfrac{6x}{(2x-1)(2x+1)} \cdot \dfrac{2x^2-3x+1}{x^2}$

$= \dfrac{6x(2x-1)(x-1)}{x^2(2x-1)(2x+1)}$

$= \dfrac{6(x-1)}{x(2x+1)}$

97. $\dfrac{1}{3} + \dfrac{1}{5} \ne \dfrac{1}{3+5}$

$\dfrac{5}{15} + \dfrac{3}{15} \ne \dfrac{1}{8}$

$\dfrac{8}{15} \ne \dfrac{1}{8}$

99. $\dfrac{3x+6y}{xy} = \dfrac{3x}{xy} + \dfrac{6y}{xy} = \dfrac{3}{y} + \dfrac{6}{x}$

101. $\dfrac{4a^2+3ab}{a^2b^2} = \dfrac{4a^2}{a^2b^2} + \dfrac{3ab}{a^2b^2} = \dfrac{4}{b^2} + \dfrac{3}{ab}$

103. a. $2x^2 + \dfrac{528}{x} = \dfrac{2x^3+528}{x}$

b.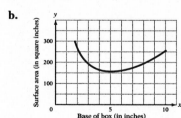

c. The point whose coordinates are (4, 164) means that when the base of the box is 4 in., 164 in^2 of cardboard will be needed.

d. The base of the box that uses the minimum amount of cardboard is 5.1 in.

e. $f(x) = 2x^2 + \dfrac{528}{x}$

$f(5.1) = 2(5.1)^2 + \dfrac{528}{5.1} \approx 155.5$ in^2

Section 6.3

Concept Review 6.3

1. Sometimes true
   The numerator of a complex fraction may contain a number, a variable, a fraction, or more than one fraction.

3. Always true

Objective 1 Exercises

3. The LCM is 3.
$$\frac{2-\frac{1}{3}}{4+\frac{11}{3}} = \frac{2-\frac{1}{3}}{4+\frac{11}{3}} \cdot \frac{3}{3} = \frac{2\cdot 3 - \frac{1}{3}\cdot 3}{4\cdot 3 + \frac{11}{3}\cdot 3} = \frac{6-1}{12+11} = \frac{5}{23}$$

5. The LCM is 6.
$$\frac{3-\frac{2}{3}}{5+\frac{5}{6}} = \frac{3-\frac{2}{3}}{5+\frac{5}{6}} \cdot \frac{6}{6} = \frac{3\cdot 6 - \frac{2}{3}\cdot 6}{5\cdot 6 + \frac{5}{6}\cdot 6} = \frac{18-4}{30+5} = \frac{14}{35} = \frac{2}{5}$$

7. The LCM of y^2 and y is y^2.
$$\frac{\frac{1}{y^2}-1}{1+\frac{1}{y}} = \frac{\frac{1}{y^2}-1}{1+\frac{1}{y}} \cdot \frac{y^2}{y^2}$$
$$= \frac{\frac{1}{y^2}\cdot y^2 - 1\cdot y^2}{1\cdot y^2 + \frac{1}{y}\cdot y^2}$$
$$= \frac{1-y^2}{y^2+y}$$
$$= \frac{(1+y)(1-y)}{y(y+1)}$$
$$= \frac{1-y}{y}$$

9. The LCM is a.
$$\frac{\frac{25}{a}-a}{5+a} = \frac{\frac{25}{a}-a}{5+a} \cdot \frac{a}{a}$$
$$= \frac{\frac{25}{a}\cdot a - a\cdot a}{5\cdot a + a\cdot a}$$
$$= \frac{25-a^2}{5a+a^2}$$
$$= \frac{(5+a)(5-a)}{a(5+a)}$$

11. The LCM of b, 2, and b^2 is $2b^2$.
$$\frac{\frac{1}{b}+\frac{1}{2}}{\frac{4}{b^2}-1} = \frac{\frac{1}{b}+\frac{1}{2}}{\frac{4}{b^2}-1} \cdot \frac{2b^2}{2b^2}$$
$$= \frac{\frac{1}{b}\cdot 2b^2 + \frac{1}{2}\cdot 2b^2}{\frac{4}{b^2}\cdot 2b^2 - 1\cdot 2b^2}$$
$$= \frac{2b+b^2}{8-2b^2}$$
$$= \frac{b(2+b)}{2(4-b^2)}$$
$$= \frac{b(2+b)}{2(2+b)(2-b)}$$
$$= \frac{b}{2(2-b)}$$

13. The LCM is $2x-3$.
$$\frac{4+\frac{12}{2x-3}}{5+\frac{15}{2x-3}} = \frac{4+\frac{12}{2x-3}}{5+\frac{15}{2x-3}} \cdot \frac{2x-3}{2x-3}$$
$$= \frac{4(2x-3)+\frac{12}{2x-3}(2x-3)}{5(2x-3)+\frac{15}{2x-3}(2x-3)}$$
$$= \frac{8x-12+12}{10x-15+15}$$
$$= \frac{8x}{10x}$$
$$= \frac{4}{5}$$

15. The LCM is $b-5$.
$$\frac{\frac{-5}{b-5}-3}{\frac{10}{b-5}+6} = \frac{\frac{-5}{b-5}-3}{\frac{10}{b-5}+6} \cdot \frac{b-5}{b-5}$$
$$= \frac{\frac{-5}{b-5}\cdot(b-5)-3(b-5)}{\frac{10}{b-5}\cdot(b-5)+6(b-5)}$$
$$= \frac{-5-3b+15}{10+6b-30}$$
$$= \frac{-3b+10}{6b-20}$$
$$= \frac{-3b+10}{2(3b-10)}$$
$$= -\frac{1}{2}$$

17. The LCM of $a-1$ and a is $a(a-1)$.
$$\frac{\frac{2a}{a-1}-\frac{3}{a}}{\frac{1}{a-1}+\frac{2}{a}} = \frac{\frac{2a}{a-1}-\frac{3}{a}}{\frac{1}{a-1}+\frac{2}{a}} \cdot \frac{a(a-1)}{a(a-1)}$$
$$= \frac{\frac{2a}{a-1}\cdot a(a-1)-\frac{3}{a}\cdot a(a-1)}{\frac{1}{a-1}\cdot a(a-1)+\frac{2}{a}\cdot a(a-1)}$$
$$= \frac{2a^2-3(a-1)}{a+2(a-1)}$$
$$= \frac{2a^2-3a+3}{a+2a-2}$$
$$= \frac{2a^2-3a+3}{3a-2}$$

19. The LCM of $3x-2$ and $9x^2-4$ is $(3x+2)(3x-2)$.

$$\frac{\frac{x}{3x-2}}{\frac{x}{9x^2-4}} = \frac{\frac{x}{3x-2}}{\frac{x}{(3x+2)(3x-2)}}$$

$$= \frac{\frac{x}{3x-2}}{\frac{x}{(3x+2)(3x-2)}} \cdot \frac{(3x+2)(3x-2)}{(3x+2)(3x-2)}$$

$$= \frac{x(3x+2)}{x}$$

$$= 3x+2$$

21. The LCM of x and x^2 is x^2.

$$\frac{1-\frac{1}{x}-\frac{6}{x^2}}{1-\frac{4}{x}+\frac{3}{x^2}} = \frac{1-\frac{1}{x}-\frac{6}{x^2}}{1-\frac{4}{x}+\frac{3}{x^2}} \cdot \frac{x^2}{x^2}$$

$$= \frac{1\cdot x^2 - \frac{1}{x}\cdot x^2 - \frac{6}{x^2}\cdot x^2}{1\cdot x^2 - \frac{4}{x}\cdot x^2 + \frac{3}{x^2}\cdot x^2}$$

$$= \frac{x^2-x-6}{x^2-4x+3}$$

$$= \frac{(x+2)(x-3)}{(x-1)(x-3)}$$

$$= \frac{x+2}{x-1}$$

23. The LCM of x^2 and x is x^2.

$$\frac{\frac{15}{x^2}-\frac{2}{x}-1}{\frac{4}{x^2}-\frac{5}{x}+4} = \frac{\frac{15}{x^2}-\frac{2}{x}-1}{\frac{4}{x^2}-\frac{5}{x}+4} \cdot \frac{x^2}{x^2}$$

$$= \frac{\frac{15}{x^2}\cdot x^2 - \frac{2}{x}\cdot x^2 - 1\cdot x^2}{\frac{4}{x^2}\cdot x^2 - \frac{5}{x}\cdot x^2 + 4\cdot x^2}$$

$$= \frac{15-2x-x^2}{4-5x+4x^2}$$

$$= -\frac{(x-3)(x+5)}{4x^2-5x+4}$$

25. The LCM is $3x+10$.

$$\frac{1-\frac{12}{3x+10}}{x-\frac{8}{3x+10}} = \frac{1-\frac{12}{3x+10}}{x-\frac{8}{3x+10}} \cdot \frac{3x+10}{3x+10}$$

$$= \frac{1\cdot(3x+10)-\frac{12}{3x+10}\cdot(3x+10)}{x(3x+10)-\frac{8}{3x+10}\cdot(3x+10)}$$

$$= \frac{3x+10-12}{3x^2+10x-8}$$

$$= \frac{3x-2}{(x+4)(3x-2)}$$

$$= \frac{1}{x+4}$$

27. The LCM is $x+2$.

$$\frac{x-5-\frac{18}{x+2}}{x+7+\frac{6}{x+2}} = \frac{x-5-\frac{18}{x+2}}{x+7+\frac{6}{x+2}} \cdot \frac{x+2}{x+2}$$

$$= \frac{(x-5)(x+2)-\frac{18}{x+2}\cdot(x+2)}{(x+7)(x+2)+\frac{6}{x+2}\cdot(x+2)}$$

$$= \frac{x^2-3x-10-18}{x^2+9x+14+6}$$

$$= \frac{x^2-3x-28}{x^2+9x+20}$$

$$= \frac{(x+4)(x-7)}{(x+4)(x+5)}$$

$$= \frac{x-7}{x+5}$$

29. The LCM is $a(a-2)$.

$$\frac{\frac{1}{a}-\frac{3}{a-2}}{\frac{2}{a}+\frac{5}{a-2}} = \frac{\frac{1}{a}-\frac{3}{a-2}}{\frac{2}{a}+\frac{5}{a-2}} \cdot \frac{a(a-2)}{a(a-2)}$$

$$= \frac{\frac{1}{a}\cdot a(a-2)-\frac{3}{a-2}\cdot a(a-2)}{\frac{2}{a}\cdot a(a-2)+\frac{5}{a-2}\cdot a(a-2)}$$

$$= \frac{a-2-3a}{2a-4+5a}$$

$$= \frac{-2a-2}{7a-4}$$

$$= \frac{-(2a+2)}{7a-4}$$

$$= -\frac{2a+2}{7a-4}$$

$$= -\frac{2(a+1)}{7a-4}$$

31. The LCM of y^2, xy, and x^2 is x^2y^2.

$$\frac{\frac{1}{y^2}-\frac{1}{xy}-\frac{2}{x^2}}{\frac{1}{y^2}-\frac{3}{xy}+\frac{2}{x^2}} = \frac{\frac{1}{y^2}-\frac{1}{xy}-\frac{2}{x^2}}{\frac{1}{y^2}-\frac{3}{xy}+\frac{2}{x^2}} \cdot \frac{x^2y^2}{x^2y^2}$$

$$= \frac{\frac{1}{y^2}\cdot x^2y^2 - \frac{1}{xy}\cdot x^2y^2 - \frac{2}{x^2}\cdot x^2y^2}{\frac{1}{y^2}\cdot x^2y^2 - \frac{3}{xy}\cdot x^2y^2 + \frac{2}{x^2}\cdot x^2y^2}$$

$$= \frac{x^2-xy-2y^2}{x^2-3xy+2y^2}$$

$$= \frac{(x+y)(x-2y)}{(x-y)(x-2y)}$$

$$= \frac{x+y}{x-y}$$

170 Chapter 6: Rational Expressions

33. The LCM is $(x+1)(x-1)$.

$$\frac{\frac{x-1}{x+1} - \frac{x+1}{x-1}}{\frac{x-1}{x+1} + \frac{x+1}{x-1}}$$

$$= \frac{\frac{x-1}{x+1} - \frac{x+1}{x-1}}{\frac{x-1}{x+1} + \frac{x+1}{x-1}} \cdot \frac{(x+1)(x-1)}{(x+1)(x-1)}$$

$$= \frac{\frac{x-1}{x+1} \cdot (x+1)(x-1) - \frac{x+1}{x-1} \cdot (x+1)(x-1)}{\frac{x-1}{x+1} \cdot (x+1)(x-1) + \frac{x+1}{x-1} \cdot (x+1)(x-1)}$$

$$= \frac{(x-1)(x-1) - (x+1)(x+1)}{(x-1)(x-1) + (x+1)(x+1)}$$

$$= \frac{x^2 - 2x + 1 - (x^2 + 2x + 1)}{x^2 - 2x + 1 + x^2 + 2x + 1}$$

$$= \frac{x^2 - 2x + 1 - x^2 - 2x - 1}{2x^2 + 2}$$

$$= \frac{-4x}{2(x^2 + 1)}$$

$$= -\frac{2x}{x^2 + 1}$$

35. The LCM is x.

$$4 - \frac{2}{2 - \frac{3}{x}} = 4 - \frac{2}{2 - \frac{3}{x}} \cdot \frac{x}{x}$$

$$= 4 - \frac{2 \cdot x}{2 \cdot x - \frac{3}{x} \cdot x}$$

$$= 4 - \frac{2x}{2x - 3}$$

The LCM is $2x - 3$.

$$4 - \frac{2x}{2x - 3} = 4 \cdot \frac{2x - 3}{2x - 3} - \frac{2x}{2x - 3}$$

$$= \frac{4(2x - 3) - 2x}{2x - 3}$$

$$= \frac{8x - 12 - 2x}{2x - 3}$$

$$= \frac{6x - 12}{2x - 3}$$

$$= \frac{6(x - 2)}{2x - 3}$$

Applying Concepts 6.3

37. $\dfrac{x^{-1}}{y^{-1}} + \dfrac{y}{x} = \dfrac{\frac{1}{x}}{\frac{1}{y}} + \dfrac{y}{x}$

$$= \frac{\frac{1}{x}}{\frac{1}{y}} \cdot \frac{xy}{xy} + \frac{y}{x}$$

$$= \frac{y}{x} + \frac{y}{x}$$

$$= \frac{y + y}{x}$$

$$= \frac{2y}{x}$$

39. $\dfrac{x - \frac{1}{x}}{1 + \frac{1}{x}} = \dfrac{x - \frac{1}{x}}{1 + \frac{1}{x}} \cdot \dfrac{x}{x}$

$$= \frac{x \cdot x - \frac{1}{x} \cdot x}{1 \cdot x + \frac{1}{x} \cdot x}$$

$$= \frac{x^2 - 1}{x + 1}$$

$$= \frac{(x+1)(x-1)}{x+1}$$

$$= x - 1$$

41. $2 - \dfrac{2}{2 - \frac{2}{c-1}} = 2 - \dfrac{2}{2 - \frac{2}{c-1}} \cdot \dfrac{c-1}{c-1}$

$$= 2 - \frac{2(c-1)}{2(c-1) - \frac{2}{c-1}(c-1)}$$

$$= 2 - \frac{2(c-1)}{2c - 2 - 2}$$

$$= 2 - \frac{2c - 2}{2c - 4}$$

$$= 2 - \frac{c - 1}{c - 2}$$

$$= \frac{2c - 4}{c - 2} - \frac{c - 1}{c - 2}$$

$$= \frac{2c - 4 - (c - 1)}{c - 2}$$

$$= \frac{c - 3}{c - 2}$$

43. $a + \dfrac{a}{2 + \frac{1}{1 - \frac{2}{a}}} = a + \dfrac{a}{2 + \frac{1}{1 - \frac{2}{a}} \cdot \frac{a}{a}}$

$$= a + \frac{a}{2 + \frac{1 \cdot a}{1 \cdot a - \frac{2}{a} \cdot a}}$$

$$= a + \frac{a}{2 + \frac{a}{a - 2}}$$

The LCM is $a - 2$.

$$a + \frac{a}{\frac{2(a-2)}{a-2} + \frac{a}{a-2}} = a + \frac{a}{\frac{2a - 4 + a}{a - 2}}$$

$$= a + \frac{a}{\frac{3a - 4}{a - 2}}$$

$$= a + \frac{a}{\frac{3a - 4}{a - 2}} \cdot \frac{a - 2}{a - 2}$$

$$= a + \frac{a(a - 2)}{3a - 4}$$

The LCM is $3a - 4$.

$$a \frac{(3-4)}{3a - 4} + \frac{a(a - 2)}{3a - 4} = \frac{3a^2 - 4a + a^2 - 2a}{3a - 4}$$

$$= \frac{4a^2 - 6a}{3a - 4}$$

$$= \frac{2a(2a - 3)}{3a - 4}$$

45. $\dfrac{\frac{1}{x+h}-\frac{1}{x}}{h} = \dfrac{\frac{1}{x+h}-\frac{1}{x}}{h} \cdot \dfrac{x(x+h)}{x(x+h)}$

$= \dfrac{\frac{1}{x+h}\cdot x(x+h) - \frac{1}{x}\cdot x(x+h)}{hx(x+h)}$

$= \dfrac{x-(x+h)}{hx(x+h)}$

$= \dfrac{-h}{hx(x+h)}$

$= -\dfrac{1}{x(x+h)}$

47. **Strategy**
The first even integer: n
The second consecutive even integer: $n+2$
The third consecutive even integer: $n+4$
Add the reciprocals of the three integers.

Solution

$\dfrac{1}{n} + \dfrac{1}{n+2} + \dfrac{1}{n+4}$

$= \dfrac{1}{n}\cdot\dfrac{(n+2)(n+4)}{(n+2)(n+4)} + \dfrac{1}{n+2}\cdot\dfrac{n(n+4)}{n(n+4)} + \dfrac{1}{n+4}\cdot\dfrac{n(n+2)}{n(n+2)}$

$= \dfrac{(n+2)(n+4) + n(n+4) + n(n+2)}{n(n+2)(n+4)}$

$= \dfrac{n^2+6n+8+n^2+4n+n^2+2n}{n^3+6n^2+8n}$

$= \dfrac{3n^2+12n+8}{n(n+2)(n+4)}$

49. a. $\dfrac{Cx}{\left(1-\dfrac{1}{(x+1)^{60}}\right)} = \dfrac{Cx}{\left(\dfrac{(x+1)^{60}-1}{(x+1)^{60}}\right)} = \dfrac{Cx(x+1)^{60}}{(x+1)^{60}-1}$

b.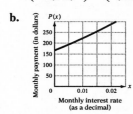

c. $12(0) = 0$
$12(0.019) = 0.228$
The interval of annual interest rates is 0% to 22.8%.

d. The ordered pair (0.006, 198.96) means that when the monthly interest rate on a car loan is 0.6%, the monthly payment on the loan is $198.96.

e. The monthly payment with a loan amount of $10,000 and an annual interest rate of 8% is $203.

Section 6.4

Concept Review 6.4

1. Never true
An equation must have an equals sign.

3. Always true

5. Always true

Objective 1 Exercises

3.
$$\frac{x}{2} + \frac{5}{6} = \frac{x}{3}$$
$$12\left(\frac{x}{2} + \frac{5}{6}\right) = 12\left(\frac{x}{3}\right)$$
$$12 \cdot \frac{x}{2} + 12 \cdot \frac{5}{6} = 4x$$
$$6x + 10 = 4x$$
$$10 = -2x$$
$$-5 = x$$
The solution is –5.

5.
$$1 - \frac{3}{y} = 4$$
$$y\left(1 - \frac{3}{y}\right) = y \cdot 4$$
$$y \cdot 1 - y \cdot \frac{3}{y} = 4y$$
$$y - 3 = 4y$$
$$-3 = 3y$$
$$-1 = y$$
The solution is –1.

7.
$$\frac{8}{2x-1} = 2$$
$$(2x-1) \cdot \frac{8}{2x-1} = (2x-1)2$$
$$8 = 4x - 2$$
$$10 = 4x$$
$$\frac{5}{2} = x$$
The solution is $\frac{5}{2}$.

9.
$$\frac{4}{x-4} = \frac{2}{x-2}$$
$$\frac{4}{x-4} \cdot (x-4)(x-2) = \frac{2}{x-2} \cdot (x-4)(x-2)$$
$$4(x-2) = 2(x-4)$$
$$4x - 8 = 2x - 8$$
$$2x - 8 = -8$$
$$2x = 0$$
$$x = 0$$
The solution is 0.

11.
$$\frac{x-2}{5} = \frac{1}{x+2}$$
$$5(x+2)\left(\frac{x-2}{5}\right) = \frac{1}{x+2} \cdot 5(x+2)$$
$$(x+2)(x-2) = 5$$
$$x^2 - 4 = 5$$
$$x^2 - 9 = 0$$
$$(x+3)(x-3) = 0$$
The solutions are –3 and 3.

13.
$$\frac{3}{x-2} = \frac{4}{x}$$
$$x(x-2) \cdot \frac{3}{x-2} = x(x-2) \cdot \frac{4}{x}$$
$$3x = (x-2)4$$
$$3x = 4x - 8$$
$$-x = -8$$
$$x = 8$$
The solution is 8.

15.
$$\frac{3}{x-4} + 2 = \frac{5}{x-4}$$
$$(x-4)\left(\frac{3}{x-4} + 2\right) = (x-4)\frac{5}{x-4}$$
$$(x-4) \cdot \frac{3}{x-4} + (x-4)2 = 5$$
$$3 + 2x - 8 = 5$$
$$2x - 5 = 5$$
$$2x = 10$$
$$x = 5$$
The solution is 5.

17.
$$\frac{8}{x-5} = \frac{3}{x}$$
$$\frac{8}{x-5} \cdot x(x-5) = \frac{3}{x} \cdot x(x-5)$$
$$8x = 3(x-5)$$
$$8x = 3x - 15$$
$$5x = -15$$
$$x = -3$$
The solution is –3.

19.
$$5 + \frac{8}{a-2} = \frac{4a}{a-2}$$
$$(a-2)\left(5 + \frac{8}{a-2}\right) = (a-2) \cdot \frac{4a}{a-2}$$
$$(a-2)5 + (a-2) \cdot \frac{8}{a-2} = 4a$$
$$5a - 10 + 8 = 4a$$
$$5a - 2 = 4a$$
$$-2 = -a$$
$$2 = a$$
2 does not check as a solution. The equation has no solution.

21.
$$\frac{x}{2} + \frac{20}{x} = 7$$
$$2x\left(\frac{x}{2} + \frac{20}{x}\right) = 7(2x)$$
$$2x\left(\frac{x}{2}\right) + 2x\left(\frac{20}{x}\right) = 14x$$
$$x^2 + 40 = 14x$$
$$x^2 - 14x + 40 = 0$$
$$(x-10)(x-4) = 0$$
$$x - 10 = 0 \qquad x - 4 = 0$$
$$x = 10 \qquad x = 4$$
The solutions are 10 and 4.

23.
$$\frac{6}{x-5} = \frac{1}{x}$$
$$\frac{6}{x-5} \cdot x(x-5) = \frac{1}{x} \cdot x(x-5)$$
$$6x = x-5$$
$$5x = -5$$
$$x = -1$$
The solution is -1.

25.
$$\frac{x}{x+2} = \frac{6}{x+5}$$
$$(x+2)(x+5)\frac{x}{x+2} = (x+2)(x+5)\frac{6}{x+5}$$
$$x(x+5) = 6(x+2)$$
$$x^2 + 5x = 6x + 12$$
$$x^2 - x - 12 = 0$$
$$(x-4)(x+3) = 0$$
$$x - 4 = 0 \qquad x + 3 = 0$$
$$x = 4 \qquad x = -3$$
The solutions are 4 and -3.

27.
$$-\frac{5}{x+7} + 1 = \frac{4}{x+7}$$
$$\frac{-5}{x+7} + 1 = \frac{4}{x+7}$$
$$(x+7)\left(\frac{-5}{x+7} + 1\right) = (x+7) \cdot \frac{4}{x+7}$$
$$(x+7) \cdot \frac{-5}{x+7} + (x+7)(1) = 4$$
$$-5 + x + 7 = 4$$
$$x + 2 = 4$$
$$x = 2$$
The solution is 2.

29.
$$\frac{2}{4y^2 - 9} + \frac{1}{2y - 3} = \frac{3}{2y + 3}$$
$$\frac{2}{(2y+3)(2y-3)} + \frac{1}{2y-3} = \frac{3}{2y+3}$$
$$(2y+3)(2y-3)\left(\frac{2}{(2y+3)(2y-3)} + \frac{1}{2y-3}\right) = (2y+3)(2y-3)\frac{3}{2y+3}$$
$$(2y+3)(2y-3)\frac{2}{(2y+3)(2y-3)} + (2y+3)(2y-3)\frac{1}{2y-3} = (2y-3)3$$
$$2 + 2y + 3 = 6y - 9$$
$$2y + 5 = 6y - 9$$
$$-4y + 5 = -9$$
$$-4y = -14$$
$$y = \frac{7}{2}$$
The solution is $\frac{7}{2}$.

31.
$$\frac{5}{x^2 - 7x + 12} = \frac{2}{x-3} + \frac{5}{x-4}$$
$$\frac{5}{(x-3)(x-4)} = \frac{2}{x-3} + \frac{5}{x-4}$$
$$(x-3)(x-4)\frac{5}{(x-3)(x-4)} = (x-3)(x-4)\left(\frac{2}{x-3} + \frac{5}{x-4}\right)$$
$$5 = (x-3)(x-4)\frac{2}{x-3} + (x-3)(x-4)\frac{5}{x-4}$$
$$5 = (x-4)2 + (x-3)5$$
$$5 = 2x - 8 + 5x - 15$$
$$5 = 7x - 23$$
$$28 = 7x$$
$$4 = x$$
4 does not check as a solution. The equation has no solution.

Objective 2 Exercises

35. Strategy
Unknown time to process the data working together: t

	Rate	Time	Part
First computer	$\frac{1}{2}$	t	$\frac{t}{2}$
Second computer	$\frac{1}{3}$	t	$\frac{t}{3}$

The sum of the part of the task completed by the first computer and the part of the task completed by the second computer is 1.
$$\frac{t}{2}+\frac{t}{3}=1$$

Solution
$$\frac{t}{2}+\frac{t}{3}=1$$
$$6\left(\frac{t}{2}+\frac{t}{3}\right)=6(1)$$
$$3t+2t=6$$
$$5t=6$$
$$t=1.2$$
With both computers, it would take 1.2 h to process the data.

37. Strategy
Unknown time to heat the water working together: t

	Rate	Time	Part
First panel	$\frac{1}{30}$	t	$\frac{t}{30}$
Second panel	$\frac{1}{45}$	t	$\frac{t}{45}$

The sum of the part of the task completed by the first panel and the part of the task completed by the second panel is 1.
$$\frac{t}{30}+\frac{t}{45}=1$$

Solution
$$\frac{t}{30}+\frac{t}{45}=1$$
$$90\left(\frac{t}{30}+\frac{t}{45}\right)=90(1)$$
$$3t+2t=90$$
$$5t=90$$
$$t=18$$
With both panels working, it would take 18 min to raise the temperature 1°.

39. Strategy
Unknown time to wire the telephone lines working together: t

	Rate	Time	Part
First member	$\frac{1}{5}$	t	$\frac{t}{5}$
Second member	$\frac{1}{7.5}$	t	$\frac{t}{7.5}$

The sum of the part of the task completed by the first member and the part of the task completed by the second member is 1.
$$\frac{t}{5}+\frac{t}{7.5}=1$$

Solution
$$\frac{t}{5}+\frac{t}{7.5}=1$$
$$15\left(\frac{t}{5}+\frac{t}{7.5}\right)=15(1)$$
$$3t+2t=15$$
$$5t=15$$
$$t=3$$
With both members working together, it would take 3 h to wire the telephone lines.

41. Strategy
Time for the new machine to package the transistors: t
Time for the old machine to package the transistors: $4t$

	Rate	Time	Part
New machine	$\frac{1}{t}$	8	$\frac{8}{t}$
Old machine	$\frac{1}{4t}$	8	$\frac{8}{4t}$

The sum of the parts of the task completed by each machine must equal 1.
$$\frac{8}{t}+\frac{8}{4t}=1$$

Solution
$$\frac{8}{t}+\frac{8}{4t}=1$$
$$4t\left(\frac{8}{t}+\frac{8}{4t}\right)=4t(1)$$
$$32+8=4t$$
$$40=4t$$
$$10=t$$
Working alone, the new machine would take 10 h to package the transistors.

43. Strategy
Time for the smaller printer to print the payroll: t

	Rate	Time	Part
Large printer	$\frac{1}{40}$	10	$\frac{10}{40}$
Smaller printer	$\frac{1}{t}$	60	$\frac{60}{t}$

The sum of the parts of the task completed by each printer must equal 1.
$$\frac{10}{40} + \frac{60}{t} = 1$$

Solution
$$\frac{10}{40} + \frac{60}{t} = 1$$
$$40t\left(\frac{10}{40} + \frac{60}{t}\right) = 40t(1)$$
$$10t + 2400 = 40t$$
$$2400 = 30t$$
$$80 = t$$
Working alone, the smaller printer would take 80 min to print the payroll.

45. Strategy
Time for the apprentice to shingle the roof: t

	Rate	Time	Part
Roofer	$\frac{1}{12}$	3	$\frac{3}{12}$
Apprentice	$\frac{1}{t}$	15	$\frac{15}{t}$

The sum of the part of the task completed by the roofer and the part of the task completed by the apprentice is 1.
$$\frac{3}{12} + \frac{15}{t} = 1$$

Solution
$$\frac{3}{12} + \frac{15}{t} = 1$$
$$12t\left(\frac{3}{12} + \frac{15}{t}\right) = 12t(1)$$
$$3t + 180 = 12t$$
$$180 = 9t$$
$$20 = t$$
It would take the apprentice 20 h to shingle the roof.

47. Strategy
Time to make a pizza if all three employees worked at the same time: t

	Rate	Time	Part
First employee	$\frac{1}{3.5}$	t	$\frac{t}{3.5}$
Second employee	$\frac{1}{2.5}$	t	$\frac{t}{2.5}$
Third employee	$\frac{1}{3.0}$	t	$\frac{t}{3.0}$

The sum of the parts of the job completed by each employee is 1.
$$\frac{t}{3.5} + \frac{t}{2.5} + \frac{t}{3.0} = 1$$

Solution
$$\frac{t}{3.5} + \frac{t}{2.5} + \frac{t}{3.0} = 1$$
$$105\left(\frac{t}{3.5} + \frac{t}{2.5} + \frac{t}{3.0}\right) = 105(1)$$
$$30t + 42t + 35t = 105$$
$$107t = 105$$
$$t \approx 0.98$$
It would take about 1 min to make the pizza if all three employees worked at the same time.

49. Strategy
Unknown time to fill the bathtub working together: t

	Rate	Time	Part
Faucets	$\frac{1}{10}$	t	$\frac{t}{10}$
Drain	$\frac{1}{15}$	t	$\frac{t}{15}$

The part of the task completed by the faucets minus the part of the task completed by the drain is 1.
$$\frac{t}{10} - \frac{t}{15} = 1$$

Solution
$$\frac{t}{10} - \frac{t}{15} = 1$$
$$30\left(\frac{t}{10} - \frac{t}{15}\right) = 30(1)$$
$$3t - 2t = 30$$
$$t = 30$$
In 30 min, the bathtub will start to overflow.

51. Strategy
Unknown time to fill the tank working together: t

	Rate	Time	Part
First inlet pipe	$\frac{1}{12}$	t	$\frac{t}{12}$
Second inlet pipe	$\frac{1}{20}$	t	$\frac{t}{20}$
Outlet pipe	$\frac{1}{10}$	t	$\frac{t}{10}$

The sum of the parts of the task completed by the inlet pipes minus the part of the task completed by the outlet pipe is 1.
$$\frac{t}{12} + \frac{t}{20} - \frac{t}{10} = 1$$

Solution
$$\frac{t}{12} + \frac{t}{20} - \frac{t}{10} = 1$$
$$60\left(\frac{t}{12} + \frac{t}{20} - \frac{t}{10}\right) = 60(1)$$
$$5t + 3t - 6t = 60$$
$$2t = 60$$
$$t = 30$$
When all three pipes are open, it will take 30 h to fill the tank.

Objective 3 Exercises

53. Strategy
Rate of the first skater: r
Rate of the second skater: $r - 3$

	Distance	Rate	Time
First skater	15	r	$\frac{15}{r}$
Second skater	12	$r-3$	$\frac{12}{r-3}$

The time of the first skater is equal to the time of the second skater.
$$\frac{15}{r} = \frac{12}{r-3}$$

Solution
$$\frac{15}{r} = \frac{12}{r-3}$$
$$r(r-3)\frac{15}{r} = r(r-3)\frac{12}{r-3}$$
$$(r-3)(15) = (r)(12)$$
$$15r - 45 = 12r$$
$$3r = 45$$
$$r = 15$$
$$r - 3 = 12$$
The first skater's rate is 15 mph, and the second skater's rate is 12 mph.

55. Strategy
Rate of the freight train: r
Rate of the passenger train: $r + 14$

	Distance	Rate	Time
Freight train	225	r	$\frac{225}{r}$
Passenger train	295	$r+14$	$\frac{295}{r+14}$

The time the freight train travels equals the time the passenger train travels.
$$\frac{225}{r} = \frac{295}{r+14}$$

Solution
$$\frac{225}{r} = \frac{295}{r+14}$$
$$r(r+14)\frac{225}{r} = r(r+14)\frac{295}{r+14}$$
$$(r+14)(225) = 295r$$
$$225r + 3150 = 295r$$
$$3150 = 70r$$
$$45 = r$$
$$r + 14 = 45 + 14 = 59$$
The rate of the freight train is 45 mph. The rate of the passenger train is 59 mph.

57. Strategy
Rate by foot: r
Rate by bicycle: $4r$

	Distance	Rate	Time
By foot	5	r	$\frac{5}{r}$
By bicycle	40	$4r$	$\frac{40}{4r}$

The total time spent walking and cycling was 5 h.
$$\frac{5}{r} + \frac{40}{4r} = 5$$

Solution
$$\frac{5}{r} + \frac{40}{4r} = 5$$
$$4r\left(\frac{5}{r} + \frac{40}{4r}\right) = 4r(5)$$
$$20 + 40 = 20r$$
$$60 = 20r$$
$$3 = r$$
$$4r = 4(3) = 12$$
The cyclist was riding 12 mph.

59. Strategy
Rate by foot: r
Rate by car: $12r$

	Distance	Rate	Time
By foot	4	r	$\frac{4}{r}$
By car	72	$12r$	$\frac{72}{12r}$

The total time spent riding and walking was 2.5 h.
$$\frac{4}{r}+\frac{72}{12r}=2.5$$

Solution
$$\frac{4}{r}+\frac{72}{12r}=2.5$$
$$12r\left(\frac{4}{r}+\frac{72}{12r}\right)=12r(2.5)$$
$$48+72=30r$$
$$120=30r$$
$$4=r$$
The motorist walks at the rate of 4 mph.

61. Strategy
Rate the executive can walk: r
Rate the executive walks using the moving sidewalk: $r+2$

	Distance	Rate	Time
Walking alone	360	r	$\frac{360}{r}$
Walking using moving sidewalk	480	$r+2$	$\frac{480}{r+2}$

It takes the executive the same time to walk 480 ft using the moving sidewalk as it takes to walk 360 ft without the moving sidewalk.
$$\frac{360}{r}=\frac{480}{r+2}$$

Solution
$$\frac{360}{r}=\frac{480}{r+2}$$
$$r(r+2)\frac{360}{r}=r(r+2)\frac{480}{r+2}$$
$$(r+2)(360)=(r)(480)$$
$$360r+720=480r$$
$$720=120r$$
$$6=r$$
The executive walks at a rate of 6 ft/s.

63. Strategy
Rate of the single-engine plane: r
Rate of the jet: $4r$

	Distance	Rate	Time
Single-engine plane	960	r	$\frac{960}{r}$
Jet	960	$4r$	$\frac{960}{4r}$

The time for the single-engine plane is 4 h more than the time for the jet.
$$\frac{960}{4r}+4=\frac{960}{r}$$

Solution
$$\frac{960}{4r}+4=\frac{960}{r}$$
$$4r\left(\frac{960}{4r}+4\right)=4r\left(\frac{960}{r}\right)$$
$$960+16r=3840$$
$$16r=2880$$
$$r=180$$
$$4r=4(180)=720$$
The rate of the single-engine plane is 180 mph.
The rate of the jet is 720 mph.

65. Strategy
Rate of the Gulf Stream: r
Rate sailing with the Gulf Stream: $28+r$
Rate sailing against the Gulf Stream: $28-r$

	Distance	Rate	Time
With Gulf Stream	170	$28+r$	$\frac{170}{28+r}$
Against Gulf Stream	110	$28-r$	$\frac{110}{28-r}$

It takes the same time to sail 170 mi with the Gulf Stream as it takes to sail 110 mi against the Gulf Stream.
$$\frac{170}{28+r}=\frac{110}{28-r}$$

Solution
$$\frac{170}{28+r}=\frac{110}{28-r}$$
$$(28-r)(28+r)\frac{170}{28+r}=(28-r)(28+r)\frac{110}{28-r}$$
$$(28-r)(170)=(28+r)(110)$$
$$4760-170r=3080+110r$$
$$1680=280r$$
$$6=r$$
The rate of the Gulf Stream is 6 mph.

178   Chapter 6: Rational Expressions

67. Strategy
Rate of the current: r

	Distance	Rate	Time
With current	20	$7+r$	$\frac{20}{7+r}$
Against current	8	$7-r$	$\frac{8}{7-r}$

The time traveling with the current equals the time traveling against the current.
$$\frac{20}{7+r} = \frac{8}{7-r}$$

Solution
$$\frac{20}{7+r} = \frac{8}{7-r}$$
$$(7+r)(7-r)\frac{20}{7+r} = (7+r)(7-r)\frac{8}{7-r}$$
$$(7-r)20 = (7+r)8$$
$$140 - 20r = 56 + 8r$$
$$140 = 56 + 28r$$
$$84 = 28r$$
$$3 = r$$
The rate of the current is 3 mph.

Applying Concepts 6.4

69. Strategy
Numerator: n
Denominator: $n + 4$
Numerator increased by 3: $n + 3$
Denominator increased by 3: $(n + 4) + 3 = n + 7$
If the numerator and denominator of the fraction are increased by 3, the new fraction is $\frac{5}{6}$.

Solution
$$\frac{n+3}{n+7} = \frac{5}{6}$$
$$6(n+7)\left(\frac{n+3}{n+7}\right) = 6(n+7)\left(\frac{5}{6}\right)$$
$$6(n+3) = (n+7)(5)$$
$$6n + 18 = 5n + 35$$
$$n + 18 = 35$$
$$n = 17$$
$$n + 4 = 17 + 4 = 21$$
The original fraction is $\frac{17}{21}$.

71. Strategy
Time to complete the job: t

	Rate	Time	Part
First printer	$\frac{1}{24}$	t	$\frac{t}{24}$
Second printer	$\frac{1}{16}$	t	$\frac{t}{16}$
Third printer	$\frac{1}{12}$	t	$\frac{t}{12}$

The sum of the parts of the task completed by each printer is 1.

Solution
$$\frac{t}{24} + \frac{t}{16} + \frac{t}{12} = 1$$
$$48\left(\frac{t}{24} + \frac{t}{16} + \frac{t}{12}\right) = 48(1)$$
$$2t + 3t + 4t = 48$$
$$9t = 48$$
$$t = \frac{48}{9} = \frac{16}{3}$$
It would take $5\frac{1}{3}$ min to print the checks.

73. Strategy
Usual speed: r
Reduced speed: $r - 5$

	Distance	Rate	Time
Usual rate	165	r	$\frac{165}{r}$
Reduced speed	165	$r-5$	$\frac{165}{r-5}$

At the decreased speed, the drive takes 15 min more than usual. $\left(\text{Note: } 15 \text{ min} = \frac{1}{4} \text{ h}\right)$

Solution
$$\frac{165}{r} = \frac{165}{r-5} - \frac{1}{4}$$
$$4r(r-5)\left(\frac{165}{r}\right) = 4r(r-5)\left(\frac{165}{r-5} - \frac{1}{4}\right)$$
$$4(r-5)(165) = 4r(165) - r(r-5)$$
$$660(r-5) = 660r - r^2 + 5r$$
$$660r - 3300 = 665r - r^2$$
$$r^2 - 5r - 3300 = 0$$
$$(r-60)(r+55) = 0$$
$$r - 60 = 0 \qquad r + 55 = 0$$
$$r = 60 \qquad r = -55$$
Because the rate cannot be negative, −55 is not a solution. The bus usually travels at 60 mph.

75. **Strategy**
    The rate running from front of parade to back of parade is the sum of the runner's rate and the rate of the parade (5 mph + 3 mph).
    The rate running from back of parade to front of parade is the difference between the runner's rate and the rate of the parade (5 mph − 3 mph).

	Distance	Rate	Time
Back to front	1	2	$\frac{1}{2}$
Front to back	1	8	$\frac{1}{8}$

    The sum of the time running from front to back and the time running from back to front is the total time t that the runner runs.
    $$\frac{1}{2}+\frac{1}{8}=t$$

    Solution
    $$\frac{1}{2}+\frac{1}{8}=t$$
    $$\frac{5}{8}=t$$

    The total time is $\frac{5}{8}$ h.

Section 6.5

Concept Review 6.5

1. Never true
   If x varies inversely as y, then when x is doubled, y is halved.

3. Never true
   If a varies jointly as b and c, then $a = kbc$.

5. Never true
   If the area of a triangle is held constant, then the length varies inversely as the width.

Objective 1 Exercises

3. **Strategy**
   To find the number of ducks in the preserve, write and solve a proportion using x to represent the number of ducks in the preserve.

   Solution
   $$\frac{60}{x}=\frac{3}{200}$$
   $$200x \cdot \frac{60}{x}=\frac{3}{200} \cdot 200x$$
   $$12{,}000 = 3x$$
   $$4000 = x$$
   There are 4000 ducks in the preserve.

5. **Strategy**
   To find the amount of additional fruit punch, write and solve a proportion using x to represent the additional amount of fruit punch. Then $x + 2$ is the total amount of fruit punch.

   Solution
   $$\frac{2}{30}=\frac{x+2}{75}$$
   $$\frac{2}{30}\cdot 150 = \frac{x+2}{75}\cdot 150$$
   $$2\cdot 5 = (x+2)2$$
   $$10 = 2x+4$$
   $$6 = 2x$$
   $$3 = x$$
   To serve 75 people, 3 additional gallons of fruit punch are necessary.

7. **Strategy**
   To find the dimensions of the room, write and solve two proportions using L to represent the length of the room and W to represent the width of the room.

   Solution
   $$\frac{\frac{1}{4}}{1}=\frac{4\frac{1}{4}}{W} \qquad \frac{\frac{1}{4}}{1}=\frac{5\frac{1}{2}}{L}$$
   $$\frac{\frac{1}{4}}{1}\cdot W = \frac{4\frac{1}{4}}{W}\cdot W \qquad \frac{\frac{1}{4}}{1}\cdot L = \frac{5\frac{1}{2}}{L}\cdot L$$
   $$\frac{1}{4}W = 4\frac{1}{4} \qquad \frac{1}{4}L = 5\frac{1}{2}$$
   $$W = 17 \qquad L = 22$$
   The dimensions of the room are 17 ft by 22 ft.

9. **Strategy**
   To find the number of defective diodes, write and solve a proportion using x to represent the number of defective diodes.

   Solution
   $$\frac{5}{4000}=\frac{x}{3200}$$
   $$\frac{5}{4000}\cdot(4000)(3200) = \frac{x}{3200}(4000)(3200)$$
   $$5(3200) = 4000x$$
   $$16{,}000 = 4000x$$
   $$4 = x$$
   In a shipment of 3200, 4 diodes would be defective.

11. Strategy
To find the additional amount of medicine, write and solve a proportion using x to represent the additional amount of medicine. Then $x + 0.75$ is the total amount of medicine.

Solution
$$\frac{0.75}{120} = \frac{x + 0.75}{200}$$
$$\frac{0.75}{120} \cdot 600 = \frac{x + 0.75}{200} \cdot 600$$
$$0.75(5) = (x + 0.75)3$$
$$3.75 = 3x + 2.25$$
$$1.5 = 3x$$
$$0.5 = x$$
An additional 0.5 oz of medicine is required.

13. Strategy
To find the additional amount of insecticide, write and solve a proportion using x to represent the additional amount of insecticide. Then $x + 6$ is the total amount of insecticide.

Solution
$$\frac{6}{15} = \frac{x + 6}{100}$$
$$\frac{6}{15} \cdot 300 = \frac{x + 6}{100} \cdot 300$$
$$6 \cdot 20 = (x + 6)3$$
$$120 = 3x + 18$$
$$102 = 3x$$
$$34 = x$$
An additional 34 oz of insecticide are required.

15. Strategy
Let x represent the person's height when standing. Then $\frac{3}{4}x$ is the person's height when kneeling.
$$\frac{3}{4}x = 4$$

Solution
$$\frac{3}{4}x = 4$$
$$\left(\frac{4}{3}\right)\frac{3}{4}x = \left(\frac{4}{3}\right)(4)$$
$$x = \frac{16}{3}$$
$$x = 5.3$$
The person is approximately 5.3 ft tall.

17. Strategy
To find the actual length of the whale, write and solve a proportion using x to represent the actual length of the whale. The length of the whale in the picture is 2 in.

Solution
$$\frac{1}{48} = \frac{2}{x}$$
$$\frac{1}{48}(48)x = \frac{2}{x}(48)x$$
$$x = 96$$
The whale is 96 ft long.

Objective 2 Exercises

21. Strategy
To find the profit:
Write the basic direct variation equation, replace the variables by the given values, and solve for k.

Write the direct variation equation, replacing k by its value. Substitute 300 for s and solve for P.

Solution
$$P = ks$$
$$2500 = k(20)$$
$$125 = k$$
$$P = 125s = 125(300) = 37,500$$
When the company sells 300 products, the profit is $37,500.

23. Strategy
To find the pressure:
Write the basic direct variation equation, replace the variables by the given values, and solve for k.

Write the direct variation equation, replacing k by its value. Substitute 30 for d and solve for p.

Solution
$$p = kd$$
$$3.6 = k(8)$$
$$0.45 = k$$
$$p = 0.45d = 0.45(30) = 13.5$$
The pressure is 13.5 pounds per square inch.

25. Strategy
To find the distance:
Write the basic direct variation equation, replace the variables by the given values, and solve for k.

Write the direct variation equation, replacing k by its value. Substitute 800 for H and solve for d.

Solution
$$d = k\sqrt{H}$$
$$19 = k\sqrt{500}$$
$$19 = k(22.36)$$
$$0.85 = k$$
$$d = 0.85\sqrt{H} = 0.85\sqrt{800} = 0.85(28.28) = 24.04$$
The horizon is 24.04 mi away from a point that is 800 ft high.

27. Strategy
To find the distance:
Write the basic direct variation equation, replace the variables by the given values, and solve for k.

Write the direct variation equation, replacing k by its value. Substitute 4 for t and solve for s.

Solution
$s = kt^2$
$5 = k(1)^2$
$5 = k$
$s = 5t^2 = 5(4)^2 = 5(16) = 80$
In 4 s, the ball will roll 80 ft.

29. Strategy
To find the length of the person's face:
Write the basic direct variation equation, replace the variables by the given values, and solve for k.

Write the direct variation equation, replacing k by its value. Substitute 1.7 for x and solve for y.

Solution
$y = kx$ $y = 6x$
$9 = k(1.5)$ $y = 6(1.7)$
$6 = k$ $y = 10.2$
The person's face is 10.2 in. long.

31. Strategy
To find the pressure:
Write the basic inverse variation equation, replace the variables by the given values, and solve for k.

Write the inverse variation equation, replacing k by its value. Substitute 150 for V and solve for P.

Solution
$P = \dfrac{k}{V}$
$25 = \dfrac{k}{400}$
$10{,}000 = k$
$P = \dfrac{10{,}000}{V} = \dfrac{10{,}000}{150} = 66\dfrac{2}{3}$
When the volume is 150 ft^3, the pressure is $66\dfrac{2}{3}$ pounds per square inch.

33. Strategy
To find the pressure:
Write the basic combined variation equation, replace the variables by the given values, and solve for k.

Write the combined variation equation, replacing k by its value. Substitute 60 for d and 1.2 for D, and solve for p.

Solution
$p = kdD$
$37.5 = k(100)(1.2)$
$37.5 = 120k$
$0.3125 = k$
$p = 0.3125dD = 0.3125(60)(1.2) = 22.5$
The pressure is 22.5 pounds per square inch.

35. Strategy
To find the repulsive force:
Write the basic inverse variation equation, replace the variables by the given values, and solve for k.

Write the inverse variation equation, replacing k by its value. Substitute 1.2 for d and solve for f.

Solution
$f = \dfrac{k}{d^2}$
$18 = \dfrac{k}{3^2}$
$18 = \dfrac{k}{9}$
$162 = k$
$f = \dfrac{162}{d^2} = \dfrac{162}{1.2^2} = \dfrac{162}{1.44} = 112.5$
The repulsive force is 112.5 lb when the distance is 1.2 in.

182 Chapter 6: Rational Expressions

37. Strategy
To find the resistance:
Write the basic combined variation equation, replace the variables by the given values, and solve for k.

Write the combined variation equation, replacing k by its value. Substitute 50 for L and 0.02 for d, and solve for R.

Solution
$$R = \frac{kL}{d^2}$$
$$9 = \frac{k(50)}{(0.05)^2}$$
$$9 = \frac{50k}{0.0025}$$
$$0.0225 = 50k$$
$$0.00045 = k$$
$$R = \frac{0.00045L}{d^2}$$
$$= \frac{0.00045(50)}{(0.02)^2}$$
$$= \frac{0.00045(50)}{0.0004}$$
$$= 56.25$$
The resistance is 56.25 ohms.

39. Strategy
To find the wind force:
Write the basic combined variation equation, replace the variables by the given values, and solve for k.

Write the combined variation equation, replacing k by its value. Substitute 10 for A and 60 for v, and solve for w.

Solution
$$w = kAv^2$$
$$45 = k(10)(30)^2$$
$$45 = k(10)(900)$$
$$45 = 9000k$$
$$0.005 = k$$
$$w = 0.005Av^2$$
$$= 0.005(10)(60)^2$$
$$= 0.005(10)(3600)$$
$$= 180$$
The wind force is 180 lb.

Applying Concepts 6.5

41. a.

b. The graph represents a linear function.

43. a.

b. The graph is the graph of a function.

45. If y doubles, x is halved.

47. If a varies directly as b and inversely as c, then c varies *directly* as b and *inversely* as a.

49. If the width of a rectangle is held constant, the area of the rectangle varies *directly* as the length.

51. a. The unit for the sales per share ratio is dollars per share.

b. $\dfrac{5269 \text{ million}}{60.49 \text{ million}} = 87.11$
$\dfrac{5879 \text{ million}}{62.17 \text{ million}} = 94.56$
$\dfrac{7213 \text{ million}}{63.60 \text{ million}} = 113.41$
$\dfrac{6972 \text{ million}}{72.72 \text{ million}} = 95.87$
$\dfrac{6881 \text{ million}}{73.91 \text{ million}} = 93.10$

For 1993 the sales per share ratio was $87.11 per share.
For 1994 the sales per share ratio was $94.56 per share.
For 1995 the sales per share ratio was $113.41 per share.
For 1996 the sales per share ratio was $95.87 per share.
For 1997 the sales per share ratio was $93.10 per share.

c. $f(x) = 0.148x^{1.4227}$
$f(94) = 0.148(94)^{1.4227} = 94.94$
$$\% \text{ error} = \left| \frac{\text{actual value} - \text{predicted value}}{\text{actual value}} \right| \times 100$$
$$\% \text{ error} = \left| \frac{94.56 - 94.94}{94.56} \right| \times 100 = 0.0040 \times 100$$
The percent error between the actual value and the predicted value is 0.40%.

Section 6.6

Concept Review 6.6

1. Always true

3. Never true
$$R_2 = \frac{R_1 R}{R_1 - R}$$

Objective 1 Exercises

1. $P = 2L + 2W$
$P - 2L = 2W$
$\dfrac{P - 2L}{2} = W$

3. $S = C - rC$
$S = C(1 - r)$
$\dfrac{S}{1 - r} = C$

5. $PV = nRT$
$\dfrac{PV}{nT} = R$

7. $F = \dfrac{Gm_1 m_2}{r^2}$
$Fr^2 = Gm_1 m_2$
$\dfrac{Fr^2}{Gm_1} = m_2$

9. $I = \dfrac{E}{R + r}$
$I(R + r) = E$
$IR + Ir = E$
$IR = E - Ir$
$R = \dfrac{E - Ir}{I}$

11. $A = \dfrac{1}{2} h(b_1 + b_2)$
$2A = h(b_1 + b_2)$
$2A = hb_1 + hb_2$
$2A - hb_1 = hb_2$
$\dfrac{2A - hb_1}{h} = b_2$

13. $\dfrac{1}{R} = \dfrac{1}{R_1} + \dfrac{1}{R_2}$
$RR_1 R_2 \left(\dfrac{1}{R}\right) = RR_1 R_2 \left(\dfrac{1}{R_1} + \dfrac{1}{R_2}\right)$
$R_1 R_2 = RR_1 R_2 \left(\dfrac{1}{R_1}\right) + RR_1 R_2 \left(\dfrac{1}{R_2}\right)$
$R_1 R_2 = RR_2 + RR_1$
$R_1 R_2 - RR_2 = RR_1$
$R_2(R_1 - R) = RR_1$
$R_2 = \dfrac{RR_1}{R_1 - R}$

15. $a_n = a_1 + (n - 1)d$
$a_n - a_1 = (n - 1)d$
$\dfrac{a_n - a_1}{n - 1} = d$

17. $S = 2WH + 2WL + 2LH$
$S - 2WL = 2WH + 2LH$
$S - 2WL = H(2W + 2L)$
$\dfrac{S - 2WL}{2W + 2L} = H$

19. $ax + by + c = 0$
$ax + by = -c$
$ax = -by - c$
$x = \dfrac{-by - c}{a}$

21. $ax + b = cx + d$
$ax + b - cx = d$
$ax - cx = d - b$
$x(a - c) = d - b$
$x = \dfrac{d - b}{a - c}$

23. $\dfrac{a}{x} = \dfrac{b}{c}$
$xc \cdot \dfrac{a}{x} = xc \cdot \dfrac{b}{c}$
$ac = xb$
$\dfrac{ac}{b} = x$

25. $\dfrac{1}{a} + \dfrac{1}{b} = \dfrac{1}{x}$
$abx \left(\dfrac{1}{a} + \dfrac{1}{b}\right) = abx \left(\dfrac{1}{x}\right)$
$bx + ax = ab$
$x(a + b) = ab$
$x = \dfrac{ab}{a + b}$

Applying Concepts 6.6

27. $\dfrac{x - y}{y} = \dfrac{x + 5}{2y}$
$2y \left(\dfrac{x - y}{y}\right) = 2y \left(\dfrac{x + 5}{2y}\right)$
$2(x - y) = x + 5$
$2x - 2y = x + 5$
$x - 2y = 5$
$x = 2y + 5$
The solution is $2y + 5$.

29.
$$\frac{x}{x+y} = \frac{2x}{4y}$$
$$(4y)(x+y)\left(\frac{x}{x+y}\right) = (4y)(x+y)\left(\frac{2x}{4y}\right)$$
$$(4y)(x) = (x+y)(2x)$$
$$4xy = 2x^2 + 2xy$$
$$2xy = 2x^2$$
$$0 = 2x^2 - 2xy$$
$$0 = 2x(x-y)$$

$2x = 0$ $x - y = 0$
$x = 0$ $x = y$

The solutions are 0 and y.

31.
$$\frac{x-y}{2x} = \frac{x-3y}{5y}$$
$$(2x)(5y)\left(\frac{x-y}{2x}\right) = (2x)(5y)\left(\frac{x-3y}{5y}\right)$$
$$(5y)(x-y) = (2x)(x-3y)$$
$$5xy - 5y^2 = 2x^2 - 6xy$$
$$-5y^2 = 2x^2 - 11xy$$
$$0 = 2x^2 - 11xy + 5y^2$$
$$0 = (2x-y)(x-5y)$$

$2x - y = 0$ $x - 5y = 0$
$2x = y$ $x = 5y$
$x = \dfrac{y}{2}$

The solutions are $\dfrac{y}{2}$ and $5y$.

33.
$$\frac{w_1}{w_2} = \frac{f_2 - f}{f - f_1}$$
$$w_2(f - f_1)\left(\frac{w_1}{w_2}\right) = \left(\frac{f_2 - f}{f - f_1}\right)w_2(f - f_1)$$
$$(f - f_1)w_1 = (f_2 - f)w_2$$
$$fw_1 - f_1 w_1 = f_2 w_2 - fw_2$$
$$fw_1 + fw_2 = f_2 w_2 + f_1 w_1$$
$$f(w_1 + w_2) = f_2 w_2 + f_1 w_1$$
$$f = \frac{f_2 w_2 + f_1 w_1}{w_1 + w_2}$$

Focus on Problem Solving

1. Statement: If a number is divisible by 8, then it is divisible by 4.
   Contrapositive: If a number is not divisible by 4, then it is not divisible by 8.
   Converse: If a number is divisible by 4, then it is divisible by 8.
   The converse is not true.

2. Statement: If $4z = 20$, then $z = 5$.
   Contrapositive: If $z \neq 5$, then $4z \neq 20$.
   Converse: If $z = 5$, then $4z = 20$.
   The converse is true. $4z = 20$ if and only if $z = 5$.

3. Statement: If p is a prime number greater than 2, then p is an odd number.
   Contrapositive: If p is not an odd number then it is not a prime number greater than 2.
   Converse: If p is an odd number, then it is a prime number greater than 2.
   The converse is not true.

4. Statement: If x is a rational number, then $x = \dfrac{a}{b}$, where a and b are integers and $b \neq 0$.
   Contrapositive: If a and b are integers, $b \neq 0$, and $x \neq \dfrac{a}{b}$, then x is not a rational number.
   Converse: If $x = \dfrac{a}{b}$, where a and b are integers and $b \neq 0$, then x is a rational number.
   The converse is true. x is a rational number if and only if $x = \dfrac{a}{b}$, where a and b are integers and $b \neq 0$.

5. If $\sqrt{x} = a$, then $a^2 = x$.
   Contrapositive: If $a^2 \neq x$, then $\sqrt{x} \neq a$.
   Converse: If $a^2 = x$, then $\sqrt{x} = a$.
   The converse is not true.

6. Statement: If the equation of a graph is $y = mx + b$, then the graph of the equation is a straight line.
   Contrapositive: If the graph of an equation is not a straight line, then the equation of the graph is not $y = mx + b$.
   Converse: If the graph of an equation is a straight line, then the equation of the graph is $y = mx + b$.
   The converse is true. The equation of a graph is $y = mx + b$ if and only if the graph of the equation is a straight line.

7. Statement: If $a = 0$ or $b = 0$, then $ab = 0$.
   Contrapositive: If $ab \neq 0$, then $a \neq 0$ and $b \neq 0$.
   Converse: If $ab = 0$, then $a = 0$ or $b = 0$.
   The converse is true. $a = 0$ or $b = 0$ if and only if $ab = 0$.

8. Statement: If the coordinates of a point are (5, 0), then the point is on the x-axis.
   Contrapositive: If a point is not on the x-axis, then the coordinates of the point are not (5, 0).
   Converse: If a point is on the x-axis, then the coordinates of the point are (5, 0).
   The converse is not true.

9. Statement: If a quadrilateral is a square, then the quadrilateral has four sides of equal length.
   Contrapositive: If a quadrilateral does not have four sides of equal length, then the quadrilateral is not a square.
   Converse: If a quadrilateral has four sides of equal length, then the quadrilateral is a square.
   The converse is not true.

Projects and Group Activities
Continued Fractions

1. $c_5 = 1 + \cfrac{1}{1+\cfrac{1}{1+\cfrac{1}{1+\cfrac{1}{1+\cfrac{1}{1+1}}}}}$

$c_5 = 1.61538$

2. $c_5 = 3 + \cfrac{1^2}{6+\cfrac{3^2}{6+\cfrac{5^2}{6+\cfrac{7^2}{6+\cfrac{9^2}{6+11^2}}}}}$

$c_5 = 3.13998$

Chapter Review Exercises

1. $P(x) = \dfrac{x}{x-3}$

$P(4) = \dfrac{4}{4-3} = \dfrac{4}{1}$

$P(4) = 4$

2. $P(x) = \dfrac{x^2 - 2}{3x^2 - 2x + 5}$

$P(-2) = \dfrac{(-2)^2 - 2}{3(-2)^2 - 2(-2) + 5} = \dfrac{4-2}{12+4+5} = \dfrac{2}{21}$

$P(-2) = \dfrac{2}{21}$

3. $g(x) = \dfrac{2x}{x-3}$

$x - 3 = 0$

$x = 3$

The domain is $\{x | x \neq 3\}$.

4. $f(x) = \dfrac{2x-7}{3x^2 + 3x - 18}$

$3x^2 + 3x - 18 = 0$

$3(x^2 + x - 6) = 0$

$3(x+3)(x-2) = 0$

$x + 3 = 0 \qquad x - 2 = 0$

$x = -3 \qquad x = 2$

The domain is $\{x | x \neq -3, x \neq 2\}$.

5. The domain must exclude values of x for which $3x^2 + 4 = 0$. This is not possible, because $3x^2 \geq 0$, and a positive number added to a number equal to or greater than zero cannot equal zero. Therefore, there are no real numbers that must be excluded from the domain of F.

The domain of $F(x)$ is $\{x | x \text{ is a real number}\}$.

6. $\dfrac{6a^{5n} + 4a^{4n} - 2a^{3n}}{2a^{3n}} = 3a^{2n} + 2a^n - 1$

7. $\dfrac{16 - x^2}{x^3 - 2x^2 - 8x} = \dfrac{(4+x)(4-x)}{x(x^2 - 2x - 8)}$

$= \dfrac{(4+x)(4-x)}{x(x-4)(x+2)}$

$= \dfrac{(4+x)(-1)}{x(1)(x+2)}$

$= -\dfrac{x+4}{x(x+2)}$

8. $\dfrac{x^3 - 27}{x^2 - 9} = \dfrac{(x-3)(x^2 + 3x + 9)}{(x+3)(x-3)} = \dfrac{x^2 + 3x + 9}{x+3}$

9. $\dfrac{a^6 b^4 + a^4 b^6}{a^5 b^4 - a^4 b^4} \cdot \dfrac{a^2 - b^2}{a^4 - b^4}$

$= \dfrac{a^4 b^4 (a^2 + b^2)}{a^4 b^4 (a-1)} \cdot \dfrac{(a+b)(a-b)}{(a^2 + b^2)(a+b)(a-b)}$

$= \dfrac{a^4 b^4 (a^2 + b^2)(a+b)(a-b)}{a^4 b^4 (a-1)(a^2 + b^2)(a+b)(a-b)}$

$= \dfrac{1}{a-1}$

10. $\dfrac{x^3 - 8}{x^3 + 2x^2 + 4x} \cdot \dfrac{x^3 + 2x^2}{x^2 - 4}$

$= \dfrac{(x-2)(x^2 + 2x + 4)}{x(x^2 + 2x + 4)} \cdot \dfrac{x^2(x+2)}{(x+2)(x-2)}$

$= \dfrac{(x-2)(x^2 + 2x + 4) \cdot x^2(x+2)}{x(x^2 + 2x + 4)(x+2)(x-2)}$

$= x$

11. $\dfrac{16 - x^2}{6x - 6} \cdot \dfrac{x^2 + 5x + 6}{x^2 - 8x + 16}$

$= \dfrac{(4+x)(4-x)}{6(x-1)} \cdot \dfrac{(x+3)(x+2)}{(x-4)(x-4)}$

$= \dfrac{(4+x)(4-x)(x+3)(x+2)}{6(x-1)(x-4)(x-4)}$

$= -\dfrac{(x+4)(x+3)(x+2)}{6(x-1)(x-4)}$

12. $\dfrac{x^{2n} - 5x^n + 4}{x^{2n} - 2x^n - 8} \div \dfrac{x^{2n} - 4x^n + 3}{x^{2n} + 8x^n + 12}$

$= \dfrac{x^{2n} - 5x^n + 4}{x^{2n} - 2x^n - 8} \cdot \dfrac{x^{2n} + 8x^n + 12}{x^{2n} - 4x^n + 3}$

$= \dfrac{(x^n - 4)(x^n - 1)}{(x^n - 4)(x^n + 2)} \cdot \dfrac{(x^n + 6)(x^n + 2)}{(x^n - 3)(x^n - 1)}$

$= \dfrac{(x^n - 4)(x^n - 1)(x^n + 6)(x^n + 2)}{(x^n - 4)(x^n + 2)(x^n - 3)(x^n - 1)}$

$= \dfrac{x^n + 6}{x^n - 3}$

13. $\dfrac{27x^3-8}{9x^3+6x^2+4x} \div \dfrac{9x^2-12x+4}{9x^2-4}$

$= \dfrac{27x^3-8}{9x^3+6x^2+4x} \cdot \dfrac{9x^2-4}{9x^2-12x+4}$

$= \dfrac{(3x-2)(9x^2+6x+4)}{x(9x^2+6x+4)} \cdot \dfrac{(3x+2)(3x-2)}{(3x-2)(3x-2)}$

$= \dfrac{(3x-2)(9x^2+6x+4)(3x+2)(3x-2)}{x(9x^2+6x+4)(3x-2)(3x-2)}$

$= \dfrac{3x+2}{x}$

14. $\dfrac{3-x}{x^2+3x+9} \div \dfrac{x^2-9}{x^3-27}$

$= \dfrac{3-x}{x^2+3x+9} \cdot \dfrac{x^3-27}{x^2-9}$

$= \dfrac{3-x}{x^2+3x+9} \cdot \dfrac{(x-3)(x^2+3x+9)}{(x+3)(x-3)}$

$= \dfrac{(3-x)(x-3)(x^2+3x+9)}{(x^2+3x+9)(x+3)(x-3)}$

$= \dfrac{3-x}{x+3}$

$= \dfrac{(-1)(x-3)}{x+3}$

$= -\dfrac{x-3}{x+3}$

15. The LCM is $24a^2b^4$.

$\dfrac{5}{3a^2b^3} + \dfrac{7}{8ab^4} = \dfrac{5}{3a^2b^3} \cdot \dfrac{8b}{8b} + \dfrac{7}{8ab^4} \cdot \dfrac{3a}{3a}$

$= \dfrac{40b+21a}{24a^2b^4}$

16. $\dfrac{3x^2+2}{x^2-4} - \dfrac{9x-x^2}{x^2-4} = \dfrac{3x^2+2-(9x-x^2)}{x^2-4}$

$= \dfrac{3x^2+2-9x+x^2}{x^2-4}$

$= \dfrac{4x^2-9x+2}{x^2-4}$

$= \dfrac{(4x-1)(x-2)}{(x+2)(x-2)}$

$= \dfrac{4x-1}{x+2}$

17. The LCM is $(3x+2)(3x-2)$.

$\dfrac{8}{9x^2-4} + \dfrac{5}{3x-2} - \dfrac{4}{3x+2} = \dfrac{8}{(3x+2)(3x-2)} + \dfrac{5}{3x-2} - \dfrac{4}{3x+2}$

$= \dfrac{8}{(3x+2)(3x-2)} + \dfrac{5}{3x-2} \cdot \dfrac{3x+2}{3x+2} - \dfrac{4}{3x+2} \cdot \dfrac{3x-2}{3x-2}$

$= \dfrac{8+5(3x+2)-4(3x-2)}{(3x+2)(3x-2)}$

$= \dfrac{8+15x+10-12x+8}{(3x+2)(3x-2)}$

$= \dfrac{3x+26}{(3x+2)(3x-2)}$

18. $3x^2 - 7x + 2 = (3x-1)(x-2)$
The LCM is $(3x-1)(x-2)$.

$$\frac{6x}{3x^2-7x+2} - \frac{2}{3x-1} + \frac{3x}{x-2} = \frac{6x}{(3x-1)(x-2)} - \frac{2}{3x-1}\cdot\frac{x-2}{x-2} + \frac{3x}{x-2}\cdot\frac{3x-1}{3x-1}$$

$$= \frac{6x - 2(x-2) + 3x(3x-1)}{(3x-1)(x-2)}$$

$$= \frac{6x - 2x + 4 + 9x^2 - 3x}{(3x-1)(x-2)}$$

$$= \frac{9x^2 + x + 4}{(3x-1)(x-2)}$$

19. The LCM is $(x-3)(x+2)$.

$$\frac{x}{x-3} - 4 - \frac{2x-5}{x+2} = \frac{x}{x-3}\cdot\frac{x+2}{x+2} - 4\cdot\frac{(x-3)(x+2)}{(x-3)(x+2)} - \frac{2x-5}{x+2}\cdot\frac{x-3}{x-3}$$

$$= \frac{x^2 + 2x - 4(x^2 - x - 6) - (2x^2 - 11x + 15)}{(x-3)(x+2)}$$

$$= \frac{x^2 + 2x - 4x^2 + 4x + 24 - 2x^2 + 11x - 15}{(x-3)(x+2)}$$

$$= \frac{-5x^2 + 17x + 9}{(x-3)(x+2)}$$

$$= \frac{-(5x^2 - 17x - 9)}{(x-3)(x+2)}$$

$$= -\frac{5x^2 - 17x - 9}{(x-3)(x+2)}$$

20. The LCM is $x - 1$.

$$\frac{x - 6 + \frac{6}{x-1}}{x + 3 - \frac{12}{x-1}} = \frac{x - 6 + \frac{6}{x-1}}{x + 3 - \frac{12}{x-1}}\cdot\frac{x-1}{x-1}$$

$$= \frac{(x-6)(x-1) + 6}{(x+3)(x-1) - 12}$$

$$= \frac{x^2 - 7x + 6 + 6}{x^2 + 2x - 3 - 12}$$

$$= \frac{x^2 - 7x + 12}{x^2 + 2x - 15}$$

$$= \frac{(x-3)(x-4)}{(x+5)(x-3)}$$

$$= \frac{x-4}{x+5}$$

21. The LCM is $x - 4$.

$$\frac{x + \frac{3}{x-4}}{3 + \frac{x}{x-4}} = \frac{x + \frac{3}{x-4}}{3 + \frac{x}{x-4}}\cdot\frac{x-4}{x-4}$$

$$= \frac{x(x-4) + 3}{3(x-4) + x}$$

$$= \frac{x^2 - 4x + 3}{3x - 12 + x}$$

$$= \frac{(x-3)(x-1)}{4x - 12}$$

$$= \frac{(x-3)(x-1)}{4(x-3)}$$

$$= \frac{x-1}{4}$$

Chapter 6: *Rational Expressions*

22.
$$\frac{5x}{2x-3}+4=\frac{3}{2x-3}$$
$$(2x-3)\left(\frac{5x}{2x-3}+4\right)=\frac{3}{2x-3}(2x-3)$$
$$5x+4(2x-3)=3$$
$$5x+8x-12=3$$
$$13x-12=3$$
$$13x=15$$
$$x=\frac{15}{13}$$

The solution is $\frac{15}{13}$.

23.
$$\frac{x}{x-3}=\frac{2x+5}{x+1}$$
$$(x-3)(x+1)\left(\frac{x}{x-3}\right)=\left(\frac{2x+5}{x+1}\right)(x-3)(x+1)$$
$$(x+1)x=(2x+5)(x-3)$$
$$x^2+x=2x^2-x-15$$
$$0=x^2-2x-15$$
$$0=(x-5)(x+3)$$

$x-5=0 \qquad x+3=0$
$x=5 \qquad x=-3$

The solutions are 5 and –3.

24.
$$\frac{6}{x-3}-\frac{1}{x+3}=\frac{51}{x^2-9}$$
$$\frac{6}{x-3}-\frac{1}{x+3}=\frac{51}{(x+3)(x-3)}$$
$$(x+3)(x-3)\left(\frac{6}{x-3}-\frac{1}{x+3}\right)=\frac{51}{(x+3)(x-3)}\cdot(x+3)(x-3)$$
$$6(x+3)-1(x-3)=51$$
$$6x+18-x+3=51$$
$$5x+21=51$$
$$5x=30$$
$$x=6$$

The solution is 6.

25.
$$\frac{30}{x^2+5x+4}+\frac{10}{x+4}=\frac{4}{x+1}$$
$$\frac{30}{(x+4)(x+1)}+\frac{10}{x+4}=\frac{4}{x+1}$$
$$(x+4)(x+1)\left(\frac{30}{(x+4)(x+1)}+\frac{10}{x+4}\right)=\frac{4}{x+1}\cdot(x+4)(x+1)$$
$$30+10(x+1)=4(x+4)$$
$$30+10x+10=4x+16$$
$$10x+40=4x+16$$
$$6x+40=16$$
$$6x=-24$$
$$x=-4$$

–4 does not check as a solution. The equation has no solution.

26.
$$I = \frac{1}{R}V$$
$$R \cdot I = R \cdot \frac{1}{R}V$$
$$RI = V$$
$$\frac{RI}{I} = \frac{V}{I}$$
$$R = \frac{V}{I}$$

27.
$$Q = \frac{N-S}{N}$$
$$Q \cdot N = \frac{N-S}{N} \cdot N$$
$$QN = N - S$$
$$QN - N = -S$$
$$N(Q-1) = -S$$
$$N = \frac{-S}{Q-1}$$
$$N = \frac{S}{1-Q}$$

28.
$$S = \frac{a}{1-r}$$
$$S(1-r) = \frac{a}{1-r}(1-r)$$
$$S - Sr = a$$
$$-Sr = a - S$$
$$r = \frac{a-S}{-S}$$
$$r = \frac{S-a}{S}$$

29. **Strategy**
To find the number of tanks of fuel, write and solve a proportion using x to represent the number of tanks of fuel.

Solution
$$\frac{4}{1800} = \frac{x}{3000}$$
$$9000 \cdot \frac{4}{1800} = \frac{x}{3000} \cdot 9000$$
$$5 \cdot 4 = 3x$$
$$20 = 3x$$
$$x = \frac{20}{3} = 6\frac{2}{3}$$

The number of tanks of fuel is $6\frac{2}{3}$.

30. **Strategy**
To find the number of miles represented, write and solve a proportion using x to represent the number of miles.

Solution
$$\frac{2.5}{10} = \frac{12}{x}$$
$$\frac{2.5}{10} \cdot 10x = \frac{12}{x} \cdot 10x$$
$$2.5x = 120$$
$$x = 48$$
The number of miles is 48.

31. **Strategy**
Unknown time for apprentice, working alone, to install fan: t

	Rate	Time	Part
Electrician	$\frac{1}{65}$	40	$\frac{40}{65}$
Apprentice	$\frac{1}{t}$	40	$\frac{40}{t}$

The sum of the part of the task completed by the electrician and the part of the task completed by the apprentice is 1.
$$\frac{40}{65} + \frac{40}{t} = 1$$

Solution
$$\frac{40}{65} + \frac{40}{t} = 1$$
$$65t\left(\frac{40}{65} + \frac{40}{t}\right) = 1 \cdot 65t$$
$$40t + 2600 = 65t$$
$$2600 = 25t$$
$$104 = t$$
The apprentice would take 104 min to complete the job alone.

32. **Strategy**
Unknown time to empty a full tub when both pipes are open: t

	Rate	Time	Part
Inlet pipe	$\frac{1}{24}$	t	$\frac{t}{24}$
Drain pipe	$\frac{1}{15}$	t	$\frac{t}{15}$

The difference between the part of the job done by the drain and the part of the job done by the inlet pipe is 1.

Solution
$$\frac{t}{15} - \frac{t}{24} = 1$$
$$120\left(\frac{t}{15} - \frac{t}{24}\right) = 1(120)$$
$$8t - 5t = 120$$
$$3t = 120$$
$$t = 40$$
It takes 40 min to empty the tub.

33. Strategy
Unknown time for 3 students to paint dormitory room working together: t

	Rate	Time	Part
1st painter	$\frac{1}{8}$	t	$\frac{t}{8}$
2nd painter	$\frac{1}{16}$	t	$\frac{t}{16}$
3rd painter	$\frac{1}{16}$	t	$\frac{t}{16}$

The sum of the parts of the task completed by 3 painters is 1.

Solution
$$\frac{t}{8}+\frac{t}{16}+\frac{t}{16}=1$$
$$16\left(\frac{t}{8}+\frac{t}{16}+\frac{t}{16}\right)=1(16)$$
$$2t+t+t=16$$
$$4t=16$$
$$t=4$$

It takes 4 h for the 3 painters to paint the dormitory room.

34. Strategy
Rate of the current: c

	Distance	Rate	Time
With the current	60	$10+c$	$\frac{60}{10+c}$
Against the current	40	$10-c$	$\frac{40}{10-c}$

The time traveling with the current is equal to the time traveling against the current.
$$\frac{60}{10+c}=\frac{40}{10-c}$$

Solution
$$\frac{60}{10+c}=\frac{40}{10-c}$$
$$(10+c)(10-c)\frac{60}{10+c}=(10+c)(10-c)\frac{40}{10-c}$$
$$(10-c)60=40(10+c)$$
$$600-60c=400+40c$$
$$200=100c$$
$$2=c$$

The rate of the current is 2 mph.

35. Strategy
Rate of cyclist: r
Rate of bus: $3r$

	Distance	Rate	Time
Bus	90	$3r$	$\frac{90}{3r}$
Cyclist	90	r	$\frac{90}{r}$

The cyclist arrives 4 h after the bus.
$$\frac{90}{3r}+4=\frac{90}{r}$$

Solution
$$\frac{90}{3r}+4=\frac{90}{r}$$
$$3r\left(\frac{90}{3r}+4\right)=\frac{90}{r}\cdot 3r$$
$$90+12r=270$$
$$12r=180$$
$$r=15$$
$$3r=3(15)=45$$

The rate of the bus is 45 mph.

36. Strategy
Rate of the car: r
Rate of the tractor: $r-15$

	Distance	Rate	Time
Car	15	r	$\frac{15}{r}$
Tractor	10	$r-15$	$\frac{10}{r-15}$

The time that the car travels equals the time that the tractor travels.
$$\frac{15}{r}=\frac{10}{r-15}$$

Solution
$$\frac{15}{r}=\frac{10}{r-15}$$
$$r(r-15)\left(\frac{15}{r}\right)=\left(\frac{10}{r-15}\right)r(r-15)$$
$$15(r-15)=10r$$
$$15r-225=10r$$
$$-225=-5r$$
$$45=r$$
$$r-15=45-15=30$$

The rate of the tractor is 30 mph.

37. Strategy
To find the pressure:
Write the basic joint variation equation, replace the variables with the given values, and solve for k.

Write the joint variation equation, replacing k with its value. Substitute 22 ft^2 for A and 20 mph for v.

Solution
$$P = kAv^2$$
$$10 = k \cdot 22 \cdot 10^2$$
$$\frac{1}{220} = k$$
$$P = \frac{1}{220}Av^2$$
$$P = \frac{1}{220} \cdot 22(20^2) = 40$$
The pressure is 40 lb.

38. Strategy
To find the illumination:
Write the basic inverse variation equation, replace the variables with the given values, and solve for k.

Write the inverse variation equation, replacing k with its value. Substitute d for 2 and solve for T.

Solution
$$I = \frac{k}{d^2}$$
$$12 = \frac{k}{(10)^2}$$
$$1200 = k$$
$$I = \frac{1200}{d^2} = \frac{1200}{(2)^2} = 300$$
The illumination is 300 lumens.

39. Strategy
To find the resistance:
Write the basic combined variation equation, replace the variables with the given values, and solve for k.

Write the combined variation equation, replacing k with its value. Substitute 8000 for l and $\frac{1}{2}$ for d, and solve for r.

Solution
$$r = \frac{kl}{d^2}$$
$$3.2 = \frac{k \cdot 16{,}000}{\left(\frac{1}{4}\right)^2}$$
$$0.0000125 = k$$
$$r = \frac{0.0000125 l}{d^2} = \frac{0.0000125(8000)}{\left(\frac{1}{2}\right)^2} = 0.4$$
The resistance is 0.4 ohm.

Chapter Test

1. $\dfrac{v^3 - 4v}{2v^2 - 5v + 2} = \dfrac{v(v^2 - 4)}{(2v - 1)(v - 2)}$
$= \dfrac{v(v + 2)(v - 2)}{(2v - 1)(v - 2)}$
$= \dfrac{v(v + 2)}{2v - 1}$

2. $\dfrac{2a^2 - 8a + 8}{4 + 4a - 3a^2} = \dfrac{2(a^2 - 4a + 4)}{(2 - a)(2 + 3a)}$
$= \dfrac{2(a - 2)(a - 2)}{(2 - a)(2 + 3a)}$
$= -\dfrac{2(a - 2)}{3a + 2}$

3. $\dfrac{3x^2 - 12}{5x - 15} \cdot \dfrac{2x^2 - 18}{x^2 + 5x + 6}$
$= \dfrac{3(x + 2)(x - 2)}{5(x - 3)} \cdot \dfrac{2(x + 3)(x - 3)}{(x + 3)(x + 2)}$
$= \dfrac{3(x + 2)(x - 2)2(x + 3)(x - 3)}{5(x - 3)(x + 3)(x + 2)}$
$= \dfrac{6(x - 2)}{5}$

4. $P(x) = \dfrac{3 - x^2}{x^3 - 2x^2 + 4}$
$P(-1) = \dfrac{3 - (-1)^2}{(-1)^3 - 2(-1)^2 + 4} = \dfrac{3 - 1}{-1 - 2 + 4} = \dfrac{2}{1}$
$P(-1) = 2$

5. $\dfrac{2x^2 - x - 3}{2x^2 - 5x + 3} \div \dfrac{3x^2 - x - 4}{x^2 - 1}$
$= \dfrac{2x^2 - x - 3}{2x^2 - 5x + 3} \cdot \dfrac{x^2 - 1}{3x^2 - x - 4}$
$= \dfrac{(2x - 3)(x + 1)}{(2x - 3)(x - 1)} \cdot \dfrac{(x + 1)(x - 1)}{(3x - 4)(x + 1)}$
$= \dfrac{(2x - 3)(x + 1)(x + 1)(x - 1)}{(2x - 3)(x - 1)(3x - 4)(x + 1)}$
$= \dfrac{x + 1}{3x - 4}$

6. $\dfrac{x^{2n} - x^n - 2}{x^{2n} + x^n} \cdot \dfrac{x^{2n} - x^n}{x^{2n} - 4}$
$= \dfrac{(x^n - 2)(x^n + 1)}{x^n(x^n + 1)} \cdot \dfrac{x^n(x^n - 1)}{(x^n + 2)(x^n - 2)}$
$= \dfrac{(x^n - 2)(x^n + 1)x^n(x^n - 1)}{x^n(x^n + 1)(x^n + 2)(x^n - 2)}$
$= \dfrac{x^n - 1}{x^n + 2}$

7. The LCM is $2x^2y^2$.

$$\frac{2}{x^2}+\frac{3}{y^2}-\frac{5}{2xy}$$

$$=\frac{2}{x^2}\cdot\frac{2y^2}{2y^2}+\frac{3}{y^2}\cdot\frac{2x^2}{2x^2}-\frac{5}{2xy}\cdot\frac{xy}{xy}$$

$$=\frac{4y^2}{2x^2y^2}+\frac{6x^2}{2x^2y^2}-\frac{5xy}{x^2y^2}$$

$$=\frac{4y^2+6x^2-5xy}{2x^2y^2}$$

8. The LCM is $(x-2)(x+2)$.

$$\frac{3x}{x-2}-3+\frac{4}{x+2}=\frac{3x}{x-2}\cdot\frac{x+2}{x+2}-3\frac{(x-2)(x+2)}{(x-2)(x+2)}+\frac{4}{x+2}\cdot\frac{x-2}{x-2}$$

$$=\frac{3x(x+2)-3(x-2)(x+2)+4(x-2)}{(x+2)(x-2)}$$

$$=\frac{3x^2+6x-3(x^2-4)+4x-8}{(x+2)(x-2)}$$

$$=\frac{3x^2+6x-3x^2+12+4x-8}{(x+2)(x-2)}$$

$$=\frac{10x+4}{(x+2)(x-2)}$$

$$=\frac{2(5x+2)}{(x+2)(x-2)}$$

9. $f(x)=\dfrac{3x^2-x+1}{x^2-9}$

$x^2-9=0$
$(x+3)(x-3)=0$
$x+3=0 \qquad x-3=0$
$\quad x=-3 \qquad \quad x=3$

The domain is $\{x|x\ne -3, 3\}$.

10. $x^2+3x-4=(x+4)(x-1)$
$x^2-1=(x+1)(x-1)$
The LCM is $(x+4)(x-1)(x+1)$.

$$\frac{x+2}{x^2+3x-4}-\frac{2x}{x^2-1}=\frac{x+2}{(x+4)(x-1)}\cdot\frac{x+1}{x+1}-\frac{2x}{(x+1)(x-1)}\cdot\frac{x+4}{x+4}$$

$$=\frac{(x+2)(x+1)-2x(x+4)}{(x+4)(x-1)(x+1)}$$

$$=\frac{x^2+3x+2-2x^2-8x}{(x+4)(x-1)(x+1)}$$

$$=\frac{-x^2-5x+2}{(x+1)(x+4)(x-1)}$$

$$=-\frac{x^2+5x-2}{(x+1)(x+4)(x-1)}$$

11. $\dfrac{1-\frac{1}{x}-\frac{12}{x^2}}{1+\frac{6}{x}+\frac{9}{x^2}} = \dfrac{1-\frac{1}{x}-\frac{12}{x^2}}{1+\frac{6}{x}+\frac{9}{x^2}} \cdot \dfrac{x^2}{x^2}$

$= \dfrac{x^2-x-12}{x^2+6x+9}$

$= \dfrac{(x-4)(x+3)}{(x+3)(x+3)}$

$= \dfrac{x-4}{x+3}$

12. $\dfrac{1-\frac{1}{x+2}}{1-\frac{3}{x+4}} = \dfrac{1-\frac{1}{x+2}}{1-\frac{3}{x+4}} \cdot \dfrac{(x+2)(x+4)}{(x+2)(x+4)}$

$= \dfrac{(x+2)(x+4)-(x+4)}{(x+2)(x+4)-3(x+2)}$

$= \dfrac{x^2+6x+8-x-4}{x^2+6x+8-3x-6}$

$= \dfrac{x^2+5x+4}{x^2+3x+2}$

$= \dfrac{(x+4)(x+1)}{(x+2)(x+1)}$

$= \dfrac{x+4}{x+2}$

13. $\dfrac{3}{x+1} = \dfrac{2}{x}$

$\dfrac{3}{x+1} \cdot x(x+1) = \dfrac{2}{x} \cdot x(x+1)$

$3x = 2(x+1)$

$3x = 2x+2$

$x = 2$

The solution is 2.

14. $\dfrac{4x}{2x-1} = 2 - \dfrac{1}{2x-1}$

$(2x-1)\dfrac{4x}{2x-1} = \left(2-\dfrac{1}{2x-1}\right)(2x-1)$

$4x = 2(2x-1)-1$

$4x = 4x-2-1$

$4x = 4x-3$

$0 = -3$

There is no solution.

15. $ax = bx+c$

$ax-bx = c$

$x(a-b) = c$

$x = \dfrac{c}{a-b}$

16. **Strategy**
Unknown time to empty the tank with both pipes open: t

	Rate	Time	Part
Inlet pipe	$\frac{1}{48}$	t	$\frac{t}{48}$
Outlet pipe	$\frac{1}{30}$	t	$\frac{t}{30}$

The difference between the part of the task completed by the outlet pipe and the part of the task completed by the inlet pipe is 1.

$\dfrac{t}{30} - \dfrac{t}{48} = 1$

Solution

$\dfrac{t}{30} - \dfrac{t}{48} = 1$

$240\left(\dfrac{t}{30} - \dfrac{t}{48}\right) = 240(1)$

$8t - 5t = 240$

$3t = 240$

$t = 80$

It will take 80 min to empty the full tank with both pipes open.

17. **Strategy**
To find the number of rolls of wallpaper, write and solve a proportion using x to represent the number of rolls.

Solution

$\dfrac{2}{45} = \dfrac{x}{315}$

$\left(\dfrac{2}{45}\right)315 = \left(\dfrac{x}{315}\right)315$

$14 = x$

The office requires 14 rolls of wallpaper.

Chapter 6: *Rational Expressions*

18. Strategy
Unknown time for both landscapers working together: t

	Rate	Time	Part
First landscaper	$\frac{1}{30}$	t	$\frac{t}{30}$
Second landscaper	$\frac{1}{15}$	t	$\frac{t}{15}$

The sum of the part of the task completed by the first landscaper and the part of the task completed by the second landscaper is 1.

$$\frac{t}{30} + \frac{t}{15} = 1$$

Solution

$$\frac{t}{30} + \frac{t}{15} = 1$$
$$30\left(\frac{t}{30} + \frac{t}{15}\right) = 1(30)$$
$$t + 2t = 30$$
$$3t = 30$$
$$t = 10$$

Working together, the landscapers can complete the task in 10 min.

19. Strategy
Rate of hiker: r
Rate of cyclist: $r + 7$

	Distance	Rate	Time
Hiker	6	r	$\frac{6}{r}$
Cyclist	20	$r + 7$	$\frac{20}{r+7}$

The time the hiker hikes equals the time the cyclist cycles.

$$\frac{6}{r} = \frac{20}{r+7}$$

Solution

$$\frac{6}{r} = \frac{20}{r+7}$$
$$\frac{6}{r}[r(r+7)] = \frac{20}{r+7}[r(r+7)]$$
$$6(r+7) = 20r$$
$$6r + 42 = 20r$$
$$42 = 14r$$
$$3 = r$$
$$r + 7 = 3 + 7 = 10$$

The rate of the cyclist is 10 mph.

20. Strategy
To find the stopping distance:
Write the general direct variation equation, replace the variables by the given values, and solve for k.

Write the direct variation equation, replacing k by its value. Substitute 30 for v and solve for s.

Solution

$$s = kv^2$$
$$170 = k(50)^2$$
$$170 = k(2500)$$
$$0.068 = k$$

$$s = kv^2$$
$$= 0.068v^2$$
$$= 0.068(30)^2$$
$$= 0.068(900)$$
$$= 61.2$$

The stopping distance for a car traveling at 30 mph is 61.2 ft.

Cumulative Review Exercises

1. $8 - 4[-3 - (-2)]^2 \div 5$
$= 8 - 4[-3 + 2]^2 \div 5$
$= 8 - 4[-1]^2 \div 5$
$= 8 - 4(1) \div 5$
$= 8 - 4 \div 5$
$= 8 - \frac{4}{5}$
$= \frac{36}{5}$

2. $\frac{2x-3}{6} - \frac{x}{9} = \frac{x-4}{3}$
$18\left(\frac{2x-3}{6} - \frac{x}{9}\right) = \left(\frac{x-4}{3}\right)18$
$3(2x-3) - 2x = 6(x-4)$
$6x - 9 - 2x = 6x - 24$
$4x - 9 = 6x - 24$
$-2x = -15$
$x = \frac{15}{2}$

3. $5 - |x-4| = 2$
$-|x-4| = -3$
$|x-4| = 3$
$x - 4 = 3 \qquad x - 4 = -3$
$x = 7 \qquad x = 1$
The solutions are 7 and 1.

4. $\frac{x}{x-3}$
$x - 3 = 0$
$x = 3$
The domain is $\{x | x \neq 3\}$.

5. $P(x) = \dfrac{x-1}{2x-3}$

$P(-2) = \dfrac{-2-1}{2(-2)-3} = \dfrac{-3}{-4-3} = \dfrac{-3}{-7}$

$P(-2) = \dfrac{3}{7}$

6. $0.000000035 = 3.5 \times 10^{-8}$

7. $\dfrac{x}{x+1} = 1$

$(x+1)\dfrac{x}{x+1} = 1(x+1)$

$x = x + 1$

$0 \neq 1$

There is no solution.

8. $(9x-1)(x-4) = 0$

$9x - 1 = 0 \qquad x - 4 = 0$
$9x = 1 \qquad\quad x = 4$
$x = \dfrac{1}{9}$

The solutions are $\dfrac{1}{9}$ and 4.

9. $\dfrac{(2a^{-2}b^3)}{(4a)^{-1}} = 2a^{-2}b^3 \cdot 4a$

$= \dfrac{2b^3 \cdot 4a}{a^2}$

$= \dfrac{8b^3 a}{a^2}$

$= \dfrac{8b^3}{a}$

10. $x - 3(1 - 2x) \geq 1 - 4(2 - 2x)$

$x - 3 + 6x \geq 1 - 8 + 8x$

$7x - 3 \geq 8x - 7$

$-x - 3 \geq -7$

$-x \geq -4$

$(-1)(-x) \leq (-1)(-4)$

$x \leq 4$

$\{x \mid x \leq 4\}$

11. $(2a^2 - 3a + 1)(-2a^2) = -4a^4 + 6a^3 - 2a^2$

12. Let $x^n = u$.

$2x^{2n} + 3x^n - 2 = 2u^2 + 3u - 2$

$= (2u - 1)(u + 2)$

$= (2x^n - 1)(x^n + 2)$

13. $x^3 y^3 - 27 = (xy)^3 - (3)^3$

$= (xy - 3)(x^2 y^2 + 3xy + 9)$

14. $\dfrac{x^4 + x^3 y - 6x^2 y^2}{x^3 - 2x^2 y} = \dfrac{x^2(x^2 + xy - 6y^2)}{x^2(x - 2y)}$

$= \dfrac{x^2(x + 3y)(x - 2y)}{x^2(x - 2y)}$

$= x + 3y$

15. $3x - 2y = 6$

$-2y = -3x + 6$

$y = \dfrac{3}{2}x - 3$

$m = \dfrac{3}{2}$

$y - y_1 = m(x - x_1)$

$y - (-1) = \dfrac{3}{2}[x - (-2)]$

$y + 1 = \dfrac{3}{2}(x + 2)$

$y + 1 = \dfrac{3}{2}x + 3$

$y = \dfrac{3}{2}x + 2$

The equation of the line is $y = \dfrac{3}{2}x + 2$.

16. $(x - x^{-1})^{-1} = \dfrac{1}{x - x^{-1}}$

$= \dfrac{1}{x - \dfrac{1}{x}}$

$= \dfrac{1}{x - \dfrac{1}{x}} \cdot \dfrac{x}{x}$

$= \dfrac{x}{x^2 - 1}$

17. $-3x + 5y = -15$

x-intercept: $(5, 0)$
y-intercept: $(0, -3)$

18. $x + y \leq 3 \qquad\quad -2x + y > 4$
$y \leq 3 - x \qquad\quad y > 4 + 2x$

19. $\dfrac{4x^3 + 2x^2 - 10x + 1}{x - 2}$

$\begin{array}{c|cccc} 2 & 4 & 2 & -10 & 1 \\ & & 8 & 20 & 20 \\ \hline & 4 & 10 & 10 & 21 \end{array}$

The simplified form is $4x^2 + 10x + 10 + \dfrac{21}{x-2}$.

196 *Chapter 6:* Rational Expressions

20. $\dfrac{16x^2 - 9y^2}{16x^2y - 12xy^2} \div \dfrac{4x^2 - xy - 3y^2}{12x^2y^2}$

    $= \dfrac{16x^2 - 9y^2}{16x^2y - 12xy^2} \cdot \dfrac{12x^2y^2}{4x^2 - xy - 3y^2}$

    $= \dfrac{(4x - 3y)(4x + 3y)}{4xy(4x - 3y)} \cdot \dfrac{12x^2y^2}{(4x + 3y)(x - y)}$

    $= \dfrac{(4x - 3y)(4x + 3y) \cdot 12x^2y^2}{4xy(4x - 3y)(4x + 3y)(x - y)}$

    $= \dfrac{3xy}{x - y}$

21. The domain must exclude values of x for which $3x^2 + 5 = 0$. This is not possible, because $3x^2 \geq 0$, and a positive number added to a number equal to or greater than zero cannot equal zero. Therefore, there are no real numbers that must be excluded from the domain of f.

    The domain of $f(x)$ is $\{x | x \text{ is a real number}\}$.

22. $3x^2 - x - 2 = (3x + 2)(x - 1)$

    $x^2 - 1 = (x + 1)(x - 1)$

    The LCM is $(3x + 2)(x + 1)(x - 1)$.

    $\dfrac{5x}{3x^2 - x - 2} - \dfrac{2x}{x^2 - 1}$

    $= \dfrac{5x}{(3x + 2)(x - 1)} \cdot \dfrac{x + 1}{x + 1} - \dfrac{2x}{(x + 1)(x - 1)} \cdot \dfrac{3x + 2}{3x + 2}$

    $= \dfrac{5x(x + 1) - 2x(x + 2)}{(3x + 2)(x - 1)(x + 1)}$

    $= \dfrac{5x^2 + 5x - 6x^2 - 4x}{(3x + 2)(x - 1)(x + 1)}$

    $= \dfrac{-x^2 + x}{(3x + 2)(x - 1)(x + 1)}$

    $= \dfrac{-x(x - 1)}{(3x + 2)(x - 1)(x + 1)}$

    $= -\dfrac{x}{(3x + 2)(x + 1)}$

23. $\begin{vmatrix} 6 & 5 \\ 2 & -3 \end{vmatrix} = 6(-3) - 5 \cdot 2 = -18 - 10 = -28$

24. $\dfrac{x - 4 + \frac{5}{x+2}}{x + 2 - \frac{1}{x+2}} = \dfrac{x - 4 + \frac{5}{x+2}}{x + 2 - \frac{1}{x+2}} \cdot \dfrac{x + 2}{x + 2}$

    $= \dfrac{(x - 4)(x + 2) + 5}{(x + 2)^2 - 1}$

    $= \dfrac{x^2 - 2x - 8 + 5}{x^2 + 4x + 4 - 1}$

    $= \dfrac{x^2 - 2x - 3}{x^2 + 4x + 3}$

    $= \dfrac{(x - 3)(x + 1)}{(x + 3)(x + 1)}$

    $= \dfrac{x - 3}{x + 3}$

25. $x + y + z = 3$
    $-2x + y + 3z = 2$
    $2x - 4y + z = -1$

    $D = \begin{vmatrix} 1 & 1 & 1 \\ -2 & 1 & 3 \\ 2 & -4 & 1 \end{vmatrix} = 27$

    $D_x = \begin{vmatrix} 3 & 1 & 1 \\ 2 & 1 & 3 \\ -1 & -4 & 1 \end{vmatrix} = 27$

    $D_y = \begin{vmatrix} 1 & 3 & 1 \\ -2 & 2 & 3 \\ 2 & -1 & 1 \end{vmatrix} = 27$

    $D_z = \begin{vmatrix} 1 & 1 & 3 \\ -2 & 1 & 2 \\ 2 & -4 & -1 \end{vmatrix} = 27$

    $x = \dfrac{D_x}{D} = \dfrac{27}{27} = 1$

    $y = \dfrac{D_y}{D} = \dfrac{27}{27} = 1$

    $z = \dfrac{D_z}{D} = \dfrac{27}{27} = 1$

    The solution is $(1, 1, 1)$.

26. $f(x) = x^2 - 3x + 3$
    $f(c) = c^2 - 3c + 3$
    $1 = c^2 - 3c + 3$
    $c^2 - 3c + 2 = 0$
    $(c - 2)(c - 1) = 0$
    $c - 2 = 0 \qquad c - 1 = 0$
    $c = 2 \qquad c = 1$

    The solutions are 1 and 2.

27. $\dfrac{2}{x - 3} = \dfrac{5}{2x - 3}$

    $\left(\dfrac{2}{x - 3}\right)(x - 3)(2x - 3) = \left(\dfrac{5}{2x - 3}\right)(x - 3)(2x - 3)$

    $2(2x - 3) = 5(x - 3)$
    $4x - 6 = 5x - 15$
    $-x - 6 = -15$
    $-x = -9$
    $x = 9$

    The solution is 9.

28.
$$\frac{3}{x^2-36} = \frac{2}{x-6} - \frac{5}{x+6}$$
$$\frac{3}{(x+6)(x-6)} = \frac{2}{x-6} - \frac{5}{x+6}$$
$$(x+6)(x-6)\left(\frac{3}{(x+6)(x-6)}\right) = \left(\frac{2}{x-6} - \frac{5}{x+6}\right)(x+6)(x-6)$$
$$3 = 2(x+6) - 5(x-6)$$
$$3 = 2x + 12 - 5x + 30$$
$$3 = -3x + 42$$
$$-39 = -3x$$
$$13 = x$$
The solution is 13.

29. $(a+5)(a^3 - 3a + 4) = a(a^3 - 3a + 4) + 5(a^3 - 3a + 4)$
$= a^4 - 3a^2 + 4a + 5a^3 - 15a + 20$
$= a^4 + 5a^3 - 3a^2 - 11a + 20$

30.
$$I = \frac{E}{R+r}$$
$$I(R+r) = \frac{E}{R+r}(R+r)$$
$$IR + Ir = E$$
$$Ir = E - IR$$
$$r = \frac{E - IR}{I}$$

31. $4x + 3y = 12$ $\quad\quad 3x + 4y = 16$
$\quad\quad 3y = -4x + 12$ $\quad\quad 4y = -3x + 16$
$\quad\quad y = -\frac{4}{3}x + 4$ $\quad\quad y = -\frac{3}{4}x + 4$

The lines are not perpendicular, since the slopes are not negative reciprocals of each other.

32.

33. Strategy
Smaller integer: x
Larger integer: $15 - x$

Five times the smaller is five more than twice the larger.
$5x = 5 + 2(15 - x)$

Solution
$5x = 5 + 2(15 - x)$
$5x = 5 + 30 - 2x$
$7x = 35$
$x = 5$
$15 - x = 10$
The smaller integer is 5 and larger integer is 10.

34. Strategy
The unknown number of pounds of almonds: x

	Amount	Cost	Total
Almonds	x	5.40	$5.40x$
Peanuts	50	2.60	$50(2.60)$
Mixture	$x + 50$	4.00	$4(x + 50)$

The sum of the values before mixing equals the value after mixing.
$5.40x + 50(2.60) = 4(x + 50)$

Solution
$$5.40x + 50(2.60) = 4(x + 50)$$
$$5.4x + 130 = 4x + 200$$
$$1.4x + 130 = 200$$
$$1.4x = 70$$
$$x = 50$$
The number of pounds of almonds is 50.

35. Strategy
To find the number of people expected to vote, write and solve a proportion using x to represent the number of people expected to vote.

Solution
$$\frac{3}{5} = \frac{x}{125,000}$$
$$\frac{3}{5} \cdot 125,000 = \frac{x}{125,000} \cdot 125,000$$
$$75,000 = x$$
The number of people expected to vote is 75,000.

36. Strategy
Time it takes older computer: $6r$
Time it takes new computer: r

	Rate	Time	Part
Older computer	$\frac{1}{6r}$	12	$\frac{12}{6r}$
New computer	$\frac{1}{r}$	12	$\frac{12}{r}$

The sum of the parts of the task completed by the older computer and the part of the task completed by the new computer is 1.
$$\frac{12}{6r} + \frac{12}{r} = 1$$

Solution
$$\frac{12}{6r} + \frac{12}{r} = 1$$
$$6r\left(\frac{12}{6r} + \frac{12}{r}\right) = (1)6r$$
$$12 + 72 = 6r$$
$$84 = 6r$$
$$14 = r$$
It takes the new computer 14 minutes to do the job working alone.

37. Strategy
Unknown rate of the wind: r

	Distance	Rate	Time
With the wind	900	$300 + r$	$\frac{900}{300+r}$
Against the wind	600	$300 - r$	$\frac{600}{300-r}$

The time traveled with the wind equals the time traveled against the wind.
$$\frac{900}{300+r} = \frac{600}{300-r}$$

Solution
$$\frac{900}{300+r} = \frac{600}{300-r}$$
$$(300+r)(300-r)\left(\frac{900}{300+r}\right) = \left(\frac{600}{300-r}\right)(300+r)(300-r)$$
$$(300-r)(900) = 600(300+r)$$
$$270{,}000 - 900r = 180{,}000 + 600r$$
$$-1500r = -90{,}000$$
$$r = 60$$
The rate of the wind is 60 mph.

38. Strategy
To find the frequency:
Write the basic inverse variation equation, replace the variables by the given values, and solve for k.

Write the inverse variation equation, replacing k by its value. Substitute 1.5 for L and solve for f.

Solution
$$f = \frac{k}{L}$$
$$60 = \frac{k}{2}$$
$$120 = k$$
$$f = \frac{120}{L} = \frac{120}{1.5} = 80$$
The frequency is 80 vibrations per minute.

Chapter 7: Rational Exponents and Radicals

Section 7.1

Concept Review 7.1

1. Sometimes true
   $\sqrt{x^2} = x$ is true for positive numbers, false for negative numbers.

3. Always true

5. Always true

Objective 1 Exercises

3. $8^{1/3} = (2^3)^{1/3} = 2$

5. $9^{3/2} = (3^2)^{3/2} = 3^3 = 27$

7. $27^{-2/3} = (3^3)^{-2/3} = 3^{-2} = \dfrac{1}{3^2} = \dfrac{1}{9}$

9. $32^{2/5} = (2^5)^{2/5} = 2^2 = 4$

11. $(-25)^{5/2}$
    The base of the exponential expression is a negative number, while the denominator of the exponent is a positive even number.
    Therefore, $(-25)^{5/2}$ is not a real number.

13. $\left(\dfrac{25}{49}\right)^{-3/2} = \left(\dfrac{5^2}{7^2}\right)^{-3/2} = \left[\left(\dfrac{5}{7}\right)^2\right]^{-3/2}$
    $= \left(\dfrac{5}{7}\right)^{-3} = \dfrac{5^{-3}}{7^{-3}} = \dfrac{7^3}{5^3} = \dfrac{343}{125}$

15. $x^{1/2} x^{1/2} = x$

17. $y^{-1/4} y^{3/4} = y^{1/2}$

19. $x^{-2/3} \cdot x^{3/4} = x^{1/12}$

21. $a^{1/3} \cdot a^{3/4} \cdot a^{-1/2} = a^{7/12}$

23. $\dfrac{a^{1/2}}{a^{3/2}} = a^{-1} = \dfrac{1}{a}$

25. $\dfrac{y^{-3/4}}{y^{1/4}} = y^{-1} = \dfrac{1}{y}$

27. $\dfrac{y^{2/3}}{y^{-5/6}} = y^{9/6} = y^{3/2}$

29. $(x^2)^{-1/2} = x^{-1} = \dfrac{1}{x}$

31. $(x^{-2/3})^6 = x^{-4} = \dfrac{1}{x^4}$

33. $(a^{-1/2})^{-2} = a$

35. $(x^{-3/8})^{-4/5} = x^{3/10}$

37. $(a^{1/2} \cdot a)^2 = (a^{3/2})^2 = a^3$

39. $(x^{-1/2} x^{3/4})^{-2} = (x^{1/4})^{-2}$
    $= x^{-1/2}$
    $= \dfrac{1}{x^{1/2}}$

41. $(y^{-1/2} y^{3/2})^{2/3} = y^{2/3}$

43. $(x^8 y^2)^{1/2} = x^4 y$

45. $(x^4 y^2 z^6)^{3/2} = x^6 y^3 z^9$

47. $(x^{-3} y^6)^{-1/3} = xy^{-2} = \dfrac{x}{y^2}$

49. $(x^{-2} y^{1/3})^{-3/4} = x^{3/2} y^{-1/4} = \dfrac{x^{3/2}}{y^{1/4}}$

51. $\left(\dfrac{x^{1/2}}{y^{-2}}\right)^4 = \dfrac{x^2}{y^{-8}} = x^2 y^8$

53. $\dfrac{x^{1/4} \cdot x^{-1/2}}{x^{2/3}} = \dfrac{x^{-1/4}}{x^{2/3}} = x^{-11/12} = \dfrac{1}{x^{11/12}}$

55. $\left(\dfrac{y^{2/3} \cdot y^{-5/6}}{y^{1/9}}\right)^9 = \left(\dfrac{y^{-1/6}}{y^{1/9}}\right)^9$
    $= (y^{-5/18})^9$
    $= y^{-5/2}$
    $= \dfrac{1}{y^{5/2}}$

57. $\left(\dfrac{b^2 \cdot b^{-3/4}}{b^{-1/2}}\right)^{-1/2} = \left(\dfrac{b^{5/4}}{b^{-1/2}}\right)^{-1/2}$
    $= (b^{7/4})^{-1/2}$
    $= b^{-7/8}$
    $= \dfrac{1}{b^{7/8}}$

59. $(a^{2/3} b^2)^6 (a^3 b^3)^{1/3} = (a^4 b^{12})(ab) = a^5 b^{13}$

61. $(16m^{-2} n^4)^{-1/2} (mn^{1/2}) = (2^4)^{-1/2} mn^{-2} \cdot mn^{1/2}$
    $= 2^{-2} m^2 n^{-3/2}$
    $= \dfrac{m^2}{2^2 n^{3/2}}$
    $= \dfrac{m^2}{4n^{3/2}}$

63. $\left(\dfrac{x^{1/2}y^{-3/4}}{y^{2/3}}\right)^{-6} = (x^{1/2}y^{-17/12})^{-6}$
$= x^{-3}y^{17/2}$
$= \dfrac{y^{17/2}}{x^3}$

65. $\left(\dfrac{2^{-6}b^{-3}}{a^{-1/2}}\right)^{-2/3} = \dfrac{2^4 b^2}{a^{1/3}} = \dfrac{16b^2}{a^{1/3}}$

67. $\dfrac{(x^{-2}y^4)^{1/2}}{(x^{1/2})^4} = \dfrac{x^{-1}y^2}{x^2} = \dfrac{y^2}{x^3}$

69. $a^{-1/4}(a^{5/4} - a^{9/4}) = a^1 - a^2 = a - a^2$

71. $y^{2/3}(y^{1/3} + y^{-2/3}) = y^1 + y^0 = y + 1$

73. $a^{1/6}(a^{5/6} - a^{-7/6}) = a^1 - a^{-1} = a - \dfrac{1}{a}$

75. $(a^{2/n})^{-5n} = a^{-10} = \dfrac{1}{a^{10}}$

77. $a^{n/2} \cdot a^{-n/3} = a^{n/6}$

79. $\dfrac{b^{m/3}}{b^m} = b^{-2m/3} = \dfrac{1}{b^{2m/3}}$

81. $(x^{5n})^{2n} = x^{10n^2}$

83. $(x^{n/2} y^{n/3})^6 = x^{3n} y^{2n}$

Objective 2 Exercises

85. $5^{1/2} = \sqrt{5}$

87. $b^{4/3} = (b^4)^{1/3} = \sqrt[3]{b^4}$

89. $(3x)^{2/3} = \sqrt[3]{(3x)^2} = \sqrt[3]{9x^2}$

91. $-3a^{2/5} = -3(a^2)^{1/5} = -3\sqrt[5]{a^2}$

93. $(x^2 y^3)^{3/4} = \sqrt[4]{(x^2 y^3)^3} = \sqrt[4]{x^6 y^9}$

95. $(a^3 b^7)^{3/2} = \sqrt{(a^3 b^7)^3}$
$= \sqrt{a^9 b^{21}}$

97. $(3x-2)^{1/3} = \sqrt[3]{3x-2}$

99. $\sqrt{14} = 14^{1/2}$

101. $\sqrt[3]{x} = x^{1/3}$

103. $\sqrt[3]{x^4} = x^{4/3}$

105. $\sqrt[5]{b^3} = b^{3/5}$

107. $\sqrt[3]{2x^2} = (2x^2)^{1/3}$

109. $-\sqrt{3x^5} = -(3x^5)^{1/2}$

111. $3x\sqrt[3]{y^2} = 3xy^{2/3}$

113. $\sqrt{a^2 + 2} = (a^2 + 2)^{1/2}$

Objective 3 Exercises

115. $\sqrt{y^{14}} = y^7$

117. $-\sqrt{a^6} = -a^3$

119. $\sqrt{a^{14} b^6} = a^7 b^3$

121. $\sqrt{121 y^{12}} = 11 y^6$

123. $\sqrt[3]{a^6 b^{12}} = a^2 b^4$

125. $-\sqrt[3]{a^9 b^9} = -a^3 b^3$

127. $\sqrt[3]{125 b^{15}} = \sqrt[3]{5^3 b^{15}} = 5b^5$

129. $\sqrt[3]{-a^6 b^9} = -a^2 b^3$

131. $\sqrt{25 x^8 y^2} = \sqrt{5^2 x^8 y^2} = 5x^4 y$

133. $\sqrt{-9 a^6 b^8}$
The square root of a negative number is not a real number, since the square of a real number must be positive. Therefore, $\sqrt{-9 a^6 b^8}$ is not a real number.

135. $\sqrt[3]{8 a^{21} b^6} = \sqrt[3]{2^3 a^{21} b^6} = 2 a^7 b^2$

137. $\sqrt[3]{-27 a^3 b^{15}} = \sqrt[3]{(-3)^3 a^3 b^{15}} = -3ab^5$

139. $\sqrt[4]{y^{12}} = y^3$

141. $\sqrt[4]{81 a^{20}} = \sqrt[4]{3^4 a^{20}} = 3a^5$

143. $-\sqrt[4]{a^{16} b^4} = -a^4 b$

145. $\sqrt[5]{a^5 b^{25}} = ab^5$

147. $\sqrt[4]{16 a^8 b^{20}} = \sqrt[4]{2^4 a^8 b^{20}} = 2a^2 b^5$

149. $\sqrt[5]{-32 x^{15} y^{20}} = \sqrt[5]{(-2)^5 x^{15} y^{20}} = -2x^3 y^4$

151. $\sqrt{\dfrac{16 x^2}{y^{14}}} = \sqrt{\dfrac{2^4 x^2}{y^{14}}} = \dfrac{2^2 x}{y^7} = \dfrac{4x}{y^7}$

153. $\sqrt[3]{\dfrac{27b^3}{a^9}} = \sqrt[3]{\dfrac{3^3 b^3}{a^9}} = \dfrac{3b}{a^3}$

155. $\sqrt{(2x+3)^2} = 2x+3$

157. $\sqrt{x^2+2x+1} = \sqrt{(x+1)^2} = x+1$

Applying Concepts 7.1

159. a. $16^{1/2} = (2^4)^{1/2} = 2^2 = 4$

   b. $16^{1/4} = (2^4)^{1/4} = 2^1 = 2$

   c. $16^{3/2} = (2^4)^{3/2} = 2^6 = 64$

   c. $16^{3/2}$ is the largest

161. a. $4^{1/2} \cdot 4^{3/2} = 4^2 = 16$

   b. $3^{4/5} \cdot 3^{6/5} = 3^2 = 9$

   c. $7^{1/4} \cdot 7^{3/4} = 7^1 = 7$

   a. $4^{1/2} \cdot 4^{3/2}$ is largest.

163. $\sqrt[3]{\sqrt{x^6}} = \sqrt[3]{x^3} = x$

165. $\sqrt[5]{\sqrt[3]{b^{15}}} = \sqrt[5]{b^5} = b$

167. $\sqrt[5]{\sqrt{a^{10}b^{20}}} = \sqrt[5]{a^5 b^{10}} = ab^2$

169. $y^p y^{2/5} = y$

   $y^p = \dfrac{y^1}{y^{2/5}}$

   $y^p = y^{3/5}$

   $p = \dfrac{3}{5}$

   When the value of p is $\tfrac{3}{5}$, the equation is true.

171. $x^p x^{-1/2} = x^{1/4}$

   $x^{p-(1/2)} = x^{1/4}$

   $p - \dfrac{1}{2} = \dfrac{1}{4}$

   $p = \dfrac{3}{4}$

   When the value of p is $\tfrac{3}{4}$, the equation is true.

Section 7.2

Concept Review 7.2

1. Sometimes true
   If a is a positive number, $\sqrt{a}$ is a real number. If a is negative, $\sqrt{a}$ is not a real number.

3. Never true
   The index of each radical must be the same in order to multiply radical expressions.

5. Always true

Objective 1 Exercises

3. $\sqrt{18} = \sqrt{3^2 \cdot 2}$
   $= \sqrt{3^2}\sqrt{2}$
   $= 3\sqrt{2}$

5. $\sqrt{98} = \sqrt{7^2 \cdot 2}$
   $= \sqrt{7^2}\sqrt{2}$
   $= 7\sqrt{2}$

7. $\sqrt[3]{72} = \sqrt[3]{2^3 \cdot 3^2}$
   $= \sqrt[3]{2^3}\sqrt[3]{3^2}$
   $= 2\sqrt[3]{9}$

9. $\sqrt[3]{16} = \sqrt[3]{2^3 \cdot 2}$
   $= \sqrt[3]{2^3}\sqrt[3]{2}$
   $= 2\sqrt[3]{2}$

11. $\sqrt{x^4 y^3 z^5} = \sqrt{x^4 y^2 z^4 (yz)}$
    $= \sqrt{x^4 y^2 z^4}\sqrt{yz}$
    $= x^2 yz^2 \sqrt{yz}$

13. $\sqrt{8a^3 b^8} = \sqrt{2^3 a^3 b^8}$
    $= \sqrt{2^2 a^2 b^8 (2a)}$
    $= \sqrt{2^2 a^2 b^8}\sqrt{2a}$
    $= 2ab^4 \sqrt{2a}$

15. $\sqrt{45 x^2 y^3 z^5} = \sqrt{3^2 \cdot 5 x^2 y^3 z^5}$
    $= \sqrt{3^2 x^2 y^2 z^4 (5yz)}$
    $= \sqrt{3^2 x^2 y^2 z^4}\sqrt{5yz}$
    $= 3xyz^2 \sqrt{5yz}$

17. $\sqrt[3]{-125 x^2 y^4} = \sqrt[3]{(-5)^3 x^2 y^4}$
    $= \sqrt[3]{(-5)^3 y^3 (x^2 y)}$
    $= \sqrt[3]{(-5)^3 y^3}\sqrt[3]{x^2 y}$
    $= -5y\sqrt[3]{x^2 y}$

19. $\sqrt[3]{-216 x^5 y^9} = \sqrt[3]{(-6)^3 x^5 y^9}$
    $= \sqrt[3]{(-6)^3 x^3 y^9 (x^2)}$
    $= \sqrt[3]{(-6)^3 x^3 y^9}\sqrt[3]{x^2}$
    $= -6xy^3 \sqrt[3]{x^2}$

21. $\sqrt[3]{a^5b^8} = \sqrt[3]{a^3b^6(a^2b^2)}$
    $= \sqrt[3]{a^3b^6}\sqrt[3]{a^2b^2}$
    $= ab^2\sqrt[3]{a^2b^2}$

Objective 2 Exercises

23. $\sqrt{2} + \sqrt{2} = 2\sqrt{2}$

25. $4\sqrt[3]{7} - \sqrt[3]{7} = 3\sqrt[3]{7}$

27. $2\sqrt{x} - 8\sqrt{x} = -6\sqrt{x}$

29. $\sqrt{8} - \sqrt{32} = \sqrt{2^3} - \sqrt{2^5}$
    $= \sqrt{2^2}\sqrt{2} - \sqrt{2^4}\sqrt{2}$
    $= 2\sqrt{2} - 2^2\sqrt{2}$
    $= 2\sqrt{2} - 4\sqrt{2}$
    $= -2\sqrt{2}$

31. $\sqrt{128x} - \sqrt{98x} = \sqrt{2^7 x} - \sqrt{2 \cdot 7^2 x}$
    $= \sqrt{2^6}\sqrt{2x} - \sqrt{7^2}\sqrt{2x}$
    $= 2^3\sqrt{2x} - 7\sqrt{2x}$
    $= 8\sqrt{2x} - 7\sqrt{2x}$
    $= \sqrt{2x}$

33. $\sqrt{27a} - \sqrt{8a} = \sqrt{3^3 a} - \sqrt{2^3 a}$
    $= \sqrt{3^2}\sqrt{3a} - \sqrt{2^2}\sqrt{2a}$
    $= 3\sqrt{3a} - 2\sqrt{2a}$

35. $2\sqrt{2x^3} + 4x\sqrt{8x} = 2\sqrt{2x^3} + 4x\sqrt{2^3 x}$
    $= 2\sqrt{x^2}\sqrt{2x} + 4x\sqrt{2^2}\sqrt{2x}$
    $= 2x\sqrt{2x} + 2 \cdot 4x\sqrt{2x}$
    $= 2x\sqrt{2x} + 8x\sqrt{2x}$
    $= 10x\sqrt{2x}$

37. $x\sqrt{75xy} - \sqrt{27x^3 y} = x\sqrt{3 \cdot 5^2 xy} - \sqrt{3^3 x^3 y}$
    $= x\sqrt{5^2}\sqrt{3xy} - \sqrt{3^2 x^2}\sqrt{3xy}$
    $= 5x\sqrt{3xy} - 3x\sqrt{3xy}$
    $= 2x\sqrt{3xy}$

39. $2\sqrt{32x^2 y^3} - xy\sqrt{98y}$
    $= 2\sqrt{2^5 x^2 y^3} - xy\sqrt{2 \cdot 7^2 y}$
    $= 2\sqrt{2^4 x^2 y^2}\sqrt{2y} - xy\sqrt{7^2}\sqrt{2y}$
    $= 2 \cdot 2^2 xy\sqrt{2y} - 7xy\sqrt{2y}$
    $= 8xy\sqrt{2y} - 7xy\sqrt{2y}$
    $= xy\sqrt{2y}$

41. $7b\sqrt{a^5 b^3} - 2ab\sqrt{a^3 b^3}$
    $= 7b\sqrt{a^4 b^2}\sqrt{ab} - 2ab\sqrt{a^2 b^2}\sqrt{ab}$
    $= 7b \cdot a^2 b\sqrt{ab} - 2ab \cdot ab\sqrt{ab}$
    $= 7a^2 b^2 \sqrt{ab} - 2a^2 b^2 \sqrt{ab}$
    $= 5a^2 b^2 \sqrt{ab}$

43. $\sqrt[3]{128} + \sqrt[3]{250} = \sqrt[3]{2^7} + \sqrt[3]{2 \cdot 5^3}$
    $= \sqrt[3]{2^6}\sqrt[3]{2} + \sqrt[3]{5^3}\sqrt[3]{2}$
    $= 2^2\sqrt[3]{2} + 5\sqrt[3]{2}$
    $= 4\sqrt[3]{2} + 5\sqrt[3]{2}$
    $= 9\sqrt[3]{2}$

45. $2\sqrt[3]{3a^4} - 3a\sqrt[3]{81a} = 2\sqrt[3]{3a^4} - 3a\sqrt[3]{3^4 a}$
    $= 2\sqrt[3]{a^3}\sqrt[3]{3a} - 3a\sqrt[3]{3^3}\sqrt[3]{3a}$
    $= 2a\sqrt[3]{3a} - 3a \cdot 3\sqrt[3]{3a}$
    $= 2a\sqrt[3]{3a} - 9a\sqrt[3]{3a}$
    $= -7a\sqrt[3]{3a}$

47. $3\sqrt[3]{x^5 y^7} - 8xy\sqrt[3]{x^2 y^4}$
    $= 3\sqrt[3]{x^3 y^6}\sqrt[3]{x^2 y} - 8xy\sqrt[3]{y^3}\sqrt[3]{x^2 y}$
    $= 3xy^2\sqrt[3]{x^2 y} - 8xy \cdot y\sqrt[3]{x^2 y}$
    $= 3xy^2\sqrt[3]{x^2 y} - 8xy^2\sqrt[3]{x^2 y}$
    $= -5xy^2\sqrt[3]{x^2 y}$

49. $2a\sqrt[4]{16ab^5} + 3b\sqrt[4]{256a^5 b}$
    $= 2a\sqrt[4]{2^4 ab^5} + 3b\sqrt[4]{2^8 a^5 b}$
    $= 2a\sqrt[4]{2^4 b^4}\sqrt[4]{ab} + 3b\sqrt[4]{2^8 a^4}\sqrt[4]{ab}$
    $= 2a \cdot 2b\sqrt[4]{ab} + 3b \cdot 2^2 a\sqrt[4]{ab}$
    $= 4ab\sqrt[4]{ab} + 12ab\sqrt[4]{ab}$
    $= 16ab\sqrt[4]{ab}$

51. $3\sqrt{108} - 2\sqrt{18} - 3\sqrt{48}$
    $= 3\sqrt{2^2 \cdot 3^3} - 2\sqrt{2 \cdot 3^2} - 3\sqrt{2^4 \cdot 3}$
    $= 3\sqrt{2^2 \cdot 3^2}\sqrt{3} - 2\sqrt{3^2}\sqrt{2} - 3\sqrt{2^4}\sqrt{3}$
    $= 3 \cdot 2 \cdot 3\sqrt{3} - 2 \cdot 3\sqrt{2} - 3 \cdot 2^2 \sqrt{3}$
    $= 18\sqrt{3} - 6\sqrt{2} - 12\sqrt{3}$
    $= 6\sqrt{3} - 6\sqrt{2}$

53. $\sqrt{4x^7y^5} + 9x^2\sqrt{x^3y^5} - 5xy\sqrt{x^5y^3} = \sqrt{2^2x^7y^5} + 9x^2\sqrt{x^3y^5} - 5xy\sqrt{x^5y^3}$
$= \sqrt{2^2x^6y^4}\sqrt{xy} + 9x^2\sqrt{x^2y^4}\sqrt{xy} - 5xy\sqrt{x^4y^2}\sqrt{xy}$
$= 2x^3y^2\sqrt{xy} + 9x^2 \cdot xy^2\sqrt{xy} - 5xy \cdot x^2y\sqrt{xy}$
$= 2x^3y^2\sqrt{xy} + 9x^3y^2\sqrt{xy} - 5x^3y^2\sqrt{xy}$
$= 6x^3y^2\sqrt{xy}$

55. $5a\sqrt{3a^3b} + 2a^2\sqrt{27ab} - 4\sqrt{75a^5b} = 5a\sqrt{3a^3b} + 2a^2\sqrt{3^3ab} - 4\sqrt{3 \cdot 5^2a^5b}$
$= 5a\sqrt{a^2}\sqrt{3ab} + 2a^2\sqrt{3^2}\sqrt{3ab} - 4\sqrt{5^2a^4}\sqrt{3ab}$
$= 5a \cdot a\sqrt{3ab} + 2a^2 \cdot 3\sqrt{3ab} - 4 \cdot 5a^2\sqrt{3ab}$
$= 5a^2\sqrt{3ab} + 6a^2\sqrt{3ab} - 20a^2\sqrt{3ab}$
$= -9a^2\sqrt{3ab}$

Objective 3 Exercises

57. $\sqrt{8}\sqrt{32} = \sqrt{256} = \sqrt{2^8} = 2^4 = 16$

59. $\sqrt[3]{4}\sqrt[3]{8} = \sqrt[3]{32} = \sqrt[3]{2^5} = \sqrt[3]{2^3}\sqrt[3]{2^2} = 2\sqrt[3]{4}$

61. $\sqrt{x^2y^5}\sqrt{xy} = \sqrt{x^3y^6} = \sqrt{x^2y^6}\sqrt{x} = xy^3\sqrt{x}$

63. $\sqrt{2x^2y}\sqrt{32xy} = \sqrt{64x^3y^2} = \sqrt{2^6x^3y^2} = \sqrt{2^6x^2y^2}\sqrt{x} = 2^3xy\sqrt{x} = 8xy\sqrt{x}$

65. $\sqrt[3]{x^2y}\sqrt[3]{16x^4y^2} = \sqrt[3]{16x^6y^3} = \sqrt[3]{2^4x^6y^3} = \sqrt[3]{2^3x^6y^3}\sqrt[3]{2} = 2x^2y\sqrt[3]{2}$

67. $\sqrt[4]{12ab^3}\sqrt[4]{4a^5b^2} = \sqrt[4]{48a^6b^5} = \sqrt[4]{2^4 \cdot 3a^6b^5} = \sqrt[4]{2^4a^4b^4}\sqrt[4]{3a^2b} = 2ab\sqrt[4]{3a^2b}$

69. $\sqrt{3}(\sqrt{27} - \sqrt{3}) = \sqrt{81} - \sqrt{9}$
$= \sqrt{3^4} - \sqrt{3^2}$
$= 3^2 - 3$
$= 9 - 3$
$= 6$

71. $\sqrt{x}(\sqrt{x} - \sqrt{2}) = \sqrt{x^2} - \sqrt{2x} = x - \sqrt{2x}$

73. $\sqrt{2x}(\sqrt{8x} - \sqrt{32}) = \sqrt{16x^2} - \sqrt{64x}$
$= \sqrt{2^4x^2} - \sqrt{2^6x}$
$= 2^2x - 2^3\sqrt{x}$
$= 4x - 8\sqrt{x}$

75. $(\sqrt{x} - 3)^2 = (\sqrt{x})^2 - 3\sqrt{x} - 3\sqrt{x} + 9$
$= x - 6\sqrt{x} + 9$

77. $(4\sqrt{5} + 2)^2 = (4\sqrt{5})^2 + 8\sqrt{5} + 8\sqrt{5} + 4$
$= 16 \cdot 5 + 16\sqrt{5} + 4$
$= 80 + 16\sqrt{5} + 4$
$= 84 + 16\sqrt{5}$

79. $2\sqrt{14xy} \cdot 4\sqrt{7x^2y} \cdot 3\sqrt{8xy^2} = 24\sqrt{784x^4y^4}$
$= 24\sqrt{2^4 \cdot 7^2x^4y^4}$
$= 24 \cdot 2^2 \cdot 7x^2y^2$
$= 672x^2y^2$

81. $\sqrt[3]{2a^2b}\sqrt[3]{4a^3b^2}\sqrt[3]{8a^5b^6} = \sqrt[3]{64a^{10}b^9}$
$= \sqrt[3]{2^6 a^{10} b^9}$
$= \sqrt[3]{2^6 a^9 b^9}\sqrt[3]{a}$
$= 2^2 a^3 b^3 \sqrt[3]{a}$
$= 4a^3 b^3 \sqrt[3]{a}$

83. $(\sqrt{5} - 5)(2\sqrt{5} + 2) = 2\sqrt{5^2} + 2\sqrt{5} - 10\sqrt{5} - 10$
$= 2 \cdot 5 - 8\sqrt{5} - 10$
$= 10 - 8\sqrt{5} - 10$
$= -8\sqrt{5}$

85. $(\sqrt{x} - y)(\sqrt{x} + y) = \sqrt{x^2} - y^2 = x - y^2$

87. $(2\sqrt{3x} - \sqrt{y})(2\sqrt{3x} + \sqrt{y}) = 4\sqrt{3^2 x^2} - \sqrt{y^2}$
$= 4 \cdot 3x - y$
$= 12x - y$

89. $(\sqrt{x} + 4)(\sqrt{x} - 7) = \sqrt{x^2} - 7\sqrt{x} + 4\sqrt{x} - 28$
$= x - 3\sqrt{x} - 28$

91. $(\sqrt[3]{x} - 4)(\sqrt[3]{x} + 5) = \sqrt[3]{x^2} + 5\sqrt[3]{x} - 4\sqrt[3]{x} - 20$
$= \sqrt[3]{x^2} + \sqrt[3]{x} - 20$

Objective 4 Exercises

93. $\dfrac{\sqrt{32x^2}}{\sqrt{2x}} = \sqrt{\dfrac{32x^2}{2x}}$
$= \sqrt{16x}$
$= \sqrt{2^4 x}$
$= \sqrt{2^4}\sqrt{x}$
$= 2^2 \sqrt{x}$
$= 4\sqrt{x}$

95. $\dfrac{\sqrt{42a^3b^5}}{\sqrt{14a^2b}} = \sqrt{\dfrac{42a^3b^5}{14a^2b}}$
$= \sqrt{3ab^4}$
$= \sqrt{b^4}\sqrt{3a}$
$= b^2 \sqrt{3a}$

97. $\dfrac{1}{\sqrt{5}} = \dfrac{1}{\sqrt{5}} \cdot \dfrac{\sqrt{5}}{\sqrt{5}} = \dfrac{\sqrt{5}}{\sqrt{5^2}} = \dfrac{\sqrt{5}}{5}$

99. $\dfrac{1}{\sqrt{2x}} = \dfrac{1}{\sqrt{2x}} \cdot \dfrac{\sqrt{2x}}{\sqrt{2x}} = \dfrac{\sqrt{2x}}{\sqrt{2^2 x^2}} = \dfrac{\sqrt{2x}}{2x}$

101. $\dfrac{5}{\sqrt{5x}} = \dfrac{5}{\sqrt{5x}} \cdot \dfrac{\sqrt{5x}}{\sqrt{5x}} = \dfrac{5\sqrt{5x}}{\sqrt{5^2 x^2}} = \dfrac{5\sqrt{5x}}{5x} = \dfrac{\sqrt{5x}}{x}$

103. $\sqrt{\dfrac{x}{5}} = \dfrac{\sqrt{x}}{\sqrt{5}} = \dfrac{\sqrt{x}}{\sqrt{5}} \cdot \dfrac{\sqrt{5}}{\sqrt{5}} = \dfrac{\sqrt{5x}}{\sqrt{5^2}} = \dfrac{\sqrt{5x}}{5}$

105. $\dfrac{3}{\sqrt[3]{2}} = \dfrac{3}{\sqrt[3]{2}} \cdot \dfrac{\sqrt[3]{2^2}}{\sqrt[3]{2^2}} = \dfrac{3\sqrt[3]{2^2}}{\sqrt[3]{2^3}} = \dfrac{3\sqrt[3]{4}}{2}$

107. $\dfrac{3}{\sqrt[3]{4x^2}} = \dfrac{3}{\sqrt[3]{2^2 x^2}} \cdot \dfrac{\sqrt[3]{2x}}{\sqrt[3]{2x}} = \dfrac{3\sqrt[3]{2x}}{\sqrt[3]{2^3 x^3}} = \dfrac{3\sqrt[3]{2x}}{2x}$

109. $\dfrac{\sqrt{40x^3 y^2}}{\sqrt{80x^2 y^3}} = \sqrt{\dfrac{40x^3 y^2}{80x^2 y^3}}$
$= \sqrt{\dfrac{x}{2y}}$
$= \dfrac{\sqrt{x}}{\sqrt{2y}} \cdot \dfrac{\sqrt{2y}}{\sqrt{2y}}$
$= \dfrac{\sqrt{2xy}}{\sqrt{2^2 y^2}}$
$= \dfrac{\sqrt{2xy}}{2y}$

111. $\dfrac{\sqrt{24a^2 b}}{\sqrt{18ab^4}} = \sqrt{\dfrac{24a^2 b}{18ab^4}}$
$= \sqrt{\dfrac{4a}{3b^3}}$
$= \dfrac{\sqrt{4a}}{\sqrt{3b^3}}$
$= \dfrac{\sqrt{2^2}\sqrt{a}}{\sqrt{b^2}\sqrt{3b}}$
$= \dfrac{2\sqrt{a}}{b\sqrt{3b}} \cdot \dfrac{\sqrt{3b}}{\sqrt{3b}}$
$= \dfrac{2\sqrt{3ab}}{b\sqrt{3^2 b^2}}$
$= \dfrac{2\sqrt{3ab}}{b \cdot 3b}$
$= \dfrac{2\sqrt{3ab}}{3b^2}$

113. $\dfrac{2}{\sqrt{5} + 2} = \dfrac{2}{\sqrt{5} + 2} \cdot \dfrac{\sqrt{5} - 2}{\sqrt{5} - 2}$
$= \dfrac{2\sqrt{5} - 4}{(\sqrt{5})^2 - 2^2}$
$= \dfrac{2\sqrt{5} - 4}{5 - 4}$
$= \dfrac{2\sqrt{5} - 4}{1}$
$= 2\sqrt{5} - 4$

115. $\dfrac{3}{\sqrt{y}-2} = \dfrac{3}{\sqrt{y}-2} \cdot \dfrac{\sqrt{y}+2}{\sqrt{y}+2}$
$= \dfrac{3\sqrt{y}+6}{(\sqrt{y})^2 - 2^2}$
$= \dfrac{3\sqrt{y}+6}{y-4}$

117. $\dfrac{\sqrt{2}-\sqrt{3}}{\sqrt{2}+\sqrt{3}} = \dfrac{\sqrt{2}-\sqrt{3}}{\sqrt{2}+\sqrt{3}} \cdot \dfrac{\sqrt{2}-\sqrt{3}}{\sqrt{2}-\sqrt{3}}$
$= \dfrac{(\sqrt{2})^2 - \sqrt{6} - \sqrt{6} + (\sqrt{3})^2}{(\sqrt{2})^2 - (\sqrt{3})^2}$
$= \dfrac{2 - 2\sqrt{6} + 3}{2 - 3}$
$= \dfrac{5 - 2\sqrt{6}}{-1}$
$= -5 + 2\sqrt{6}$

119. $\dfrac{4-\sqrt{2}}{2-\sqrt{3}} = \dfrac{4-\sqrt{2}}{2-\sqrt{3}} \cdot \dfrac{2+\sqrt{3}}{2+\sqrt{3}}$
$= \dfrac{8 + 4\sqrt{3} - 2\sqrt{2} - \sqrt{6}}{(2)^2 - (\sqrt{3})^2}$
$= \dfrac{8 + 4\sqrt{3} - 2\sqrt{2} - \sqrt{6}}{4 - 3}$
$= \dfrac{8 + 4\sqrt{3} - 2\sqrt{2} - \sqrt{6}}{1}$
$= 8 + 4\sqrt{3} - 2\sqrt{2} - \sqrt{6}$

121. $\dfrac{\sqrt{3}-\sqrt{5}}{\sqrt{2}+\sqrt{5}} = \dfrac{\sqrt{3}-\sqrt{5}}{\sqrt{2}+\sqrt{5}} \cdot \dfrac{\sqrt{2}-\sqrt{5}}{\sqrt{2}-\sqrt{5}}$
$= \dfrac{\sqrt{6} - \sqrt{15} - \sqrt{10} + \sqrt{5^2}}{(\sqrt{2})^2 - (\sqrt{5})^2}$
$= \dfrac{\sqrt{6} - \sqrt{15} - \sqrt{10} + 5}{2 - 5}$
$= \dfrac{\sqrt{6} - \sqrt{15} - \sqrt{10} + 5}{-3}$
$= \dfrac{\sqrt{15} + \sqrt{10} - \sqrt{6} - 5}{3}$

123. $\dfrac{3}{\sqrt[4]{8x^3}} = \dfrac{3}{\sqrt[4]{2^3 x^3}} \cdot \dfrac{\sqrt[4]{2x}}{\sqrt[4]{2x}} = \dfrac{3\sqrt[4]{2x}}{\sqrt[4]{2^4 x^4}} = \dfrac{3\sqrt[4]{2x}}{2x}$

125. $\dfrac{4}{\sqrt[5]{16a^2}} = \dfrac{4}{\sqrt[5]{2^4 a^2}} \cdot \dfrac{\sqrt[5]{2a^3}}{\sqrt[5]{2a^3}}$
$= \dfrac{4\sqrt[5]{2a^3}}{\sqrt[5]{2^5 a^5}}$
$= \dfrac{4\sqrt[5]{2a^3}}{2a}$
$= \dfrac{2\sqrt[5]{2a^3}}{a}$

127. $\dfrac{2x}{\sqrt[5]{64x^3}} = \dfrac{2x}{\sqrt[5]{2^6 x^3}} \cdot \dfrac{\sqrt[5]{2^4 x^2}}{\sqrt[5]{2^4 x^2}}$
$= \dfrac{2x\sqrt[5]{2^4 x^2}}{\sqrt[5]{2^{10} x^5}}$
$= \dfrac{2x\sqrt[5]{16x^2}}{2^2 x}$
$= \dfrac{\sqrt[5]{16x^2}}{2}$

129. $\dfrac{\sqrt{a}+a\sqrt{b}}{\sqrt{a}-a\sqrt{b}} = \dfrac{\sqrt{a}+a\sqrt{b}}{\sqrt{a}-a\sqrt{b}} \cdot \dfrac{\sqrt{a}+a\sqrt{b}}{\sqrt{a}+a\sqrt{b}}$
$= \dfrac{\sqrt{a^2} + a\sqrt{ab} + a\sqrt{ab} + a^2\sqrt{b^2}}{(\sqrt{a})^2 - (a\sqrt{b})^2}$
$= \dfrac{a + 2a\sqrt{ab} + a^2 b}{a - a^2 b}$
$= \dfrac{a(1 + 2\sqrt{ab} + ab)}{a(1 - ab)}$
$= \dfrac{1 + 2\sqrt{ab} + ab}{1 - ab}$

131. $\dfrac{3\sqrt{xy} + 2\sqrt{xy}}{\sqrt{x} - \sqrt{y}} = \dfrac{5\sqrt{xy}}{\sqrt{x} - \sqrt{y}}$
$= \dfrac{5\sqrt{xy}}{\sqrt{x} - \sqrt{y}} \cdot \dfrac{\sqrt{x} + \sqrt{y}}{\sqrt{x} + \sqrt{y}}$
$= \dfrac{5\sqrt{x^2 y} + 5\sqrt{xy^2}}{(\sqrt{x})^2 - (\sqrt{y})^2}$
$= \dfrac{5\sqrt{x^2}\sqrt{y} + 5\sqrt{y^2}\sqrt{x}}{x - y}$
$= \dfrac{5x\sqrt{y} + 5y\sqrt{x}}{x - y}$

Applying Concepts 7.2

133. $(\sqrt{8} - \sqrt{2})^3 = (\sqrt{2^3} - \sqrt{2})^3 = (2\sqrt{2} - \sqrt{2})^3$
$= (\sqrt{2})^3 = \sqrt{2} \cdot \sqrt{2} \cdot \sqrt{2}$
$= 2\sqrt{2}$

135. $(\sqrt{2} - 3)^3 = (\sqrt{2} - 3)(\sqrt{2} - 3)(\sqrt{2} - 3)$
$= (2 - 6\sqrt{2} + 9)(\sqrt{2} - 3)$
$= (-6\sqrt{2} + 11)(\sqrt{2} - 3)$
$= -12 + 18\sqrt{2} + 11\sqrt{2} - 33$
$= 29\sqrt{2} - 45$

137. $\dfrac{3}{\sqrt{y+1}+1} = \dfrac{3}{\sqrt{y+1}+1} \cdot \dfrac{\sqrt{y+1}-1}{\sqrt{y+1}-1}$

$= \dfrac{3\sqrt{y+1}-3}{\left(\sqrt{y+1}\right)^2 - 1^2}$

$= \dfrac{3\sqrt{y+1}-3}{y+1-1}$

$= \dfrac{3\sqrt{y+1}-3}{y}$

139. $\dfrac{\sqrt[3]{(x+y)^2}}{\sqrt{x+y}} = \dfrac{(x+y)^{2/3}}{(x+y)^{1/2}}$

$= (x+y)^{1/6}$

$= \sqrt[6]{x+y}$

141. $\sqrt[4]{2y}\sqrt{x+3} = (2y)^{1/4}(x+3)^{1/2}$

$= (2y)^{1/4}(x+3)^{2/4}$

$= \sqrt[4]{2y}\sqrt[4]{(x+3)^2}$

$= \sqrt[4]{2y(x+3)^2}$

143. $\sqrt{a}\sqrt[3]{a+3} = a^{1/2}(a+3)^{1/3}$

$= a^{3/6}(a+3)^{2/6}$

$= \sqrt[6]{a^3}\sqrt[6]{(a+3)^2}$

$= \sqrt[6]{a^3(a+3)^2}$

145. $\sqrt{16^{1/2}} = (16^{1/2})^{1/2} = 16^{1/4} = 2$

147. $\sqrt[4]{32^{-4/5}} = (32^{-4/5})^{1/4}$

$= 32^{-1/5}$

$= (2^5)^{-1/5}$

$= 2^{-1}$

$= \dfrac{1}{2}$

149. a. $D(x) = 2.79x^{0.3}$

$D(x) = 2.79x^{3/10}$

$D(x) = 2.79\sqrt[10]{x^3}$

b. For the year 1993, $x = 5$.

$D(5) = 2.79\sqrt[10]{5^3}$
$D(5) \approx 4.52$
The model predicted a debt of $4.52 trillion by 1993.

c. For the year 2004, $x = 16$.

$D(16) = 2.79\sqrt[10]{16^3}$
$D(16) \approx 6.41$
The model predicts a debt of $6.41 trillion by the year 2004.

Section 7.3

Concept Review 7.3

1. Always true
3. Always true

Objective 1 Exercises

3. $\sqrt{-4} = i\sqrt{4} = i\sqrt{2^2} = 2i$

5. $\sqrt{-98} = i\sqrt{98} = i\sqrt{2\cdot 7^2} = 7i\sqrt{2}$

7. $\sqrt{-27} = i\sqrt{27} = i\sqrt{3^2\cdot 3} = 3i\sqrt{3}$

9. $\sqrt{16} + \sqrt{-4} = \sqrt{16} + i\sqrt{4} = \sqrt{2^4} + i\sqrt{2^2} = 4 + 2i$

11. $\sqrt{12} - \sqrt{-18} = \sqrt{12} - i\sqrt{18}$
$= \sqrt{2^2\cdot 3} - i\sqrt{3^2\cdot 2}$
$= 2\sqrt{3} - 3i\sqrt{2}$

13. $\sqrt{160} - \sqrt{-147} = \sqrt{160} - i\sqrt{147}$
$= \sqrt{2^4\cdot 2\cdot 5} - i\sqrt{7^2\cdot 3}$
$= 4\sqrt{10} - 7i\sqrt{3}$

15. $\sqrt{-4a^2} = i\sqrt{4a^2} = i\sqrt{2^2 a^2} = 2ai$

17. $\sqrt{-49x^{12}} = i\sqrt{49x^{12}} = i\sqrt{7^2 x^{12}} = 7x^6 i$

19. $\sqrt{-144a^3 b^5} = i\sqrt{2^4\cdot 3^2 a^3 b^5}$
$= 2^2\cdot 3iab^2\sqrt{ab}$
$= 12ab^2 i\sqrt{ab}$

21. $\sqrt{4a} + \sqrt{-12a^2} = \sqrt{2^2 a} + i\sqrt{2^2\cdot 3a^2}$
$= 2\sqrt{a} + 2ai\sqrt{3}$

23. $\sqrt{18b^5} - \sqrt{-27b^3} = \sqrt{2\cdot 3^2 b^5} - i\sqrt{3\cdot 3^2 b^3}$
$= 3b^2\sqrt{2b} - 3bi\sqrt{3b}$

25. $\sqrt{-50x^3 y^3} + x\sqrt{25x^4 y^3}$
$= i\sqrt{5^2\cdot 2x^3 y^3} + x\sqrt{5^2 x^4 y^3}$
$= 5ixy\sqrt{2xy} + 5x^3 y\sqrt{y}$
$= 5x^3 y\sqrt{y} + 5xyi\sqrt{2xy}$

27. $\sqrt{-49a^5 b^2} - ab\sqrt{-25a^3}$
$= i\sqrt{7^2 a^5 b^2} - iab\sqrt{5^2 a^3}$
$= 7ia^2 b\sqrt{a} - 5ia^2 b\sqrt{a}$
$= 2a^2 bi\sqrt{a}$

29. $\sqrt{12a^3} + \sqrt{-27b^3}$
$= \sqrt{2^2\cdot 3a^3} + i\sqrt{3^2\cdot 3b^3}$
$= 2a\sqrt{3a} + 3bi\sqrt{3b}$

Objective 2 Exercises

31. $(6-9i)+(4+2i) = 10-7i$

33. $(3-5i)+(8-2i) = 11-7i$

35. $(5-\sqrt{-25})-(11-\sqrt{-36})$
$= (5-i\sqrt{25})-(11-i\sqrt{36})$
$= (5-i\sqrt{5^2})-(11-i\sqrt{2^2 \cdot 3^2})$
$= (5-5i)-(11-6i)$
$= -6+i$

37. $(5-\sqrt{-12})-(9+\sqrt{-108})$
$= (5-i\sqrt{12})-(9+i\sqrt{108})$
$= (5-i\sqrt{2^2 \cdot 3})-(9+i\sqrt{2^2 \cdot 3^2 \cdot 3})$
$= (5-2i\sqrt{3})-(9+6i\sqrt{3})$
$= -4-8i\sqrt{3}$

39. $(\sqrt{40}-\sqrt{-98})-(\sqrt{90}+\sqrt{-32})$
$= (\sqrt{40}-i\sqrt{98})-(\sqrt{90}+i\sqrt{32})$
$= (\sqrt{2^2 \cdot 2 \cdot 5}-i\sqrt{7^2 \cdot 2})-(\sqrt{3^2 \cdot 2 \cdot 5}+i\sqrt{2^4 \cdot 2})$
$= (2\sqrt{10}-7i\sqrt{2})-(3\sqrt{10}+4i\sqrt{2})$
$= -\sqrt{10}-11i\sqrt{2}$

41. $(6-8i)+4i = 6-4i$

43. $(8-3i)+(-8+3i) = 0$

45. $(4+6i)+7 = (4+6i)+(7+0i) = 11+6i$

Objective 3 Exercises

47. $(-6i)(-4i) = 24i^2 = 24(-1) = -24$

49. $\sqrt{-5}\sqrt{-45} = i\sqrt{5} \cdot i\sqrt{45}$
$= i^2 \sqrt{225}$
$= -\sqrt{3^2 \cdot 5^2}$
$= -15$

51. $\sqrt{-5}\sqrt{-10} = i\sqrt{5} \cdot i\sqrt{10} = i^2 \sqrt{50}$
$= -\sqrt{5^2 \cdot 2} = -5\sqrt{2}$

53. $-3i(4-5i) = -12i+15i^2$
$= -12i+15(-1)$
$= -15-12i$

55. $\sqrt{-3}(\sqrt{12}-\sqrt{-6}) = i\sqrt{3}(\sqrt{12}-i\sqrt{6})$
$= i\sqrt{36}-i^2\sqrt{18}$
$= i\sqrt{2^2 \cdot 3^2}+\sqrt{3^2 \cdot 2}$
$= 6i+3\sqrt{2}$
$= 3\sqrt{2}+6i$

57. $(2-4i)(2-i) = 4-2i-8i+4i^2$
$= 4-10i+4i^2$
$= 4-10i+4(-1)$
$= -10i$

59. $(4-7i)(2+3i) = 8+12i-14i-21i^2$
$= 8-2i-21i^2$
$= 8-2i-21(-1)$
$= 29-2i$

61. $\left(\dfrac{4}{5}-\dfrac{2}{5}i\right)\left(1+\dfrac{1}{2}i\right) = \dfrac{4}{5}+\dfrac{2}{5}i-\dfrac{2}{5}i-\dfrac{1}{5}i^2$
$= \dfrac{4}{5}-\dfrac{1}{5}i^2$
$= \dfrac{4}{5}-\dfrac{1}{5}(-1)$
$= \dfrac{4}{5}+\dfrac{1}{5}$
$= 1$

63. $(2-i)\left(\dfrac{2}{5}+\dfrac{1}{5}i\right) = \dfrac{4}{5}+\dfrac{2}{5}i-\dfrac{2}{5}i-\dfrac{1}{5}i^2$
$= \dfrac{4}{5}-\dfrac{1}{5}i^2$
$= \dfrac{4}{5}-\dfrac{1}{5}(-1)$
$= \dfrac{4}{5}+\dfrac{1}{5}$
$= 1$

65. $(8-5i)(8+5i) = 8^2+5^2 = 64+25 = 89$

67. $(7-i)(7+i) = 7^2+1^2 = 49+1 = 50$

Objective 4 Exercises

69. $\dfrac{4}{5i} = \dfrac{4}{5i} \cdot \dfrac{i}{i} = \dfrac{4i}{5i^2} = \dfrac{4i}{5(-1)} = \dfrac{4i}{-5} = -\dfrac{4}{5}i$

71. $\dfrac{16+5i}{-3i} = \dfrac{16+5i}{-3i} \cdot \dfrac{i}{i}$
$= \dfrac{16i+5i^2}{-3i^2}$
$= \dfrac{16i+5(-1)}{-3(-1)}$
$= \dfrac{-5+16i}{3}$
$= -\dfrac{5}{3}+\dfrac{16}{3}i$

73. $\dfrac{6}{5+2i} = \dfrac{6}{5+2i} \cdot \dfrac{5-2i}{5-2i}$
$= \dfrac{30-12i}{25+4}$
$= \dfrac{30-12i}{29}$
$= \dfrac{30}{29}-\dfrac{12}{29}i$

75. $\dfrac{5}{4-i} = \dfrac{5}{4-i} \cdot \dfrac{4+i}{4+i}$
$= \dfrac{20+5i}{16+1}$
$= \dfrac{20+5i}{17}$
$= \dfrac{20}{17} + \dfrac{5}{17}i$

77. $\dfrac{2+12i}{5+i} = \dfrac{2+12i}{5+i} \cdot \dfrac{5-i}{5-i}$
$= \dfrac{10-2i+60i-12i^2}{25+1}$
$= \dfrac{10+58i-12(-1)}{26}$
$= \dfrac{22+58i}{26}$
$= \dfrac{11+29i}{13}$
$= \dfrac{11}{13} + \dfrac{29}{13}i$

79. $\dfrac{\sqrt{-2}}{\sqrt{12}-\sqrt{-8}} = \dfrac{i\sqrt{2}}{\sqrt{12}-i\sqrt{8}}$
$= \dfrac{i\sqrt{2}}{\sqrt{2^2 \cdot 3} - i\sqrt{2^2 \cdot 2}}$
$= \dfrac{i\sqrt{2}}{2\sqrt{3}-2i\sqrt{2}} \cdot \dfrac{2\sqrt{3}+2i\sqrt{2}}{2\sqrt{3}+2i\sqrt{2}}$
$= \dfrac{2i^2\sqrt{6}+2i^2\sqrt{2^2}}{(2\sqrt{3})^2+(2\sqrt{2})^2}$
$= \dfrac{2i\sqrt{6}-2\cdot 2}{12+8}$
$= \dfrac{-4+2i\sqrt{6}}{20}$
$= \dfrac{-2+i\sqrt{6}}{10}$
$= -\dfrac{1}{5} + \dfrac{\sqrt{6}}{10}i$

81. $\dfrac{3+5i}{1-i} = \dfrac{3+5i}{1-i} \cdot \dfrac{1+i}{1+i}$
$= \dfrac{3+3i+5i+5i^2}{1+1}$
$= \dfrac{3+8i+5i^2}{2}$
$= \dfrac{3+8i+5(-1)}{2}$
$= \dfrac{-2+8i}{2}$
$= -1+4i$

Applying Concepts 7.3

83. $i^6 = i^4 \cdot i^2 = 1(-1) = -1$

85. $i^{57} = i^{56} \cdot i = 1 \cdot i = i$

87. $i^{-6} = \dfrac{1}{i^6} = \dfrac{1}{i^4 \cdot i^2} = \dfrac{1}{1 \cdot -1} = \dfrac{1}{-1} = -1$

89. $i^{-58} = \dfrac{1}{i^{58}} = \dfrac{1}{i^{56} \cdot i^2} = \dfrac{1}{1(-1)} = \dfrac{1}{-1} = -1$

91. $y^2+1 = y^2+1^2 = (y+i)(y-i)$

93. $x^2+25 = x^2+5^2 = (x+5i)(x-5i)$

95. a.

$x^2-2x-10=0$	
$(1+3i)^2 - 2(1+3i) - 10$	0
$1+6i+9i^2-2-6i-10$	0
$9i^2-11$	0
$-9-11$	0
$-20 \ne 0$	

No, $1+3i$ is not a solution of $x^2-2x-10=0$.

b.

$x^2-2x-10=0$	
$(1-3i)^2 - 2(1-3i) - 10$	0
$1-6i+9i^2-2+6i-10$	0
$9i^2-11$	0
$-9-11$	0
$-20 \ne 0$	

No, $1-3i$ is not a solution of $x^2-2x-10=0$.

97. $\sqrt{-i} = \sqrt{(-1)i}$
$= \sqrt{-1}\sqrt{i}$
$= i\sqrt{i}$
$= i\left(\dfrac{\sqrt{2}}{2} + i\dfrac{\sqrt{2}}{2}\right)$
$= i\dfrac{\sqrt{2}}{2} + i^2\dfrac{\sqrt{2}}{2}$
$= i\dfrac{\sqrt{2}}{2} - 1\left(\dfrac{\sqrt{2}}{2}\right)$
$= -\dfrac{\sqrt{2}}{2} + i\dfrac{\sqrt{2}}{2}$

Section 7.4

Concept Review 7.4

1. Always true

3. Never true
The domain is $\{x | x \text{ is a real number}\}$.

210 Chapter 7: Rational Exponents and Radicals

Objective 1 Exercises

3. The domain of $f(x) = 2x^{1/3}$ is $\{x | x \text{ is a real number}\}$.

5. $x + 1 \geq 0$
   $x \geq -1$
   The domain of $g(x) = -2\sqrt{x+1}$ is $\{x | x \geq -1\}$.

7. $x \geq 0$
   The domain of $f(x) = 2x\sqrt{x} - 3$ is $\{x | x \geq 0\}$.

9. $x^{3/4} = \sqrt[4]{x^3}$
   $x^3 \geq 0$
   $x \geq 0$
   The domain of $C(x) = -3x^{3/4} + 1$ is $\{x | x \geq 0\}$.

11. $(3x - 6)^{1/2} = \sqrt{3x - 6}$
    $3x - 6 \geq 0$
    $3x \geq 6$
    $x \geq 2$
    The domain of $F(x) = 4(3x - 6)^{1/2}$ is $\{x | x \geq 2\}$.

13. The domain of $g(x) = 2(2x - 10)^{2/3}$ is $(-\infty, \infty)$.

15. $12 - 4x \geq 0$
    $-4x \geq -12$
    $x \leq 3$
    The domain of $V(x) = x - \sqrt{12 - 4x}$ is $(-\infty, 3]$.

17. $(x - 2)^2 \geq 0$
    $x - 2 \geq 0$
    $x \geq 2$
    The domain of $h(x) = 3\sqrt[4]{(x-2)^3}$ is $[2, \infty)$.

19. $(4 - 6x)^{1/2} = \sqrt{4 - 6x}$
    $4 - 6x \geq 0$
    $-6x \geq -4$
    $x \leq \frac{2}{3}$
    The domain of $f(x) = x - (4 - 6x)^{1/2}$ is $\left(-\infty, \frac{2}{3}\right]$.

Objective 2 Exercises

21.

23.

25.

27.

29.

31.

33.

35.

37.

39.

41.

Section 7.5

Concept Review 7.5

1. Sometimes true
   $(-2)^2 = 2^2$, but -2 does not equal 2.

3. Sometimes true
   The Pythagorean Therorem is valid only for right triangles.

Objective 1 Exercises

3. $\sqrt{x} = 5$
   $(\sqrt{x})^2 = 5^2$
   $x = 25$

   Check:

   $\sqrt{x} = 5$

   | $\sqrt{25}$ | 5 |

   $5 = 5$

   The solution is 25.

5. $\sqrt[3]{a} = 3$
   $(\sqrt[3]{a})^3 = 3$
   $a = 27$

   Check:

   $\sqrt[3]{a} = 3$

   | $\sqrt[3]{27}$ | 3 |

   $3 = 3$

   The solution is 27.

7. $\sqrt{3x} = 12$
   $(\sqrt{3x})^2 = 12^2$
   $3x = 144$
   $x = 48$

   Check:

   $\sqrt{3x} = 12$

   | $\sqrt{3(48)}$ | 12 |
   | $\sqrt{144}$ | 12 |

   $12 = 12$

   The solution is 48.

9. $\sqrt[3]{4x} = -2$
   $(\sqrt[3]{4x})^3 = (-2)^3$
   $4x = -8$
   $x = -2$

   Check:

   $\sqrt[3]{4x} = -2$

   | $\sqrt[3]{4(-2)}$ | -2 |
   | $\sqrt[3]{-8}$ | -2 |

   $-2 = -2$

   The solution is -2.

11. $\sqrt{2x} = -4$
    $(\sqrt{2x})^2 = (-4)^2$
    $2x = 16$
    $x = 8$

    Check:

    $\sqrt{2x} = -4$

    | $\sqrt{2(8)}$ | -4 |
    | $\sqrt{16}$ | -4 |

    $4 \neq -4$

    8 does not check as a solution. The equation has no solution.

13. $\sqrt{3x-2} = 5$
    $(\sqrt{3x-2})^2 = 5^2$
    $3x - 2 = 25$
    $3x = 27$
    $x = 9$

    Check:

    $\sqrt{3x-2} = 5$

    | $\sqrt{3(9)-2}$ | 5 |
    | $\sqrt{27-2}$ | 5 |
    | $\sqrt{25}$ | 5 |

    $5 = 5$

    The solution is 9.

15. $\sqrt{3-2x} = 7$
    $(\sqrt{3-2x})^2 = 7^2$
    $3 - 2x = 49$
    $-2x = 46$
    $x = -23$

    Check:

    $\sqrt{3-2x} = 7$

    | $\sqrt{3-2(-23)}$ | 7 |
    | $\sqrt{3+46}$ | 7 |
    | $\sqrt{49}$ | 7 |

    $7 = 7$

    The solution is -23.

17. $7 = \sqrt{1-3x}$
$(7)^2 = (\sqrt{1-3x})^2$
$49 = 1 - 3x$
$48 = -3x$
$-16 = x$

Check:

$7 = \sqrt{1-3x}$	
7	$\sqrt{1-3(-16)}$
7	$\sqrt{1+48}$
7	$\sqrt{49}$
$7 = 7$	

The solution is -16.

19. $\sqrt[3]{4x-1} = 2$
$(\sqrt[3]{4x-1})^3 = 2^3$
$4x - 1 = 8$
$4x = 9$
$x = \dfrac{9}{4}$

Check:

$\sqrt[3]{4x-1} = 2$	
$\sqrt[3]{4\left(\dfrac{9}{4}\right)-1}$	2
$\sqrt[3]{9-1}$	2
$\sqrt[3]{8}$	2
$2 = 2$	

The solution is $\dfrac{9}{4}$.

21. $\sqrt[3]{1-2x} = -3$
$(\sqrt[3]{1-2x})^3 = (-3)^3$
$1 - 2x = -27$
$-2x = -28$
$x = 14$

Check:

$\sqrt[3]{1-2x} = -3$	
$\sqrt[3]{1-2(14)}$	-3
$\sqrt[3]{1-28}$	-3
$\sqrt[3]{-27}$	-3
$-3 = -3$	

The solution is 14.

23. $\sqrt[3]{9x+1} = 4$
$(\sqrt[3]{9x+1})^3 = 4^3$
$9x + 1 = 64$
$9x = 63$
$x = 7$

Check:

$\sqrt[3]{9x+1} = 4$	
$\sqrt[3]{9(7)+1}$	4
$\sqrt[3]{63+1}$	4
$\sqrt[3]{64}$	4
$4 = 4$	

The solution is 7.

25. $\sqrt{4x-3} - 5 = 0$
$\sqrt{4x-3} = 5$
$(\sqrt{4x-3})^2 = 5^2$
$4x - 3 = 25$
$4x = 28$
$x = 7$

Check:

$\sqrt{4x-3} - 5 = 0$	
$\sqrt{4(7)-3} - 5$	0
$\sqrt{28-3} - 5$	0
$\sqrt{25} - 5$	0
$5 - 5$	0
$0 = 0$	

The solution is 7.

27. $\sqrt[3]{x-3} + 5 = 0$
$\sqrt[3]{x-3} = -5$
$(\sqrt[3]{x-3})^3 = (-5)^3$
$x - 3 = -125$
$x = -122$

Check:

$\sqrt[3]{x-3} + 5 = 0$	
$\sqrt[3]{-122-3} + 5$	0
$\sqrt[3]{-125} + 5$	0
$-5 + 5$	0
$0 = 0$	

The solution is -122.

29. $\sqrt[3]{2x-6} = 4$
$(\sqrt[3]{2x-6})^3 = 4^3$
$2x - 6 = 64$
$2x = 70$
$x = 35$

Check:

$\sqrt[3]{2x-6} = 4$	
$\sqrt[3]{2(35)-6}$	4
$\sqrt[3]{70-6}$	4
$\sqrt[3]{64}$	4
$4 = 4$	

The solution is 35.

31. $\sqrt[4]{2x-9} = 3$
$(\sqrt[4]{2x-9})^4 = 3^4$
$2x - 9 = 81$
$2x = 90$
$x = 45$

Check:

$\sqrt[4]{2x-9} = 3$	
$\sqrt[4]{2(45)-9}$	3
$\sqrt[4]{90-9}$	3
$\sqrt[4]{81}$	3
$3 = 3$	

The solution is 45.

33. $\sqrt{3x-5} - 5 = 3$
$\sqrt{3x-5} = 8$
$(\sqrt{3x-5})^2 = 8^2$
$3x - 5 = 64$
$3x = 69$
$x = 23$

Check:

$\sqrt{3x-5} - 5 = 3$	
$\sqrt{3(23)-5} - 5$	3
$\sqrt{69-5} - 5$	3
$\sqrt{64} - 5$	3
$8 - 5$	3
$3 = 3$	

The solution is 23.

35. $\sqrt[3]{x-4} + 7 = 5$
$\sqrt[3]{x-4} = -2$
$(\sqrt[3]{x-4})^3 = (-2)^3$
$x - 4 = -8$
$x = -4$

Check:

$\sqrt[3]{x-4} + 7 = 5$	
$\sqrt[3]{-4-4} + 7$	5
$\sqrt[3]{-8} + 7$	5
$-2 + 7$	5
$5 = 5$	

The solution is –4.

37. $\sqrt{3x-5} - 2 = 3$
$(\sqrt{3x-5})^2 = (5)^2$
$3x - 5 = 25$
$3x = 30$
$x = 10$

Check:

$\sqrt{3x-5} - 2 = 3$	
$\sqrt{3(10)-5} - 2$	3
$\sqrt{30-5} - 2$	3
$\sqrt{25} - 2$	3
$5 - 2$	3
$3 = 3$	

The solution is 10.

39. $\sqrt{7x+2} - 10 = -7$
$(\sqrt{7x+2})^2 = (3)^2$
$7x + 2 = 9$
$7x = 7$
$x = 1$

Check:

$\sqrt{7x+2} - 10 = -7$	
$\sqrt{7(1)+2} - 10$	–7
$\sqrt{7+2} - 10$	–7
$\sqrt{9} - 10$	–7
$3 - 10$	–7
$-7 = -7$	

The solution is 1.

41. $\sqrt[3]{1-3x}+5=3$
$(\sqrt[3]{1-3x})^3=(-2)^3$
$1-3x=-8$
$-3x=-9$
$x=3$

Check:

$\sqrt[3]{1-3x}+5=3$	
$\sqrt[3]{1-3(3)}+5$	3
$\sqrt[3]{1-9}+5$	3
$\sqrt[3]{-8}+5$	3
$-2+5$	3
	$3=3$

The solution is 3.

43. $7-\sqrt{3x+1}=-1$
$(-\sqrt{3x+1})^2=(-8)^2$
$3x+1=64$
$3x=63$
$x=21$

Check:

$7-\sqrt{3x+1}=-1$	
$7-\sqrt{3(21)+1}$	-1
$7-\sqrt{63+1}$	-1
$7-\sqrt{64}$	-1
$7-8$	-1
	$-1=-1$

The solution is 21.

45. $\sqrt{2x+4}=3-\sqrt{2x}$
$(\sqrt{2x+4})^2=(3-\sqrt{2x})^2$
$2x+4=9-6\sqrt{2x}+2x$
$6\sqrt{2x}=5$
$(6\sqrt{2x})^2=(5)^2$
$36(2x)=25$
$72x=25$
$x=\dfrac{25}{72}$

Check:

$\sqrt{2x+4}=3-\sqrt{2x}$	
$\sqrt{2\left(\dfrac{25}{72}\right)+4}$	$3-\sqrt{2\left(\dfrac{25}{72}\right)}$
$\sqrt{\dfrac{25}{36}+4}$	$3-\sqrt{\dfrac{25}{36}}$
$\sqrt{\dfrac{169}{36}}$	$3-\dfrac{5}{6}$
$\dfrac{13}{6}$	$=\dfrac{13}{6}$

The solution is $\dfrac{25}{72}$.

47. $\sqrt{x^2-4x-1}+3=x$
$\sqrt{x^2-4x-1}=x-3$
$(\sqrt{x^2-4x-1})^2=(x-3)^2$
$x^2-4x-1=x^2-6x+9$
$2x=10$
$x=5$

Check:

$\sqrt{x^2-4x-1}+3=x$	
$\sqrt{5^2-4(5)-1}+3$	5
$\sqrt{25-20-1}+3$	5
$\sqrt{4}+3$	5
$2+3=5$	

The solution is 5.

49.
$$\sqrt{x^2 - 2x + 1} = 3$$
$$(\sqrt{x^2 - 2x + 1})^2 = 3^2$$
$$x^2 - 2x + 1 = 9$$
$$x^2 - 2x - 8 = 0$$
$$(x - 4)(x + 2) = 0$$
$$x - 4 = 0 \quad x + 2 = 0$$
$$x = 4 \quad\quad x = -2$$

Check:

$\sqrt{x^2 - 2x + 1} = 3$		$\sqrt{x^2 - 2x + 1} = 3$	
$\sqrt{4^2 - 2(4) + 1}$	3	$\sqrt{(-2)^2 - 2(-2) + 1}$	3
$\sqrt{16 - 8 + 1}$	3	$\sqrt{4 + 4 + 1}$	3
$\sqrt{9}$	3	$\sqrt{9}$	3
	3 = 3		3 = 3

The solutions are 4 and −2.

51.
$$\sqrt{4x + 1} - \sqrt{2x + 4} = 1$$
$$(\sqrt{4x + 1})^2 = (1 + \sqrt{2x + 4})^2$$
$$4x + 1 = 1 + 2\sqrt{2x + 4} + 2x + 4$$
$$4x + 1 = 5 + 2x + 2\sqrt{2x + 4}$$
$$(2x - 4)^2 = (2\sqrt{2x + 4})^2$$
$$4x^2 - 16x + 16 = 4(2x + 4)$$
$$4x^2 - 16x + 16 = 8x + 16$$
$$4x^2 - 24x = 0$$
$$4x(x - 6) = 0$$
$$x = 0 \quad x - 6 = 0$$
$$\quad\quad\quad x = 6$$

Check:

$\sqrt{4x + 1} - \sqrt{2x + 4} = 1$		$\sqrt{4x + 1} - \sqrt{2x + 4} = 1$	
$\sqrt{4(0) + 1} - \sqrt{2(0) + 4}$	1	$\sqrt{4(6) + 1} - \sqrt{2(6) + 4}$	1
$\sqrt{1} - \sqrt{4}$	1	$\sqrt{24 + 1} - \sqrt{12 + 4}$	1
$1 - 2$	1	$\sqrt{25} - \sqrt{16}$	1
	$-1 \neq 1$	$5 - 4$	1
			1 = 1

The solution is 6.

Chapter 7: Rational Exponents and Radicals

53.
$$\sqrt{5x+4} - \sqrt{3x+1} = 1$$
$$(\sqrt{5x+4})^2 = (1+\sqrt{3x+1})^2$$
$$5x+4 = 1 + 2\sqrt{3x+1} + 3x+1$$
$$5x+4 = 2 + 3x + 2\sqrt{3x+1}$$
$$(2x+2)^2 = (2\sqrt{3x+1})^2$$
$$4x^2 + 8x + 4 = 4(3x+1)$$
$$4x^2 + 8x + 4 = 12x + 4$$
$$4x^2 - 4x = 0$$
$$4x(x-1) = 0$$
$$x = 0 \quad x - 1 = 0$$
$$x = 1$$

Check:

$\sqrt{5x+4} - \sqrt{3x+1} = 1$	
$\sqrt{5(0)+4} - \sqrt{3(0)+1}$	1
$\sqrt{4} - \sqrt{1}$	1
$2 - 1$	1
$1 = 1$	

$\sqrt{5x+4} - \sqrt{3x+1} = 1$	
$\sqrt{5(1)+4} - \sqrt{3(1)+1}$	1
$\sqrt{5+4} - \sqrt{3+1}$	1
$\sqrt{9} - \sqrt{4}$	1
$3 - 2$	1
$1 = 1$	

The solutions are 0 and 1.

55.
$$\sqrt[4]{x^2+x-1} - 1 = 0$$
$$\sqrt[4]{x^2+x-1} = 1$$
$$(\sqrt[4]{x^2+x-1})^4 = 1^4$$
$$x^2 + x - 1 = 1$$
$$x^2 + x - 2 = 0$$
$$(x+2)(x-1) = 0$$
$$x+2 = 0 \quad x-1 = 0$$
$$x = -2 \quad x = 1$$

Check:

$\sqrt[4]{x^2+x-1} - 1 = 0$	
$\sqrt[4]{(-2)^2+(-2)-1} - 1$	0
$\sqrt[4]{4-2-1} - 1$	0
$\sqrt[4]{1} - 1$	0
$1 - 1$	0
$0 = 0$	

$\sqrt[4]{x^2+x-1} - 1 = 0$	
$\sqrt[4]{1^2+1-1} - 1$	0
$\sqrt[4]{1+1-1} - 1$	0
$\sqrt[4]{1} - 1$	0
$1 - 1$	0
$0 = 0$	

The solutions are −2 and 1.

57. $3\sqrt{x-2}+2=x$
$3\sqrt{x-2}=x-2$
$(3\sqrt{x-2})^2=(x-2)^2$
$9(x-2)=x^2-4x+4$
$9x-18=x^2-4x+4$
$0=x^2-13x+22$
$0=(x-2)(x-11)$
$x-2=0 \quad x-11=0$
$x=2 \quad\quad x=11$

Check:

$3\sqrt{x-2}+2=x$	
$3\sqrt{2-2}+2$	2
$3\sqrt{0}+2$	2
$3(0)+2$	2
$0+2$	2
$2=2$	

$3\sqrt{x-2}+2=x$	
$3\sqrt{11-2}+2$	11
$3\sqrt{9}+2$	11
$3(3)+2$	11
$9+2$	11
$11=11$	

The solutions are 2 and 11.

59. $x+2\sqrt{x+1}=7$
$2\sqrt{x+1}=7-x$
$(2\sqrt{x+1})^2=(7-x)^2$
$4(x+1)=49-14x+x^2$
$4x+4=49-14x+x^2$
$0=x^2-18x+45$
$0=(x-3)(x-15)$
$x-3=0 \quad x-15=0$
$x=3 \quad\quad x=15$

Check:

$x+2\sqrt{x+1}=7$	
$3+2\sqrt{3+1}$	7
$3+2\sqrt{4}$	7
$3+2(2)$	7
$3+4$	7
$7=7$	

$x+2\sqrt{x+1}=7$	
$15+2\sqrt{15+1}$	7
$15+2\sqrt{16}$	7
$15+2(4)$	7
$15+8$	7
$23\neq 7$	

The solution is 3.

Objective 2 Exercises

61. Strategy
To find the air pressure, replace v in the equation by 64 and solve for p.

Solution
$$v = 6.3\sqrt{1013 - p}$$
$$64 = 6.3\sqrt{1013 - p}$$
$$\frac{64}{6.3} = \sqrt{1013 - p}$$
$$\left(\frac{64}{5.3}\right)^2 = 1013 - p$$
$$p = 1013 - \left(\frac{64}{6.3}\right)^2$$
$$p = 909.8$$
The air pressure is 909.8 mb.
As air pressure decreases, the wind speed increases.

63. Strategy
To find the mean distance of Tethys from Saturn, replace T in the equation by 1.89 and solve for d.

Solution
$$T = 0.373\sqrt{d^3}$$
$$1.89 = 0.373\sqrt{d^3}$$
$$\frac{1.89}{0.373} = d^{3/2}$$
$$\left(\frac{1.89}{0.373}\right)^{2/3} = d$$
$$2.95 = d$$
Tethys is 295,000 km from Saturn.

65. Strategy
To find the weight, replace M in the equation by 60,000 and solve for W.

Solution
$$M = 126.4\sqrt[4]{W^3}$$
$$60,000 = 126.4\sqrt[4]{W^3}$$
$$\frac{60,000}{126.4} = W^{3/4}$$
$$\left(\frac{60,000}{126.4}\right)^{4/3} = W$$
$$3700 = W$$
The elephant weighs 3700 lb.

67. Strategy
To find the distance, use the Pythagorean Theorem. The hypotenuse is the length of the ladder. The distance along the ground from the building to the ladder is the unknown leg.

Solution
$$c^2 = a^2 + b^2$$
$$26^2 = 24^2 + b^2$$
$$676 = 576 + b^2$$
$$100 = b^2$$
$$\sqrt{100} = \sqrt{b^2}$$
$$10 = b$$
The distance is 10 ft.

69. Strategy
To find the distance above the water, replace d in the equation by the given value and solve for h.

Solution
$$d = \sqrt{1.5h}$$
$$3.5 = \sqrt{1.5h}$$
$$(3.5)^2 = (\sqrt{1.5h})^2$$
$$(3.5)^2 = 1.5h$$
$$\frac{(3.5)^2}{1.5} = h$$
$$8.17 = h$$
The periscope must be 8.17 ft above the water.

71. Strategy
To find the distance, replace the variable v in the equation by 120 and solve for d.

Solution
$$v = 8\sqrt{d}$$
$$120 = 8\sqrt{d}$$
$$\frac{120}{8} = \sqrt{d}$$
$$\left(\frac{120}{8}\right)^2 = \left(\sqrt{d}\right)^2$$
$$225 = d$$
The distance is 225 ft.

73. Strategy
To find the distance, replace the variables v and a in the equation by their given values and solve for s.

Solution
$$v = \sqrt{2as}$$
$$48 = \sqrt{24s}$$
$$(48)^2 = (\sqrt{24s})^2$$
$$2304 = 24s$$
$$96 = s$$
The distance is 96 ft.

75. Strategy
To find the length of the pendulum, replace T in the equation by the given value and solve for L.

Solution
$$T = 2\pi\sqrt{\frac{L}{32}}$$
$$2.4 = 2(3.14)\sqrt{\frac{L}{32}}$$
$$\frac{2.4}{2(3.14)} = \sqrt{\frac{L}{32}}$$
$$\left(\frac{2.4}{2(3.14)}\right)^2 = \left(\sqrt{\frac{L}{32}}\right)^2$$
$$\left(\frac{2.4}{6.28}\right)^2 = \frac{L}{32}$$
$$4.67 = L$$

The length of the pendulum is 4.67 ft.

Applying Concepts 7.5

77. $x^{2/3} = 9$
$(x^{2/3})^{3/2} = 9^{3/2}$
$x = (\sqrt{9})^3$
$x = 3^3$
$x = 27$

79. $v = \sqrt{64d}$
$v^2 = (\sqrt{64d})^2$
$v^2 = 64d$
$\dfrac{v^2}{64} = d$

81. $V = \pi r^2 h$
$\dfrac{V}{\pi h} = r^2$
$\sqrt{\dfrac{V}{\pi h}} = \sqrt{r^2}$
$\sqrt{\dfrac{V}{\pi h}} = r$
$r = \dfrac{\sqrt{V}}{\sqrt{\pi h}} \cdot \dfrac{\sqrt{\pi h}}{\sqrt{\pi h}} = \dfrac{\sqrt{V\pi h}}{\pi h}$

83. $\sqrt{3x-2} = \sqrt{2x-3} + \sqrt{x-1}$
$(\sqrt{3x-2})^2 = (\sqrt{2x-3} + \sqrt{x-1})^2$
$3x - 2 = 2x - 3 + 2\sqrt{2x-3}\sqrt{x-1} + x - 1$
$2 = 2\sqrt{2x-3}\sqrt{x-1}$
$1 = \sqrt{2x-3}\sqrt{x-1}$
$1^2 = (\sqrt{2x-3}\sqrt{x-1})^2$
$1 = (2x-3)(x-1)$
$1 = 2x^2 - 5x + 3$
$0 = 2x^2 - 5x + 2$
$0 = (2x-1)(x-2)$

$2x - 1 = 0 \quad x - 2 = 0$
$\quad 2x = 1 \qquad x = 2$
$\quad x = \dfrac{1}{2}$

Check:
$\sqrt{3x-2} = \sqrt{2x-3} + \sqrt{x-1}$

$\sqrt{3\left(\frac{1}{2}\right)-2}$	$\sqrt{2\left(\frac{1}{2}\right)-3} + \sqrt{\frac{1}{2}-1}$
$\sqrt{-\frac{1}{2}}$	$\neq \sqrt{-2} + \sqrt{-\frac{1}{2}}$

$\sqrt{3x-2} = \sqrt{2x-3} + \sqrt{x-1}$

$\sqrt{3(2)-2}$	$\sqrt{2(2)-3} + \sqrt{2-1}$
$\sqrt{4}$	$\sqrt{1} + \sqrt{1}$
2	$1 + 1$

$2 = 2$

The solution is 2.

85. Strategy
The greatest distance between the two corners is the hypotenuse of the triangle whose legs are the diagonal of the base of the box (d) and the height of the box (3 in.). The diagonal of the base (d) is the hypotenuse of the triangle whose lengths are the sides of the base (4 in. by 6 in.)
To find the greatest distance
(1) use the Pythagorean Theorem with $c = d$ (diagonal of the base) and $a = 4$, $b = 6$.
(2) use the Pythagorean Theorem with $c =$ greatest distance, $a = d$, and $b = 3$.

Solution
$d^2 = 4^2 + 6^2$
$d^2 = 16 + 36$
$d^2 = 52$
$d = \sqrt{52}$
$c^2 = a^2 + b^2$
$c^2 = d^2 + 3^2$
$c^2 = \left(\sqrt{52}\right)^2 + 9$
$c^2 = 52 + 9$
$c = \sqrt{61} \approx 7.81$
The greatest distance is
$\sqrt{61}$ in. ≈ 7.81 in.

87. a. $x^2 + 3 = 7$
$x^2 = 4$
$x = \pm 2$
x is an integer, a rational number, and a real number.

b. $x^2 + 1 = 0$
$x^2 = -1$
$x = \pm i$
x is an imaginary number.

c. $\dfrac{5}{8}x = \dfrac{2}{3}$
$\dfrac{8}{5} \cdot \dfrac{5}{8}x = \dfrac{8}{5} \cdot \dfrac{2}{3}$
$x = \dfrac{16}{15}$
x is a rational number and a real number.

d. $x^2 + 1 = 9$
$x^2 = 8$
$x = \pm 2\sqrt{2}$
x is a irrational number and a real number.

e. $x^{3/4} = 8$
$\left(x^{3/4}\right)^{4/3} = (8)^{4/3}$
$x = \left(\sqrt[3]{8}\right)^4 = 2^4 = 16$
x is an integer, a rational number, and a real number.

f. $\sqrt[3]{x} = -27$
$(\sqrt[3]{x})^3 = (-27)^3$
$x = -19,683$
x is an integer, a rational number, and a real number.

89. Strategy
Find the hypotenuses of the triangles in order, letting each become a leg of the next triangle.

Triangle 1: $a^2 = b^2 + c^2$
$a^2 = 1^2 + 1^2$
$a^2 = 2$
$a = \sqrt{2}$

Triangle 2: Let a become b. Then the new a is
$a^2 = (\sqrt{2})^2 + 1^2$
$a^2 = 2 + 1$
$a^2 = 3$
$a = \sqrt{3}$

Triangle 3: Let a become b. Then the new a is
$a^2 = (\sqrt{3})^2 + 1^2$
$a^2 = 3 + 1$
$a^2 = 4$
$a = 2$

Triangle 4: Let a become b. Then the new a is
$a^2 = (2)^2 + (1)^2$
$a^2 = 4 + 1$
$a^2 = 5$
$a = \sqrt{5}$

For the last triangle, let a become b. Then x is
$x^2 = (\sqrt{5})^2 + 1^2$
$x^2 = 5 + 1$
$x^2 = 6$
$x = \sqrt{6}$

Projects and Group Activities

Graphing Compex Numbers

1. $m_1 = \dfrac{4-0}{3-0} = \dfrac{4}{3}$

2. $m_2 = \dfrac{3-0}{-4-0} = -\dfrac{3}{4}$

3. $m_1 m_2 = \dfrac{4}{3}\left(-\dfrac{3}{4}\right) = -1$

   Since the product of the slopes is –1, the lines are perpendicular.

4. $d_1 = \sqrt{(-4-0)^2 + (3-0)^2} = \sqrt{16+9} = 5$
   $d_2 = \sqrt{(3-0)^2 + (4-0)^2} = \sqrt{9+16} = 5$

Solving Radical Equations with a Graphing Calculator

1. $\sqrt{x+0.3} = 1.3$
   $y = \sqrt{x+0.3} - 1.3$

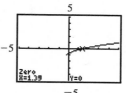

   The solution is 1.390.

2. $\sqrt[3]{x+1.2} = -1.1$
   $y = \sqrt[3]{x+1.2} + 1.1$

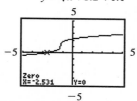

   The solution is –2.531.

3. $\sqrt[4]{3x-1.5} = 1.4$
   $y = \sqrt[4]{3x-1.5} - 1.4$

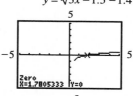

   The solution is 1.781.

Chapter Review Exercises

1. $81^{-1/4} = (3^4)^{-1/4} = 3^{-1} = \dfrac{1}{3}$

2. $\dfrac{x^{-3/2}}{x^{7/2}} = x^{-10/2} = x^{-5} = \dfrac{1}{x^5}$

3. $(a^{16})^{-5/8} = a^{-10} = \dfrac{1}{a^{10}}$

4. $(16x^{-4}y^{12})(100x^6 y^{-2})^{1/2}$
   $= 16x^{-4}y^{12} \cdot (10^2)^{1/2} x^3 y^{-1}$
   $= 160 x^{-1} y^{11}$
   $= \dfrac{160 y^{11}}{x}$

5. $3x^{3/4} = 3\sqrt[4]{x^3}$

6. $7y\sqrt[3]{x^2} = 7x^{2/3}y$

7. $\sqrt[4]{81a^8 b^{12}} = \sqrt[4]{3^4 a^8 b^{12}} = 3a^2 b^3$

8. $-\sqrt{49x^6y^{16}} = -\sqrt{7^2x^6y^{16}} = -7x^3y^8$

9. $\sqrt[3]{-8a^6b^{12}} = \sqrt[3]{(-2)^3a^6b^{12}} = -2a^2b^4$

10. $\sqrt{18a^3b^6} = \sqrt{3^2a^2b^6(2a)} = 3ab^3\sqrt{2a}$

11. $\sqrt[5]{-64a^8b^{12}} = \sqrt[5]{(-2)^5a^5b^{10}(2a^3b^2)}$
    $= -2ab^2\sqrt[5]{2a^3b^2}$

12. $\sqrt[4]{x^6y^8z^{10}} = \sqrt[4]{x^4y^8z^8 \cdot x^2z^2}$
    $= xy^2z^2\sqrt[4]{x^2z^2}$
    $= xy^2z^2\sqrt{xz}$

13. $\sqrt{54} + \sqrt{24} = \sqrt{3^2 \cdot 6} + \sqrt{2^2 \cdot 6}$
    $= 3\sqrt{6} + 2\sqrt{6}$
    $= 5\sqrt{6}$

14. $\sqrt{48x^5y} - x\sqrt{80x^3y}$
    $= \sqrt{4^2x^4(3xy)} - x\sqrt{4^2x^2(5xy)}$
    $= 4x^2\sqrt{3xy} - 4x^2\sqrt{5xy}$

15. $\sqrt{50a^4b^3} - ab\sqrt{18a^2b}$
    $= \sqrt{5^2a^4b^2(2b)} - ab\sqrt{3^2a^2(2b)}$
    $= 5a^2b\sqrt{2b} - 3a^2b\sqrt{2b}$
    $= 2a^2b\sqrt{2b}$

16. $4x\sqrt{12x^2y} + \sqrt{3x^4y} - x^2\sqrt{27y}$
    $= 4x\sqrt{2^2x^2(3y)} + \sqrt{x^4(3y)} - x^2\sqrt{3^2(3y)}$
    $= 8x^2\sqrt{3y} + x^2\sqrt{3y} - 3x^2\sqrt{3y}$
    $= 6x^2\sqrt{3y}$

17. $\sqrt{32}\sqrt{50} = \sqrt{1600} = \sqrt{40^2} = 40$

18. $\sqrt[3]{16x^4y}\sqrt[3]{4xy^5} = \sqrt[3]{64x^5y^6}$
    $= \sqrt[3]{4^3x^3y^6(x^2)}$
    $= 4xy^2\sqrt[3]{x^2}$

19. $\sqrt{3x}(3+\sqrt{3x}) = 3\sqrt{3x} + (\sqrt{3x})^2$
    $= 3\sqrt{3x} + 3x$
    $= 3x + 3\sqrt{3x}$

20. $(5-\sqrt{6})^2 = 25 - 10\sqrt{6} + \sqrt{6}^2$
    $= 25 - 10\sqrt{6} + 6$
    $= 31 - 10\sqrt{6}$

21. $(\sqrt{3}+8)(\sqrt{3}-2) = \sqrt{3^2} + 6\sqrt{3} - 16$
    $= 3 + 6\sqrt{3} - 16$
    $= -13 + 6\sqrt{3}$

22. $\dfrac{\sqrt{125x^6}}{\sqrt{5x^3}} = \sqrt{\dfrac{125x^6}{5x^3}}$
    $= \sqrt{25x^3}$
    $= \sqrt{5^2x^2(x)}$
    $= 5x\sqrt{x}$

23. $\dfrac{8}{\sqrt{3y}} = \dfrac{8}{\sqrt{3y}} \cdot \dfrac{\sqrt{3y}}{\sqrt{3y}} = \dfrac{8\sqrt{3y}}{\sqrt{3^2y^2}} = \dfrac{8\sqrt{3y}}{3y}$

24. $\dfrac{x+2}{\sqrt{x}+\sqrt{2}} = \dfrac{x+2}{\sqrt{x}+\sqrt{2}} \cdot \dfrac{\sqrt{x}-\sqrt{2}}{\sqrt{x}-\sqrt{2}}$
    $= \dfrac{x\sqrt{x} - x\sqrt{2} + 2\sqrt{x} - 2\sqrt{2}}{\sqrt{x^2} - \sqrt{2^2}}$
    $= \dfrac{x\sqrt{x} - x\sqrt{2} + 2\sqrt{x} - 2\sqrt{2}}{x-2}$

25. $\dfrac{\sqrt{x}+\sqrt{y}}{\sqrt{x}-\sqrt{y}} = \dfrac{\sqrt{x}+\sqrt{y}}{\sqrt{x}-\sqrt{y}} \cdot \dfrac{\sqrt{x}+\sqrt{y}}{\sqrt{x}+\sqrt{y}}$
    $= \dfrac{\sqrt{x^2} + \sqrt{xy} + \sqrt{xy} + \sqrt{y^2}}{\sqrt{x^2} - \sqrt{y^2}}$
    $= \dfrac{x + 2\sqrt{xy} + y}{x-y}$

26. $\sqrt{-36} = \sqrt{-1}\sqrt{36} = i\sqrt{6^2} = 6i$

27. $\sqrt{-50} = i\sqrt{50} = i\sqrt{5^2 \cdot 2} = 5i\sqrt{2}$

28. $\sqrt{49} - \sqrt{-16} = \sqrt{7^2} - \sqrt{-1}\sqrt{16}$
    $= 7 - i\sqrt{4^2}$
    $= 7 - 4i$

29. $\sqrt{200} + \sqrt{-12} = \sqrt{10^2 \cdot 2} + \sqrt{-1}\sqrt{2^2}\sqrt{3}$
    $= 10\sqrt{2} + 2i\sqrt{3}$

30. $(5+2i) + (4-3i) = (5+4) + (2+(-3))i$
    $= 9 - i$

31. $(-8+3i) - (4-7i)$
    $= (-8+3i) + (-4+7i)$
    $= (-8+(-4)) + (3+7)i$
    $= -12 + 10i$

32. $(9-\sqrt{-16}) + (5+\sqrt{-36})$
    $= (9-4i) + (5+6i)$
    $= (9+5) + (-4+6)i$
    $= 14 + 2i$

33. $(\sqrt{50} + \sqrt{-72}) - (\sqrt{162} - \sqrt{-8})$
    $= (\sqrt{5^2 \cdot 2} + i\sqrt{6^2 \cdot 2}) - (\sqrt{9^2 \cdot 2} - i\sqrt{2^2 \cdot 2})$
    $= (5\sqrt{2} + 6i\sqrt{2}) - (9\sqrt{2} - 2i\sqrt{2})$
    $= -4\sqrt{2} + 8i\sqrt{2}$

34. $(3-9i) + 7 = 10 - 9i$

35. $(8i)(2i) = 16i^2 = 16(-1) = -16$

36. $i(3-7i) = 3i - 7i^2 = 3i - 7(-1) = 7 + 3i$

37. $\sqrt{-12}\sqrt{-6} = i\sqrt{12} \cdot i\sqrt{6}$
$= i^2\sqrt{72}$
$= (-1)\sqrt{6^2 \cdot 2}$
$= -6\sqrt{2}$

38. $(6-5i)(4+3i) = 24 + 18i - 20i - 15i^2$
$= 24 - 2i - 15(-1)$
$= 24 + 15 - 2i$
$= 39 - 2i$

39. $\dfrac{-6}{i} = -\dfrac{6}{i} \cdot \dfrac{i}{i} = \dfrac{-6i}{i^2} = \dfrac{-6i}{-1} = 6i$

40. $\dfrac{5+2i}{3i} = \dfrac{5+2i}{3i} \cdot \dfrac{-3i}{-3i}$
$= \dfrac{-15i - 6i^2}{-9i^2}$
$= \dfrac{-15i - 6(-1)}{-9(-1)}$
$= \dfrac{-15i + 6}{9}$
$= \dfrac{6}{9} - \dfrac{15}{9}i$
$= \dfrac{2}{3} - \dfrac{5}{3}i$

41. $\dfrac{7}{2-i} = \dfrac{7}{2-i} \cdot \dfrac{2+i}{2+i}$
$= \dfrac{14+7i}{2^2+1}$
$= \dfrac{14+7i}{4+1}$
$= \dfrac{14+7i}{5}$
$= \dfrac{14}{5} + \dfrac{7}{5}i$

42. $\dfrac{\sqrt{16}}{\sqrt{4} - \sqrt{-4}} = \dfrac{4}{2-2i}$
$= \dfrac{4}{2-2i} \cdot \dfrac{2+2i}{2+2i}$
$= \dfrac{8+8i}{4-4i^2}$
$= \dfrac{8+8i}{8}$
$= 1 + i$

43. $\dfrac{5+9i}{1-i} = \dfrac{5+9i}{1-i} \cdot \dfrac{1+i}{1+i}$
$= \dfrac{5+5i+9i+9i^2}{1+1}$
$= \dfrac{5+14i+9(-1)}{2}$
$= \dfrac{5-9+14i}{2}$
$= \dfrac{-4+14i}{2}$
$= -2 + 7i$

44. $\sqrt[3]{9x} = -6$
$(\sqrt[3]{9x})^3 = (-6)^3$
$9x = -216$
$x = -24$

Check:

$$\begin{array}{c|c} \sqrt[3]{9x} = -6 \\ \hline \sqrt[3]{9(-24)} & -6 \\ \sqrt[3]{-216} & -6 \\ -6 = -6 \end{array}$$

The solution is –24.

45. The function f contains an even root. The radicand must be greater than or equal to zero.
$3x - 2 \geq 0$
$3x \geq 2$
$x \geq \dfrac{2}{3}$

The domain is $\left\{x \mid x \geq \dfrac{2}{3}\right\}$.

46. The function f contains an odd root. The radicand may be positive or negative.

The domain is $\{x \mid x \text{ is a real number}\}$.

47.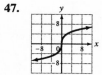

48.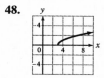

224 *Chapter 7: Rational Exponents and Radicals*

49. $\sqrt[3]{3x-5} = 2$
$(\sqrt[3]{3x-5})^3 = 2^3$
$3x - 5 = 8$
$3x = 13$
$x = \dfrac{13}{3}$

Check:

$\sqrt[3]{3x-5} = 2$	
$\sqrt[3]{3 \cdot \dfrac{13}{3} - 5}$	2
$\sqrt[3]{13-5}$	2
$\sqrt[3]{8}$	2
$2 = 2$	

The solution is $\dfrac{13}{3}$.

50. $\sqrt{4x+9} + 10 = 11$
$\sqrt{4x+9} = 1$
$(\sqrt{4x+9})^2 = 1^2$
$4x + 9 = 1$
$4x = -8$
$x = -2$

Check:

$\sqrt{4x+9} + 10 = 11$	
$\sqrt{4(-2)+9} + 10$	11
$\sqrt{1} + 10$	11
$1 + 10$	11
$11 = 11$	

The solution is -2.

51. Strategy
To find the width of the rectangle, use the Pythagorean Theorem. The unknown width is one leg, the length is the other leg, and the diagonal is the hypotenuse.

Solution
$a^2 + b^2 = c^2$
$a^2 + (12)^2 = (13)^2$
$a^2 + 144 = 169$
$a^2 = 25$
$a = 5$
The width is 5 in.

52. Strategy
To find the amount of power, replace v in the equation with the given value and solve for p.

Solution
$v = 4.05\sqrt[3]{P}$
$20 = 4.05\sqrt[3]{P}$
$\dfrac{20}{4.05} \approx \sqrt[3]{P}$
$\left(\dfrac{20}{4.05}\right)^3 = (\sqrt[3]{P})^3$
$120 \approx P$
The amount of power is 120 watts.

53. Strategy
To find the distance required, replace v and a in the equation with the given values and solve for s.

Solution
$v = \sqrt{2as}$
$88 = \sqrt{2 \cdot 16s}$
$7744 = 32s$
$242 = s$
The distance required is 242 feet.

54. Strategy
To find the distance, use the Pythagorean Theorem. The hypotenuse is the length of the ladder (12 ft). One leg is the height on the building that the ladder reaches (10 ft). The distance from the bottom of the ladder to the building is the other leg.

Solution
$c^2 = a^2 + b^2$
$12^2 = 10^2 + b^2$
$144 = 100 + b^2$
$44 = b^2$
$44^{1/2} = (b^2)^{1/2}$
$\sqrt{44} = b$
$6.63 = b$

The distance is 6.63 feet.

Chapter Test

1. $\dfrac{r^{2/3} r^{-1}}{r^{-1/2}} = \dfrac{r^{-1/3}}{r^{-1/2}} = r^{1/6}$

2. $\dfrac{(2x^{1/3} y^{-2/3})^6}{(x^{-4} y^8)^{1/4}} = \dfrac{2^6 x^2 y^{-4}}{x^{-1} y^2} = 2^6 x^3 y^{-6} = \dfrac{64x^3}{y^6}$

3. $\left(\dfrac{4a^4}{b^2}\right)^{-3/2} = \dfrac{4^{-3/2} a^{-6}}{b^{-3}}$
$= (2^2)^{-3/2} a^{-6} b^3$
$= 2^{-3} a^{-6} b^3$
$= \dfrac{b^3}{8a^6}$

4. $3y^{2/5} = 3\sqrt[5]{y^2}$

5. $\dfrac{1}{2}\sqrt[4]{x^3} = \dfrac{1}{2}x^{3/4}$

6. The function f contains an even root. The radicand must be greater than or equal to zero.
$$4 - x \geq 0$$
$$-x \geq -4$$
$$x \leq 4$$
The domain is $\{x | x \leq 4\}$.

7. The function f contains an odd root. The radicand may be positive or negative.

   The domain is $(-\infty, \infty)$.

8. $\sqrt[3]{27a^4b^3c^7} = \sqrt[3]{3^3 a^3 b^3 c^6 (ac)} = 3abc^2\sqrt[3]{ac}$

9. $\sqrt{18a^3} + a\sqrt{50a} = \sqrt{3^2 a^2(2a)} + a\sqrt{5^2(2a)}$
$= 3a\sqrt{2a} + 5a\sqrt{2a}$
$= 8a\sqrt{2a}$

10. $\sqrt[3]{54x^7y^3} - x\sqrt[3]{128x^4y^3} - x^2\sqrt[3]{2xy^3}$
$= \sqrt[3]{3^3 x^6 y^3 (2x)} - x\sqrt[3]{4^3 x^3 y^3 (2x)} - x^2\sqrt[3]{y^3(2x)}$
$= 3x^2y\sqrt[3]{2x} - 4x^2y\sqrt[3]{2x} - x^2y\sqrt[3]{2x}$
$= -2x^2y\sqrt[3]{2x}$

11. $\sqrt{3x}(\sqrt{x} - \sqrt{25x}) = \sqrt{3x^2} - \sqrt{75x^2}$
$= \sqrt{x^2(3)} - \sqrt{5^2 x^2(3)}$
$= x\sqrt{3} - 5x\sqrt{3}$
$= -4x\sqrt{3}$

12. $(2\sqrt{3} + 4)(3\sqrt{3} - 1) = 6\sqrt{3^2} - 2\sqrt{3} + 12\sqrt{3} - 4$
$= 18 + 10\sqrt{3} - 4$
$= 14 + 10\sqrt{3}$

13. $(\sqrt{a} - 3\sqrt{b})(2\sqrt{a} + 5\sqrt{b})$
$= 2\sqrt{a^2} + 5\sqrt{ab} - 6\sqrt{ab} - 15\sqrt{b^2}$
$= 2a - \sqrt{ab} - 15b$

14. $(2\sqrt{x} + \sqrt{y})^2 = 4\sqrt{x^2} + 4\sqrt{xy} + \sqrt{y^2}$
$= 4x + 4\sqrt{xy} + y$

15. $\dfrac{\sqrt{32x^5y}}{\sqrt{2xy^3}} = \sqrt{\dfrac{32x^5y}{2xy^3}} = \sqrt{\dfrac{16x^4}{y^2}} = \sqrt{\dfrac{4^2 x^4}{y^2}} = \dfrac{4x^2}{y}$

16. $\dfrac{4 - 2\sqrt{5}}{2 - \sqrt{5}} = \dfrac{4 - 2\sqrt{5}}{2 - \sqrt{5}} \cdot \dfrac{2 + \sqrt{5}}{2 + \sqrt{5}}$
$= \dfrac{8 + 4\sqrt{5} - 4\sqrt{5} - 2\sqrt{5^2}}{2^2 - \sqrt{5^2}}$
$= \dfrac{8 - 2 \cdot 5}{4 - 5}$
$= \dfrac{8 - 10}{-1}$
$= \dfrac{-2}{-1}$
$= 2$

17. $\dfrac{\sqrt{x}}{\sqrt{x} - \sqrt{y}} = \dfrac{\sqrt{x}}{\sqrt{x} - \sqrt{y}} \cdot \dfrac{\sqrt{x} + \sqrt{y}}{\sqrt{x} + \sqrt{y}}$
$= \dfrac{\sqrt{x^2} + \sqrt{xy}}{\sqrt{x^2} - \sqrt{y^2}}$
$= \dfrac{x + \sqrt{xy}}{x - y}$

18. $(\sqrt{-8})(\sqrt{-2}) = i\sqrt{8} \cdot i\sqrt{2} = i^2\sqrt{16} = -1 \cdot 4 = -4$

19. $(5 - 2i) - (8 - 4i) = -3 + 2i$

20. $(2 + 5i)(4 - 2i) = 8 - 4i + 20i - 10i^2$
$= 8 + 16i - 10(-1)$
$= 8 + 16i + 10$
$= 18 + 16i$

21. $\dfrac{2 + 3i}{1 - 2i} = \dfrac{2 + 3i}{1 - 2i} \cdot \dfrac{1 + 2i}{1 + 2i}$
$= \dfrac{2 + 4i + 3i + 6i^2}{1 + 4}$
$= \dfrac{2 + 7i + 6(-1)}{5}$
$= \dfrac{2 - 6 + 7i}{5} = \dfrac{-4 + 7i}{5}$
$= -\dfrac{4}{5} + \dfrac{7}{5}i$

22. $(2 + i) + (2 - i) = 4$

23. $\sqrt{x+12} - \sqrt{x} = 2$
    $\sqrt{x+12} = 2 + \sqrt{x}$
    $(\sqrt{x+12})^2 = (2+\sqrt{x})^2$
    $x + 12 = 4 + 4\sqrt{x} + x$
    $12 = 4 + 4\sqrt{x}$
    $8 = 4\sqrt{x}$
    $2 = \sqrt{x}$
    $2^2 = (\sqrt{x})^2$
    $4 = x$

    Check:

$\sqrt{x+12} - \sqrt{x} = 2$	
$\sqrt{4+12} - \sqrt{4}$	2
$\sqrt{16} - \sqrt{4}$	2
$4 - 2$	2
$2 = 2$	

    The solution is 4.

24. $\sqrt[3]{2x-2} + 4 = 2$
    $\sqrt[3]{2x-2} = -2$
    $(\sqrt[3]{2x-2})^3 = (-2)^3$
    $2x - 2 = -8$
    $2x = -6$
    $x = -3$

    Check:

$\sqrt[3]{2x-2} + 4 = 2$	
$\sqrt[3]{2(-3)-2} + 4$	2
$\sqrt[3]{-8} + 4$	2
$-2 + 4$	2
$2 = 2$	

    The solution is -3.

25. **Strategy**
    To find the distance, replace v in the formula with the given value and solve for d.

    Solution
    $v = \sqrt{64d}$
    $192 = \sqrt{64d}$
    $(192)^2 = (\sqrt{64d})^2$
    $36,864 = 64d$
    $576 = d$
    The distance is 576 feet.

Cumulative Review Exercises

1. The Distributive Property

2. $2x - 3[x - 2(x-4) + 2x]$
   $= 2x - 3[x - 2x + 8 + 2x]$
   $= 2x - 3[x + 8]$
   $= 2x - 3x - 24$
   $= -x - 24$

3. $A \cap B = \emptyset$

4. $\sqrt[3]{2x-5} + 3 = 6$
   $\sqrt[3]{2x-5} = 3$
   $(\sqrt[3]{2x-5})^3 = 3^3$
   $2x - 5 = 27$
   $2x = 32$
   $x = 16$

   Check:

$\sqrt[3]{2x-5} + 3 = 6$	
$\sqrt[3]{2(16)-5} + 3$	6
$\sqrt[3]{32-5} + 3$	6
$\sqrt[3]{27} + 3$	6
$6 = 6$	

   The solution is 16.

5. $5 - \dfrac{2}{3}x = 4$
   $5 - \dfrac{2}{3}x - 5 = 4 - 5$
   $-\dfrac{2}{3}x = -1$
   $\left(-\dfrac{3}{2}\right)\left(-\dfrac{2}{3}\right)x = -1\left(-\dfrac{3}{2}\right)$
   $x = \dfrac{3}{2}$
   The solution is $\dfrac{3}{2}$.

6. $2[4 - 2(3 - 2x)] = 4(1 - x)$
   $2[4 - 6 + 4x] = 4 - 4x$
   $2[-2 + 4x] = 4 - 4x$
   $-4 + 8x = 4 - 4x$
   $-4 + 8x + 4x = 4 - 4x + 4x$
   $12x - 4 = 4$
   $12x - 4 + 4 = 4 + 4$
   $12x = 8$
   $\left(\dfrac{1}{12}\right)12x = \dfrac{1}{12}(8)$
   $x = \dfrac{2}{3}$
   The solution is $\dfrac{2}{3}$.

7. $3x - 4 \leq 8x + 1$
   $-5x - 4 \leq 1$
   $-5x \leq 5$
   $x \geq -1$
   $\{x | x \geq -1\}$

8. $5 < 2x - 3 < 7$
$5 + 3 < 2x - 3 + 3 < 7 + 3$
$8 < 2x < 10$
$\frac{1}{2} \cdot 8 < \frac{1}{2} \cdot 2x < \frac{1}{2} \cdot 10$
$4 < x < 5$
$\{x | 4 < x < 5\}$

9. $|7 - 3x| > 1$
$7 - 3x < -1$ or $7 - 3x > 1$
$-3x < -8 \qquad -3x > -6$
$x > \frac{8}{3} \qquad x < 2$
$\{x | x > \frac{8}{3}\} \quad \{x | x < 2\}$
$\{x | x > \frac{8}{3}\} \cup \{x | x < 2\} = \{x | x < 2 \text{ or } x > \frac{8}{3}\}$

10. $64a^2 - b^2 = (8a)^2 - b^2 = (8a + b)(8a - b)$

11. $x^5 + 2x^3 - 3x = x(x^4 + 2x^2 - 3)$
$= x(x^2 + 3)(x^2 - 1)$
$= x(x^2 + 3)(x + 1)(x - 1)$

12. $3x^2 + 13x - 10 = 0$
$(3x - 2)(x + 5) = 0$
$3x - 2 = 0 \quad x + 5 = 0$
$3x = 2 \qquad x = -5$
$x = \frac{2}{3}$
The solutions are $\frac{2}{3}$ and -5.

13.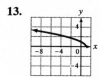

14. $x - 2y = 4 \qquad 2x + y = 4$
$-2y = -x + 4 \qquad y = -2x + 4$
$y = \frac{1}{2}x - 2$
$m_1 m_2 = \frac{1}{2}(-2) = -1$
Yes, the lines are perpendicular.

15. $(3^{-1}x^3 y^{-5})(3^{-1}y^{-2})^{-2} = (3^{-1}x^3 y^{-5})(3^2 y^4)$
$= \frac{x^3}{3y^5}(3^2 y^4)$
$= \frac{3x^3}{y}$

16. $\left(\frac{x^{-1/2} y^{3/4}}{y^{-5/4}}\right)^4 = \left(\frac{y^2}{x^{1/2}}\right)^4 = \frac{y^8}{x^2}$

17. $\sqrt{20x^3} - x\sqrt{45x} = \sqrt{2^2 x^2 \cdot 5x} - x\sqrt{3^2 \cdot 5x}$
$= \sqrt{2^2 x^2}\sqrt{5x} - x\sqrt{3^2}\sqrt{5x}$
$= 2x\sqrt{5x} - 3x\sqrt{5x}$
$= -x\sqrt{5x}$

18. $(\sqrt{5} - 3)(\sqrt{5} - 2) = 5 - 2\sqrt{5} - 3\sqrt{5} + 6$
$= 11 - 5\sqrt{5}$

19. $\frac{\sqrt[3]{4x^5 y^4}}{\sqrt[3]{8x^2 y^5}} = \sqrt[3]{\frac{4x^5 y^4}{8x^2 y^5}}$
$= \sqrt[3]{\frac{x^3}{2y}}$
$= \frac{\sqrt[3]{x^3}}{\sqrt[3]{2y}}$
$= \frac{x}{\sqrt[3]{2y}} \cdot \frac{\sqrt[3]{4y^2}}{\sqrt[3]{4y^2}}$
$= \frac{x\sqrt[3]{4y^2}}{\sqrt[3]{2^3 y^3}}$
$= \frac{x\sqrt[3]{4y^2}}{2y}$

20. $\frac{3i}{2-i} = \frac{3i}{2-i} \cdot \frac{2+i}{2+i}$
$= \frac{6i + 3i^2}{2^2 - i^2}$
$= \frac{6i + 3(-1)}{4 - (-1)}$
$= \frac{-3 + 6i}{5}$
$= -\frac{3}{5} + \frac{6}{5}i$

21.

22. The function g contains an odd root. The radicand may be positive or negative.
The domain is $\{x | x$ is a real number$\}$.

23. $f(x) = 3x^2 - 2x + 1$
$f(-3) = 3(-3)^2 - 2(-3) + 1$
$f(-3) = 27 + 6 + 1$
$f(-3) = 34$
The value of $f(-3)$ is 34.

24. First find the slope of the line.
$$m = \frac{y_2 - y_1}{x_2 - x_1} = \frac{2-3}{-1-2} = \frac{-1}{-3} = \frac{1}{3}$$
Use the point-slope form to find an equation of the line.
$$y - y_1 = m(x - x_1)$$
$$y - 3 = \frac{1}{3}(x - 2)$$
$$y - 3 = \frac{1}{3}x - \frac{2}{3}$$
$$y = \frac{1}{3}x + \frac{7}{3}$$
An equation of the line is $y = \frac{1}{3}x + \frac{7}{3}$.

25. $\begin{vmatrix} 1 & 2 & -3 \\ 0 & -1 & 2 \\ 3 & 1 & -2 \end{vmatrix} = 1 \cdot \begin{vmatrix} -1 & 2 \\ 1 & -2 \end{vmatrix} - 2 \begin{vmatrix} 0 & 2 \\ 3 & -2 \end{vmatrix} - 3 \begin{vmatrix} 0 & -1 \\ 3 & 1 \end{vmatrix}$
$$= 1 \cdot 0 - 2(-6) - 3 \cdot 3$$
$$= 3$$

26. $\quad 2x - y = 4$
$\quad -2x + 3y = 5$
$$D = \begin{vmatrix} 2 & -1 \\ -2 & 3 \end{vmatrix} = 4$$
$$D_x = \begin{vmatrix} 4 & -1 \\ 5 & 3 \end{vmatrix} = 17$$
$$D_y = \begin{vmatrix} 2 & 4 \\ -2 & 5 \end{vmatrix} = 18$$
$$x = \frac{D_x}{D} = \frac{17}{4}$$
$$y = \frac{D_y}{D} = \frac{18}{4} = \frac{9}{2}$$
The solution is $\left(\frac{17}{4}, \frac{9}{2}\right)$.

27. Find the y-intercept at $x = 0$.
$\quad 3(0) - 2y = -6$
$\qquad\quad y = 3$ The y-intercept is $(0, 3)$.
To find the slope, find the x-intercept and use it to get the slope
$\quad 3x - 2(0) = -6$
$\qquad\qquad x = -2$ The x-intercept is $(-2, 0)$.
$$m = \frac{y_2 - y_1}{x_2 - x_1}$$
$$m = \frac{3 - 0}{0 - (-2)} = \frac{3}{2}$$
The slope is $\frac{3}{2}$, and the y-intercept is $(0, 3)$.

28. $3x + 2y \leq 4$
$\quad 2y \leq -3x + 4$
$\quad\; y \leq -\frac{3}{2}x + 2$

Sketch the solid line $y = -\frac{3}{2}x + 2$. Shade below the solid line.

29. Strategy
Number of 18¢ stamps: x
Number of 13¢ stamps: $30 - x$

Stamps	Number	Value	Total Value
18¢	x	18	$18x$
13¢	$30 - x$	13	$13(30 - x)$

The sum of the total values of each type of stamp equals the total value of the stamps (485¢).
$$18x + 13(30 - x) = 485$$

Solution
$$18x + 13(30 - x) = 485$$
$$18x + 390 - 13x = 485$$
$$5x + 390 = 485$$
$$5x = 95$$
$$x = 19$$
There are nineteen 18¢ stamps.

30. Strategy
Amount invested at 8.4%: x

	Principal	Rate	Interest
Amount invested at 7.2%	2500	0.072	0.072(2500)
Amount invested at 8.4%	x	0.084	$0.084x$

The total amount of interest earned is $516.
$$0.072(2500) + 0.084x = 516$$

Solution
$$0.072(2500) + 0.084x = 516$$
$$180 + 0.084x = 516$$
$$0.084x = 336$$
$$x = 4000$$
The additional investment must be $4000.

31. Strategy
Length of the rectangle: x
Width of the rectangle: $x - 6$
Use the equation for the area of a rectangle, $A = L \cdot W$.

Solution
$A = L \cdot W$
$72 = x(x - 6)$
$72 = x^2 - 6x$
$0 = x^2 - 6x - 72$
$0 = (x - 12)(x + 6)$
$x - 12 = 0 \quad x + 6 = 0$
$\quad x = 12 \quad\quad x = -6$
The length cannot be negative, so -6 is not a solution.
$x - 6 = 12 - 6 = 6$
The length is 12 ft and the width is 6 ft.

32. Strategy
Unknown rate of the car: x
Unknown rate of the plane: $5x$

	Distance	Rate	Time
Car	25	x	$\dfrac{25}{x}$
Plane	625	$5x$	$\dfrac{625}{5x}$

The total time of the trip was 3 h.
$\dfrac{25}{x} + \dfrac{625}{5x} = 3$

Solution
$5x\left(\dfrac{25}{x} + \dfrac{625}{5x}\right) = 3(5x)$
$125 + 625 = 15x$
$750 = 15x$
$50 = x$
$250 = 5x$
The rate of the plane is 250 mph.

33. Strategy
To find the time it takes light to travel from the earth to the moon, use the formula $RT = D$, substituting for R and D and solving for T.

Solution
$RT = D$
$1.86 \times 10^5 \cdot T = 232{,}500$
$1.86 \times 10^5 \cdot T = 2.325 \times 10^5$
$T = 1.25 \times 10^0$
$T = 1.25$
The time is 1.25 seconds.

34. Strategy
To find the height of the periscope, replace d in the equation by the given value and solve for h.

Solution
$d = \sqrt{1.5h}$
$7 = \sqrt{1.5h}$
$7^2 = 1.5h$
$\dfrac{7^2}{1.5} = h$
$32.7 = h$
The height of the periscope is 32.7 ft.

35. Slope $m = \dfrac{y_2 - y_1}{x_2 - x_1} = \dfrac{400 - 0}{5000 - 0} = \dfrac{400}{5000} = 0.08$
The slope represents the simple interest on the investment. The interest rate is 8%.

Chapter 8: Quadratic Equations and Inequalities

Section 8.1

Concept Review 8.1

1. Sometimes true
   A quadratic equation may have two real roots, one real root, or two imaginary roots.

3. Never true
   The Principle of Zero Products states that at least one of the factors must be zero if the product is zero. In this case the product is 8.

Objective 1 Exercises

3. $x^2 - 4x = 0$
   $x(x - 4) = 0$
   $x = 0 \qquad x - 4 = 0$
   $\qquad\qquad\quad x = 4$
   The solutions are 0 and 4.

5. $t^2 - 25 = 0$
   $(t - 5)(t + 5) = 0$
   $t - 5 = 0 \qquad t + 5 = 0$
   $t = 5 \qquad\quad t = -5$
   The solutions are 5 and –5.

7. $s^2 - s - 6 = 0$
   $(s - 3)(s + 2) = 0$
   $s - 3 = 0 \qquad s + 2 = 0$
   $s = 3 \qquad\quad s = -2$
   The solutions are 3 and –2.

9. $y^2 - 6y + 9 = 0$
   $(y - 3)(y - 3) = 0$
   $y - 3 = 0 \qquad y - 3 = 0$
   $y = 3 \qquad\quad y = 3$
   The solution is 3.

11. $9z^2 - 18z = 0$
    $9z(z - 2) = 0$
    $9z = 0 \qquad z - 2 = 0$
    $z = 0 \qquad\quad z = 2$
    The solutions are 0 and 2.

13. $r^2 - 3r = 10$
    $r^2 - 3r - 10 = 0$
    $(r - 5)(r + 2) = 0$
    $r - 5 = 0 \qquad r + 2 = 0$
    $r = 5 \qquad\quad r = -2$
    The solutions are 5 and –2.

15. $v^2 + 10 = 7v$
    $v^2 - 7v + 10 = 0$
    $(v - 2)(v - 5) = 0$
    $v - 2 = 0 \qquad v - 5 = 0$
    $v = 2 \qquad\quad v = 5$
    The solutions are 2 and 5.

17. $2x^2 - 9x - 18 = 0$
    $(x - 6)(2x + 3) = 0$
    $x - 6 = 0 \qquad 2x + 3 = 0$
    $x = 6 \qquad\quad 2x = -3$
    $\qquad\qquad\qquad x = -\dfrac{3}{2}$
    The solutions are 6 and $-\dfrac{3}{2}$.

19. $4z^2 - 9z + 2 = 0$
    $(z - 2)(4z - 1) = 0$
    $z - 2 = 0 \qquad 4z - 1 = 0$
    $z = 2 \qquad\quad 4z = 1$
    $\qquad\qquad\qquad z = \dfrac{1}{4}$
    The solutions are 2 and $\dfrac{1}{4}$.

21. $3w^2 + 11w = 4$
    $3w^2 + 11w - 4 = 0$
    $(3w - 1)(w + 4) = 0$
    $3w - 1 = 0 \qquad w + 4 = 0$
    $3w = 1 \qquad\quad w = -4$
    $w = \dfrac{1}{3}$
    The solutions are $\dfrac{1}{3}$ and –4.

23. $6x^2 = 23x + 18$
    $6x^2 - 23x - 18 = 0$
    $(2x - 9)(3x + 2) = 0$
    $2x - 9 = 0 \qquad 3x + 2 = 0$
    $2x = 9 \qquad\quad 3x = -2$
    $x = \dfrac{9}{2} \qquad\quad x = -\dfrac{2}{3}$
    The solutions are $\dfrac{9}{2}$ and $-\dfrac{2}{3}$.

25. $4 - 15u - 4u^2 = 0$
    $(1 - 4u)(4 + u) = 0$
    $1 - 4u = 0 \qquad 4 + u = 0$
    $-4u = -1 \qquad\quad u = -4$
    $u = \dfrac{1}{4}$
    The solutions are $\dfrac{1}{4}$ and –4.

27. $x + 18 = x(x - 6)$
    $x + 18 = x^2 - 6x$
    $0 = x^2 - 7x - 18$
    $0 = (x - 9)(x + 2)$
    $x - 9 = 0 \qquad x + 2 = 0$
    $x = 9 \qquad\quad x = -2$
    The solutions are 9 and –2.

29.
$$4s(s+3) = s-6$$
$$4s^2 + 12s = s-6$$
$$4s^2 + 11s + 6 = 0$$
$$(s+2)(4s+3) = 0$$
$$s+2 = 0 \qquad 4s+3 = 0$$
$$s = -2 \qquad 4s = -3$$
$$s = -\frac{3}{4}$$
The solutions are -2 and $-\frac{3}{4}$.

31.
$$u^2 - 2u + 4 = (2u-3)(u+2)$$
$$u^2 - 2u + 4 = 2u^2 + u - 6$$
$$0 = u^2 + 3u - 10$$
$$0 = (u-2)(u+5)$$
$$u - 2 = 0 \qquad u + 5 = 0$$
$$u = 2 \qquad u = -5$$
The solutions are 2 and -5.

33.
$$(3x-4)(x+4) = x^2 - 3x - 28$$
$$3x^2 + 8x - 16 = x^2 - 3x - 28$$
$$2x^2 + 11x + 12 = 0$$
$$(x+4)(2x+3) = 0$$
$$x+4 = 0 \qquad 2x+3 = 0$$
$$x = -4 \qquad 2x = -3$$
$$x = -\frac{3}{2}$$
The solutions are -4 and $-\frac{3}{2}$.

35.
$$x^2 - 9bx + 14b^2 = 0$$
$$(x-2b)(x-7b) = 0$$
$$x - 2b = 0 \qquad x - 7b = 0$$
$$x = 2b \qquad x = 7b$$
The solutions are $2b$ and $7b$.

37.
$$x^2 - 6cx - 7c^2 = 0$$
$$(x-7c)(x+c) = 0$$
$$x - 7c = 0 \qquad x + c = 0$$
$$x = 7c \qquad x = -c$$
The solutions are $7c$ and $-c$.

39.
$$2x^2 + 3bx + b^2 = 0$$
$$(2x+b)(x+b) = 0$$
$$2x + b = 0 \qquad x + b = 0$$
$$2x = -b \qquad x = -b$$
$$x = -\frac{b}{2}$$
The solutions are $-\frac{b}{2}$ and $-b$.

41.
$$3x^2 - 14ax + 8a^2 = 0$$
$$(x-4a)(3x-2a) = 0$$
$$x - 4a = 0 \qquad 3x - 2a = 0$$
$$x = 4a \qquad 3x = 2a$$
$$x = \frac{2a}{3}$$
The solutions are $4a$ and $\frac{2a}{3}$.

Objective 2 Exercises

43.
$$(x-r_1)(x-r_2) = 0$$
$$(x-2)(x-5) = 0$$
$$x^2 - 7x + 10 = 0$$

45.
$$(x-r_1)(x-r_2) = 0$$
$$[x-(-2)][x-(-4)] = 0$$
$$(x+2)(x+4) = 0$$
$$x^2 + 6x + 8 = 0$$

47.
$$(x-r_1)(x-r_2) = 0$$
$$(x-6)[x-(-1)] = 0$$
$$(x-6)(x+1) = 0$$
$$x^2 - 5x - 6 = 0$$

49.
$$(x-r_1)(x-r_2) = 0$$
$$(x-3)[x-(-3)] = 0$$
$$(x-3)(x+3) = 0$$
$$x^2 - 9 = 0$$

51.
$$(x-r_1)(x-r_2) = 0$$
$$(x-4)(x-4) = 0$$
$$x^2 - 8x + 16 = 0$$

53.
$$(x-r_1)(x-r_2) = 0$$
$$(x-0)(x-5) = 0$$
$$x(x-5) = 0$$
$$x^2 - 5x = 0$$

55.
$$(x-r_1)(x-r_2) = 0$$
$$(x-0)(x-3) = 0$$
$$x(x-3) = 0$$
$$x^2 - 3x = 0$$

57.
$$(x-r_1)(x-r_2) = 0$$
$$(x-3)\left(x-\frac{1}{2}\right) = 0$$
$$x^2 - \frac{7}{2}x + \frac{3}{2} = 0$$
$$2\left(x^2 - \frac{7}{2}x + \frac{3}{2}\right) = 2 \cdot 0$$
$$2x^2 - 7x + 3 = 0$$

59.
$(x-r_1)(x-r_2)=0$
$\left[x-\left(-\frac{3}{4}\right)\right](x-2)=0$
$\left(x+\frac{3}{4}\right)(x-2)=0$
$x^2-\frac{5}{4}x-\frac{3}{2}=0$
$4\left(x^2-\frac{5}{4}x-\frac{3}{2}\right)=4\cdot 0$
$4x^2-5x-6=0$

61.
$(x-r_1)(x-r_2)=0$
$\left[x-\left(-\frac{5}{3}\right)\right][x-(-2)]=0$
$\left(x+\frac{5}{3}\right)(x+2)=0$
$x^2+\frac{11}{3}x+\frac{10}{3}=0$
$3\left(x^2+\frac{11}{3}x+\frac{10}{3}\right)=3\cdot 0$
$3x^2+11x+10=0$

63.
$(x-r_1)(x-r_2)=0$
$\left[x-\left(-\frac{2}{3}\right)\right]\left(x-\frac{2}{3}\right)=0$
$\left(x+\frac{2}{3}\right)\left(x-\frac{2}{3}\right)=0$
$x^2-\frac{4}{9}=0$
$9\left(x^2-\frac{4}{9}\right)=9\cdot 0$
$9x^2-4=0$

65.
$(x-r_1)(x-r_2)=0$
$\left(x-\frac{1}{2}\right)\left(x-\frac{1}{3}\right)=0$
$x^2-\frac{5}{6}x+\frac{1}{6}=0$
$6\left(x^2-\frac{5}{6}x+\frac{1}{6}\right)=6\cdot 0$
$6x^2-5x+1=0$

67.
$(x-r_1)(x-r_2)=0$
$\left(x-\frac{6}{5}\right)\left[x-\left(-\frac{1}{2}\right)\right]=0$
$\left(x-\frac{6}{5}\right)\left(x+\frac{1}{2}\right)=0$
$x^2-\frac{7}{10}x-\frac{3}{5}=0$
$10\left(x^2-\frac{7}{10}x-\frac{3}{5}\right)=10\cdot 0$
$10x^2-7x-6=0$

69.
$(x-r_1)(x-r_2)=0$
$\left[x-\left(-\frac{1}{4}\right)\right]\left[x-\left(-\frac{1}{2}\right)\right]=0$
$\left(x+\frac{1}{4}\right)\left(x+\frac{1}{2}\right)=0$
$x^2+\frac{3}{4}x+\frac{1}{8}=0$
$8\left(x^2+\frac{3}{4}x+\frac{1}{8}\right)=8\cdot 0$
$8x^2+6x+1=0$

71.
$(x-r_1)(x-r_2)=0$
$\left(x-\frac{3}{5}\right)\left[x-\left(-\frac{1}{10}\right)\right]=0$
$\left(x-\frac{3}{5}\right)\left(x+\frac{1}{10}\right)=0$
$x^2-\frac{1}{2}x-\frac{3}{50}=0$
$50\left(x^2-\frac{1}{2}x-\frac{3}{50}\right)=50\cdot 0$
$50x^2-25x-3=0$

Objective 3 Exercises

73.
$y^2=49$
$\sqrt{y^2}=\sqrt{49}$
$y=\pm\sqrt{49}=\pm 7$
The solutions are 7 and −7.

75.
$z^2=-4$
$\sqrt{z^2}=\sqrt{-4}$
$z=\pm\sqrt{-4}=\pm 2i$
The solutions are $2i$ and $-2i$.

77.
$s^2-4=0$
$s^2=4$
$\sqrt{s^2}=\sqrt{4}$
$s=\pm\sqrt{4}=\pm 2$
The solutions are 2 and −2.

79.
$4x^2-81=0$
$4x^2=81$
$x^2=\frac{81}{4}$
$\sqrt{x^2}=\sqrt{\frac{81}{4}}$
$x=\pm\sqrt{\frac{81}{4}}=\pm\frac{9}{2}$

The solutions are $\frac{9}{2}$ and $-\frac{9}{2}$.

81. $y^2 + 49 = 0$
$y^2 = -49$
$\sqrt{y^2} = \sqrt{-49}$
$y = \pm\sqrt{-49} = \pm 7i$
The solutions are $7i$ and $-7i$.

83. $v^2 - 48 = 0$
$v^2 = 48$
$\sqrt{v^2} = \sqrt{48}$
$v = \pm\sqrt{48} = \pm 4\sqrt{3}$
The solutions are $4\sqrt{3}$ and $-4\sqrt{3}$.

85. $r^2 - 75 = 0$
$r^2 = 75$
$\sqrt{r^2} = \sqrt{75}$
$r = \pm\sqrt{75} = \pm 5\sqrt{3}$
The solutions are $5\sqrt{3}$ and $-5\sqrt{3}$.

87. $z^2 + 18 = 0$
$z^2 = -18$
$\sqrt{z^2} = \sqrt{-18}$
$z = \pm\sqrt{-18} = \pm 3i\sqrt{2}$
The solutions are $3i\sqrt{2}$ and $-3i\sqrt{2}$.

89. $(x-1)^2 = 36$
$\sqrt{(x-1)^2} = \sqrt{36}$
$x - 1 = \pm\sqrt{36} = \pm 6$
$x - 1 = 6 \qquad x - 1 = -6$
$x = 7 \qquad x = -5$
The solutions are 7 and -5.

91. $3(y+3)^2 = 27$
$(y+3)^2 = 9$
$\sqrt{(y+3)^2} = \sqrt{9}$
$y + 3 = \pm\sqrt{9} = \pm 3$
$y + 3 = 3 \qquad y + 3 = -3$
$y = 0 \qquad y = -6$
The solutions are 0 and -6.

93. $5(z+2)^2 = 125$
$(z+2)^2 = 25$
$\sqrt{(z+2)^2} = \sqrt{25}$
$z + 2 = \pm\sqrt{25} = \pm 5$
$z + 2 = 5 \qquad z + 2 = -5$
$z = 3 \qquad z = -7$
The solutions are 3 and -7.

95. $(x+5)^2 = -25$
$\sqrt{(x+5)^2} = \sqrt{-25}$
$x + 5 = \pm\sqrt{-25} = \pm 5i$
$x + 5 = 5i \qquad x + 5 = -5i$
$x = -5 + 5i \qquad x = -5 - 5i$
The solutions are $-5 + 5i$ and $-5 - 5i$.

97. $3(x-4)^2 = -12$
$(x-4)^2 = -4$
$\sqrt{(x-4)^2} = \sqrt{-4}$
$x - 4 = \pm\sqrt{-4} = \pm 2i$
$x - 4 = 2i \qquad x - 4 = -2i$
$x = 4 + 2i \qquad x = 4 - 2i$
The solutions are $4 + 2i$ and $4 - 2i$.

99. $3(x-9)^2 = -27$
$(x-9)^2 = -9$
$\sqrt{(x-9)^2} = \sqrt{-9}$
$x - 9 = \pm\sqrt{-9} = \pm 3i$
$x - 9 = 3i \qquad x - 9 = -3i$
$x = 9 + 3i \qquad x = 9 - 3i$
The solutions are $9 + 3i$ and $9 - 3i$.

101. $\left(v - \dfrac{1}{2}\right)^2 = \dfrac{1}{4}$
$\sqrt{\left(v - \dfrac{1}{2}\right)^2} = \sqrt{\dfrac{1}{4}}$
$v - \dfrac{1}{2} = \pm\sqrt{\dfrac{1}{4}} = \pm\dfrac{1}{2}$
$v - \dfrac{1}{2} = \dfrac{1}{2} \qquad v - \dfrac{1}{2} = -\dfrac{1}{2}$
$v = 1 \qquad v = 0$
The solutions are 1 and 0.

103. $\left(x - \dfrac{2}{5}\right)^2 = \dfrac{9}{25}$
$\sqrt{\left(x - \dfrac{2}{5}\right)^2} = \sqrt{\dfrac{9}{25}}$
$x - \dfrac{2}{5} = \pm\sqrt{\dfrac{9}{25}} = \pm\dfrac{3}{5}$
$x - \dfrac{2}{5} = \dfrac{3}{5} \qquad x - \dfrac{2}{5} = -\dfrac{3}{5}$
$x = \dfrac{2}{5} + \dfrac{3}{5} = \dfrac{5}{5} \qquad x = \dfrac{2}{5} - \dfrac{3}{5}$
$x = 1 \qquad x = -\dfrac{1}{5}$
The solutions are 1 and $-\dfrac{1}{5}$.

105. $\left(a+\dfrac{3}{4}\right)^2 = \dfrac{9}{16}$

$\sqrt{\left(a+\dfrac{3}{4}\right)^2} = \sqrt{\dfrac{9}{16}}$

$a+\dfrac{3}{4} = \pm\sqrt{\dfrac{9}{16}} = \pm\dfrac{3}{4}$

$a+\dfrac{3}{4} = \dfrac{3}{4} \qquad a+\dfrac{3}{4} = -\dfrac{3}{4}$

$a = \dfrac{3}{4}-\dfrac{3}{4} \qquad a = -\dfrac{3}{4}-\dfrac{3}{4} = -\dfrac{6}{4}$

$a = 0 \qquad\qquad a = -\dfrac{3}{2}$

The solutions are 0 and $-\dfrac{3}{2}$.

107. $3\left(x-\dfrac{5}{3}\right)^2 = \dfrac{4}{3}$

$\left(x-\dfrac{5}{3}\right)^2 = \dfrac{4}{9}$

$\sqrt{\left(x-\dfrac{5}{3}\right)^2} = \sqrt{\dfrac{4}{9}}$

$x-\dfrac{5}{3} = \pm\sqrt{\dfrac{4}{9}} = \pm\dfrac{2}{3}$

$x-\dfrac{5}{3} = \dfrac{2}{3} \qquad x-\dfrac{5}{3} = -\dfrac{2}{3}$

$x = \dfrac{5}{3}+\dfrac{2}{3} \qquad x = \dfrac{5}{3}-\dfrac{2}{3} = \dfrac{3}{3}$

$x = \dfrac{7}{3} \qquad\qquad x = 1$

The solutions are $\dfrac{7}{3}$ and 1.

109. $(x+5)^2 - 6 = 0$

$(x+5)^2 = 6$

$\sqrt{(x+5)^2} = \sqrt{6}$

$x+5 = \pm\sqrt{6}$

$x+5 = \sqrt{6} \qquad x+5 = -\sqrt{6}$

$x = -5+\sqrt{6} \qquad x = -5-\sqrt{6}$

The solutions are $-5+\sqrt{6}$ and $-5-\sqrt{6}$.

111. $(s-2)^2 - 24 = 0$

$(s-2)^2 = 24$

$\sqrt{(s-2)^2} = \sqrt{24}$

$s-2 = \pm\sqrt{24} = \pm 2\sqrt{6}$

$s-2 = 2\sqrt{6} \qquad s-2 = -2\sqrt{6}$

$s = 2+2\sqrt{6} \qquad s = 2-2\sqrt{6}$

The solutions are $2+2\sqrt{6}$ and $2-2\sqrt{6}$.

113. $(z+1)^2 + 12 = 0$

$(z+1)^2 = -12$

$\sqrt{(z+1)^2} = \sqrt{-12}$

$z+1 = \pm\sqrt{-12} = \pm 2i\sqrt{3}$

$z+1 = 2i\sqrt{3} \qquad z+1 = -2i\sqrt{3}$

$z = -1+2i\sqrt{3} \qquad z = -1-2i\sqrt{3}$

The solutions are $-1+2i\sqrt{3}$ and $-1-2i\sqrt{3}$.

115. $(v-3)^2 + 45 = 0$

$(v-3)^2 = -45$

$\sqrt{(v-3)^2} = \sqrt{-45}$

$v-3 = \pm\sqrt{-45} = \pm 3i\sqrt{5}$

$v-3 = 3i\sqrt{5} \qquad v-3 = -3i\sqrt{5}$

$v = 3+3i\sqrt{5} \qquad v = 3-3i\sqrt{5}$

The solutions are $3+3i\sqrt{5}$ and $3-3i\sqrt{5}$.

117. $\left(u+\dfrac{2}{3}\right)^2 - 18 = 0$

$\left(u+\dfrac{2}{3}\right)^2 = 18$

$\sqrt{\left(u+\dfrac{2}{3}\right)^2} = \sqrt{18}$

$u+\dfrac{2}{3} = \pm\sqrt{18} = \pm 3\sqrt{2}$

$u+\dfrac{2}{3} = 3\sqrt{2} \qquad u+\dfrac{2}{3} = -3\sqrt{2}$

$u = -\dfrac{2}{3}+3\sqrt{2} \qquad u = -\dfrac{2}{3}-3\sqrt{2}$

$u = \dfrac{-2+9\sqrt{2}}{3} \qquad u = \dfrac{-2-9\sqrt{2}}{3}$

The solutions are $\dfrac{-2+9\sqrt{2}}{3}$ and $\dfrac{-2-9\sqrt{2}}{3}$.

119. $\left(x+\dfrac{1}{2}\right)^2 + 40 = 0$

$\left(x+\dfrac{1}{2}\right)^2 = -40$

$\sqrt{\left(x+\dfrac{1}{2}\right)^2} = \sqrt{-40}$

$x+\dfrac{1}{2} = \pm\sqrt{-40} = \pm 2i\sqrt{10}$

$x+\dfrac{1}{2} = 2i\sqrt{10} \qquad x+\dfrac{1}{2} = -2i\sqrt{10}$

$x = -\dfrac{1}{2}+2i\sqrt{10} \qquad x = -\dfrac{1}{2}-2i\sqrt{10}$

The solutions are

$-\dfrac{1}{2}+2i\sqrt{10}$ and $-\dfrac{1}{2}-2i\sqrt{10}$.

Applying Concepts 8.1

121. $(x - r_1)(x - r_2) = 0$
$(x - \sqrt{2})[x - (-\sqrt{2})] = 0$
$(x - \sqrt{2})(x + \sqrt{2}) = 0$
$x^2 - 2 = 0$

123. $(x - r_1)(x - r_2) = 0$
$(x - i)[x - (-i)] = 0$
$(x - i)(x + i) = 0$
$x^2 - i^2 = 0$
$x^2 + 1 = 0$

125. $(x - r_1)(x - r_2) = 0$
$(x - 2\sqrt{2})[x - (-2\sqrt{2})] = 0$
$(x - 2\sqrt{2})(x + 2\sqrt{2}) = 0$
$x^2 - 4(2) = 0$
$x^2 - 8 = 0$

127. $(x - r_1)(x - r_2) = 0$
$(x - 2\sqrt{3})[x - (-2\sqrt{3})] = 0$
$(x - 2\sqrt{3})(x + 2\sqrt{3}) = 0$
$x^2 - 4(3) = 0$
$x^2 - 12 = 0$

129. $(x - r_1)(x - r_2) = 0$
$(x - 2i\sqrt{3})[x - (-2i\sqrt{3})] = 0$
$(x - 2i\sqrt{3})(x + 2i\sqrt{3}) = 0$
$x^2 - 4i^2(3) = 0$
$x^2 - 12i^2 = 0$
$x^2 + 12 = 0$

131. $5y^2 x^2 = 125 z^2$
$y^2 x^2 = 25 z^2$
$y^2 x^2 - 25 z^2 = 0$
$(xy + 5z)(xy - 5z) = 0$
$xy + 5z = 0 \qquad xy - 5z = 0$
$xy = -5z \qquad xy = 5z$
$x = -\dfrac{5z}{y} \qquad x = \dfrac{5z}{y}$

The solutions are $-\dfrac{5z}{y}$ and $\dfrac{5z}{y}$.

133. $2(x - y)^2 - 8 = 0$
$(x - y)^2 - 4 = 0$
$(x - y)^2 = 4$
$\sqrt{(x - y)^2} = \sqrt{4}$
$x - y = \pm 2$
$x = y \pm 2$

The solutions are $y + 2$ and $y - 2$.

135. $(x - 4)^2 = (x + 2)^2$
$x^2 - 8x + 16 = x^2 + 4x + 4$
$x^2 - x^2 - 8x + 16 = x^2 - x^2 + 4x + 4$
$-8x + 16 = 4x + 4$
$-12x + 16 = 4$
$-12x = -12$
$x = 1$

The solution is 1.

137. $ax^2 + c = 0$
$ax^2 = -c$
$x^2 = -\dfrac{c}{a}$
$x = \pm \sqrt{-\dfrac{c}{a}}$
$= \pm \sqrt{\dfrac{c}{a}} i$
$= \pm \dfrac{\sqrt{c}}{\sqrt{a}} i$
$= \pm \dfrac{\sqrt{c}}{\sqrt{a}} i \cdot \dfrac{\sqrt{a}}{\sqrt{a}}$
$= \pm \dfrac{\sqrt{ac}}{\sqrt{a^2}} i$
$= \pm \dfrac{\sqrt{ac}}{a}$

The solutions are $\dfrac{\sqrt{ac}}{a} i$ and $-\dfrac{\sqrt{ac}}{a} i$.

Section 8.2

Concept Review 8.2

1. Always true

3. Never true
   If $b^2 > 4ac$, then $b^2 - 4ac > 0$ and the quadratic equation has two real roots.

5. Never true
   $\left[\left(\dfrac{1}{2}\right)5\right]^2 = \dfrac{25}{4}$, so the last term is not correct.

Objective 1 Exercises

3. $x^2 - 4x - 5 = 0$
$x^2 - 4x = 5$
Complete the square.
$x^2 - 4x + 4 = 5 + 4$
$(x-2)^2 = 9$
$\sqrt{(x-2)^2} = \sqrt{9}$
$x - 2 = \pm\sqrt{9} = \pm 3$
$x - 2 = 3 \qquad x - 2 = -3$
$x = 5 \qquad x = -1$
The solutions are 5 and −1.

5. $v^2 + 8v - 9 = 0$
$v^2 + 8v = 9$
Complete the square.
$v^2 + 8v + 16 = 9 + 16$
$(v+4)^2 = 25$
$\sqrt{(v+4)^2} = \sqrt{25}$
$v + 4 = \pm\sqrt{25} = \pm 5$
$v + 4 = 5 \qquad v + 4 = -5$
$v = 1 \qquad v = -9$
The solutions are 1 and −9.

7. $z^2 - 6z + 9 = 0$
$z^2 - 6z = -9$
Complete the square.
$z^2 - 6z + 9 = -9 + 9$
$(z-3)^2 = 0$
$\sqrt{(z-3)^2} = \sqrt{0}$
$z - 3 = 0$
$z = 3$
The solution is 3.

9. $r^2 + 4r - 7 = 0$
$r^2 + 4r = 7$
Complete the square.
$r^2 + 4r + 4 = 7 + 4$
$(r+2)^2 = 11$
$\sqrt{(r+2)^2} = \sqrt{11}$
$r + 2 = \pm\sqrt{11}$
$r + 2 = \sqrt{11} \qquad r + 2 = -\sqrt{11}$
$r = -2 + \sqrt{11} \qquad r = -2 - \sqrt{11}$
The solutions are $-2 + \sqrt{11}$ and $-2 - \sqrt{11}$.

11. $x^2 - 6x + 7 = 0$
$x^2 - 6x = -7$
Complete the square.
$x^2 - 6x + 9 = -7 + 9$
$(x-3)^2 = 2$
$\sqrt{(x-3)^2} = \sqrt{2}$
$x - 3 = \pm\sqrt{2}$
$x - 3 = \sqrt{2} \qquad x - 3 = -\sqrt{2}$
$x = 3 + \sqrt{2} \qquad x = 3 - \sqrt{2}$
The solutions are $3 + \sqrt{2}$ and $3 - \sqrt{2}$.

13. $z^2 - 2z + 2 = 0$
$z^2 - 2z = -2$
Complete the square.
$z^2 - 2z + 1 = -2 + 1$
$(z-1)^2 = -1$
$\sqrt{(z-1)^2} = \sqrt{-1}$
$z - 1 = \pm i$
$z - 1 = i \qquad z - 1 = -i$
$z = 1 + i \qquad z = 1 - i$
The solutions are $1 + i$ and $1 - i$.

15. $t^2 - t - 1 = 0$
$t^2 - t = 1$
Complete the square.
$t^2 - t + \dfrac{1}{4} = 1 + \dfrac{1}{4}$
$\left(t - \dfrac{1}{2}\right)^2 = \dfrac{5}{4}$
$\sqrt{\left(t - \dfrac{1}{2}\right)^2} = \sqrt{\dfrac{5}{4}}$
$t - \dfrac{1}{2} = \pm\dfrac{\sqrt{5}}{2}$
$t - \dfrac{1}{2} = \dfrac{\sqrt{5}}{2} \qquad t - \dfrac{1}{2} = -\dfrac{\sqrt{5}}{2}$
$t = \dfrac{1}{2} + \dfrac{\sqrt{5}}{2} \qquad t = \dfrac{1}{2} - \dfrac{\sqrt{5}}{2}$
The solutions are $\dfrac{1 + \sqrt{5}}{2}$ and $\dfrac{1 - \sqrt{5}}{2}$.

17. $y^2 - 6y = 4$
Complete the square.
$y^2 - 6y + 9 = 4 + 9$
$(y-3)^2 = 13$
$\sqrt{(y-3)^2} = \sqrt{13}$
$y - 3 = \pm\sqrt{13}$
$y - 3 = \sqrt{13} \qquad y - 3 = -\sqrt{13}$
$y = 3 + \sqrt{13} \qquad y = 3 - \sqrt{13}$
The solutions are $3 + \sqrt{13}$ and $3 - \sqrt{13}$.

19. $x^2 = 8x - 15$
$x^2 - 8x = -15$
Complete the square.
$x^2 - 8x + 16 = -15 + 16$
$(x-4)^2 = 1$
$\sqrt{(x-4)^2} = \sqrt{1}$
$x - 4 = \pm 1$
$x - 4 = 1 \qquad x - 4 = -1$
$x = 5 \qquad\qquad x = 3$
The solutions are 5 and 3.

21. $v^2 = 4v - 13$
$v^2 - 4v = -13$
Complete the square.
$v^2 - 4v + 4 = -13 + 4$
$(v-2)^2 = -9$
$\sqrt{(v-2)^2} = \sqrt{-9}$
$v - 2 = \pm 3i$
$v - 2 = 3i \qquad v - 2 = -3i$
$v = 2 + 3i \qquad v = 2 - 3i$
The solutions are $2 + 3i$ and $2 - 3i$.

23. $p^2 + 6p = -13$
Complete the square.
$p^2 + 6p + 9 = -13 + 9$
$(p+3)^2 = -4$
$\sqrt{(p+3)^2} = \sqrt{-4}$
$p + 3 = \pm 2i$
$p + 3 = 2i \qquad p + 3 = -2i$
$p = -3 + 2i \qquad p = -3 - 2i$
The solutions are $-3 + 2i$ and $-3 - 2i$.

25. $y^2 - 2y = 17$
Complete the square.
$y^2 - 2y + 1 = 17 + 1$
$(y-1)^2 = 18$
$\sqrt{(y-1)^2} = \sqrt{18}$
$y - 1 = \pm 3\sqrt{2}$
$y - 1 = 3\sqrt{2} \qquad y - 1 = -3\sqrt{2}$
$y = 1 + 3\sqrt{2} \qquad y = 1 - 3\sqrt{2}$
The solutions are $1 + 3\sqrt{2}$ and $1 - 3\sqrt{2}$.

27. $z^2 = z + 4$
$z^2 - z = 4$
Complete the square.
$z^2 - z + \frac{1}{4} = 4 + \frac{1}{4}$
$\left(z - \frac{1}{2}\right)^2 = \frac{17}{4}$
$\sqrt{\left(z - \frac{1}{2}\right)^2} = \sqrt{\frac{17}{4}}$
$z - \frac{1}{2} = \pm \frac{\sqrt{17}}{2}$
$z - \frac{1}{2} = \frac{\sqrt{17}}{2} \qquad z - \frac{1}{2} = -\frac{\sqrt{17}}{2}$
$z = \frac{1}{2} + \frac{\sqrt{17}}{2} \qquad z = \frac{1}{2} - \frac{\sqrt{17}}{2}$
The solutions are $\frac{1 + \sqrt{17}}{2}$ and $\frac{1 - \sqrt{17}}{2}$.

29. $x^2 + 13 = 2x$
$x^2 - 2x = -13$
Complete the square.
$x^2 - 2x + 1 = -13 + 1$
$(x-1)^2 = -12$
$\sqrt{(x-1)^2} = \sqrt{-12}$
$x - 1 = \pm 2i\sqrt{3}$
$x - 1 = 2i\sqrt{3} \qquad x - 1 = -2i\sqrt{3}$
$x = 1 + 2i\sqrt{3} \qquad x = 1 - 2i\sqrt{3}$
The solutions are $1 + 2i\sqrt{3}$ and $1 - 2i\sqrt{3}$.

31. $2y^2 + 3y + 1 = 0$
$2y^2 + 3y = -1$
$\frac{1}{2}(2y^2 + 3y) = \frac{1}{2}(-1)$
$y^2 + \frac{3}{2}y = -\frac{1}{2}$
Complete the square.
$y^2 + \frac{3}{2}y + \frac{9}{16} = -\frac{1}{2} + \frac{9}{16}$
$\left(y + \frac{3}{4}\right)^2 = \frac{1}{16}$
$\sqrt{\left(y + \frac{3}{4}\right)^2} = \sqrt{\frac{1}{16}}$
$y + \frac{3}{4} = \pm \frac{1}{4}$
$y + \frac{3}{4} = \frac{1}{4} \qquad\qquad y + \frac{3}{4} = -\frac{1}{4}$
$y = -\frac{2}{4} = -\frac{1}{2} \qquad y = -\frac{4}{4} = -1$
The solutions are $-\frac{1}{2}$ and -1.

33. $4r^2 - 8r = -3$

$\frac{1}{4}(4r^2 - 8r) = \frac{1}{4}(-3)$

$r^2 - 2r = -\frac{3}{4}$

Complete the square.

$r^2 - 2r + 1 = -\frac{3}{4} + 1$

$(r-1)^2 = \frac{1}{4}$

$\sqrt{(r-1)^2} = \sqrt{\frac{1}{4}}$

$r - 1 = \pm\frac{1}{2}$

$r - 1 = \frac{1}{2} \qquad r - 1 = -\frac{1}{2}$

$r = \frac{3}{2} \qquad r = \frac{1}{2}$

The solutions are $\frac{3}{2}$ and $\frac{1}{2}$.

35. $6y^2 - 5y = 4$

$\frac{1}{6}(6y^2 - 5y) = \frac{1}{6}(4)$

$y^2 - \frac{5}{6}y = \frac{2}{3}$

Complete the square.

$y^2 - \frac{5}{6}y + \frac{25}{144} = \frac{2}{3} + \frac{25}{144}$

$\left(y - \frac{5}{12}\right)^2 = \frac{121}{144}$

$\sqrt{\left(y - \frac{5}{12}\right)^2} = \sqrt{\frac{121}{144}}$

$y - \frac{5}{12} = \pm\frac{11}{12}$

$y - \frac{5}{12} = \frac{11}{12} \qquad y - \frac{5}{12} = -\frac{11}{12}$

$y = \frac{16}{12} = \frac{4}{3} \qquad y = -\frac{6}{12} = -\frac{1}{2}$

The solutions are $\frac{4}{3}$ and $-\frac{1}{2}$.

37. $4x^2 - 4x + 5 = 0$

$4x^2 - 4x = -5$

$\frac{1}{4}(4x^2 - 4x) = \frac{1}{4}(-5)$

$x^2 - x = -\frac{5}{4}$

Complete the square.

$x^2 - x + \frac{1}{4} = -\frac{5}{4} + \frac{1}{4}$

$\left(x - \frac{1}{2}\right)^2 = -1$

$\sqrt{\left(x - \frac{1}{2}\right)^2} = \sqrt{-1}$

$x - \frac{1}{2} = \pm i$

$x - \frac{1}{2} = i \qquad x - \frac{1}{2} = -i$

$x = \frac{1}{2} + i \qquad x = \frac{1}{2} - i$

The solutions are $\frac{1}{2} + i$ and $\frac{1}{2} - i$.

39. $9x^2 - 6x + 2 = 0$

$9x^2 - 6x = -2$

$\frac{1}{9}(9x^2 - 6x) = \frac{1}{9}(-2)$

$x^2 - \frac{2}{3}x = -\frac{2}{9}$

Complete the square.

$x^2 - \frac{2}{3}x + \frac{1}{9} = -\frac{2}{9} + \frac{1}{9}$

$\left(x - \frac{1}{3}\right)^2 = -\frac{1}{9}$

$\sqrt{\left(x - \frac{1}{3}\right)^2} = \sqrt{-\frac{1}{9}}$

$x - \frac{1}{3} = \pm\frac{1}{3}i$

$x - \frac{1}{3} = \frac{1}{3}i \qquad x - \frac{1}{3} = -\frac{1}{3}i$

$x = \frac{1}{3} + \frac{1}{3}i \qquad x = \frac{1}{3} - \frac{1}{3}i$

The solutions are $\frac{1}{3} + \frac{1}{3}i$ and $\frac{1}{3} - \frac{1}{3}i$.

41.
$$2s^2 = 4s + 5$$
$$2s^2 - 4s = 5$$
$$\frac{1}{2}(2s^2 - 4s) = \frac{1}{2}(5)$$
$$s^2 - 2s = \frac{5}{2}$$
Complete the square.
$$s^2 - 2s + 1 = \frac{5}{2} + 1$$
$$(s-1)^2 = \frac{7}{2}$$
$$\sqrt{(s-1)^2} = \sqrt{\frac{7}{2}}$$
$$s - 1 = \pm\sqrt{\frac{7}{2}} = \pm\frac{\sqrt{14}}{2}$$
$$s - 1 = \frac{\sqrt{14}}{2} \qquad s - 1 = -\frac{\sqrt{14}}{2}$$
$$s = \frac{2}{2} + \frac{\sqrt{14}}{2} \qquad x = \frac{2}{2} - \frac{\sqrt{14}}{2}$$
The solutions are $\frac{2+\sqrt{14}}{2}$ and $\frac{2-\sqrt{14}}{2}$.

43.
$$2r^2 = 3 - r$$
$$2r^2 + r = 3$$
$$\frac{1}{2}(2r^2 + r) = \frac{1}{2}(3)$$
$$r^2 + \frac{1}{2}r = \frac{3}{2}$$
Complete the square.
$$r^2 + \frac{1}{2}r + \frac{1}{16} = \frac{3}{2} + \frac{1}{16}$$
$$\left(r + \frac{1}{4}\right)^2 = \frac{25}{16}$$
$$\sqrt{\left(r + \frac{1}{4}\right)^2} = \sqrt{\frac{25}{16}}$$
$$r + \frac{1}{4} = \pm\frac{5}{4}$$
$$r + \frac{1}{4} = \frac{5}{4} \qquad r + \frac{1}{4} = -\frac{5}{4}$$
$$r = \frac{4}{4} = 1 \qquad r = -\frac{6}{4} = -\frac{3}{2}$$
The solutions are 1 and $-\frac{3}{2}$.

45.
$$y - 2 = (y-3)(y+2)$$
$$y - 2 = y^2 - y - 6$$
$$y^2 - 2y = 4$$
Complete the square.
$$y^2 - 2y + 1 = 4 + 1$$
$$(y-1)^2 = 5$$
$$\sqrt{(y-1)^2} = \sqrt{5}$$
$$y - 1 = \pm\sqrt{5}$$
$$y - 1 = \sqrt{5} \qquad y - 1 = -\sqrt{5}$$
$$y = 1 + \sqrt{5} \qquad y = 1 - \sqrt{5}$$
The solutions are $1+\sqrt{5}$ and $1-\sqrt{5}$.

47.
$$6t - 2 = (2t-3)(t-1)$$
$$6t - 2 = 2t^2 - 5t + 3$$
$$2t^2 - 11t = -5$$
$$\frac{1}{2}(2t^2 - 11t) = \frac{1}{2}(-5)$$
$$t^2 - \frac{11}{2}t = -\frac{5}{2}$$
Complete the square.
$$t^2 - \frac{11}{2}t + \frac{121}{16} = -\frac{5}{2} + \frac{121}{16}$$
$$\left(t - \frac{11}{4}\right)^2 = \frac{81}{16}$$
$$\sqrt{\left(t - \frac{11}{4}\right)^2} = \sqrt{\frac{81}{16}}$$
$$t - \frac{11}{4} = \pm\frac{9}{4}$$
$$t - \frac{11}{4} = \frac{9}{4} \qquad t - \frac{11}{4} = -\frac{9}{4}$$
$$t = \frac{20}{4} = 5 \qquad t = \frac{2}{4} = \frac{1}{2}$$
The solutions are 5 and $\frac{1}{2}$.

49.
$$(x-4)(x+1) = x - 3$$
$$x^2 - 3x - 4 = x - 3$$
$$x^2 - 4x = 1$$
Complete the square.
$$x^2 - 4x + 4 = 1 + 4$$
$$(x-2)^2 = 5$$
$$\sqrt{(x-2)^2} = \sqrt{5}$$
$$x - 2 = \pm\sqrt{5}$$
$$x - 2 = \sqrt{5} \qquad x - 2 = -\sqrt{5}$$
$$x = 2 + \sqrt{5} \qquad x = 2 - \sqrt{5}$$
The solutions are $2+\sqrt{5}$ and $2-\sqrt{5}$.

51. $z^2 + 2z = 4$
Complete the square.
$z^2 + 2z + 1 = 4 + 1$
$(z+1)^2 = 5$
$\sqrt{(z+1)^2} = \sqrt{5}$
$z + 1 = \pm 2.236$

$z + 1 \approx 2.236 \qquad z + 1 \approx -2.236$
$z \approx -1 + 2.236 \qquad z \approx -1 - 2.236$
$z \approx 1.236 \qquad z \approx -3.236$

The solutions are approximately 1.236 and −3.236.

53. $2x^2 = 4x - 1$
$2x^2 - 4x = -1$
$\frac{1}{2}(2x^2 - 4x) = \frac{1}{2}(-1)$
$x^2 - 2x = -\frac{1}{2}$
Complete the square.
$x^2 - 2x + 1 = -\frac{1}{2} + 1$
$(x-1)^2 = \frac{1}{2}$
$\sqrt{(x-1)^2} = \sqrt{\frac{1}{2}}$
$x - 1 \approx \pm 0.707$

$x - 1 \approx 0.707 \qquad x - 1 \approx -0.707$
$x \approx 1 + 0.707 \qquad x \approx 1 - 0.707$
$x \approx 1.707 \qquad x \approx 0.293$

The solutions are approximately 1.707 and 0.293.

55. $4z^2 + 2z - 1 = 0$
$4z^2 + 2z = 1$
$\frac{1}{4}(4z^2 + 2z) = \frac{1}{4}(1)$
$z^2 + \frac{1}{2}z = \frac{1}{4}$
Complete the square.
$z^2 + \frac{1}{2}z + \frac{1}{16} = \frac{1}{4} + \frac{1}{16}$
$\left(z + \frac{1}{4}\right)^2 = \frac{5}{16}$
$\sqrt{\left(z + \frac{1}{4}\right)^2} = \sqrt{\frac{5}{16}}$
$z + \frac{1}{4} \approx \pm 0.559$

$z + \frac{1}{4} \approx 0.559 \qquad z + \frac{1}{4} \approx -0.559$
$z \approx -\frac{1}{4} + 0.559 \qquad z \approx -\frac{1}{4} - 0.559$
$z \approx 0.309 \qquad z \approx -0.809$

The solutions are approximately 0.309 and −0.809.

Objective 2 Exercises

59. $x^2 - 3x - 10 = 0$
$a = 1, b = -3, c = -10$
$x = \dfrac{-b \pm \sqrt{b^2 - 4ac}}{2a}$
$= \dfrac{-(-3) \pm \sqrt{(-3)^2 - 4(1)(-10)}}{2(1)}$
$= \dfrac{3 \pm \sqrt{9 + 40}}{2}$
$= \dfrac{3 \pm \sqrt{49}}{2}$
$= \dfrac{3 \pm 7}{2}$

$x = \dfrac{3+7}{2} \qquad x = \dfrac{3-7}{2}$
$= \dfrac{10}{2} \qquad = \dfrac{-4}{2}$
$= 5 \qquad = -2$

The solutions are 5 and −2.

61. $y^2 + 5y - 36 = 0$
$a = 1, b = 5, c = -36$
$y = \dfrac{-b \pm \sqrt{b^2 - 4ac}}{2a}$
$= \dfrac{-5 \pm \sqrt{(5)^2 - 4(1)(-36)}}{2(1)}$
$= \dfrac{-5 \pm \sqrt{25 + 144}}{2}$
$= \dfrac{-5 \pm \sqrt{169}}{2}$
$= \dfrac{-5 \pm 13}{2}$

$y = \dfrac{-5+13}{2} \qquad y = \dfrac{-5-13}{2}$
$= \dfrac{8}{2} \qquad = \dfrac{-18}{2}$
$= 4 \qquad = -9$

The solutions are 4 and −9.

63. $w^2 = 8w + 72$
$w^2 - 8w - 72 = 0$
$a = 1, b = -8, c = -72$
$w = \dfrac{-b \pm \sqrt{b^2 - 4ac}}{2a}$
$= \dfrac{-(-8) \pm \sqrt{(-8)^2 - 4(1)(-72)}}{2(1)}$
$= \dfrac{8 \pm \sqrt{64 + 288}}{2}$
$= \dfrac{8 \pm \sqrt{352}}{2}$
$= \dfrac{8 \pm 4\sqrt{22}}{2}$
$= 4 \pm 2\sqrt{22}$

The solutions are $4 + 2\sqrt{22}$ and $4 - 2\sqrt{22}$.

65. $v^2 = 24 - 5v$
$v^2 + 5v - 24 = 0$
$a = 1, b = 5, c = -24$
$v = \dfrac{-b \pm \sqrt{b^2 - 4ac}}{2a}$
$= \dfrac{-5 \pm \sqrt{(5)^2 - 4(1)(-24)}}{2(1)}$
$= \dfrac{-5 \pm \sqrt{25 + 96}}{2}$
$= \dfrac{-5 \pm \sqrt{121}}{2}$
$= \dfrac{-5 \pm 11}{2}$

$v = \dfrac{-5 + 11}{2}$ $v = \dfrac{-5 - 11}{2}$
$= \dfrac{6}{2}$ $= \dfrac{-16}{2}$
$= 3$ $= -8$

The solutions are 3 and –8.

67. $2y^2 + 5y - 3 = 0$
$a = 2, b = 5, c = -3$
$y = \dfrac{-b \pm \sqrt{b^2 - 4ac}}{2a}$
$= \dfrac{-5 \pm \sqrt{(5)^2 - 4(2)(-3)}}{2(2)}$
$= \dfrac{-5 \pm \sqrt{25 + 24}}{4}$
$= \dfrac{-5 \pm \sqrt{49}}{4}$
$= \dfrac{-5 \pm 7}{4}$

$y = \dfrac{-5 + 7}{4}$ $y = \dfrac{-5 - 7}{4}$
$= \dfrac{2}{4}$ $= \dfrac{-12}{4}$
$= \dfrac{1}{2}$ $= -3$

The solutions are $\dfrac{1}{2}$ and –3.

69. $8s^2 = 10s + 3$
$8s^2 - 10s - 3 = 0$
$a = 8, b = -10, c = -3$
$s = \dfrac{-b \pm \sqrt{b^2 - 4ac}}{2a}$
$= \dfrac{-(-10) \pm \sqrt{(-10)^2 - 4(8)(-3)}}{2(8)}$
$= \dfrac{10 \pm \sqrt{100 + 96}}{16}$
$= \dfrac{10 \pm \sqrt{196}}{16}$
$= \dfrac{10 \pm 14}{16}$

$s = \dfrac{10 + 14}{16}$ $s = \dfrac{10 - 14}{16}$
$= \dfrac{24}{16}$ $= \dfrac{-4}{16}$
$= \dfrac{3}{2}$ $= -\dfrac{1}{4}$

The solutions are $\dfrac{3}{2}$ and $-\dfrac{1}{4}$.

71. $v^2 - 2v - 7 = 0$
$a = 1, b = -2, c = -7$
$v = \dfrac{-b \pm \sqrt{b^2 - 4ac}}{2a}$
$= \dfrac{-(-2) \pm \sqrt{(-2)^2 - 4(1)(-7)}}{2(1)}$
$= \dfrac{2 \pm \sqrt{4 + 28}}{2}$
$= \dfrac{2 \pm \sqrt{32}}{2}$
$= \dfrac{2 \pm 4\sqrt{2}}{2}$
$= 1 \pm 2\sqrt{2}$

The solutions are $1 + 2\sqrt{2}$ and $1 - 2\sqrt{2}$.

73. $y^2 - 8y - 20 = 0$
$a = 1, b = -8, c = -20$

$$y = \frac{-b \pm \sqrt{b^2 - 4ac}}{2a}$$

$$= \frac{-(-8) \pm \sqrt{(-8)^2 - 4(1)(-20)}}{2(1)}$$

$$= \frac{8 \pm \sqrt{64 + 80}}{2}$$

$$= \frac{8 \pm \sqrt{144}}{2}$$

$$= \frac{8 \pm 12}{2}$$

$y = \dfrac{8+12}{2}$ $\quad y = \dfrac{8-12}{2}$

$= \dfrac{20}{2}$ $\quad\quad = \dfrac{-4}{2}$

$= 10$ $\quad\quad\quad = -2$

The solutions are 10 and –2.

75. $\quad v^2 = 12v - 24$
$v^2 - 12v + 24 = 0$
$a = 1, b = -12, c = 24$

$$v = \frac{-b \pm \sqrt{b^2 - 4ac}}{2a}$$

$$= \frac{-(-12) \pm \sqrt{(-12)^2 - 4(1)(24)}}{2(1)}$$

$$= \frac{12 \pm \sqrt{144 - 96}}{2}$$

$$= \frac{12 \pm \sqrt{48}}{2}$$

$$= \frac{12 \pm 4\sqrt{3}}{2}$$

$$= 6 \pm 2\sqrt{3}$$

The solutions are $6 + 2\sqrt{3}$ and $6 - 2\sqrt{3}$.

77. $4x^2 - 4x - 7 = 0$
$a = 4, b = -4, c = -7$

$$x = \frac{-b \pm \sqrt{b^2 - 4ac}}{2a}$$

$$= \frac{-(-4) \pm \sqrt{(-4)^2 - 4(4)(-7)}}{2(4)}$$

$$= \frac{4 \pm \sqrt{16 + 112}}{8}$$

$$= \frac{4 \pm \sqrt{128}}{8}$$

$$= \frac{4 \pm 8\sqrt{2}}{8}$$

$$= \frac{1 \pm 2\sqrt{2}}{2}$$

The solutions are $\dfrac{1+2\sqrt{2}}{2}$ and $\dfrac{1-2\sqrt{2}}{2}$.

79. $2s^2 - 3s + 1 = 0$
$a = 2, b = -3, c = 1$

$$s = \frac{-b \pm \sqrt{b^2 - 4ac}}{2a}$$

$$= \frac{-(-3) \pm \sqrt{(-3)^2 - 4(2)(1)}}{2(2)}$$

$$= \frac{3 \pm \sqrt{9 - 8}}{4}$$

$$= \frac{3 \pm \sqrt{1}}{4}$$

$$= \frac{3 \pm 1}{4}$$

$s = \dfrac{3+1}{4}$ $\quad s = \dfrac{3-1}{4}$

$= \dfrac{4}{4}$ $\quad\quad = \dfrac{2}{4}$

$= 1$ $\quad\quad\quad = \dfrac{1}{2}$

The solutions are 1 and $\dfrac{1}{2}$.

81. $3x^2 + 10x + 6 = 0$
$a = 3, b = 10, c = 6$

$$x = \frac{-b \pm \sqrt{b^2 - 4ac}}{2a}$$

$$= \frac{-10 \pm \sqrt{(10)^2 - 4(3)(6)}}{2(3)}$$

$$= \frac{-10 \pm \sqrt{100 - 72}}{6}$$

$$= \frac{-10 \pm \sqrt{28}}{6}$$

$$= \frac{-10 \pm 2\sqrt{7}}{6}$$

$$= \frac{-5 \pm \sqrt{7}}{3}$$

The solutions are $\dfrac{-5+\sqrt{7}}{3}$ and $\dfrac{-5-\sqrt{7}}{3}$.

83.
$$6w^2 = 19w - 10$$
$$6w^2 - 19w + 10 = 0$$
$$a = 6, b = -19, c = 10$$
$$w = \frac{-b \pm \sqrt{b^2 - 4ac}}{2a}$$
$$= \frac{-(-19) \pm \sqrt{(-19)^2 - 4(6)(10)}}{2(6)}$$
$$= \frac{19 \pm \sqrt{361 - 240}}{12}$$
$$= \frac{19 \pm \sqrt{121}}{12}$$
$$= \frac{19 \pm 11}{12}$$

$$w = \frac{19 + 11}{12} \qquad w = \frac{19 - 11}{12}$$
$$= \frac{30}{12} \qquad = \frac{8}{12}$$
$$= \frac{5}{2} \qquad = \frac{2}{3}$$

The solutions are $\frac{5}{2}$ and $\frac{2}{3}$.

85.
$$p^2 - 4p + 5 = 0$$
$$a = 1, b = -4, c = 5$$
$$p = \frac{-b \pm \sqrt{b^2 - 4ac}}{2a}$$
$$= \frac{-(-4) \pm \sqrt{(-4)^2 - 4(1)(5)}}{2(1)}$$
$$= \frac{4 \pm \sqrt{16 - 20}}{2}$$
$$= \frac{4 \pm \sqrt{-4}}{2}$$
$$= \frac{4 \pm 2i}{2}$$
$$= 2 \pm i$$

The solutions are $2 + i$ and $2 - i$.

87.
$$x^2 + 6x + 13 = 0$$
$$a = 1, b = 6, c = 13$$
$$x = \frac{-b \pm \sqrt{b^2 - 4ac}}{2a}$$
$$= \frac{-6 \pm \sqrt{(6)^2 - 4(1)(13)}}{2(1)}$$
$$= \frac{-6 \pm \sqrt{36 - 52}}{2}$$
$$= \frac{-6 \pm \sqrt{-16}}{2}$$
$$= \frac{-6 \pm 4i}{2}$$
$$= -3 \pm 2i$$

The solutions are $-3 + 2i$ and $-3 - 2i$.

89.
$$t^2 - 6t + 10 = 0$$
$$a = 1, b = -6, c = 10$$
$$t = \frac{-b \pm \sqrt{b^2 - 4ac}}{2a}$$
$$= \frac{-(-6) \pm \sqrt{(-6)^2 - 4(1)(10)}}{2(1)}$$
$$= \frac{6 \pm \sqrt{36 - 40}}{2}$$
$$= \frac{6 \pm \sqrt{-4}}{2}$$
$$= \frac{6 \pm 2i}{2}$$
$$= 3 \pm i$$

The solutions are $3 + i$ and $3 - i$.

91.
$$4v^2 + 8v + 3 = 0$$
$$a = 4, b = 8, c = 3$$
$$v = \frac{-b \pm \sqrt{b^2 - 4ac}}{2a}$$
$$= \frac{-8 \pm \sqrt{(8)^2 - 4(4)(3)}}{2(4)}$$
$$= \frac{-8 \pm \sqrt{64 - 48}}{8}$$
$$= \frac{-8 \pm \sqrt{16}}{8}$$
$$= \frac{-8 \pm 4}{8}$$

$$v = \frac{-8 + 4}{8} \qquad v = \frac{-8 - 4}{8}$$
$$= \frac{-4}{8} \qquad = \frac{-12}{8}$$
$$= -\frac{1}{2} \qquad = -\frac{3}{2}$$

The solutions are $-\frac{1}{2}$ and $-\frac{3}{2}$.

93.
$$2y^2 + 2y + 13 = 0$$
$$a = 2, b = 2, c = 13$$
$$y = \frac{-b \pm \sqrt{b^2 - 4ac}}{2a}$$
$$= \frac{-2 \pm \sqrt{(2)^2 - 4(2)(13)}}{2(2)}$$
$$= \frac{-2 \pm \sqrt{4 - 104}}{4}$$
$$= \frac{-2 \pm \sqrt{-100}}{4}$$
$$= \frac{-2 \pm 10i}{4}$$
$$= \frac{-1 \pm 5i}{2}$$

The solutions are $-\frac{1}{2} + \frac{5}{2}i$ and $-\frac{1}{2} - \frac{5}{2}i$.

95. $3v^2 + 6v + 1 = 0$
$a = 3, b = 6, c = 1$

$$x = \frac{-b \pm \sqrt{b^2 - 4ac}}{2a}$$
$$= \frac{-6 \pm \sqrt{(6)^2 - 4(3)(1)}}{2(3)}$$
$$= \frac{-6 \pm \sqrt{36 - 12}}{6}$$
$$= \frac{-6 \pm \sqrt{24}}{6}$$
$$= \frac{-6 \pm 2\sqrt{6}}{6}$$
$$= \frac{-3 \pm \sqrt{6}}{3}$$

The solutions are $\frac{-3 + \sqrt{6}}{3}$ and $\frac{-3 - \sqrt{6}}{3}$.

97. $3y^2 = 6y - 5$
$3y^2 - 6y + 5 = 0$
$a = 3, b = -6, c = 5$

$$y = \frac{-b \pm \sqrt{b^2 - 4ac}}{2a}$$
$$= \frac{-(-6) \pm \sqrt{(-6)^2 - 4(3)(5)}}{2(3)}$$
$$= \frac{6 \pm \sqrt{36 - 60}}{6}$$
$$= \frac{6 \pm \sqrt{-24}}{6}$$
$$= \frac{6 \pm 2i\sqrt{6}}{6}$$
$$= \frac{3 \pm i\sqrt{6}}{3}$$

The solutions are $1 + \frac{\sqrt{6}}{3}i$ and $1 - \frac{\sqrt{6}}{3}i$.

99. $10y(y + 4) = 15y - 15$
$10y^2 + 40y = 15y - 15$
$10y^2 + 25y + 15 = 0$
$5(2y^2 + 5y + 3) = 0$
$2y^2 + 5y + 3 = 0$
$a = 2, b = 5, c = 3$

$$y = \frac{-b \pm \sqrt{b^2 - 4ac}}{2a}$$
$$= \frac{-5 \pm \sqrt{(5)^2 - 4(2)(3)}}{2(2)}$$
$$= \frac{-5 \pm \sqrt{25 - 24}}{4}$$
$$= \frac{-5 \pm \sqrt{1}}{4}$$
$$= \frac{-5 \pm 1}{4}$$

$y = \frac{-5 + 1}{4}$ $y = \frac{-5 - 1}{4}$
$= \frac{-4}{4}$ $= \frac{-6}{4}$
$= -1$ $= -\frac{3}{2}$

The solutions are -1 and $-\frac{3}{2}$.

101. $(2t + 1)(t - 3) = 9$
$2t^2 - 5t - 3 = 9$
$2t^2 - 5t - 12 = 0$
$a = 2, b = -5, c = -12$

$$t = \frac{-b \pm \sqrt{b^2 - 4ac}}{2a}$$
$$= \frac{-(-5) \pm \sqrt{(-5)^2 - 4(2)(-12)}}{2(2)}$$
$$= \frac{5 \pm \sqrt{25 + 96}}{4}$$
$$= \frac{5 \pm \sqrt{121}}{4}$$
$$= \frac{5 \pm 11}{4}$$

$t = \frac{5 + 11}{4}$ $t = \frac{5 - 11}{4}$
$= \frac{16}{4}$ $= \frac{-6}{4}$
$= 4$ $= -\frac{3}{2}$

The solutions are 4 and $-\frac{3}{2}$.

103. $3y^2 + y + 1 = 0$
$a = 3, b = 1, c = 1$
$b^2 - 4ac$
$1^2 - 4(3)(1) = 1 - 12 = -11$
$-11 < 0$
Since the discriminant is less than zero, the equation has two complex number solutions.

105. $4x^2 + 20x + 25 = 0$
$a = 4, b = 20, c = 25$
$b^2 - 4ac$
$20^2 - 4(4)(25) = 400 - 400 = 0$
Since the discriminant is equal to zero, the equation has one real number solution, a double root.

107. $3w^2 + 3w - 2 = 0$
$a = 3, b = 3, c = -2$
$b^2 - 4ac$
$3^2 - 4(3)(-2) = 9 + 24 = 33$
$33 > 0$
Since the discriminant is greater than zero, the equation has two real number solutions that are not equal.

109. $2t^2 + 9t + 3 = 0$
$a = 2, b = 9, c = 3$
$b^2 - 4ac$
$9^2 - 4(2)(3) = 81 - 24 = 57$
$57 > 0$
Since the discriminant is greater than zero, the equation has two real number solutions that are not equal.

111. $x^2 + 6x - 6 = 0$
$a = 1, b = 6, c = -6$
$$x = \frac{-b \pm \sqrt{b^2 - 4ac}}{2a}$$
$$= \frac{-6 \pm \sqrt{(6)^2 - 4(1)(-6)}}{2(1)}$$
$$= \frac{-6 \pm \sqrt{36 + 24}}{2}$$
$$= \frac{-6 \pm \sqrt{60}}{2}$$
$$= \frac{-6 \pm 2\sqrt{15}}{2}$$
$$= -3 \pm \sqrt{15}$$
$\approx -3 \pm 3.873$
$x \approx -3 + 3.873 \qquad x \approx -3 - 3.873$
$\approx 0.873 \qquad\qquad \approx -6.873$
The solutions are approximately 0.873 and −6.873.

113. $r^2 - 2r - 4 = 0$
$a = 1, b = -2, c = -4$
$$r = \frac{-b \pm \sqrt{b^2 - 4ac}}{2a}$$
$$= \frac{-(-2) \pm \sqrt{(-2)^2 - 4(1)(-4)}}{2(1)}$$
$$= \frac{2 \pm \sqrt{4 + 16}}{2}$$
$$= \frac{2 \pm \sqrt{20}}{2}$$
$$= \frac{2 \pm 2\sqrt{5}}{2}$$
$$= 1 \pm \sqrt{5}$$
$\approx 1 \pm 2.236$
$r \approx 1 + 2.236 \qquad r \approx 1 - 2.236$
$\approx 3.236 \qquad\qquad \approx -1.236$
The solutions are approximately 3.236 and −1.236.

115. $3t^2 = 7t + 1$
$3t^2 - 7t - 1 = 0$
$a = 3, b = -7, c = -1$
$$t = \frac{-b \pm \sqrt{b^2 - 4ac}}{2a}$$
$$= \frac{-(-7) \pm \sqrt{(-7)^2 - 4(3)(-1)}}{2(3)}$$
$$= \frac{7 \pm \sqrt{49 + 12}}{6}$$
$$= \frac{7 \pm \sqrt{61}}{6}$$
$$\approx \frac{7 \pm 7.810}{6}$$
$t \approx \frac{7 + 7.810}{6} \qquad t \approx \frac{7 - 7.810}{6}$
$\approx \frac{14.810}{6} \qquad\qquad \approx \frac{-0.810}{6}$
$\approx 2.468 \qquad\qquad \approx -0.135$
The solutions are approximately 2.468 and −0.135.

Chapter 8: Quadratic Equations and Inequalities

Applying Concepts 8.2

117. $\sqrt{2}y^2 + 3y - 2\sqrt{2} = 0$
$a = \sqrt{2},\ b = 3,\ c = -2\sqrt{2}$
$y = \dfrac{-b \pm \sqrt{b^2 - 4ac}}{2a}$
$= \dfrac{-3 \pm \sqrt{(3)^2 - 4(\sqrt{2})(-2\sqrt{2})}}{2(\sqrt{2})}$
$= \dfrac{-3 \pm \sqrt{9 + 16}}{2\sqrt{2}}$
$= \dfrac{-3 \pm \sqrt{25}}{2\sqrt{2}} = \dfrac{-3 \pm 5}{2\sqrt{2}}$
$y = \dfrac{-3 + 5}{2\sqrt{2}}$
$= \dfrac{2}{2\sqrt{2}} \cdot \dfrac{\sqrt{2}}{\sqrt{2}}$
$= \dfrac{2\sqrt{2}}{4}$
$= \dfrac{\sqrt{2}}{2}$
$= \dfrac{-3 - 5}{2\sqrt{2}}$
$= \dfrac{-8}{2\sqrt{2}}$
$= -\dfrac{4}{\sqrt{2}}$
$= -\dfrac{4}{\sqrt{2}} \cdot \dfrac{\sqrt{2}}{\sqrt{2}}$
$= -\dfrac{4\sqrt{2}}{2}$
$= -2\sqrt{2}$

The solutions are $\dfrac{\sqrt{2}}{2}$ and $-2\sqrt{2}$.

119. $\sqrt{2}x^2 + 5x - 3\sqrt{2} = 0$
$a = \sqrt{2},\ b = 5,\ c = -3\sqrt{2}$
$x = \dfrac{-b \pm \sqrt{b^2 - 4ac}}{2a}$
$= \dfrac{-5 \pm \sqrt{(5)^2 - 4(\sqrt{2})(-3\sqrt{2})}}{2(\sqrt{2})}$
$= \dfrac{-5 \pm \sqrt{25 + 24}}{2\sqrt{2}}$
$= \dfrac{-5 \pm \sqrt{49}}{2\sqrt{2}}$
$= \dfrac{-5 \pm 7}{2\sqrt{2}}$
$x = \dfrac{-5 + 7}{2\sqrt{2}}$
$= \dfrac{2}{2\sqrt{2}}$
$= \dfrac{1}{\sqrt{2}}$
$= \dfrac{1}{\sqrt{2}} \cdot \dfrac{\sqrt{2}}{\sqrt{2}}$
$= \dfrac{\sqrt{2}}{2}$
$= \dfrac{-5 - 7}{2\sqrt{2}}$
$= \dfrac{-12}{2\sqrt{2}}$
$= \dfrac{-6}{\sqrt{2}}$
$= -\dfrac{6}{\sqrt{2}} \cdot \dfrac{\sqrt{2}}{\sqrt{2}}$
$= -\dfrac{6\sqrt{2}}{2}$
$= -3\sqrt{2}$

The solutions are $\dfrac{\sqrt{2}}{2}$ and $-3\sqrt{2}$.

121. $t^2 - t\sqrt{3} + 1 = 0$
$a = 1,\ b = -\sqrt{3},\ c = 1$
$t = \dfrac{-b \pm \sqrt{b^2 - 4ac}}{2a}$
$= \dfrac{-(-\sqrt{3}) \pm \sqrt{(-\sqrt{3})^2 - 4(1)(1)}}{2(1)}$
$= \dfrac{\sqrt{3} \pm \sqrt{3 - 4}}{2}$
$= \dfrac{\sqrt{3} \pm \sqrt{-1}}{2}$
$= \dfrac{\sqrt{3} \pm i}{2}$
$= \dfrac{\sqrt{3}}{2} \pm \dfrac{1}{2}i$

The solutions are $\dfrac{\sqrt{3}}{2} + \dfrac{1}{2}i$ and $\dfrac{\sqrt{3}}{2} - \dfrac{1}{2}i$.

123. $x^2 - ax - 2a^2 = 0$
$(x - 2a)(x + a) = 0$
$x - 2a = 0 \qquad x + a = 0$
$\quad x = 2a \qquad\quad x = -a$
The solutions are $2a$ and $-a$.

125. $2x^2 + 3ax - 2a^2 = 0$
$(2x - a)(x + 2a) = 0$
$2x - a = 0 \qquad x + 2a = 0$
$\quad 2x = a \qquad\quad x = -2a$
$\quad x = \dfrac{a}{2}$
The solutions are $\dfrac{a}{2}$ and $-2a$.

127. $x^2 - 2x - y = 0$
$a = 1, b = -2, c = -y$
$x = \dfrac{-b \pm \sqrt{b^2 - 4ac}}{2a}$
$= \dfrac{-(-2) \pm \sqrt{(-2)^2 - 4(1)(-y)}}{2(1)}$
$= \dfrac{2 \pm \sqrt{4 + 4y}}{2}$
$= \dfrac{2 \pm 2\sqrt{1 + y}}{2}$
$= 1 \pm \sqrt{1 + y}$
The solutions are $1 + \sqrt{y+1}$ and $1 - \sqrt{y+1}$.

129. $x^2 - 6x + p = 0$
$a = 1, b = -6, c = p$
$b^2 - 4ac > 0$
$(-6)^2 - 4(1)(p) > 0$
$36 - 4p > 0$
$-4p > -36$
$p < 9$
$\{p | p < 9, \; p \in \text{real numbers}\}$

131. $x^2 - 2x + p = 0$
$a = 1, b = -2, c = p$
$b^2 - 4ac < 0$
$(-2)^2 - 4(1)(p) < 0$
$4 - 4p < 0$
$-4p < -4$
$p > 1$
$\{p | p > 1, \; p \in \text{real numbers}\}$

133. $x^2 + bx - 1 = 0$
$b^2 - 4ac = b^2 - 4(1)(-1) = b^2 + 4$
Because b^2 is greater than zero regardless of the value of b, $b^2 + 4$ is greater than zero regardless of the value of b.

Therefore, $x^2 + bx - 1 = 0$ has real number solutions regardless of the value of b.

135. Strategy
To find the time it takes for the ball to hit the ground, use the value for height ($h = 0$) and solve for t.

Solution
$$h = -16t^2 + 70t + 4$$
$$0 = -16t^2 + 70t + 4$$
$$-4 = -16t^2 + 70t$$
$$-\dfrac{1}{16}(-4) = -\dfrac{1}{16}(16t^2 + 70t)$$
$$\dfrac{1}{4} = t^2 - \dfrac{35}{8}t$$
Complete the square.
$$\dfrac{1}{4} + \dfrac{1225}{256} = t^2 - \dfrac{35}{8}t + \dfrac{1225}{256}$$
$$\dfrac{1289}{256} = \left(t - \dfrac{35}{16}\right)^2$$
$$\sqrt{\dfrac{1289}{256}} = \sqrt{\left(t - \dfrac{35}{16}\right)^2}$$
$$\pm\sqrt{\dfrac{1289}{256}} = t - \dfrac{35}{16}$$

$t - \dfrac{35}{16} = \sqrt{\dfrac{1289}{256}} \qquad t - \dfrac{35}{16} = -\sqrt{\dfrac{1289}{256}}$

$t = \dfrac{35}{16} + \sqrt{\dfrac{1289}{256}} \qquad t = \dfrac{35}{16} - \sqrt{\dfrac{1289}{256}}$

$t \approx 4.431 \qquad\qquad\qquad t \approx -0.0564$

The solution $t = -0.0564$ is not possible because it represents a time before the ball is thrown.
The ball takes about 4.43 seconds to hit the ground.

137. **a.** $\qquad x^2 - s_a x - s_j s_s f = 0$
$x^2 - 0.97x - (0.34)(0.97)(0.24) = 0$
$x^2 - 0.97x - 0.079152 = 0$
$x = \dfrac{0.97 \pm \sqrt{(-0.97)^2 - 4(1)(-0.079152)}}{2(1)}$
$x \approx 1.05$ and $x \approx -0.08$
The larger of the two roots is 1.05.
$1.05 > 1$
The model predicts that the population will increase.

b. $\qquad x^2 - s_a x - s_j s_s f = 0$
$x^2 - 0.94x - (0.11)(0.71)(0.24) = 0$
$x^2 - 0.94x - 0.018744 = 0$
$x = \dfrac{0.94 \pm \sqrt{(-0.94)^2 - 4(1)(-0.018744)}}{2(1)}$
$x \approx 0.96$ and $x \approx -0.02$
The larger of the two roots is 0.96.
$0.96 < 1$
The model predicts that the population will decrease.

Section 8.3

Concept Review 8.3

1. Always true

3. Always true

Objective 1 Exercises

3. $x^4 - 13x^2 + 36 = 0$
$(x^2)^2 - 13(x^2) + 36 = 0$
$u^2 - 13u + 36 = 0$
$(u-4)(u-9) = 0$
$u - 4 = 0 \quad u - 9 = 0$
$u = 4 \quad u = 9$
Replace u by x^2.
$x^2 = 4 \quad x^2 = 9$
$\sqrt{x^2} = \sqrt{4} \quad \sqrt{x^2} = \sqrt{9}$
$x = \pm 2 \quad x = \pm 3$
The solutions are 2, −2, 3, and −3.

5. $z^4 - 6z^2 + 8 = 0$
$(z^2)^2 - 6(z^2) + 8 = 0$
$u^2 - 6u + 8 = 0$
$(u-4)(u-2) = 0$
$u - 4 = 0 \quad u - 2 = 0$
$u = 4 \quad u = 2$
Replace u by z^2.
$z^2 = 4 \quad z^2 = 2$
$\sqrt{z^2} = \sqrt{4} \quad \sqrt{z^2} = \sqrt{2}$
$z = \pm 2 \quad z = \pm\sqrt{2}$
The solutions are 2, −2, $\sqrt{2}$, and $-\sqrt{2}$.

7. $p - 3p^{1/2} + 2 = 0$
$(p^{1/2})^2 - 3(p^{1/2}) + 2 = 0$
$u^2 - 3u + 2 = 0$
$(u-1)(u-2) = 0$
$u - 1 = 0 \quad u - 2 = 0$
$u = 1 \quad u = 2$
Replace u by $p^{1/2}$.
$p^{1/2} = 1 \quad p^{1/2} = 2$
$(p^{1/2})^2 = 1^2 \quad (p^{1/2})^2 = 2^2$
$p = 1 \quad p = 4$
The solutions are 1 and 4.

9. $x - x^{1/2} - 12 = 0$
$(x^{1/2})^2 - (x^{1/2}) - 12 = 0$
$u^2 - u - 12 = 0$
$(u+3)(u-4) = 0$
$u + 3 = 0 \quad u - 4 = 0$
$u = -3 \quad u = 4$
Replace u by $x^{1/2}$.
$x^{1/2} = -3 \quad x^{1/2} = 4$
$(x^{1/2})^2 = (-3)^2 \quad (x^{1/2})^2 = 4^2$
$x = 9 \quad x = 16$
16 checks as a solution.
9 does not check as a solution.
The solution is 16.

11. $z^4 + 3z^2 - 4 = 0$
$(z^2)^2 + 3(z^2) - 4 = 0$
$u^2 + 3u - 4 = 0$
$(u+4)(u-1) = 0$
$u + 4 = 0 \quad u - 1 = 0$
$u = -4 \quad u = 1$
Replace u by z^2.
$z^2 = -4 \quad z^2 = 1$
$\sqrt{z^2} = \sqrt{-4} \quad \sqrt{z^2} = \sqrt{1}$
$z = \pm 2i \quad z = \pm 1$
The solutions are $2i$, $-2i$, 1, and −1.

13. $x^4 + 12x^2 - 64 = 0$
$(x^2)^2 + 12(x^2) - 64 = 0$
$u^2 + 12u - 64 = 0$
$(u+16)(u-4) = 0$
$u + 16 = 0 \quad u - 4 = 0$
$u = -16 \quad u = 4$
Replace u by x^2.
$x^2 = -16 \quad x^2 = 4$
$\sqrt{x^2} = \sqrt{-16} \quad \sqrt{x^2} = \sqrt{4}$
$x = \pm 4i \quad x = \pm 2$
The solutions are $4i$, $-4i$, 2, and −2.

15. $p + 2p^{1/2} - 24 = 0$
$(p^{1/2})^2 + 2(p^{1/2}) - 24 = 0$
$u^2 + 2u - 24 = 0$
$(u+6)(u-4) = 0$
$u + 6 = 0 \quad u - 4 = 0$
$u = -6 \quad u = 4$
Replace u by $p^{1/2}$.
$p^{1/2} = -6 \quad p^{1/2} = 4$
$(p^{1/2})^2 = (-6)^2 \quad (p^{1/2})^2 = 4^2$
$p = 36 \quad p = 16$
16 checks as a solution.
36 does not check as a solution.
The solution is 16.

17. $y^{2/3} - 9y^{1/3} + 8 = 0$
$(y^{1/3})^2 - 9(y^{1/3}) + 8 = 0$
$u^2 - 9u + 8 = 0$
$(u-1)(u-8) = 0$
$u - 1 = 0 \qquad u - 8 = 0$
$u = 1 \qquad\quad u = 8$
Replace u by $y^{1/3}$.
$y^{1/3} = 1 \qquad\qquad y^{1/3} = 8$
$(y^{1/3})^3 = 1^3 \qquad (y^{1/3})^3 = 8^3$
$y = 1 \qquad\qquad\quad y = 512$
The solutions are 1 and 512.

19. $x^6 - 9x^3 + 8 = 0$
$(x^3)^2 - 9(x^3) + 8 = 0$
$u^2 - 9u + 8 = 0$
$(u-8)(u-1) = 0$
$u - 8 = 0 \qquad\qquad u - 1 = 0$
Replace u by x^3.
$\qquad\qquad x^3 - 8 = 0 \qquad\qquad\qquad x^3 - 1 = 0$
$(x-2)(x^2 + 2x + 4) = 0 \quad (x-1)(x^2 + x + 1) = 0$

$x - 2 = 0 \qquad x^2 + 2x + 4 = 0 \qquad\qquad x - 1 = 0 \qquad x^2 + x + 1 = 0$
$x = 2 \qquad\qquad x = \dfrac{-2 \pm \sqrt{2^2 - 4(1)(4)}}{2(1)} \qquad x = 1 \qquad\qquad x = \dfrac{-1 \pm \sqrt{1^2 - 4(1)(1)}}{2(1)}$
$\qquad\qquad\qquad = \dfrac{-2 \pm \sqrt{-12}}{2} \qquad\qquad\qquad\qquad\qquad = \dfrac{-1 \pm \sqrt{-3}}{2}$
$\qquad\qquad\qquad = \dfrac{-2 \pm 2i\sqrt{3}}{2} \qquad\qquad\qquad\qquad\qquad = \dfrac{-1 \pm i\sqrt{3}}{2}$
$\qquad\qquad\qquad = -1 \pm i\sqrt{3}$

The solutions are 2, 1, $-1 + i\sqrt{3}$, $-1 - i\sqrt{3}$, $-\dfrac{1}{2} + \dfrac{\sqrt{3}}{2}i$, and $-\dfrac{1}{2} - \dfrac{\sqrt{3}}{2}i$.

21. $z^8 - 17z^4 + 16 = 0$
$(z^4)^2 - 17(z^4) + 16 = 0$
$u^2 - 17u + 16 = 0$
$(u-16)(u-1) = 0$
$u - 16 = 0 \qquad\qquad u - 1 = 0$
Replace u by z^4.
$z^4 - 16 = 0 \qquad z^4 - 1 = 0$
$(z^2)^2 - 16 = 0 \quad (z^2)^2 - 1 = 0$
$v^2 - 16 = 0 \qquad v^2 - 1 = 0$

$(v+4)(v-4) = 0 \qquad (v+1)(v-1) = 0$
$v + 4 = 0 \qquad v - 4 = 0 \qquad v + 1 = 0 \qquad v - 1 = 0$
$v = -4 \qquad\quad v = 4 \qquad\quad v = -1 \qquad\quad v = 1$
Replace v by z^2.
$z^2 = -4 \qquad\quad z^2 = 4 \qquad\quad z^2 = -1 \qquad\quad z^2 = 1$
$\sqrt{z^2} = \sqrt{-4} \quad \sqrt{z^2} = \sqrt{4} \quad \sqrt{z^2} = \sqrt{-1} \quad \sqrt{z^2} = 1$
$z = \pm 2i \qquad\quad z = \pm 2 \qquad\quad z = \pm i \qquad\quad z = \pm 1$
The solutions are $-2, 2, 2i, -2i, -1, 1, i,$ and $-i$.

23. $p^{2/3} + 2p^{1/3} - 8 = 0$
$(p^{1/3})^2 + 2(p^{1/3}) - 8 = 0$
$u^2 + 2u - 8 = 0$
$(u+4)(u-2) = 0$
$u + 4 = 0 \qquad u - 2 = 0$
$u = -4 \qquad u = 2$
Replace u by $p^{1/3}$.
$p^{1/3} = -4 \qquad\qquad p^{1/3} = 2$
$(p^{1/3})^3 = (-4)^3 \qquad (p^{1/3})^3 = 2^3$
$p = -64 \qquad\qquad p = 8$
The solutions are –64 and 8.

25. $2x - 3x^{1/2} + 1 = 0$
$2(x^{1/2})^2 - 3(x^{1/2}) + 1 = 0$
$2u^2 - 3u + 1 = 0$
$(2u - 1)(u - 1) = 0$
$2u - 1 = 0 \qquad u - 1 = 0$
$2u = 1 \qquad\quad u = 1$
$u = \dfrac{1}{2}$
Replace u by $x^{1/2}$.
$x^{1/2} = \dfrac{1}{2} \qquad\qquad x^{1/2} = 1$
$(x^{1/2})^2 = \left(\dfrac{1}{2}\right)^2 \qquad (x^{1/2})^2 = 1^2$
$x = \dfrac{1}{4} \qquad\qquad\qquad x = 1$
The solutions are $\dfrac{1}{4}$ and 1.

Objective 2 Exercises

27. $\sqrt{x+1} + x = 5$
$\sqrt{x+1} = 5 - x$
$(\sqrt{x+1})^2 = (5-x)^2$
$x + 1 = 25 - 10x + x^2$
$0 = 24 - 11x + x^2$
$0 = (3-x)(8-x)$
$3 - x = 0 \qquad 8 - x = 0$
$3 = x \qquad\quad 8 = x$
3 checks as a solution.
8 does not check as a solution.
The solution is 3.

29. $x = \sqrt{x} + 6$
$x - 6 = \sqrt{x}$
$(x-6)^2 = (\sqrt{x})^2$
$x^2 - 12x + 36 = x$
$x^2 - 13x + 36 = 0$
$(x-4)(x-9) = 0$
$x - 4 = 0 \qquad x - 9 = 0$
$x = 4 \qquad\quad x = 9$
9 checks as a solution.
4 does not check as a solution.
The solution is 9.

31. $\sqrt{3w+3} = w + 1$
$(\sqrt{3w+3})^2 = (w+1)^2$
$3w + 3 = w^2 + 2w + 1$
$0 = w^2 - w - 2$
$0 = (w-2)(w+1)$
$w - 2 = 0 \qquad w + 1 = 0$
$w = 2 \qquad\quad w = -1$
2 and –1 check as solutions.
The solutions are 2 and –1.

33. $\sqrt{4y+1} - y = 1$
$\sqrt{4y+1} = y + 1$
$(\sqrt{4y+1})^2 = (y+1)^2$
$4y + 1 = y^2 + 2y + 1$
$0 = y^2 - 2y$
$0 = y(y-2)$
$y = 0 \qquad y - 2 = 0$
$\qquad\qquad y = 2$
0 and 2 check as solutions.
The solutions are 0 and 2.

35. $\sqrt{10x+5} - 2x = 1$
$\sqrt{10x+5} = 2x + 1$
$(\sqrt{10x+5})^2 = (2x+1)^2$
$10x + 5 = 4x^2 + 4x + 1$
$0 = 4x^2 - 6x - 4$
$0 = 2(2x^2 - 3x - 2)$
$0 = 2(2x+1)(x-2)$
$2x + 1 = 0 \qquad x - 2 = 0$
$2x = -1 \qquad\quad x = 2$
$x = -\dfrac{1}{2}$
$-\dfrac{1}{2}$ and 2 check as solutions.
The solutions are $-\dfrac{1}{2}$ and 2.

37. $\sqrt{p+11} = 1 - p$
$(\sqrt{p+11})^2 = (1-p)^2$
$p + 11 = 1 - 2p + p^2$
$0 = -10 - 3p + p^2$
$0 = p^2 - 3p - 10$
$0 = (p-5)(p+2)$
$p - 5 = 0 \qquad p + 2 = 0$
$p = 5 \qquad\quad p = -2$
–2 checks as a solution.
5 does not check as a solution.
The solution is –2.

39.
$$\sqrt{x-1} - \sqrt{x} = -1$$
$$\sqrt{x-1} = \sqrt{x} - 1$$
$$(\sqrt{x-1})^2 = (\sqrt{x} - 1)^2$$
$$x - 1 = x - 2\sqrt{x} + 1$$
$$2\sqrt{x} = 2$$
$$\sqrt{x} = 1$$
$$(\sqrt{x})^2 = 1^2$$
$$x = 1$$
1 checks as a solution.
The solution is 1.

41.
$$\sqrt{2x-1} = 1 - \sqrt{x-1}$$
$$(\sqrt{2x-1})^2 = (1 - \sqrt{x-1})^2$$
$$2x - 1 = 1 - 2\sqrt{x-1} + x - 1$$
$$2\sqrt{x-1} = -x + 1$$
$$(2\sqrt{x-1})^2 = (-x+1)^2$$
$$4(x-1) = x^2 - 2x + 1$$
$$4x - 4 = x^2 - 2x + 1$$
$$0 = x^2 - 6x + 5$$
$$0 = (x-5)(x-1)$$
$$x - 5 = 0 \qquad x - 1 = 0$$
$$x = 5 \qquad x = 1$$
1 checks as a solution.
5 does not check as a solution.
The solution is 1.

43.
$$\sqrt{t+3} + \sqrt{2t+7} = 1$$
$$\sqrt{2t+7} = 1 - \sqrt{t+3}$$
$$(\sqrt{2t+7})^2 = (1 - \sqrt{t+3})^2$$
$$2t + 7 = 1 - 2\sqrt{t+3} + t + 3$$
$$t + 3 = -2\sqrt{t+3}$$
$$(t+3)^2 = (-2\sqrt{t+3})^2$$
$$t^2 + 6t + 9 = 4(t+3)$$
$$t^2 + 6t + 9 = 4t + 12$$
$$t^2 + 2t - 3 = 0$$
$$(t+3)(t-1) = 0$$
$$t + 3 = 0 \qquad t - 1 = 0$$
$$t = -3 \qquad t = 1$$
−3 checks as a solution.
1 does not check as a solution.
The solution is −3.

Objective 3 Exercises

45.
$$x = \frac{10}{x-9}$$
$$(x-9)x = (x-9)\frac{10}{x-9}$$
$$x^2 - 9x = 10$$
$$x^2 - 9x - 10 = 0$$
$$(x-10)(x+1) = 0$$
$$x - 10 = 0 \qquad x + 1 = 0$$
$$x = 10 \qquad x = -1$$
The solutions are 10 and −1.

47.
$$\frac{t}{t+1} = \frac{-2}{t-1}$$
$$(t-1)(t+1)\frac{t}{t+1} = (t-1)(t+1)\frac{-2}{t-1}$$
$$(t-1)t = (t+1)(-2)$$
$$t^2 - t = -2t - 2$$
$$t^2 + t + 2 = 0$$
$$t = \frac{-b \pm \sqrt{b^2 - 4ac}}{2a}$$
$$= \frac{-1 \pm \sqrt{1^2 - 4(1)(2)}}{2(1)}$$
$$= \frac{-1 \pm \sqrt{1-8}}{2} = \frac{-1 \pm \sqrt{-7}}{2} = \frac{-1 \pm i\sqrt{7}}{2}$$
The solutions are $-\frac{1}{2} + \frac{\sqrt{7}}{2}i$ and $-\frac{1}{2} - \frac{\sqrt{7}}{2}i$.

49.
$$\frac{y-1}{y+2} + y = 1$$
$$(y+2)\left(\frac{y-1}{y+2} + y\right) = (y+2)1$$
$$(y+2)\frac{y-1}{y+2} + (y+2)y = y + 2$$
$$y - 1 + y^2 + 2y = y + 2$$
$$y^2 + 3y - 1 = y + 2$$
$$y^2 + 2y - 3 = 0$$
$$(y+3)(y-1) = 0$$
$$y + 3 = 0 \qquad y - 1 = 0$$
$$y = -3 \qquad y = 1$$
The solutions are −3 and 1.

51.
$$\frac{3r+2}{r+2} - 2r = 1$$
$$(r+2)\left(\frac{3r+2}{r+2} - 2r\right) = (r+2)1$$
$$(r+2)\frac{3r+2}{r+2} - (r+2)2r = r + 2$$
$$3r + 2 - 2r^2 - 4r = r + 2$$
$$-2r^2 - r + 2 = r + 2$$
$$-2r^2 - 2r = 0$$
$$-2r(r+1) = 0$$
$$-2r = 0 \qquad r + 1 = 0$$
$$r = 0 \qquad r = -1$$
The solutions are 0 and −1.

53.
$$\frac{2}{2x+1}+\frac{1}{x}=3$$
$$x(2x+1)\left(\frac{2}{2x+1}+\frac{1}{x}\right)=x(2x+1)3$$
$$x(2x+1)\frac{2}{2x+1}+x(2x+1)\frac{1}{x}=3x(2x+1)$$
$$2x+2x+1=6x^2+3x$$
$$4x+1=6x^2+3x$$
$$0=6x^2-x-1$$
$$0=(2x-1)(3x+1)$$

$2x-1=0 \qquad 3x+1=0$
$2x=1 \qquad\quad 3x=-1$
$x=\frac{1}{2} \qquad\quad x=-\frac{1}{3}$

The solutions are $\frac{1}{2}$ and $-\frac{1}{3}$.

55.
$$\frac{16}{z-2}+\frac{16}{z+2}=6$$
$$(z-2)(z+2)\left(\frac{16}{z-2}+\frac{16}{z+2}\right)=(z-2)(z+2)6$$
$$(z-2)(z+2)\frac{16}{z-2}+(z-2)(z+2)\frac{16}{z+2}=(z^2-4)6$$
$$(z+2)16+(z-2)16=6z^2-24$$
$$16z+32+16z-32=6z^2-24$$
$$32z=6z^2-24$$
$$0=6z^2-32z-24$$
$$0=2(3z^2-16z-12)$$
$$0=2(3z+2)(z-6)$$

$3z+2=0 \qquad z-6=0$
$3z=-2 \qquad\quad z=6$
$z=-\frac{2}{3}$

The solutions are $-\frac{2}{3}$ and 6.

57.
$$\frac{t}{t-2}+\frac{2}{t-1}=4$$
$$(t-2)(t-1)\left(\frac{t}{t-2}+\frac{2}{t-1}\right)=(t-2)(t-1)4$$
$$(t-2)(t-1)\frac{t}{t-2}+(t-2)(t-1)\frac{2}{t-1}=(t^2-3t+2)4$$
$$(t-1)t+(t-2)2=4t^2-12t+8$$
$$t^2-t+2t-4=4t^2-12t+8$$
$$t^2+t-4=4t^2-12t+8$$
$$0=3t^2-13t+12$$
$$0=(3t-4)(t-3)$$

$3t-4=0 \qquad t-3=0$
$3t=4 \qquad\quad t=3$
$t=\frac{4}{3}$

The solutions are $\frac{4}{3}$ and 3.

59.
$$\frac{5}{2p-1} + \frac{4}{p+1} = 2$$
$$(2p-1)(p+1)\left(\frac{5}{2p-1} + \frac{4}{p+1}\right) = 2(p-1)(p+1)2$$
$$(2p-1)(p+1)\frac{5}{2p-1} + (2p-1)(p+1)\frac{4}{p+1} = (2p^2 + p - 1)2$$
$$(p+1)5 + (2p-1)4 = 4p^2 + 2p - 2$$
$$5p + 5 + 8p - 4 = 4p^2 + 2p - 2$$
$$13p + 1 = 4p^2 + 2p - 2$$
$$0 = 4p^2 - 11p - 3$$
$$0 = (4p+1)(p-3)$$

$4p + 1 = 0 \qquad p - 3 = 0$
$4p = -1 \qquad\quad p = 3$
$p = -\dfrac{1}{4}$

The solutions are $-\dfrac{1}{4}$ and 3.

61.
$$\frac{2v}{v+2} + \frac{3}{v+4} = 1$$
$$(v+2)(v+4)\left(\frac{2v}{v+2} + \frac{3}{v+4}\right) = (v+2)(v+4)1$$
$$(v+2)(v+4)\frac{2v}{v+2} + (v+2)(v+4)\frac{3}{v+4} = v^2 + 6v + 8$$
$$(v+4)2v + (v+2)3 = v^2 + 6v + 8$$
$$2v^2 + 8v + 3v + 6 = v^2 + 6v + 8$$
$$2v^2 + 11v + 6 = v^2 + 6v + 8$$
$$v^2 + 5v - 2 = 0$$
$$v = \frac{-b \pm \sqrt{b^2 - 4ac}}{2a} = \frac{-5 \pm \sqrt{5^2 - 4(1)(-2)}}{2(1)} = \frac{-5 \pm \sqrt{25 + 8}}{2} = \frac{-5 \pm \sqrt{33}}{2}$$

The solutions are $\dfrac{-5 + \sqrt{33}}{2}$ and $\dfrac{-5 - \sqrt{33}}{2}$.

Applying Concepts 8.3

63. $\dfrac{x^2}{4} + \dfrac{x}{2} = 6$

$4\left(\dfrac{x^2}{4} + \dfrac{x}{2}\right) = 4(6)$

$x^2 + 2x = 24$

$x^2 + 2x - 24 = 0$

$(x+6)(x-4) = 0$

$x + 6 = 0 \qquad x - 4 = 0$
$\quad x = -6 \qquad \quad x = 4$

The solutions are -6 and 4.

65. $\dfrac{x+2}{3} + \dfrac{2}{x-2} = 3$

$3(x-2)\left(\dfrac{x+2}{3} + \dfrac{2}{x-2}\right) = 3(x-2)3$

$(x-2)(x+2) + 3\cdot 2 = 9(x-2)$

$x^2 - 4 + 6 = 9x - 18$

$x^2 + 2 = 9x - 18$

$x^2 - 9x + 20 = 0$

$(x-4)(x-5) = 0$

$x - 4 = 0 \qquad x - 5 = 0$
$\quad x = 4 \qquad \quad x = 5$

The solutions are 4 and 5.

67. $\dfrac{x^4}{3} - \dfrac{8x^2}{3} = 3$

$3\left(\dfrac{x^4}{3} - \dfrac{8x^2}{3}\right) = 3(3)$

$x^4 - 8x^2 = 9$

$x^4 - 8x^2 - 9 = 0$

$(x^2)^2 - 8(x^2) - 9 = 0$

$u^2 - 8u - 9 = 0$

$(u-9)(u+1) = 0$

$u - 9 = 0 \qquad u + 1 = 0$
$\quad u = 9 \qquad \quad u = -1$

Replace u by x^2.

$x^2 = 9 \qquad \qquad x^2 = -1$

$\sqrt{x^2} = \sqrt{9} \qquad \sqrt{x^2} = \sqrt{-1}$

$x = \pm 3 \qquad \qquad x = \pm i$

The solutions are $3, -3, i,$ and $-i$.

69. $\dfrac{x^4}{8} + \dfrac{x^2}{4} = 3$

$8\left(\dfrac{x^4}{8} + \dfrac{x^2}{4}\right) = 8(3)$

$x^4 + 2x^2 = 24$

$x^4 + 2x^2 - 24 = 0$

$(x^2)^2 + 2(x^2) - 24 = 0$

$u^2 + 2u - 24 = 0$

$(u+6)(u-4) = 0$

$u + 6 = 0 \qquad u - 4 = 0$
$\quad u = -6 \qquad \quad u = 4$

Replace u by x^2.

$x^2 = -6 \qquad \qquad x^2 = 4$

$\sqrt{x^2} = \sqrt{-6} \qquad \sqrt{x^2} = \sqrt{4}$

$x = \pm i\sqrt{6} \qquad \quad x = \pm 2$

The solutions are $2, -2, i\sqrt{6},$ and $-i\sqrt{6}$.

71. $\sqrt{x^4 + 4} = 2x$

$(\sqrt{x^4+4})^2 = (2x)^2$

$x^4 + 4 = 4x^2$

$x^4 - 4x^2 + 4 = 0$

$(x^2)^2 - 4(x^2) + 4 = 0$

$u^2 - 4u + 4 = 0$

$(u-2)(u-2) = 0$

$u - 2 = 0 \quad u - 2 = 0$
$\quad u = 2 \qquad \quad u = 2$

Replace u by x^2.

$x^2 = 2$

$\sqrt{x^2} = \sqrt{2}$

$x = \pm\sqrt{2}$

$\sqrt{2}$ checks as a solution.

$-\sqrt{2}$ does not check as a solution.

The solution is $\sqrt{2}$.

73.
$(\sqrt{x}+3)^2 - 4\sqrt{x} - 17 = 0$
$(\sqrt{x}+3)^2 - 4\sqrt{x} - 12 - 5 = 0$
$(\sqrt{x}+3)^2 - 4(\sqrt{x}+3) - 5 = 0$
Let $u = \sqrt{x}+3$.
$u^2 - 4u - 5 = 0$
$(u-5)(u+1) = 0$
$u - 5 = 0 \qquad u + 1 = 0$
$u = 5 \qquad\qquad u = -1$
Replace u by $\sqrt{x}+3$.

$\sqrt{x}+3 = 5 \qquad\qquad \sqrt{x}+3 = -1$
$\sqrt{x} = 2 \qquad\qquad \sqrt{x} = -4$
$(\sqrt{x})^2 = 2^2 \qquad\qquad (\sqrt{x})^2 = (-4)^2$
$x = 4 \qquad\qquad\qquad x = 16$

16 does not check in the original equation.
The solution is 4.

Section 8.4

Concept Review 8.4

1. Sometimes true
If $x = 2$,
$x^2 = 2^2$ and $(x+2)^2 = 4^2$.
These are the squares of two consecutive even integers.

3. Always true

Objective 1 Exercises

1. Strategy
This is a geometry problem.
The width of the rectangle: x
The length of the rectangle: $2x + 8$

The area of the rectangle is 640 ft^2. Use the equation for the area of a rectangle ($A = L \cdot W$).

Solution
$A = L \cdot W$
$640 = (2x+8)x$
$640 = 2x^2 + 8x$
$0 = 2x^2 + 8x - 640$
$0 = 2(x^2 + 4x - 320)$
$0 = 2(x+20)(x-16)$
$x + 20 = 0 \qquad x - 16 = 0$
$x = -20 \qquad\quad x = 16$
Since the width of the rectangle cannot be negative, -20 cannot be a solution.
$2x + 8 = 2(16) + 8 = 32 + 8 = 40$
The width of the rectangle is 16 ft.
The length of the rectangle is 40 ft.

3. Strategy
This is a geometry problem.
The width of the rectangle: x
The length of the rectangle: $3x - 2$

The area of the rectangle is 65 ft^2. Use the equation for the area of a rectangle ($A = L \cdot W$).

Solution
$A = L \cdot W$
$65 = (3x - 2)x$
$65 = 3x^2 - 2x$
$0 = 3x^2 - 2x - 65$
$0 = (3x+13)(x-5)$
$3x + 13 = 0 \qquad x - 5 = 0$
$x = -\dfrac{13}{3} \qquad\quad x = 5$

Since the width cannot be negative, $-\dfrac{13}{3}$ cannot be a solution.
$3x + 2 = 3(5) - 2 = 15 - 2 = 13$
The length is of the rectangle is 13 ft.
The width of the rectangle is 5 ft.

5. Strategy
This is a uniform motion problem.
Rate of truck on return trip: r

	Distance	Rate	Time
With load	550	$r-5$	$\dfrac{550}{r-5}$
Without load	550	r	$\dfrac{550}{r}$

The total time of the trip was 21 h.

Solution
$\dfrac{550}{r-5} + \dfrac{550}{r} = 21$

$r(r-5)\left(\dfrac{550}{r-5} + \dfrac{550}{r}\right) = r(r-5)21$

$550r + 550(r-5) = (r^2 - 5r)21$
$550r + 550r - 2750 = 21r^2 - 105r$
$1100r - 2750 = 21r^2 - 105r$
$0 = 21r^2 - 1205 + 2750$
$0 = (21r - 50)(r - 55)$
$21r - 50 = 0 \qquad r - 55 = 0$
$21r = 0 \qquad\qquad r = 55$
$r = \dfrac{50}{21}$

$r - 5 = \dfrac{50}{21} = \dfrac{50 - 105}{21} = -\dfrac{55}{21}$ or
$r - 5 = 55 - 5 = 50$

$\dfrac{50}{21}$ cannot be a solution because then the rate of the trip with the load would be negative. The rate of the truck on the return trip was 55 mph.

7. **Strategy**
To find the time for a projectile to return to Earth, substitute the values for height ($s = 0$) and initial velocity ($v_0 = 200$ ft/s) and solve for t.

Solution
$$s = v_0 t - 16t^2$$
$$0 = 200t - 16t^2$$
$$0 = 8t(25 - 2t)$$
$$\begin{array}{ll} 8t = 0 & 25 - 2t = 0 \\ t = 0 & -2t = -25 \\ & t = 12.5 \end{array}$$
The solution $t = 0$ is not appropriate because the projectile has not yet left Earth. The rocket takes 12.5 s to return to Earth.

9. **Strategy**
To find the maximum speed, substitute for distance ($d = 150$) and solve for v.

Solution
$$d = 0.019v^2 + 0.69v$$
$$150 = 0.019v^2 + 0.69v$$
$$0 = 0.019v^2 + 0.69v - 150$$
$$v = \frac{-b \pm \sqrt{b^2 - 4ac}}{2a}$$
$$= \frac{-0.69 \pm \sqrt{(0.69)^2 - 4(0.019)(-150)}}{2(0.019)}$$
$$= \frac{-0.69 \pm \sqrt{11.8761}}{0.038}$$
$$= 72.5 \text{ or } -108.85$$
Since the speed cannot be negative, -108.85 cannot be a solution.
The maximum speed a driver can be going and still be able to stop within 150 m is 72.5 km/h.

11. **Strategy**
To find when the rocket will be 300 ft above the ground, substitute for height ($h = 300$) and solve for t.

Solution
$$h = -16t^2 + 200t$$
$$300 = -16t^2 + 200t$$
$$0 = -16t^2 + 200t - 300$$
$$t = \frac{-b \pm \sqrt{b^2 - 4ac}}{2a}$$
$$t = \frac{-200 \pm \sqrt{(200)^2 - 4(-16)(-300)}}{2(-16)}$$
$$t = \frac{-200 \pm \sqrt{20,800}}{-32}$$
$$t = 1.74 \text{ or } 10.76$$
The rocket will be 300 ft above the ground 1.74 s and 10.76 s after the launch.

13. **Strategy**
This is a work problem.
Time for the smaller pipe to fill the tank: t
Time for the larger pipe to fill the tank: $t - 6$

	Rate	Time	Part
Smaller pipe	$\dfrac{1}{t}$	4	$\dfrac{4}{t}$
Larger pipe	$\dfrac{1}{t-6}$	4	$\dfrac{4}{t-6}$

The sum of the parts of the task completed must equal 1.
$$\frac{4}{t} + \frac{4}{t-6} = 1$$

Solution
$$\frac{4}{t} + \frac{4}{t-6} = 1$$
$$t(t-6)\left(\frac{4}{t} + \frac{4}{t-6}\right) = t(t-6)1$$
$$(t-6)4 + 4t = t^2 - 6t$$
$$4t - 24 + 4t = t^2 - 6t$$
$$8t - 24 = t^2 - 6t$$
$$0 = t^2 - 14t + 24$$
$$0 = (t - 12)(t - 2)$$
$$\begin{array}{ll} t - 12 = 0 & t - 2 = 0 \\ t = 12 & t = 2 \\ t - 6 = 12 - 6 = 6 & t - 6 = 2 - 6 = -4 \end{array}$$
The solution -4 is not possible, since time cannot be a negative number.
It would take the larger pipe 6 min to fill the tank.
It would take the smaller pipe 12 min to fill the tank.

15. **Strategy**
    This is a distance-rate problem.
    Rate of the cruise ship for the first 40 mi: r
    Rate of the cruise ship for the next 60 mi: $r + 5$

	Distance	Rate	Time
First 40 mi	40	r	$\frac{40}{r}$
Next 60 mi	60	$r+5$	$\frac{60}{r+5}$

    The total time of travel was 8 h.
    $$\frac{40}{r} + \frac{60}{r+5} = 8$$

    Solution
    $$\frac{40}{r} + \frac{60}{r+5} = 8$$
    $$r(r+5)\left(\frac{40}{r} + \frac{60}{r+5}\right) = r(r+5)8$$
    $$(r+5)40 + 60r = 8r(r+5)$$
    $$40r + 200 + 60r = 8r^2 + 40r$$
    $$100r + 200 = 8r^2 + 40r$$
    $$0 = 8r^2 - 60r - 200$$
    $$0 = 4(2r^2 - 15r - 50)$$
    $$0 = 4(2r+5)(r-10)$$

    $2r + 5 = 0 \qquad r - 10 = 0$
    $2r = -5 \qquad\quad r = 10$
    $r = -\frac{5}{2}$

    The solution $-\frac{5}{2}$ is not possible, since rate cannot be a negative number. The rate of the cruise ship for the first 40 mi was 10 mph.

17. **Strategy**
This is a distance-rate problem.
Rate of the wind: w

	Distance	Rate	Time
With wind	240	$100 + w$	$\dfrac{240}{100+w}$
Against wind	240	$100 - w$	$\dfrac{240}{100-w}$

The time with the wind is 1 h less than the time against the wind.
$$\frac{240}{100+w} = \frac{240}{100-w} - 1$$

Solution
$$\frac{240}{100+w} = \frac{240}{100-w} - 1$$
$$(100+w)(100-w)\left(\frac{240}{100+w}\right) = (100+w)(100-w)\left(\frac{240}{100-w} - 1\right)$$
$$240(100-w) = 240(100+w) - (100+w)(100-w)$$
$$24,000 - 240w = 24,000 + 240w - 10,000 + w^2$$
$$0 = w^2 + 480w - 10,000$$
$$0 = (w+500)(w-20)$$

$w + 500 = 0 \qquad w - 20 = 0$
$w = -500 \qquad w = 20$

The solution –500 is not possible, since rate cannot be a negative number.
The rate of the wind is 20 mph.

19. Strategy
  This is a distance-rate problem.
  Rate of crew in calm water: x

	Distance	Rate	Time
With current	16	$x+2$	$\dfrac{16}{x+2}$
Against current	16	$x-2$	$\dfrac{16}{x-2}$

The total trip took 6 h.
$\dfrac{16}{x+2} + \dfrac{16}{x-2} = 6$

Solution
$$\dfrac{16}{x+2} + \dfrac{16}{x-2} = 6$$
$$(x+2)(x-2)\left(\dfrac{16}{x+2} + \dfrac{16}{x-2}\right) = (x+2)(x-2)6$$
$$16(x-2) + 16(x+2) = 6(x^2 - 4)$$
$$16x - 32 + 16x + 32 = 6x^2 - 24$$
$$32x = 6x^2 - 24$$
$$0 = 6x^2 - 32x - 24$$
$$0 = 2(3x^2 - 16x - 12)$$
$$0 = 2(3x + 2)(x - 6)$$

$3x + 2 = 0 \qquad x - 6 = 0$
$x = -\dfrac{2}{3} \qquad\quad x = 6$

The solution $x = -\dfrac{2}{3}$ is not possible, since rate cannot be a negative number.
The rate of the crew in calm water is 6 mph.

Applying Concepts 8.4

21. Strategy
  Unknown number: n
  Twice the reciprocal of the number: $2 \cdot \dfrac{1}{n} = \dfrac{2}{n}$
  The sum of the number and twice its reciprocal is $\dfrac{33}{4}$.

Solution
$$n + \dfrac{2}{n} = \dfrac{33}{4}$$
$$4n\left(n + \dfrac{2}{n}\right) = 4n\left(\dfrac{33}{4}\right)$$
$$4n^2 + 8 = 33n$$
$$4n^2 - 33n + 8 = 0$$
$$(4n - 1)(n - 8) = 0$$

$4n - 1 = 0 \qquad n - 8 = 0$
$4n = 1 \qquad\qquad n = 8$
$n = \dfrac{1}{4}$

The number is $\dfrac{1}{4}$ or 8.

23. Strategy
  This is an integer problem.
  The first integer: n
  Next consecutive integer: $n + 1$
  The difference of the cubes of the integers is 127.

Solution
$$(n+1)^3 - n^3 = 127$$
$$(n+1)(n+1)(n+1) - n^3 = 127$$
$$(n^2 + 2n + 1)(n+1) - n^3 = 127$$
$$(n^3 + n^2 + 2n^2 + 2n + n + 1) - n^3 = 127$$
$$(n^3 + 3n^2 + 3n + 1) - n^3 = 127$$
$$n^3 + 3n^2 + 3n + 1 - n^3 = 127$$
$$3n^2 + 3n + 1 = 127$$
$$3n^2 + 3n - 126 = 0$$
$$3(n^2 + n - 42) = 0$$
$$n^2 + n - 42 = 0$$
$$(n+7)(n-6) = 0$$

$n + 7 = 0 \qquad n - 6 = 0$
$n = -7 \qquad\quad n = 6$
For $n = -7$, $n + 1 = -7 + 1 = -6$.
For $n = 6$, $n + 1 = 6 + 1 = 7$.
The integers are –7 and –6, or 6 and 7.

25. Strategy
This is a geometry problem.
Width of rectangular piece of cardboard: w
Length of rectangular piece of cardboard: $w + 8$
Width of open box: $w - 2(2) = w - 4$
Length of open box: $w + 8 - 2(2) = w + 4$
Height of open box: 2
Use the equation $V = lwh$.

Solution
$$V = lwh$$
$$256 = (w+4)(w-4)(2)$$
$$256 = (w^2 - 16)(2)$$
$$256 = 2w^2 - 32$$
$$0 = 2w^2 - 288$$
$$0 = 2(w^2 - 144)$$
$$0 = (w+12)(w-12)$$
$$w + 12 = 0 \qquad w - 12 = 0$$
$$w = -12 \qquad w = 12$$
The width cannot be negative, so –12 is not a solution.
$w - 4 = 12 - 4 = 8$
$w + 4 = 12 + 4 = 16$
The width is 8 cm.
The length is 16 cm.
The height is 2 cm.

27. To find the year in which more than 150 million units are shipped, substitute 150 for $G(x)$ and solve for x.
Here $x = 1$ corresponds to 1997.
$$G(x) = \frac{1}{2}x^2 + 15.5x + 78$$
$$150 = \frac{1}{2}x^2 + 15.5x + 78$$
$$0 = \frac{1}{2}x^2 + 15.5x - 72$$
$$x = \frac{-b \pm \sqrt{b^2 - 4ac}}{2a}$$
$$x = \frac{-15.5 \pm \sqrt{(15.5)^2 - 4\left(\frac{1}{2}\right)(-72)}}{2\left(\frac{1}{2}\right)}$$
$$x = \frac{-15.5 \pm \sqrt{384.25}}{1}$$
$$x \approx 4.10 \text{ or } x \approx -35.10$$
The solution –35.10 is not possible.
Since $x = 1$ refers to 1997, $G(1)$ refers to the number of millions of units shipped in 1997. More than 150 million units will be shipped when $x > 4.1$. When $x = 5$, the year will be $1997 + 5 - 1 = 2001$.

29. Strategy
This is a geometry problem.
Use the Pythagorean formula $(a^2 + b^2 = c^2)$, with the legs being $a = 1.5$, $b = 3.5$, and the hypotenuse being $c = x + 1.5$.

Solution
$$a^2 + b^2 = c^2$$
$$(1.5)^2 + (3.5)^2 = (x+1.5)^2$$
$$14.5 = (x+1.5)^2$$
$$\pm 3.8 \approx x + 1.5$$
$$3.8 = x + 1.5 \qquad -3.8 = x + 1.5$$
$$2.3 = x \qquad -5.3 = x$$
The solution $x = -5.3$ is not possible since distance cannot be negative.
The bottom of the scoop of ice cream is 2.3 in. from the bottom of the cone.

Section 8.5

Concept Review 8.5

1. Sometimes true

The end points of $x^2 - 4 \le 0$ are included in the solution set.

3. Sometimes true

The solution set of $x^2 \le 0$ is $\{0\}$.

Objective 1 Exercises

3. $(x-4)(x+2) > 0$

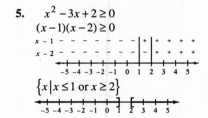

$\{x \mid x < -2 \text{ or } x > 4\}$

5. $x^2 - 3x + 2 \ge 0$
$(x-1)(x-2) \ge 0$

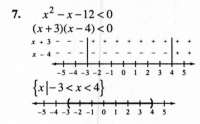

$\{x \mid x \le 1 \text{ or } x \ge 2\}$

7. $x^2 - x - 12 < 0$
$(x+3)(x-4) < 0$

$\{x \mid -3 < x < 4\}$

9. $(x-1)(x+2)(x-3) < 0$

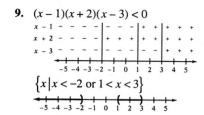

$\{x \mid x < -2 \text{ or } 1 < x < 3\}$

11. $x^2 - 16 > 0$
$(x-4)(x+4) > 0$

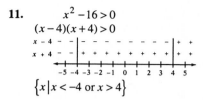

$\{x \mid x < -4 \text{ or } x > 4\}$

13. $x^2 - 4x + 4 > 0$
$(x-2)(x-2) > 0$

```
x - 2  - - - - - - - - | + + + +
x - 2  - - - - - - - - | + + + +
      -5 -4 -3 -2 -1 0 1 2 3 4 5
```

$\{x \mid x < 2 \text{ or } x > 2\}$

15. $x^2 - 9x \le 36$
$x^2 - 9x - 36 \le 0$
$(x+3)(x-12) \le 0$

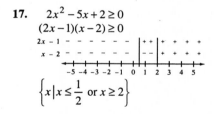

$\{x \mid -3 \le x \le 12\}$

17. $2x^2 - 5x + 2 \ge 0$
$(2x-1)(x-2) \ge 0$

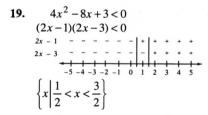

$\left\{x \mid x \le \dfrac{1}{2} \text{ or } x \ge 2\right\}$

19. $4x^2 - 8x + 3 < 0$
$(2x-1)(2x-3) < 0$

```
2x - 1  - - - - - - | + + + + +
2x - 3  - - - - - - | - + + + +
       -5 -4 -3 -2 -1 0 1 2 3 4 5
```

$\left\{x \mid \dfrac{1}{2} < x < \dfrac{3}{2}\right\}$

21. $(x-6)(x+3)(x-2) \le 0$

```
x - 6   - - - - - | - - | - - + +
x + 3   - - - | + + + + | + + + +
x - 2   - - - - - | - + | + + + +
       -10 -8 -6 -4 -2 0 2 4 6 8 10
```

$\{x \mid x \le -3 \text{ or } 2 \le x \le 6\}$

23. $(2x-1)(x-4)(2x+3) > 0$

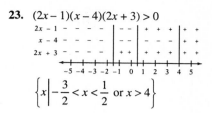

$\left\{x \mid -\dfrac{3}{2} < x < \dfrac{1}{2} \text{ or } x > 4\right\}$

25. $x^3 + 3x^2 - x - 3 \le 0$
$x^2(x+3) - 1(x+3) \le 0$
$(x+3)(x^2-1) \le 0$
$(x+3)(x+1)(x-1) \le 0$

```
x + 3   - - - | + + | + + | + + + +
x + 1   - - - - - | + + | + + + +
x - 1   - - - - - | - - | + + + +
       -5 -4 -3 -2 -1 0 1 2 3 4 5
```

$\{x \mid x \le -3 \text{ or } -1 \le x \le 1\}$

27. $x^3 - x^2 - 4x + 4 \ge 0$
$x^2(x-1) - 4(x-1) \ge 0$
$(x-1)(x^2-4) \ge 0$
$(x-1)(x+2)(x-2) \ge 0$

```
x - 1   - - - - - | - + | + + + +
x + 2   - - | + + | + + | + + + +
x - 2   - - - - - | - - | + + + +
       -5 -4 -3 -2 -1 0 1 2 3 4 5
```

$\{x \mid -2 \le x \le 1 \text{ or } x \ge 2\}$

Objective 2 Exercises

29. $\dfrac{x-4}{x+2} > 0$

```
x - 4   - - - | - - - - - | + +
x + 2   - - - | + + + + + | + +
       -5 -4 -3 -2 -1 0 1 2 3 4 5
```

$\{x \mid x < -2 \text{ or } x > 4\}$

31. $\dfrac{x-3}{x+1} \le 0$

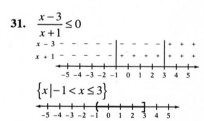

$\{x \mid -1 < x \le 3\}$

33. $\dfrac{(x-1)(x+2)}{x-3} \le 0$

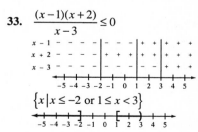

$\{x \mid x \le -2 \text{ or } 1 \le x < 3\}$

35. $\dfrac{3x}{x-2} > 1$

$\dfrac{3x}{x-2} - 1 > 0$

$\dfrac{3x}{x-2} - \dfrac{x-2}{x-2} > 0$

$\dfrac{2x+2}{x-2} > 0$

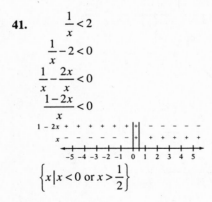

$\{x \mid x < -1 \text{ or } x > 2\}$

37. $\dfrac{2}{x+1} \geq 2$

$\dfrac{2}{x+1} - 2 \geq 0$

$\dfrac{2}{x+1} - \dfrac{2x+2}{x+1} \geq 0$

$\dfrac{-2x}{x+1} \geq 0$

$\{x \mid -1 < x \leq 0\}$

39. $\dfrac{x}{(x-1)(x+2)} \geq 0$

$\{x \mid -2 < x \leq 0 \text{ or } x > 1\}$

41. $\dfrac{1}{x} < 2$

$\dfrac{1}{x} - 2 < 0$

$\dfrac{1}{x} - \dfrac{2x}{x} < 0$

$\dfrac{1-2x}{x} < 0$

$\left\{x \mid x < 0 \text{ or } x > \dfrac{1}{2}\right\}$

Applying Concepts 8.5

43. $(x+2)(x-3)(x+1)(x+4) > 0$

$\{x \mid x < -4 \text{ or } -2 < x < -1 \text{ or } x > 3\}$

45. $(x^2+2x-8)(x^2-2x-3) < 0$

$(x+4)(x-2)(x-3)(x+1) < 0$

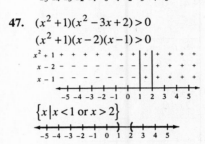

$\{x \mid -4 < x < -1 \text{ or } 2 < x < 3\}$

47. $(x^2+1)(x^2-3x+2) > 0$

$(x^2+1)(x-2)(x-1) > 0$

$\{x \mid x < 1 \text{ or } x > 2\}$

49. $\dfrac{x^2(3-x)(2x+1)}{(x+4)(x+2)} \geq 0$

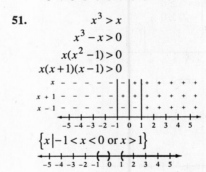

$\left\{x \mid -4 < x < -2 \text{ or } -\dfrac{1}{2} \leq x \leq 3\right\}$

51. $x^3 > x$

$x^3 - x > 0$

$x(x^2-1) > 0$

$x(x+1)(x-1) > 0$

$\{x \mid -1 < x < 0 \text{ or } x > 1\}$

53.
$$3 - \frac{1}{x} \le 2$$
$$3x - \frac{1}{x} - 2 \le 0$$
$$\frac{3x^2 - 1 - 2x}{x} \le 0$$
$$\frac{3x^2 - 2x - 1}{x} \le 0$$
$$\frac{(3x+1)(x-1)}{x} \le 0$$

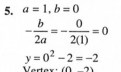

$$\left\{ x \mid x \le -\frac{1}{3} \text{ or } 0 < x \le 1 \right\}$$

Section 8.6

Concept Review 8.6

1. Sometimes true
   If the vertex is on an axis, the axis of symmetry will intersect the origin.

3. Sometimes true
   A parabola may have one, two or no x-intercepts.

5. Always true

Objective 1 Exercises

3. $a = 1, b = 0$
$-\frac{b}{2a} = -\frac{0}{2(1)} = 0$
$y = 0^2 = 0$
Vertex: (0, 0)
Axis of symmetry: $x = 0$

5. $a = 1, b = 0$
$-\frac{b}{2a} = -\frac{0}{2(1)} = 0$
$y = 0^2 - 2 = -2$
Vertex: (0, -2)
Axis of symmetry: $x = 0$

7. $a = -1, b = 0$
$-\frac{b}{2a} = -\frac{0}{2(-1)} = 0$
$y = -0^2 + 3 = 3$
Vertex: (0, 3)
Axis of symmetry: $x = 0$

9. $a = \frac{1}{2}, b = 0$
$-\frac{b}{2a} = -\frac{0}{2\left(\frac{1}{2}\right)} = 0$
$y = \frac{1}{2}(0)^2 = 0$
Vertex: (0, 0)
Axis of symmetry: $x = 0$

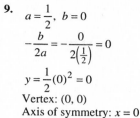 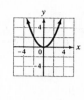

11. $a = 2, b = 0$
$-\frac{b}{2a} = -\frac{0}{2(2)} = 0$
$y = 2(0)^2 - 1 = -1$
Vertex: (0, -1)
Axis of symmetry: $x = 0$

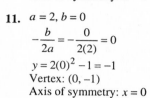

13. $a = 1, b = -2$
$-\frac{b}{2a} = -\frac{-2}{2(1)} = 1$
$y = 1^2 - 2(1) = -1$
Vertex: (1, -1)
Axis of symmetry: $x = 1$

15. $a = -2, b = 4$
$-\frac{b}{2a} = -\frac{4}{2(-2)} = 1$
$y = -2(1)^2 + 4(1) = 2$
Vertex: (1, 2)
Axis of symmetry: $x = 1$

17. $a = 1, b = -1$
$-\frac{b}{2a} = -\frac{-1}{2(1)} = \frac{1}{2}$
$y = \left(\frac{1}{2}\right)^2 - \frac{1}{2} - 2 = -\frac{9}{4}$
Vertex: $\left(\frac{1}{2}, -\frac{9}{4}\right)$
Axis of symmetry: $x = \frac{1}{2}$

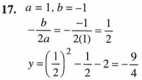

19. $a = 2, b = -1$
$-\frac{b}{2a} = -\frac{-1}{2(2)} = \frac{1}{4}$
$y = 2\left(\frac{1}{4}\right)^2 - \frac{1}{4} - 5 = -\frac{41}{8}$
Vertex: $\left(\frac{1}{4}, -\frac{41}{8}\right)$
Axis of symmetry: $x = \frac{1}{4}$

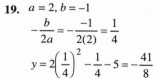

21.

Domain: $\{x \mid x \in \text{real numbers}\}$
Range: $\{y \mid y \ge -7\}$

23.

Domain: $\{x \mid x \in \text{real numbers}\}$

Range: $\left\{y \mid y \leq \dfrac{25}{8}\right\}$

25.

Domain: $\{x \mid x \in \text{real numbers}\}$

Range: $\{y \mid y \geq 0\}$

27.

Domain: $\{x \mid x \in \text{real numbers}\}$

Range: $\{y \mid y \geq -7\}$

29.

Domain: $\{x \mid x \in \text{real numbers}\}$

Range: $\{y \mid y \leq -1\}$

Objective 2 Exercises

33. $y = x^2 - 4$
$0 = x^2 - 4$
$0 = (x-2)(x+2)$
$x - 2 = 0 \quad\quad x + 2 = 0$
$\quad x = 2 \quad\quad\quad x = -2$
The x-intercepts are $(2, 0)$ and $(-2, 0)$.

35. $y = 2x^2 - 4x$
$0 = 2x^2 - 4x$
$0 = 2x(x - 2)$
$2x = 0 \quad\quad x - 2 = 0$
$\quad x = 0 \quad\quad\quad x = 2$
The x-intercepts are $(0, 0)$ and $(2, 0)$.

37. $y = x^2 - x - 2$
$0 = x^2 - x - 2$
$0 = (x - 2)(x + 1)$
$x - 2 = 0 \quad\quad x + 1 = 0$
$\quad x = 2 \quad\quad\quad x = -1$
The x-intercepts are $(2, 0)$ and $(-1, 0)$.

39. $y = 2x^2 - 5x - 3$
$0 = 2x^2 - 5x - 3$
$0 = (2x + 1)(x - 3)$
$2x + 1 = 0 \quad\quad x - 3 = 0$
$\quad 2x = -1 \quad\quad\quad x = 3$
$\quad\quad x = -\dfrac{1}{2}$

The x-intercepts are $\left(-\dfrac{1}{2},\ 0\right)$ and $(3, 0)$.

41. $y = 3x^2 - 19x - 14$
$0 = 3x^2 - 19x - 14$
$0 = (3x + 2)(x - 7)$
$3x + 2 = 0 \quad\quad x - 7 = 0$
$\quad 3x = -2 \quad\quad\quad x = 7$
$\quad\quad x = -\dfrac{2}{3}$

The x-intercepts are $\left(-\dfrac{2}{3},\ 0\right)$ and $(7, 0)$.

43. $y = 3x^2 - 19x + 20$
$0 = 3x^2 - 19x + 20$
$0 = (3x - 4)(x - 5)$
$3x - 4 = 0 \quad\quad x - 5 = 0$
$\quad 3x = 4 \quad\quad\quad x = 5$
$\quad\quad x = \dfrac{4}{3}$

The x-intercepts are $\left(\dfrac{4}{3},\ 0\right)$ and $(5, 0)$.

45. $y = 9x^2 - 12x + 4$
$0 = 9x^2 - 12x + 4$
$0 = (3x - 2)(3x - 2)$
$3x - 2 = 0 \quad\quad 3x - 2 = 0$
$\quad 3x = 2 \quad\quad\quad 3x = 2$
$\quad\quad x = \dfrac{2}{3} \quad\quad\quad x = \dfrac{2}{3}$

The x-intercept is $\left(\dfrac{2}{3},\ 0\right)$.

47. $y = 9x^2 - 2$
$0 = 9x^2 - 2$
$2 = 9x^2$
$\dfrac{2}{9} = x^2$
$\sqrt{\dfrac{2}{9}} = \sqrt{x^2}$
$\pm \dfrac{\sqrt{2}}{3} = x$

The x-intercepts are $\left(\dfrac{\sqrt{2}}{3},\ 0\right)$ and $\left(-\dfrac{\sqrt{2}}{3},\ 0\right)$.

49. $y = 4x^2 - 4x - 15$
$0 = 4x^2 - 4x - 15$
$0 = (2x+3)(2x-5)$
$2x+3 = 0 \quad (2x-5) = 0$
$2x = -3 \quad\quad 2x = 5$
$x = -\dfrac{3}{2} \quad\quad x = \dfrac{5}{2}$

The x-intercepts are $\left(-\dfrac{3}{2},\ 0\right)$ and $\left(\dfrac{5}{2}, 0\right)$.

51. $y = x^2 + 4x - 3$
$0 = x^2 + 4x - 3$
$a = 1, b = 4, c = -3$
$x = \dfrac{-b \pm \sqrt{b^2 - 4ac}}{2a}$
$= \dfrac{-4 \pm \sqrt{4^2 - 4(1)(-3)}}{2(1)}$
$= \dfrac{-4 \pm \sqrt{16+12}}{2}$
$= \dfrac{-4 \pm \sqrt{28}}{2}$
$= \dfrac{-4 \pm 2\sqrt{7}}{2}$
$= -2 \pm \sqrt{7}$

The x-intercepts are $(-2+\sqrt{7},\ 0)$ and $(-2-\sqrt{7},\ 0)$.

53. $y = -x^2 - 4x - 5$
$0 = -x^2 - 4x - 5$
$a = -1, b = -4, c = -5$
$x = \dfrac{-b \pm \sqrt{b^2 - 4ac}}{2a}$
$= \dfrac{-(-4) \pm \sqrt{(-4)^2 - 4(-1)(-5)}}{2(-1)}$
$= \dfrac{4 \pm \sqrt{16 - 20}}{-2}$
$= \dfrac{4 \pm \sqrt{-4}}{-2}$
$= \dfrac{4 \pm 2i}{-2}$
$= -2 \pm i$

The equation has no real solutions.
The parabola has no x-intercepts.

55. $y = -x^2 - 2x + 1$
$0 = -x^2 - 2x + 1$
$0 = x^2 + 2x - 1$
$a = 1, b = 2, c = -1$
$x = \dfrac{-b \pm \sqrt{b^2 - 4ac}}{2a}$
$= \dfrac{-2 \pm \sqrt{2^2 - 4(1)(-1)}}{2(1)}$
$= \dfrac{-2 \pm \sqrt{4+4}}{2}$
$= \dfrac{-2 \pm \sqrt{8}}{2}$
$= \dfrac{-2 \pm 2\sqrt{2}}{2}$
$= -1 \pm \sqrt{2}$

The x-intercepts are $(-1+\sqrt{2},\ 0)$ and $(-1-\sqrt{2},\ 0)$.

57.
To the nearest tenth, the zeros of $f(x) = x^2 + 3x - 1$ are -3.3 and 0.3.

59.
To the nearest tenth, the zeros of $f(x) = 2x^2 - 3x - 7$ are -1.3 and 2.8.

61.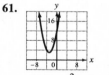
$f(x) = x^2 + 6x + 12$ has no x-intercepts.

63. $y = 2x^2 + x + 1$
$a = 2, b = 1, c = 1$
$b^2 - 4ac$
$1^2 - 4(2)(1) = 1 - 8 = -7$
$-7 < 0$
Since the discriminant is less than zero, the parabola has no x-intercepts.

65. $y = -x^2 - x + 3$
$a = -1, b = -1, c = 3$
$b^2 - 4ac$
$(-1)^2 - 4(-1)(3) = 1 + 12 = 13$
$13 > 0$
Since the discriminant is greater than zero, the parabola has two x-intercepts.

67. $y = x^2 - 8x + 16$
$a = 1, b = -8, c = 16$
$b^2 - 4ac$
$(-8)^2 - 4(1)(16) = 64 - 64 = 0$
Since the discriminant is equal to zero, the parabola has one x-intercept.

69. $y = -3x^2 - x - 2$
$a = -3, b = -1, c = -2$
$b^2 - 4ac$
$(-1)^2 - 4(-3)(-2) = 1 - 24 = -23$
$-23 < 0$
Since the discriminant is less than zero, the parabola has no x-intercepts.

71. $y = 4x^2 - x - 2$
$a = 4, b = -1, c = -2$
$b^2 - 4ac$
$(-1)^2 - 4(4)(-2) = 1 + 32 = 33$
$33 > 0$
Since the discriminant is greater than zero, the parabola has two x-intercepts.

73. $y = -2x^2 - x - 5$
$a = -2, b = -1, c = -5$
$b^2 - 4ac$
$(-1)^2 - 4(-2)(-5) = 1 - 40 = -39$
$-39 < 0$
Since the discriminant is less than zero, the parabola has no x-intercepts.

75. $y = x^2 + 8x + 16$
$a = 1, b = 8, c = 16$
$b^2 - 4ac$
$8^2 - 4(1)(16) = 64 - 64 = 0$
Since the discriminant is equal to zero, the parabola has one x-intercept.

77. $y = x^2 + x - 3$
$a = 1, b = 1, c = -3$
$b^2 - 4ac$
$1^2 - 4(1)(-3) = 1 + 12 = 13$
$13 > 0$
Since the discriminant is greater than zero, the parabola has two x-intercepts.

Applying Concepts 8.6

79. $y = x^2 + 2x + k$
$1 = (-3)^2 + 2(-3) + k$
$1 = 9 - 6 + k$
$1 = 3 + k$
$-2 = k$
The value of k is -2.

81. $y = 3x^2 + kx - 6$
$4 = 3(-2)^2 + k(-2) - 6$
$4 = 3(4) - 2k - 6$
$4 = 12 - 2k - 6$
$4 = 6 - 2k$
$0 = 2 - 2k$
$2k = 2$
$k = 1$
The value of k is 1.

83. Substitute 5 for y in the equation and solve for x.
$5 = 3x^2 - 2x - 1$
$0 = 3x^2 - 2x - 6$
$a = 3, b = -2, c = -6$
$x = \dfrac{-b \pm \sqrt{b^2 - 4ac}}{2a}$
$= \dfrac{-(-2) \pm \sqrt{(-2)^2 - 4(3)(-6)}}{2(3)}$
$= \dfrac{2 \pm \sqrt{4 + 72}}{6}$
$= \dfrac{2 \pm \sqrt{76}}{6}$
$= \dfrac{2 \pm 2\sqrt{19}}{6}$
$= \dfrac{1 \pm \sqrt{19}}{3}$
$\dfrac{1 - \sqrt{19}}{3}$ is not a possible solution because $\left(\dfrac{1 - \sqrt{19}}{3}, 5\right)$ is in quadrant II.
The solution is $\dfrac{1 + \sqrt{19}}{3}$.

85. When the coefficient of x^2 is increased, the parabola becomes narrower.

87. When the constant term is increased, the parabola is higher on the rectangular coordinate system.

89. $y = x^2 - 4x + 7$
$= (x^2 - 4x) + 7$
$= (x^2 - 4x + 4) - 4 + 7$
$= (x - 2)^2 + 3$
$x - 2 = 0$
$x = 2$
$y = (x - 2)^2 + 3$
$= (2 - 2)^2 + 3$
$= 3$
The vertex is (2, 3).

91. $y = x^2 + x + 2$
$= (x^2 + x) + 2$
$= \left(x^2 + x + \dfrac{1}{4}\right) - \dfrac{1}{4} + 2$
$= \left(x + \dfrac{1}{2}\right)^2 + \dfrac{7}{4}$

$x + \dfrac{1}{2} = 0$
$x = -\dfrac{1}{2}$

$y = \left(x + \dfrac{1}{2}\right)^2 + \dfrac{7}{4}$
$= \left(-\dfrac{1}{2} + \dfrac{1}{2}\right)^2 + \dfrac{7}{4}$
$= \dfrac{7}{4}$

The vertex is $\left(-\dfrac{1}{2}, \dfrac{7}{4}\right)$.

93. $y = a(x - h)^2 + k$
Vertex (1, 2): $h = 1$, $k = 2$
$P(2, 5)$: $x = 2$, $y = 5$
Substitute into the equation:
$5 = a(2 - 1)^2 + 2$
$5 = a(1)^2 + 2$
$5 = a + 2$
$3 = a$
$y = 3(x - 1)^2 + 2$
$y = 3(x^2 - 2x + 1) + 2$
$y = 3x^2 - 6x + 3 + 2$
$y = 3x^2 - 6x + 5$

Section 8.7

Concept Review 8.7

1. Sometimes true
   If the parabola opens down, it has a maximum value. The range of the parabola would be from negative infinity to the maximum value. Thus, there would be no minimum value.

3. Always true

5. Sometimes true
   If a parabola opens down and the parabola is below the x-axis, the maximum value of the parabola is a negative number.

Objective 1 Exercises

3. $f(x) = x^2 - 2x + 3$
$x = -\dfrac{b}{2a} = -\dfrac{-2}{2(1)} = 1$
$f(x) = x^2 - 2x + 3$
$f(1) = 1^2 - 2(1) + 3$
$= 1 - 2 + 3 = 2$
Since a is positive, the function has a minimum value. The minimum value of the function is 2.

5. $f(x) = -2x^2 + 4x - 3$
$x = -\dfrac{b}{2a} = -\dfrac{4}{2(-2)} = 1$
$f(x) = -2x^2 + 4x - 3$
$f(1) = -2(1)^2 + 4(1) - 3$
$= -2 + 4 - 3 = -1$
Since a is negative, the function has a maximum value. The maximum value of the function is -1.

7. $f(x) = 2x^2 + 4x$
$x = -\dfrac{b}{2a} = -\dfrac{4}{2(2)} = -1$
$f(x) = 2x^2 + 4x$
$f(-1) = 2(-1)^2 + 4(-1) = 2 - 4 = -2$
Since a is positive, the function has a minimum value. The minimum value of the function is -2.

9. $f(x) = -2x^2 + 4x - 5$
$x = -\dfrac{b}{2a} = -\dfrac{4}{2(-2)} = 1$
$f(x) = -2x^2 + 4x - 5$
$f(1) = -2(1)^2 + 4(1) - 5 = -2 + 4 - 5 = -3$
Since a is negative, the function has a maximum value. The maximum value of the function is -3.

11. $f(x) = 2x^2 + 3x - 8$
$x = -\dfrac{b}{2a} = -\dfrac{3}{2(2)} = -\dfrac{3}{4}$
$f(x) = 2x^2 + 3x - 8$
$f\left(-\dfrac{3}{4}\right) = 2\left(-\dfrac{3}{4}\right)^2 + 3\left(-\dfrac{3}{4}\right) - 8$
$= \dfrac{9}{8} - \dfrac{9}{4} - 8$
$= -\dfrac{73}{8}$
Since a is positive, the function has a minimum value. The minimum value of the function is $-\dfrac{73}{8}$.

13. $f(x) = 3x^2 + 3x - 2$
$$x = -\frac{b}{2a} = -\frac{3}{2(3)} = -\frac{1}{2}$$
$$f(x) = 3x^2 + 3x - 2$$
$$f\left(-\frac{1}{2}\right) = 3\left(-\frac{1}{2}\right)^2 + 3\left(-\frac{1}{2}\right) - 2$$
$$= \frac{3}{4} - \frac{3}{2} - 2$$
$$= -\frac{11}{4}$$
Since a is positive, the function has a minimum value. The minimum value of the function is $-\frac{11}{4}$.

15. $f(x) = -3x^2 + 4x - 2$
$$x = -\frac{b}{2a} = -\frac{4}{2(-3)} = \frac{2}{3}$$
$$f(x) = -3x^2 + 4x - 2$$
$$f\left(\frac{2}{3}\right) = -3\left(\frac{2}{3}\right)^2 + 4\left(\frac{2}{3}\right) - 2$$
$$= -\frac{4}{3} + \frac{8}{3} - 2$$
$$= -\frac{2}{3}$$
Since a is negative, the function has a maximum value. The maximum value of the function is $-\frac{2}{3}$.

17. $f(x) = 3x^2 + 5x + 2$
$$x = -\frac{b}{2a} = -\frac{5}{2(3)} = -\frac{5}{6}$$
$$f(x) = 3x^2 + 5x + 2$$
$$f\left(-\frac{5}{6}\right) = 3\left(-\frac{5}{6}\right)^2 + 5\left(-\frac{5}{6}\right) + 2$$
$$= \frac{25}{12} - \frac{25}{6} + 2$$
$$= -\frac{1}{12}$$
Since a is positive, the function has a minimum value. The minimum value of the function is $-\frac{1}{12}$.

19. **Strategy**
To find the time it takes the ball to reach its maximum height, find the t-coordinate of the vertex.
To find the maximum height, evaluate the function at the t-coordinate of the vertex.

Solution
$$t = -\frac{b}{2a} = -\frac{80}{2(-16)} = 2.5$$
The ball reaches its maximum height in 2.5 s.
$$s(t) = -16t^2 + 80t + 50$$
$$s(2.5) = -16(2.5)^2 + 80(2.5) + 50$$
$$= -100 + 200 + 50$$
$$= 150$$
The maximum height is 150 ft.

21. **Strategy**
To find the number of lenses to minimize the cost, find the x-coordinate of the vertex.

Solution
$$x = -\frac{b}{2a} = -\frac{-20}{2(0.1)} = 100$$
To minimize cost, 100 lenses should be produced.

23. **Strategy**
To find the distance from one end of the bridge where the cable is at its minimum height, find the x-coordinate of the vertex.
To find the minimum height, evaluate the function at the x-coordinate of the vertex.

Solution
$$x = -\frac{b}{2a} = -\frac{-0.8}{2(0.025)} = 16$$
The cable is at its minimum height 16 ft from one end of the bridge.
$$h(x) = 0.025x^2 - 0.8x + 25$$
$$h(16) = 0.025(16)^2 - 0.8(16) + 25$$
$$= 6.4 - 12.8 + 25$$
$$= 18.6$$
The minimum height is 18.6 ft.

25. **Strategy**
To find out whether Saul will be able to catch the orange, find the maximum height of the orange. Find the t-coordinate of the vertex, and evaluate the function at that value.

Solution
$$h(t) = -16t^2 + 32t + 4$$
$$t = -\frac{b}{2a} = -\frac{32}{2(-16)} = 1$$
$$h(1) = -16(1)^2 + 32(1) + 4 = -16 + 32 + 4 = 20$$
The maximum height of the orange is 20 ft. Since Saul's arms are only 18 ft above ground, the orange will be high enough for Saul to catch.

27. Strategy
To find the maximum height, find the t-coordinate of the vertex and evaluate the function at that value.

Solution
$h(t) = -16t^2 + 90t + 15$
$t = -\dfrac{b}{2a} = -\dfrac{90}{2(-16)} = 2.8125$
$h(2.8125) = -16(2.8125)^2 + 90(2.8125) + 15$
$= 141.6$
The maximum height of the waterspout is 141.6 ft.

29. Strategy
To find the speed that will yield the maximum fuel efficiency, find the v-coordinate of the vertex. To find the maximum fuel efficiency, evaluate the function at the v-coordinate of the vertex.

Solution
$v = -\dfrac{b}{2a} = -\dfrac{1.476}{2(-0.018)} = 41$
The speed that will yield the maximum fuel efficiency is 41 mph.
$E(v) = -0.018v^2 + 1.476v + 3.4$
$E(41) = -0.018(41)^2 + 1.476(41) + 3.4$
$= 33.658$
The maximum fuel efficiency is 33.658 mi/gal.

Applying Concepts 8.7

31. The minimum value of the function $f(x) = x^4 - 2x^2 + 4$ is 3.0.

33. The maximum value of the function $f(x) = -x^6 + x^4 - x^3 + x$ is 0.5.

35. a.

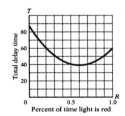

b. R is the percent of time that the light is red in the horizontal direction. Since $1 = 100\%$, and the percent of time the light is red in the horizontal direction cannot be more than 100%, R cannot be greater than 1. Since $0 = 0\%$, and the percent of time the light is red in the horizontal direction cannot be less than 0%, R cannot be less than 0. Therefore, the graph is drawn only for $0 \le R \le 1$.

c. $T = \left(\dfrac{100 + 150}{2}\right)R^2 + [(0.08(100)$
$\qquad -1.08(150)]R + 0.58(150)$
$T = 125R^2 - 154R + 87$
$R = -\dfrac{b}{2a} = -\dfrac{-154}{2(125)} = 0.616$

The traffic light should remain red in the horizontal direction approximately 62% of the time.

Focus on Problem Solving

1. The sum of the first n odd integers is n^2. If this conjecture is true, then the sum of
$1 + 3 + 5 + 7 + 9 + 11 + 13 + 15 = 8^2 = 64$.

2. The next figure should be ⌐⌐.

3. All even numbers are divisible by 2. Because 14,386 is an even number, it is divisible by 2.

4. Inductive reasoning

5. Deductive reasoning

6. Inductive reasoning proceeds from patterns or particular facts of individual cases to a general conjecture that may or may not be true. Deductive reasoning uses established facts, definitions, and general principles to prove particular statements.

Projects and Group Activities

Completing the Square

Solve $x^2 + 4x = 12$ by completing the square.
Begin with a line of unknown length, x, and form a square with x as a side.
Draw a rectangle with width x and length 2 at the right of the square and another such rectangle below the square.
Complete the square with sides $x + 2$.
The area of the large square is $(x+2)^2$.
From the diagram, we see that
$x^2 + 4x + 4 = (x+2)^2$
From the original equation, $x^2 + 4x = 12$. Thus
$12 + 4 = (x+2)^2$
$\quad 16 = (x+2)^2$
$\quad\; 4 = x + 2$
$\quad\; 2 = x$
The positive solution is 2.

Chapter Review Exercises

1. $2x^2 - 3x = 0$
$x(2x - 3) = 0$
$x = 0 \qquad 2x - 3 = 0$
$\qquad\qquad 2x = 3$
$\qquad\qquad x = \dfrac{3}{2}$

The solutions are 0 and $\dfrac{3}{2}$.

2. $6x^2 + 9xc = 6c^2$
$6x^2 + 9xc - 6c^2 = 0$
$3(2x^2 + 3cx - 2c^2) = 0$
$3(2x - c)(x + 2c) = 0$
$2x - c = 0 \qquad x + 2c = 0$
$2x = c \qquad\quad x = -2c$
$x = \dfrac{c}{2}$

The solutions are $\dfrac{c}{2}$ and $-2c$.

3. $x^2 = 48$
$\sqrt{x^2} = \sqrt{48}$
$x = \pm\sqrt{48} = \pm 4\sqrt{3}$

The solutions are $4\sqrt{3}$ and $-4\sqrt{3}$.

4. $\left(x + \dfrac{1}{2}\right)^2 + 4 = 0$
$\left(x + \dfrac{1}{2}\right)^2 = -4$
$\sqrt{\left(x + \dfrac{1}{2}\right)^2} = \sqrt{-4}$
$x + \dfrac{1}{2} = \pm\sqrt{-4} = \pm 2i$

$x + \dfrac{1}{2} = 2i \qquad\quad x + \dfrac{1}{2} = -2i$
$x = -\dfrac{1}{2} + 2i \qquad x = -\dfrac{1}{2} - 2i$

The solutions are $-\dfrac{1}{2} + 2i$ and $-\dfrac{1}{2} - 2i$.

5. $-\dfrac{b}{2a} = -\dfrac{-7}{2(1)} = \dfrac{7}{2}$
$f(x) = x^2 - 7x + 8$
$f\left(\dfrac{7}{2}\right) = \left(\dfrac{7}{2}\right)^2 - 7\left(\dfrac{7}{2}\right) + 8$
$= \dfrac{49}{4} - \dfrac{49}{2} + 8$
$= -\dfrac{17}{4}$

The minimum value of the function is $-\dfrac{17}{4}$.

6. $-\dfrac{b}{2a} = -\dfrac{4}{2(-2)} = 1$
$f(x) = -2x^2 + 4x + 1$
$f(1) = -2(1)^2 + 4(1) + 1$
$= -2 + 4 + 1$
$= 3$

The maximum value of the function is 3.

7. $(x - r_1)(x - r_2) = 0$
$\left(x - \dfrac{1}{3}\right)[x - (-3)] = 0$
$\left(x - \dfrac{1}{3}\right)(x + 3) = 0$
$x^2 + \dfrac{8}{3}x - 1 = 0$
$3\left(x^2 + \dfrac{8}{3}x - 1\right) = 3 \cdot 0$
$3x^2 + 8x - 3 = 0$

8. $2x^2 + 9x = 5$
$2x^2 + 9x - 5 = 0$
$(2x - 1)(x + 5) = 0$
$2x - 1 = 0 \qquad x + 5 = 0$
$2x = 1 \qquad\quad x = -5$
$x = \dfrac{1}{2}$

The solutions are $\dfrac{1}{2}$ and -5.

9. $2(x + 1)^2 - 36 = 0$
$2(x + 1)^2 = 36$
$(x + 1)^2 = 18$
$\sqrt{(x + 1)^2} = \sqrt{18}$
$x + 1 = \pm 3\sqrt{2}$
$x + 1 = 3\sqrt{2} \qquad\quad x + 1 = -3\sqrt{2}$
$x = -1 + 3\sqrt{2} \qquad x = -1 - 3\sqrt{2}$

The solutions are $-1 + 3\sqrt{2}$ and $-1 - 3\sqrt{2}$.

10. $x^2 + 6x + 10 = 0$
$a = 1, b = 6, c = 10$
$x = \dfrac{-b \pm \sqrt{b^2 - 4ac}}{2a}$
$= \dfrac{-6 \pm \sqrt{6^2 - 4(1)(10)}}{2(1)}$
$= \dfrac{-6 \pm \sqrt{36 - 40}}{2}$
$= \dfrac{-6 \pm \sqrt{-4}}{2}$
$= \dfrac{-6 \pm 2i}{2}$
$= -3 \pm i$

The solutions are $-3 + i$ and $-3 - i$.

11.
$$\frac{2}{x-4} + 3 = \frac{x}{2x-3}$$
$$(x-4)(2x-3)\left(\frac{2}{x-4} + 3\right) = (x-4)(2x-3)\frac{x}{2x-3}$$
$$(2x-3)2 + 3(x-4)(2x-3) = (x-4)x$$
$$4x - 6 + 3(2x^2 - 11x + 12) = x^2 - 4x$$
$$4x - 6 + 6x^2 - 33x + 36 = x^2 - 4x$$
$$6x^2 - 29x + 30 = x^2 - 4x$$
$$5x^2 - 25x + 30 = 0$$
$$5(x^2 - 5x + 6) = 0$$
$$5(x-2)(x-3) = 0$$
$$x - 2 = 0 \qquad x - 3 = 0$$
$$x = 2 \qquad x = 3$$
The solutions are 2 and 3.

12.
$$x^4 - 6x^2 + 8 = 0$$
$$(x^2)^2 - 6(x^2) + 8 = 0$$
$$u^2 - 6u + 8 = 0$$
$$(u-2)(u-4) = 0$$
$$u - 2 = 0 \qquad u - 4 = 0$$
$$u = 2 \qquad u = 4$$
Replace u by x^2.
$$x^2 = 2 \qquad x^2 = 4$$
$$\sqrt{x^2} = \sqrt{2} \qquad \sqrt{x^2} = \sqrt{4}$$
$$x = \pm\sqrt{2} \qquad x = \pm 2$$
The solutions are $\sqrt{2}$, $-\sqrt{2}$, 2, and -2.

13.
$$\sqrt{2x-1} + \sqrt{2x} = 3$$
$$\sqrt{2x-1} = 3 - \sqrt{2x}$$
$$(\sqrt{2x-1})^2 = (3 - \sqrt{2x})^2$$
$$2x - 1 = 9 - 6\sqrt{2x} + 2x$$
$$-10 = -6\sqrt{2x}$$
$$\frac{-10}{-6} = \frac{-6\sqrt{2x}}{-6}$$
$$\frac{5}{3} = \sqrt{2x}$$
$$\left(\frac{5}{3}\right)^2 = (\sqrt{2x})^2$$
$$\frac{25}{9} = 2x$$
$$\frac{1}{2}\left(\frac{25}{9}\right) = \frac{1}{2}(2x)$$
$$\frac{25}{18} = x$$
The solution is $\frac{25}{18}$.

14.
$$2x^{2/3} + 3x^{1/3} - 2 = 0$$
$$2(x^{1/3})^2 + 3(x^{1/3}) - 2 = 0$$
$$2u^2 + 3u - 2 = 0$$
$$(2u - 1)(u + 2) = 0$$
$$2u - 1 = 0 \qquad u + 2 = 0$$
$$2u = 1 \qquad u = -2$$
$$u = \frac{1}{2}$$
Replace u by $x^{1/3}$.
$$x^{1/3} = \frac{1}{2} \qquad x^{1/3} = -2$$
$$(x^{1/3})^3 = \left(\frac{1}{2}\right)^3 \qquad (x^{1/3})^3 = (-2)^3$$
$$x = \frac{1}{8} \qquad x = -8$$
The solutions are $\frac{1}{8}$ and -8.

15.
$$\sqrt{3x-2} + 4 = 3x$$
$$\sqrt{3x-2} = 3x - 4$$
$$(\sqrt{3x-2})^2 = (3x-4)^2$$
$$3x - 2 = 9x^2 - 24x + 16$$
$$0 = 9x^2 - 27x + 18$$
$$0 = 9(x^2 - 3x + 2)$$
$$0 = 9(x-2)(x-1)$$
$$x - 2 = 0 \qquad x - 1 = 0$$
$$x = 2 \qquad x = 1$$
1 does not check as a solution.
The solution is 2.

16.
$$x^2 - 6x - 2 = 0$$
$$a = 1, b = -6, c = -2$$
$$x = \frac{-b \pm \sqrt{b^2 - 4ac}}{2a}$$
$$= \frac{6 \pm \sqrt{(-6)^2 - 4(1)(-2)}}{2(1)}$$
$$= \frac{6 \pm \sqrt{36 + 8}}{2}$$
$$= \frac{6 \pm \sqrt{44}}{2}$$
$$= \frac{6 \pm 2\sqrt{11}}{2}$$
$$= 3 \pm \sqrt{11}$$
The solutions are $3 + \sqrt{11}$ and $3 - \sqrt{11}$.

17.
$$\frac{2x}{x-4} + \frac{6}{x+1} = 11$$
$$(x-4)(x+1)\left(\frac{2x}{x-4} + \frac{6}{x+1}\right) = (x-4)(x+1)11$$
$$(x-4)(x+1)\frac{2x}{x-4} + (x-4)(x+1)\frac{6}{x+1} = 11(x-4)(x+1)$$
$$2x(x+1) + 6(x-4) = 11(x^2 - 3x - 4)$$
$$2x^2 + 2x + 6x - 24 = 11x^2 - 33x - 44$$
$$2x^2 + 8x - 24 = 11x^2 - 33x - 44$$
$$0 = 9x^2 - 41x - 20$$
$$0 = (9x + 4)(x - 5)$$

$9x + 4 = 0$ $x - 5 = 0$
$9x = -4$ $x = 5$
$x = -\frac{4}{9}$

The solutions are $-\frac{4}{9}$ and 5.

18. $2x^2 - 2x = 1$
$2x^2 - 2x - 1 = 0$
$a = 2, b = -2, c = -1$
$$x = \frac{-b \pm \sqrt{b^2 - 4ac}}{2a}$$
$$= \frac{2 \pm \sqrt{(-2)^2 - 4(2)(-1)}}{2(2)}$$
$$= \frac{2 \pm \sqrt{4 + 8}}{4}$$
$$= \frac{2 \pm \sqrt{12}}{4}$$
$$= \frac{2 \pm 2\sqrt{3}}{4}$$
$$= \frac{1 \pm \sqrt{3}}{2}$$

The solutions are $\frac{1 + \sqrt{3}}{2}$ and $\frac{1 - \sqrt{3}}{2}$.

19. $2x = 4 - 3\sqrt{x - 1}$
$2x - 4 = -3\sqrt{x - 1}$
$(2x - 4)^2 = \left(-3\sqrt{x - 1}\right)^2$
$4x^2 - 16x + 16 = 9(x - 1)$
$4x^2 - 16x + 16 = 9x - 9$
$4x^2 - 25x + 25 = 0$
$(4x - 5)(x - 5) = 0$

$4x - 5 = 0$ $x - 5 = 0$
$4x = 5$ $x = 5$
$x = \frac{5}{4}$

5 does not check as a solution. The solution is $\frac{5}{4}$.

20.
$$3x = \frac{9}{x-2}$$
$$3x(x-2) = \frac{9}{x-2}(x-2)$$
$$3x^2 - 6x = 9$$
$$3x^2 - 6x - 9 = 0$$
$$3(x^2 - 2x - 3) = 0$$
$$3(x-3)(x+1) = 0$$
$$x - 3 = 0 \qquad x + 1 = 0$$
$$x = 3 \qquad x = -1$$
The solutions are 3 and -1.

21.
$$\frac{3x+7}{x+2} + x = 3$$
$$(x+2)\left(\frac{3x+7}{x+2} + x\right) = (x+2)3$$
$$3x + 7 + x(x+2) = 3x + 6$$
$$3x + 7 + x^2 + 2x = 3x + 6$$
$$x^2 + 5x + 7 = 3x + 6$$
$$x^2 + 2x + 1 = 0$$
$$(x+1)^2 = 0$$
$$\sqrt{(x+1)^2} = \sqrt{0}$$
$$x + 1 = 0$$
$$x = -1$$
The solution is -1.

22.
$$\frac{x-2}{2x+3} - \frac{x-4}{x} = 2$$
$$x(2x+3)\left(\frac{x-2}{2x+3} - \frac{x-4}{x}\right) = x(2x+3)2$$
$$x(x-2) - (2x+3)(x-4) = 2x(2x+3)$$
$$x^2 - 2x - (2x^2 - 5x - 12) = 4x^2 + 6x$$
$$x^2 - 2x - 2x^2 + 5x + 12 = 4x^2 + 6x$$
$$-x^2 + 3x + 12 = 4x^2 + 6x$$
$$0 = 5x^2 + 3x - 12$$
$a = 5, \ b = 3, \ c = -12$
$$x = \frac{-b \pm \sqrt{b^2 - 4ac}}{2a}$$
$$= \frac{-3 \pm \sqrt{3^2 - 4(5)(-12)}}{2(5)}$$
$$= \frac{-3 \pm \sqrt{9 + 240}}{10}$$
$$= \frac{-3 \pm \sqrt{249}}{10}$$
The solutions are $\frac{-3 + \sqrt{249}}{10}$ and $\frac{-3 - \sqrt{249}}{10}$.

23.
$$1 - \frac{x+4}{2-x} = \frac{x-3}{x+2}$$
$$(x+2)(2-x)\left(1 - \frac{x+4}{2-x}\right) = (x+2)(2-x)\frac{x-3}{x+2}$$
$$(x+2)(2-x) - (x+2)(x+4) = (2-x)(x-3)$$
$$4 - x^2 - \left(x^2 + 6x + 8\right) = -x^2 + 5x - 6$$
$$4 - x^2 - x^2 - 6x - 8 = -x^2 + 5x - 6$$
$$-2x^2 - 6x - 4 = -x^2 + 5x - 6$$
$$0 = x^2 + 11x - 2$$

$a = 1$, $b = 11$, $c = -2$

$$x = \frac{-b \pm \sqrt{b^2 - 4ac}}{2a}$$
$$= \frac{-11 \pm \sqrt{11^2 - 4(1)(-2)}}{2(1)}$$
$$= \frac{-11 \pm \sqrt{121 + 8}}{2}$$
$$= \frac{-11 \pm \sqrt{129}}{2}$$

The solutions are $\frac{-11 + \sqrt{129}}{2}$ and $\frac{-11 - \sqrt{129}}{2}$.

24. $y = -x^2 + 6x - 5$

The axis of symmetry is the line with equation $x = -\frac{b}{2a}$.

$a = -1$, $b = 6$

$$x = -\frac{6}{2(-1)} = -\frac{6}{-2} = 3$$

The axis of symmetry is $x = 3$.

25. $y = -x^2 + 3x - 2$

The x-coordinate of the vertex is $-\frac{b}{2a}$.

$a = -1$, $b = 3$

$$-\frac{b}{2a} = -\frac{3}{2(-1)} = -\frac{3}{-2} = \frac{3}{2}$$

$$y = -x^2 + 3x - 2$$
$$= -\left(\frac{3}{2}\right)^2 + 3\left(\frac{3}{2}\right) - 2$$
$$= \frac{1}{4}$$

The vertex is $\left(\frac{3}{2}, \frac{1}{4}\right)$.

26. $y = -2x^2 + 2x - 3$

$a = -2$, $b = 2$, $c = -3$

$b^2 - 4ac$

$(2)^2 - 4(-2)(-3) = 4 - 24 = -20$

$-20 < 0$

Since the discriminant is less than zero, the parabola has no x-intercepts.

27. $y = 3x^2 - 2x - 4$
$a = 3, \ b = -2, \ c = -4$
$b^2 - 4ac$
$(-2)^2 - 4(3)(-4) = 4 + 48 = 52$
$50 > 0$
Since the discriminant is greater than zero, the parabola has two x-intercepts.

28. $y = 4x^2 + 12x + 4$
$0 = 4x^2 + 12x + 4$
$0 = 4(x^2 + 3x + 1)$
$a = 1, \ b = 3, \ c = 1$
$x = \dfrac{-b \pm \sqrt{b^2 - 4ac}}{2a}$
$= \dfrac{-3 \pm \sqrt{3^2 - 4(1)(1)}}{2(1)}$
$= \dfrac{-3 \pm \sqrt{9 - 4}}{2}$
$= \dfrac{-3 \pm \sqrt{5}}{2}$
The x-intercepts are $\left(\dfrac{-3 + \sqrt{5}}{2}, 0\right)$ and $\left(\dfrac{-3 - \sqrt{5}}{2}, 0\right)$.

29. $y = -2x^2 - 3x + 2$
$0 = -2x^2 - 3x + 2$
$0 = (-2x + 1)(x + 2)$
$-2x + 1 = 0 \qquad x + 2 = 0$
$-2x = -1 \qquad x = -2$
$x = \dfrac{1}{2}$
The x-intercepts are $\left(\dfrac{1}{2}, 0\right)$ and $(-2, 0)$.

30. $f(x) = 3x^2 + 2x + 2$
$0 = 3x^2 + 2x + 2$
$a = 3, \ b = 2, \ c = -2$
$x = \dfrac{-b \pm \sqrt{b^2 - 4ac}}{2a}$
$= \dfrac{-2 \pm \sqrt{2^2 - 4(3)(2)}}{2(3)}$
$= \dfrac{-2 \pm \sqrt{4 - 24}}{6}$
$= \dfrac{-2 \pm \sqrt{-20}}{6}$
$= \dfrac{-2 \pm 2i\sqrt{5}}{6}$
$= \dfrac{-1 \pm i\sqrt{5}}{3}$
The zeros are $-\dfrac{1}{3} + \dfrac{\sqrt{5}}{3}i$ and $-\dfrac{1}{3} - \dfrac{\sqrt{5}}{3}i$.

31. $(x + 3)(2x - 5) < 0$
$\left\{x \mid -3 < x < \dfrac{5}{2}\right\}$

32. $(x - 2)(x + 4)(2x + 3) \leq 0$
$\left\{x \mid x \leq -4 \text{ or } -\dfrac{3}{2} \leq x \leq 2\right\}$

33. $\dfrac{x - 2}{2x - 3} \geq 0$
$\left\{x \mid x < \dfrac{3}{2} \text{ or } x \geq 2\right\}$

34. $\dfrac{(2x - 1)(x + 3)}{x - 4} \leq 0$
$\left\{x \mid x \leq -3 \text{ or } \dfrac{1}{2} \leq x < 4\right\}$

35. $-\dfrac{b}{2a} = -\dfrac{2}{2(1)} = -1$
$y = (-1)^2 + 2(-1) - 4 = -5$
The vertex is $(-1, -5)$.
The axis of symmetry is $x = -1$.

The domain is $\{x \mid x \in \text{real numbers}\}$
The range is $\{y \mid y \geq -5\}$.

36. $-\dfrac{b}{2a} = -\dfrac{-2}{2(1)} = 1$
$y = 1^2 - 2(1) + 3 = 2$
The vertex is $(1, 2)$.
The axis of symmetry is $x = 1$.

Chapter 8: Quadratic Equations and Inequalities

37. Strategy
This is a geometry problem.
The width of the rectangle: x
The length of the rectangle: $2x + 2$
The area of the rectangle is 60 cm^2.
Use the equation for the area of the rectangle $(A = L \cdot W)$.

Solution
$A = L \cdot W$
$60 = x(2x + 2)$
$60 = 2x^2 + 2x$
$0 = 2x^2 + 2x - 60$
$0 = 2(x^2 + x - 30)$
$0 = 2(x + 6)(x - 5)$
$x + 6 = 0 \qquad x - 5 = 0$
$x = -6 \qquad x = 5$
Since the width cannot be negative, -6 cannot be a solution.
$2x + 2 = 2(5) + 2 = 10 + 2 = 12$
The width of the rectangle is 5 cm.
The length of the rectangle is 12 cm.

38. Strategy
This is an integer problem.
The first integer: x
The second consecutive even integer: $x + 2$
The third consecutive even integer: $x + 4$
The sum of the squares of the three consecutive integers is 56.
$x^2 + (x + 2)^2 + (x + 4)^2 = 56$

Solution
$x^2 + (x + 2)^2 + (x + 4)^2 = 56$
$x^2 + x^2 + 4x + 4 + x^2 + 8x + 16 = 56$
$3x^2 + 12x - 36 = 0$
$3(x^2 + 4 - 12) = 0$
$3(x + 6)(x - 2) = 0$
$x + 6 = 0 \qquad x - 2 = 0$
$x = -6 \qquad x = 2$
$x = 2, \; x + 2 = 4, \; x + 4 = 6$
$x = -6, \; x + 2 = -4, \; x + 4 = -2$
The integers are 2, 4, and 6 or -6, -4, and -2.

39. Strategy
This is a work problem.
Time for new computer to print payroll: x
Time for older computer to print payroll: $x + 12$

	Rate	Time	Part
New computer	$\dfrac{1}{x}$	8	$\dfrac{8}{x}$
Older computer	$\dfrac{1}{x+12}$	8	$\dfrac{8}{x+12}$

The sum of the parts of the task completed must be 1.
$\dfrac{8}{x} + \dfrac{8}{x+12} = 1$

Solution
$\dfrac{8}{x} + \dfrac{8}{x+12} = 1$
$x(x+12)\left(\dfrac{8}{x} + \dfrac{8}{x+12}\right) = x(x+12)(1)$
$8(x+12) + 8x = x(x+12)$
$8x + 96 + 8x = x^2 + 12x$
$16x + 96 = x^2 + 12x$
$0 = x^2 - 4x - 96$
$0 = (x - 12)(x + 8)$
$x - 12 = 0 \qquad x + 8 = 0$
$x = 12 \qquad x = -8$
The solution -8 is not possible, since time cannot be a negative number. Working alone, the new computer can print the payroll in 12 min.

40. Strategy
This is a distance-rate problem.
Rate of the first car: r
Rate of the second car: $r + 10$

	Distance	Rate	Time
First car	200	r	$\dfrac{200}{r}$
Second car	200	$r+10$	$\dfrac{200}{r+10}$

The second car's time is one hour less than the time of the first car.
$$\frac{200}{r+10} = \frac{200}{r} - 1$$

Solution
$$\frac{200}{r+10} = \frac{200}{r} - 1$$
$$r(r+10)\left(\frac{200}{r+10}\right) = r(r+10)\left(\frac{200}{r} - 1\right)$$
$$200r = 200(r+10) - r(r+10)$$
$$200r = 200r + 2000 - r^2 - 10r$$
$$r^2 + 10r - 2000 = 0$$
$$(r+50)(r-40) = 0$$

$r + 50 = 0 \qquad r - 40 = 0$
$r = -50 \qquad r = 40$

The solution -50 is not possible, since rate cannot be a negative number.
$r + 10 = 40 + 10 = 50$
The rate of the first car is 40 mph.
The rate of the second car is 50 mph.

Chapter Test

1.
$$2x^2 + x = 6$$
$$2x^2 + x - 6 = 0$$
$$(2x - 3)(x + 2) = 0$$

$2x - 3 = 0 \qquad x + 2 = 0$
$2x = 3 \qquad x = -2$
$x = \dfrac{3}{2}$

The solutions are $\dfrac{3}{2}$ and -2.

2.
$$12x^2 + 7x - 12 = 0$$
$$(3x + 4)(4x - 3) = 0$$

$3x + 4 = 0 \qquad 4x - 3 = 0$
$3x = -4 \qquad 4x = 3$
$x = -\dfrac{4}{3} \qquad x = \dfrac{3}{4}$

The solutions are $-\dfrac{4}{3}$ and $\dfrac{3}{4}$.

3. $f(x) = -x^2 + 8x - 7$
$$x = -\frac{b}{2a} = -\frac{8}{2(-1)} = 4$$
$$f(4) = -4^2 + 8(4) - 7 = 9.$$
The maximum value of the function is 9.

4.
$$(x - r_1)(x - r_2) = 0$$
$$\left[x - \left(-\frac{1}{3}\right)\right](x - 3) = 0$$
$$\left(x + \frac{1}{3}\right)(x - 3) = 0$$
$$x^2 - 3x + \frac{1}{3}x - 1 = 0$$
$$x^2 - \frac{8}{3}x - 1 = 0$$
$$3\left(x^2 - \frac{8}{3}x - 1\right) = 3 \cdot 0$$
$$3x^2 - 8x - 3 = 0$$

5. $2(x+3)^2 - 36 = 0$
$$2(x+3)^2 = 36$$
$$(x+3)^2 = 18$$
$$\sqrt{(x+3)^2} = \sqrt{18}$$
$$x + 3 = \pm 3\sqrt{2}$$
$$x = -3 \pm 3\sqrt{2}$$
The solutions are $-3 + 3\sqrt{2}$ and $-3 - 3\sqrt{2}$.

6. $x^2 + 4x - 1 = 0$
$$x^2 + 4x = 1$$
Complete the square.
$$x^2 + 4x + 4 = 1 + 4$$
$$(x+2)^2 = 5$$
$$\sqrt{(x+2)^2} = \sqrt{5}$$
$$x + 2 = \pm\sqrt{5}$$
$$x = -2 \pm \sqrt{5}$$
The solutions are $-2 + \sqrt{5}$ and $-2 - \sqrt{5}$.

7. $g(x) = x^2 + 3x - 8$
$$0 = x^2 + 3x - 8$$
$a = 1, \ b = 3, \ c = -8$
$$x = \frac{-b \pm \sqrt{b^2 - 4ac}}{2a}$$
$$= \frac{-3 \pm \sqrt{3^2 - 4(1)(-8)}}{2(1)}$$
$$= \frac{-3 \pm \sqrt{9 + 32}}{2}$$
$$= \frac{-3 \pm \sqrt{41}}{2}$$
The zeros of $g(x)$ are $\dfrac{-3 + \sqrt{41}}{2}$ and $\dfrac{-3 - \sqrt{41}}{2}$.

8. $3x^2 - x + 8 = 0$
$a = 3, \ b = -1, \ c = 8$
$x = \dfrac{-b \pm \sqrt{b^2 - 4ac}}{2a}$
$= \dfrac{-(-1) \pm \sqrt{(-1)^2 - 4(3)(8)}}{2(3)}$
$= \dfrac{1 \pm \sqrt{1 - 96}}{6}$
$= \dfrac{1 \pm \sqrt{-95}}{6}$
$= \dfrac{1 \pm i\sqrt{95}}{6}$

The solutions are $= \dfrac{1}{6} + \dfrac{\sqrt{95}}{6}i$ and $= \dfrac{1}{6} - \dfrac{\sqrt{95}}{6}i$.

9. $\dfrac{2x}{x-1} + \dfrac{3}{x+2} = 1$
$(x-1)(x+2)\left(\dfrac{2x}{x-1} + \dfrac{3}{x+2}\right) = (x-1)(x+2)1$
$2x(x+2) + 3(x-1) = x^2 + x - 2$
$2x^2 + 4x + 3x - 3 = x^2 + x - 2$
$2x^2 + 7x - 3 = x^2 + x - 2$
$x^2 + 6x - 1 = 0$
$a = 1, \ b = 6, \ c = -1$
$x = \dfrac{-b \pm \sqrt{b^2 - 4ac}}{2a}$
$= \dfrac{-6 \pm \sqrt{6^2 - 4(1)(-1)}}{2(1)}$
$= \dfrac{-6 \pm \sqrt{36 + 4}}{2}$
$= \dfrac{-6 \pm \sqrt{40}}{2}$
$= \dfrac{-6 \pm 2\sqrt{10}}{2}$
$= -3 \pm \sqrt{10}$

The solutions are $-3 + \sqrt{10}$ and $-3 - \sqrt{10}$.

10. $2x + 7x^{1/2} - 4 = 0$
$2\left(x^{1/2}\right)^2 + 7x^{1/2} - 4 = 0$
$2u^2 + 7u - 4 = 0$
$(2u - 1)(u + 4) = 0$

$2u - 1 = 0 \qquad u + 4 = 0$
$2u = 1 \qquad\qquad u = -4$
$u = \dfrac{1}{2}$

Replace u by $x^{1/2}$.

$x^{1/2} = \dfrac{1}{2} \qquad\qquad x^{1/2} = -4$
$\left(x^{1/2}\right)^2 = \left(\dfrac{1}{2}\right)^2 \qquad \left(x^{1/2}\right)^2 = (-4)^2$
$x = \dfrac{1}{4} \qquad\qquad\qquad x = 16$

16 does not check as a solution.
$\dfrac{1}{4}$ does check as a solution.

The solution is $\dfrac{1}{4}$.

11. $x^4 - 11x^2 + 18 = 0$
$\left(x^2\right)^2 - 11x^2 + 18 = 0$
$u^2 - 11u + 18 = 0$
$(u - 9)(u - 2) = 0$

$u - 9 = 0 \qquad u - 2 = 0$
$u = 9 \qquad\qquad u = 2$

Replace u by x^2.

$x^2 = 9 \qquad\qquad x^2 = 2$
$\sqrt{x^2} = \sqrt{9} \qquad \sqrt{x^2} = \sqrt{2}$
$x = \pm 3 \qquad\qquad x = \pm\sqrt{2}$

The solutions are $3, -3, \sqrt{2}, -\sqrt{2}$.

12. $\sqrt{2x+1} + 5 = 2x$
$\sqrt{2x+1} = 2x - 5$
$\left(\sqrt{2x+1}\right)^2 = (2x-5)^2$
$2x + 1 = 4x^2 - 20x + 25$
$0 = 4x^2 - 22x + 24$
$0 = 2\left(2x^2 - 11x + 12\right)$
$0 = 2(2x - 3)(x - 4)$

$2x - 3 = 0 \qquad x - 4 = 0$
$2x = 3 \qquad\qquad x = 4$
$x = \dfrac{3}{2}$

$\dfrac{3}{2}$ does not check as a solution.
4 does check as a solution.
The solution is 4.

13. $\sqrt{x-2} = \sqrt{x} - 2$
$\left(\sqrt{x-2}\right)^2 = \left(\sqrt{x} - 2\right)^2$
$x - 2 = x - 4\sqrt{x} + 4$
$-6 = -4\sqrt{x}$
$-\dfrac{1}{4}(-6) = \left(-\dfrac{1}{4}\right)(-4\sqrt{x})$
$\dfrac{3}{2} = \sqrt{x}$
$\dfrac{9}{4} = x$

$\dfrac{9}{4}$ does not check as a solution.
The equation has no solution.

14. $b^2 - 4ac = 2^2 - 4(3)(-4)$
$= 4 + 48$
$= 52$
The discriminant is positive.
The parabola has two x-intercepts.

15. $y = 2x^2 + 5x - 12$
$0 = 2x^2 + 5x - 12$
$0 = (2x - 3)(x + 4)$
$2x - 3 = 0 \qquad x + 4 = 0$
$2x = 3 \qquad x = -4$
$x = \dfrac{3}{2}$

The x-intercepts of the parabola are $\left(\dfrac{3}{2}, 0\right)$ and $(-4, 0)$.

16. $y = 2x^2 + 6x + 3$
The equation of the axis of symmetry is $x = -\dfrac{b}{2a}$.
$a = 2, \; b = 6$
$x = -\dfrac{b}{2a} = -\dfrac{6}{2(2)} = -\dfrac{6}{4} = -\dfrac{3}{2}$
The axis of symmetry is $x = -\dfrac{3}{2}$.

17. $-\dfrac{b}{2a} = -\dfrac{1}{2\left(\frac{1}{2}\right)} = -1$

$y = \dfrac{1}{2}(-1)^2 - 1 - 4 = -4.5$
The vertex is $(-1, -4.5)$.
The axis of symmetry is $x = -1$.

The domain is $\{x \mid x \in \text{real numbers}\}$.
The range is $\{y \mid y \geq -4.5\}$.

18. $\dfrac{2x - 3}{x + 4} \leq 0$

$\{x \mid -4 < x \leq \dfrac{3}{2}\}$

19. **Strategy**
This is a geometry problem.
The height of the triangle: x
The base of the triangle: $3x + 3$
The area of the triangle is 30 ft^2. Use the formula for the area of a triangle $\left(A = \dfrac{1}{2}bh\right)$.

Solution
$A = \dfrac{1}{2}bh$
$30 = \dfrac{1}{2}(3x + 3)x$
$60 = 3x^2 + 3x$
$0 = 3x^2 + 3x - 60$
$0 = 3(x^2 + x - 20)$
$0 = 3(x + 5)(x - 4)$
$x + 5 = 0 \qquad x - 4 = 0$
$x = -5 \qquad x = 4$
The solution -5 is not possible, since height cannot be a negative number.
$x = 4, \; 3x + 3 = 3(4) + 3 = 15$
The height of the triangle is 4 ft.
The base of the triangle is 15 ft.

20. Strategy
This is a distance-rate problem.
The rate of the canoe in calm water: x

	Distance	Rate	Time
With current	6	$x+2$	$\dfrac{6}{x+2}$
Against current	6	$x-2$	$\dfrac{6}{x-2}$

The total traveling time is 4 h.
$$\frac{6}{x+2}+\frac{6}{x-2}=4$$

Solution
$$\frac{6}{x+2}+\frac{6}{x-2}=4$$
$$(x+2)(x-2)\left(\frac{6}{x+2}+\frac{6}{x-2}\right)=(x+2)(x-2)4$$
$$6(x-2)+6(x+2)=4(x^2-4)$$
$$6x-12+6x+12=4x^2-16$$
$$12x=4x^2-16$$
$$0=4x^2-12x-16$$
$$0=4(x^2-3x-4)$$
$$0=4(x-4)(x+1)$$

$x-4=0 \qquad x+1=0$
$x=4 \qquad\quad x=-1$

The solution -1 is not possible, since the rate cannot be a negative number.
The rate of the canoe in calm water is 4 mph.

Cumulative Review Exercises

1. $2a^2 - b^2 \div c^2$
$2(3)^2 - (-4)^2 \div (-2)^2$
$= 2(9) - 16 \div 4$
$= 18 - 16 \div 4$
$= 18 - 4$
$= 14$

2. $\dfrac{2x-3}{4} - \dfrac{x+4}{6} = \dfrac{3x-2}{8}$
$24\left(\dfrac{2x-3}{4} - \dfrac{x+4}{6}\right) = 24\left(\dfrac{3x-2}{8}\right)$
$6(2x-3) - 4(x+4) = 3(3x-2)$
$12x - 18 - 4x - 16 = 9x - 6$
$8x - 34 = 9x - 6$
$-x - 34 = -6$
$-x = 28$
$x = -28$
The solution is -28.

3. $P_1(3, -4)$, $P_2(-1, 2)$
$m = \dfrac{y_2 - y_1}{x_2 - x_1} = \dfrac{2-(-4)}{-1-3} = \dfrac{2+4}{-4} = \dfrac{6}{-4} = -\dfrac{3}{2}$

4. $x - y = 1$
$y = x - 1$
$m = 1 \quad (x_1, y_1) = (1, 2)$
$y - y_1 = m(x - x_1)$
$y - 2 = 1(x - 1)$
$y - 2 = x - 1$
$y = x + 1$

5. $-3x^3y + 6x^2y^2 - 9xy^3$
$-3xy(x^2 - 2xy + 3y^2)$

6. $6x^2 - 7x - 20 = (2x - 5)(3x + 4)$

7. $a^n x + a^n y - 2x - 2y = a^n(x+y) - 2(x+y)$
$= (x+y)(a^n - 2)$

8.
$$3x - 4 \overline{\smash{\big)}\, 3x^3 - 13x^2 + 10}$$
$\underline{3x^3 - 4x^2}$
$-9x^2$
$\underline{-9x^2 + 12x}$
$-12x + 10$
$\underline{-12x + 16}$
-6

$(3x^3 - 13x^2 + 10) \div (3x - 4) = x^2 - 3x - 4 - \dfrac{6}{3x-4}$

9. $\dfrac{x^2 + 2x + 1}{8x^2 + 8x} \cdot \dfrac{4x^3 - 4x^2}{x^2 - 1}$
$= \dfrac{(x+1)(x+1)}{8x(x+1)} \cdot \dfrac{4x^2(x-1)}{(x+1)(x-1)}$
$= \dfrac{(x+1)(x+1)4x^2(x-1)}{8x(x+1)(x+1)(x-1)}$
$= \dfrac{x}{2}$

10. Distance between points is
$\sqrt{(x_2 - x_1)^2 + (y_2 - y_1)^2}$
Distance $= \sqrt{[2-(-2)]^2 + (5-3)^2}$
$= \sqrt{4^2 + 2^2}$
$= \sqrt{20}$
$= \sqrt{4 \cdot 5}$
$= 2\sqrt{5}$
The distance between the points is $2\sqrt{5}$

11. $S = \dfrac{n}{2}(a+b)$

$2S = 2\dfrac{n}{2}(a+b)$

$2S = n(a+b)$

$2S = an + bn$

$2S - an = bn$

$\dfrac{2S - an}{n} = b$

12. $-2i(7 - 4i) = -14i + 8i^2$
$= -14i + 8(-1)$
$= -8 - 14i$

13. $a^{-1/2}\left(a^{1/2} - a^{3/2}\right) = a^{-1/2+1/2} - a^{-1/2+3/2}$
$= a^0 - a^1$
$= 1 - a$

14. $\dfrac{\sqrt[3]{8x^4y^5}}{\sqrt[3]{16xy^6}} = \sqrt[3]{\dfrac{8x^4y^5}{16xy^6}}$

$= \sqrt[3]{\dfrac{x^3}{2y}}$

$= \sqrt[3]{\dfrac{x^3}{2y}}\sqrt[3]{\dfrac{4y^2}{4y^2}}$

$= \sqrt[3]{\dfrac{4y^2 x^3}{8y^3}}$

$= \sqrt[3]{\dfrac{x^3\left(4y^2\right)}{8y^3}}$

$= \dfrac{x\sqrt[3]{4y^2}}{2y}$

15. $\dfrac{x}{x+2} - \dfrac{4x}{x+3} = 1$

$(x+2)(x+3)\left(\dfrac{x}{x+2} - \dfrac{4x}{x+3}\right) = (x+2)(x+3)1$

$x(x+3) - 4x(x+2) = (x+2)(x+3)$

$x^2 + 3x - 4x^2 - 8x = x^2 + 5x + 6$

$-3x^2 - 5x = x^2 + 5x + 6$

$0 = 4x^2 + 10x + 6$

$0 = 2\left(2x^2 + 5x + 3\right)$

$0 = 2(2x+3)(x+1)$

$2x + 3 = 0 \qquad x + 1 = 0$
$2x = -3 \qquad x = -1$
$x = -\dfrac{3}{2}$

The solutions are $-\dfrac{3}{2}$ and -1.

282 Chapter 8: Quadratic Equations and Inequalities

16.
$$\frac{x}{2x+3} - \frac{3}{4x^2-9} = \frac{x}{2x-3}$$
$$\frac{x}{2x+3} - \frac{3}{(2x+3)(2x-3)} = \frac{x}{2x-3}$$
$$(2x+3)(2x-3)\left[\frac{x}{2x+3} - \frac{3}{(2x+3)(2x-3)}\right] = (2x+3)(2x-3)\frac{x}{2x-3}$$
$$(2x-3)x - 3 = (2x+3)x$$
$$2x^2 - 3x - 3 = 2x^2 + 3x$$
$$-3 = 6x$$
$$-\frac{1}{2} = x$$

Check: $\dfrac{-\frac{1}{2}}{2\left(-\frac{1}{2}\right)+3} - \dfrac{3}{4\left(-\frac{1}{2}\right)^2-9} = \dfrac{-\frac{1}{2}}{2\left(-\frac{1}{2}\right)-3}$

$$\frac{-\frac{1}{2}}{2} - \frac{3}{-8} = \frac{-\frac{1}{2}}{-4}$$
$$\frac{1}{8} = \frac{1}{8}$$

The solution is $x = -\dfrac{1}{2}$.

17. $x^4 - 6x^2 + 8 = 0$
$\left(x^2\right)^2 - 6x^2 + 8 = 0$
$u^2 - 6u + 8 = 0$
$(u-4)(u-2) = 0$

$u - 4 = 0 \qquad u - 2 = 0$
$u = 4 \qquad\quad u = 2$

Replace u by x^2
$x^2 = 4 \qquad\qquad x^2 = 2$
$\sqrt{x^2} = \sqrt{4} \qquad \sqrt{x^2} = \sqrt{2}$
$x = \pm 2 \qquad\quad x = \pm\sqrt{2}$

The solutions are 2, –2, $\sqrt{2}$, and $-\sqrt{2}$.

18. $\sqrt{3x+1} - 1 = x$
$\sqrt{3x+1} = x + 1$
$\left(\sqrt{3x+1}\right)^2 = (x+1)^2$
$3x + 1 = x^2 + 2x + 1$
$0 = x^2 - x$
$0 = x(x-1)$

$x = 0 \qquad x - 1 = 0$
$\qquad\qquad\quad x = 1$

0 and 1 both check as solutions. The solutions are 0 and 1.

19. $|3x - 2| < 8$
$-8 < 3x - 2 < 8$
$-8 + 2 < 3x - 2 + 2 < 8 + 2$
$-6 < 3x < 10$
$\dfrac{1}{3} \cdot (-6) < \dfrac{1}{3} \cdot (3x) < \dfrac{1}{3} \cdot 10$
$-2 < x < \dfrac{10}{3}$
$\left\{x \mid -2 < x < \dfrac{10}{3}\right\}$

20.
$$6x - 5y = 15$$
$$6x - 5(0) = 15$$
$$6x = 15$$
$$x = \frac{15}{6} = \frac{5}{2}$$
The x-intercept is $\left(\frac{5}{2},\ 0\right)$.

$$6x - 5y = 15$$
$$6(0) - 5y = 15$$
$$-5y = 15$$
$$y = -3$$
The y-intercept is $(0, -3)$.

21. Solve each inequality.

$$x + y \le 3 \qquad\qquad 2x - y < 4$$
$$y \le 3 - x \qquad\qquad -y < 4 - 2x$$
$$\qquad\qquad\qquad\qquad y > -4 + 2x$$

22.
$$x + y + z = 2$$
$$-x + 2y - 3z = -9$$
$$x - 2y - 2z = -1$$

$$D = \begin{vmatrix} 1 & 1 & 1 \\ -1 & 2 & -3 \\ 1 & -2 & -2 \end{vmatrix} = \begin{vmatrix} 2 & -3 \\ -2 & -2 \end{vmatrix} - \begin{vmatrix} -1 & -3 \\ 1 & -2 \end{vmatrix} + \begin{vmatrix} -1 & 2 \\ 1 & -2 \end{vmatrix} = (-4 - 6) - [2 - (-3)] + (2 - 2) = -10 - 5 + 0 = -15$$

$$D_x = \begin{vmatrix} 2 & 1 & 1 \\ -9 & 2 & -3 \\ -1 & -2 & -2 \end{vmatrix} = 2\begin{vmatrix} 2 & -3 \\ -2 & -2 \end{vmatrix} - \begin{vmatrix} -9 & -3 \\ -1 & -2 \end{vmatrix} + \begin{vmatrix} -9 & 2 \\ -1 & -2 \end{vmatrix} = 2(-4 - 6) - (18 - 3) + [18 - (-2)] = -20 - 15 + 20 = -15$$

$$D_y = \begin{vmatrix} 1 & 2 & 1 \\ -1 & -9 & -3 \\ 1 & -1 & -2 \end{vmatrix} = \begin{vmatrix} -9 & -3 \\ -1 & -2 \end{vmatrix} - 2\begin{vmatrix} -1 & -3 \\ 1 & -2 \end{vmatrix} + \begin{vmatrix} -1 & -9 \\ 1 & -1 \end{vmatrix} = (18 - 3) - 2[2 - (-3)] + [1 - (-9)] = 15 - 10 + 10 = 15$$

$$D_z = \begin{vmatrix} 1 & 1 & 2 \\ -1 & 2 & -9 \\ 1 & -2 & -1 \end{vmatrix} = \begin{vmatrix} 2 & -9 \\ -2 & -1 \end{vmatrix} - \begin{vmatrix} -1 & -9 \\ 1 & -1 \end{vmatrix} + 2\begin{vmatrix} -1 & 2 \\ 1 & -2 \end{vmatrix} = (-2 - 18) - [1 - (-9)] + 2(2 - 2) = -20 - 10 + 0 = -30$$

$$x = \frac{D_x}{D} = \frac{-15}{-15} = 1$$
$$y = \frac{D_y}{D} = \frac{15}{-15} = -1$$
$$z = \frac{D_z}{D} = \frac{-30}{-15} = 2$$
The solution is $(1, -1, 2)$.

23.
$$f(x) = \frac{2x - 3}{x^2 - 1}$$
$$f(-2) = \frac{2(-2) - 3}{(-2)^2 - 1} = -\frac{7}{3}$$

24. $f(x) = \dfrac{x-2}{x^2 - 2x - 15}$

$f(x) = \dfrac{x-2}{(x-5)(x+3)}$

$0 = (x-5)(x+3)$

$x - 5 = 0 \qquad x + 3 = 0$
$x = 5 \qquad\quad x = -3$

The domain of $f(x)$ is $\{x \mid x \neq 5 \text{ and } x \neq -3\}$.

25. $x^3 + x^2 - 6x < 0$

$x(x^2 + x - 6) < 0$

$x(x+3)(x-2) < 0$

$\{x \mid x < -3 \text{ or } 0 < x < 2\}$

26. $\dfrac{(x-1)(x-5)}{x+3} \geq 0$

$\{x \mid -3 < x \leq 1 \text{ or } x \geq 5\}$

27. **Strategy**
Let p represent the length of the piston rod, T the tolerance, and m the given length.
Solve the absolute value inequality $|m - p| \leq T$ for m.

Solution

$|m - p| \leq T$

$\left| m - 9\dfrac{3}{8} \right| \leq \dfrac{1}{64}$

$-\dfrac{1}{64} \leq m - 9\dfrac{3}{8} \leq \dfrac{1}{64}$

$-\dfrac{1}{64} + 9\dfrac{3}{8} \leq m \leq \dfrac{1}{64} + 9\dfrac{3}{8}$

$9\dfrac{23}{64} \leq m \leq 9\dfrac{25}{64}$

The lower limit is $9\dfrac{23}{64}$ in.
The upper limit is $9\dfrac{25}{64}$ in.

28. $A = \dfrac{1}{2} b \cdot h$

$= \dfrac{1}{2}(x+8)(2x-4)$

$= \dfrac{1}{2}(2x^2 + 12x - 32)$

$= (x^2 + 6x - 16)$ ft^2

29. $2x^2 + 4x + 3 = 0$

$a = 2, \ b = 4, \ c = 3$

$b^2 - 4ac = 4^2 - 4(2)(3)$

$16 - 24 = -8$

$-8 < 0$

Since the discriminant is less than zero, the equation has two complex number solutions.

30. $m = \dfrac{y_2 - y_1}{x_2 - x_1}$

$= \dfrac{0 - 250{,}000}{30 - 0}$

$= \dfrac{-250{,}000}{30}$

$= -\dfrac{25{,}000}{3}$

The slope represents the amount in dollars that the building depreciates each year
$\left(-\dfrac{25{,}000}{3} = \$8333 \right)$.

Chapter 9: Functions and Relations

Section 9.1

Concept Review 9.1

1. Sometimes true
   For the absolute value function $f(x) = |x| - 3$, the value of $f(1) = -2$.

3. Sometimes true
   The range of $f(x) = x^2$ is $\{y | y \geq 0\}$.

5. Always true

Objective 1 Exercises

3. A vertical line intersects the graph at most once. The graph is the graph of a function.

5. A vertical line exists that intersects the graph more than once. The graph is not the graph of a function.

7. A vertical line intersects the graph at most once. The graph is the graph of a function.

9.

   The domain is $\{x | x \in \text{real numbers}\}$.
   The range is $\{y | y \in \text{real numbers}\}$.

11.

    The domain is $\{x | x \in \text{real numbers}\}$.
    The range is $\{y | y \geq -1\}$.

13.

    The domain is $\{x | x \in \text{real numbers}\}$.
    The range is $\{y | y \geq 0\}$.

15.

    The domain is $\{x | x \in \text{real numbers}\}$.
    The range is $\{y | y \in \text{real numbers}\}$.

17.

    The domain is $\{x | x \geq -1\}$.
    The range is $\{y | y \geq 0\}$.

19.

    The domain is $\{x | x \in \text{real numbers}\}$.
    The range is $\{y | y \geq 0\}$.

21.

    The domain is $\{x | x \in \text{real numbers}\}$.
    The range is $\{y | y \geq 1\}$.

23.

    The domain is $\{x | x \in \text{real numbers}\}$.
    The range is $\{y | y \in \text{real numbers}\}$.

25.

    The domain is $\{x | x \geq -2\}$.
    The range is $\{y | y \leq 0\}$.

27.

The domain is $\{x|x \in \text{real numbers}\}$.
The range is $\{y|y \geq -5\}$.

29.

The domain is $\{x|x \in \text{real numbers}\}$.
The range is $\{y|y \geq -1\}$.

31.

The domain is $\{x|x \in \text{real numbers}\}$.
The range is $\{y|y \in \text{real numbers}\}$.

33.

The domain is $\{x|x \in \text{real numbers}\}$.
The range is $\{y|y \in \text{real numbers}\}$.

35.

The domain is $\{x|x \in \text{real numbers}\}$.
The range is $\{y|y \geq -5\}$.

37.

The domain is $\{x|x \in \text{real numbers}\}$.
The range is $\{y|y \leq 0\}$.

39.

The domain is $\{x|x \in \text{real numbers}\}$.
The range is $\{y|y \in \text{real numbers}\}$.

Applying Concepts 9.1

41. a. $f(x) = x$ is a function.

b. $f(x) = \left|\dfrac{x}{2}\right|$ is a function.

c. $\{(3, 1), (1, 3), (3, 0), (0, 3)\}$ is not a function because the number 3 in the domain is paired with different values in the range.

43. $f(x) = \sqrt{x-2}$
$f(a) = 4 = \sqrt{a-2}$
$4^2 = \left(\sqrt{a-2}\right)^2$
$16 = a - 2$
$18 = a$

45. $f(14, 35)$: The greatest common divisor is 7.
$g(14, 35)$: The least common multiple is 70.
$f(14, 35) + g(14, 35) = 7 + 70 = 77$

47. $f(x) = -|x+3|$
$f(x)$ is greatest when $x + 3 = 0$
$x = -3$

49. a. $x^3 = 0$
$\sqrt[3]{x} = \sqrt[3]{0}$
$x = 0$

b. The x-intercept of the graph of $f(x) = x^3$ is $(0, 0)$.

51.

The graph has one turning point.

Section 9.2

Concept Review 9.2

1. Always true

3. Always true

5. Never true
If c_1 and c_2 are positive constants, the graph of $y = f(x + c_1) + c_2$ is the graph of $y = f(x)$ translated to the left c_1 units and shifted upward c_2 units.

Objective 1 Exercises

3.

5.

7.

9.

11.

13.

15.

17.

19.

21.

23.

25.

27.

Applying Concepts 9.2

29.

31.

33.

35.

37.

Section 9.3

Concept Review 9.3

1. Always true

3. Never true
$(f \circ g)(x) = 2(x+4) + 1$

Objective 1 Exercises

1. $(f-g)(2) = f(2) - g(2)$
$= [2(2)^2 - 3] - [-2(2) + 4]$
$= 5 - 0$
$= 5$
$(f-g)(2) = 5$

3. $(f+g)(0) = f(0) + g(0)$
$= [2(0)^2 - 3] + [-2(0) + 4]$
$= -3 + 4$
$= 1$
$(f+g)(0) = 1$

5. $(f \cdot g)(2) = f(2) \cdot g(2)$
$= [2(2)^2 - 3] \cdot [-2(2) + 4]$
$= 5 \cdot 0$
$= 0$
$(f \cdot g)(2) = 0$

7. $\left(\dfrac{f}{g}\right)(4) = \dfrac{f(4)}{g(4)}$
$= \dfrac{2(4)^2 - 3}{-2(4) + 4}$
$= -\dfrac{29}{4}$
$\dfrac{f}{g}(4) = -\dfrac{29}{4}$

9. $\left(\dfrac{g}{f}\right)(-3) = \dfrac{g(-3)}{f(-3)}$
$= \dfrac{-2(-3) + 4}{2(-3)^2 - 3}$
$= \dfrac{10}{15}$
$= \dfrac{2}{3}$
$\left(\dfrac{g}{f}\right)(-3) = \dfrac{2}{3}$

11. $(f+g)(1) = f(1) + g(1)$
$= [2(1)^2 + 3(1) - 1] + [2(1) - 4]$
$= [2 + 3 - 1] + [2 - 4]$
$= 4 - 2$
$= 2$
$(f+g)(1) = 2$

13. $(f-g)(4) = f(4) - g(4)$
$= [2(4)^2 + 3(4) - 1] - [2(4) - 4]$
$= [32 + 12 - 1] - [8 - 4]$
$= 43 - 4$
$= 39$
$(f-g)(4) = 39$

15. $(f \cdot g)(1) = f(1) \cdot g(1)$
$= [2(1)^2 + 3(1) - 1] \cdot [2(1) - 4]$
$= [2 + 3 - 1] \cdot [2 - 4]$
$= 4(-2)$
$= -8$
$(f \cdot g)(1) = -8$

17. $\left(\dfrac{f}{g}\right)(-3) = \dfrac{f(-3)}{g(-3)}$
$= \dfrac{2(-3)^2 + 3(-3) - 1}{2(-3) - 4}$
$= \dfrac{18 - 9 - 1}{-6 - 4}$
$= \dfrac{8}{-10}$
$= -\dfrac{4}{5}$
$\left(\dfrac{f}{g}\right)(-3) = -\dfrac{4}{5}$

19. $(f-g)(2) = f(2) - g(2)$
$= [2^2 + 3(2) - 5] - [2^3 - 2(2) + 3]$
$= [4 + 6 - 5] - [8 - 4 + 3]$
$= 5 - 7$
$= -2$
$(f-g)(-2) = -2$

21. $\left(\dfrac{f}{g}\right)(-2) = \dfrac{f(-2)}{g(-2)}$
$= \dfrac{(-2)^2 + 3(-2) - 5}{(-2)^3 - 2(-2) + 3}$
$= \dfrac{4 - 6 - 5}{-8 + 4 + 3}$
$= \dfrac{-7}{-1} = 7$
$\left(\dfrac{f}{g}\right)(-2) = 7$

Objective 2 Exercises

25. $f(x) = 2x - 3$
$f(0) = 2(0) - 3 = 0 - 3 = -3$
$g(x) = 4x - 1$
$g[f(0)] = g(-3) = 4(-3) - 1 = -12 - 1 = -13$
$g[f(0)] = -13$

27. $f(x) = 2x - 3$
$f(-2) = 2(-2) - 3 = -4 - 3 = -7$
$g(x) = 4x - 1$
$g[f(-2)] = g(-7) = 4(-7) - 1 = -28 - 1 = -29$
$g[f(-2)] = -29$

29. $f(x) = 2x - 3$
$g(x) = 4x - 1$
$g[f(x)] = g(2x - 3)$
$= 4(2x - 3) - 1$
$= 8x - 12 - 1$
$= 8x - 13$
$g[f(x)] = 8x - 13$

31. $g(x) = x^2 + 3$
$g(0) = 0^2 + 3 = 3$
$h(x) = x - 2$
$h[g(0)] = h(3) = 3 - 2 = 1$
$h[g(0)] = 1$

33. $g(x) = x^2 + 3$
$g(-2) = (-2)^2 + 3 = 4 + 3 = 7$
$h(x) = x - 2$
$h[g(-2)] = h(7) = 7 - 2 = 5$
$h[g(-2)] = 5$

35. $g(x) = x^2 + 3$
$h(x) = x - 2$
$h[g(x)] = h(x^2 + 3) = x^2 + 3 - 2 = x^2 + 1$
$h[g(x)] = x^2 + 1$

37. $f(x) = x^2 + x + 1$
$f(0) = 0^2 + 0 + 1 = 0 + 0 + 1 = 1$
$h(x) = 3x + 2$
$h[f(0)] = h(1) = 3(1) + 2 = 3 + 2 = 5$
$h[f(0)] = 5$

39. $f(x) = x^2 + x + 1$
$f(-2) = (-2)^2 - 2 + 1 = 4 - 2 + 1 = 3$
$h(x) = 3x + 2$
$h[f(-2)] = h(3) = 3(3) + 2 = 9 + 2 = 11$
$h[f(-2)] = 11$

41. $f(x) = x^2 + x + 1$
$h(x) = 3x + 2$
$h[f(x)] = h(x^2 + x + 1) = 3(x^2 + x + 1) + 2$
$= 3x^2 + 3x + 3 + 2$
$= 3x^2 + 3x + 5$
$h[f(x)] = 3x^2 + 3x + 5$

43. $g(x) = x^3$
$g(-1) = (-1)^3 = -1$
$f(x) = x - 2$
$f[g(-1)] = f(-1) = -1 - 2 = -3$
$f[g(-1)] = -3$

45. $f(x) = x - 2$
$f(-1) = -1 - 2 = -3$
$g(x) = x^3$
$g[f(-1)] = g(-3) = (-3)^3 = -27$
$g[f(-1)] = -27$

47. $f(x) = x - 2,\ g(x) = x^3$
$g[f(x)] = g(x - 2) = (x - 2)^3$
$g[f(x)] = x^3 - 6x^2 - 12x - 8$

Applying Concepts 9.3

49. $g(3 + h) - g(3) = (3 + h)^2 - 1 - [(3)^2 - 1]$
$= 9 + 6h + h^2 - 1 - 8$
$g(3 + h) - g(3) = h^2 + 6h$

51. $\dfrac{g(1+h) - g(1)}{h} = \dfrac{[(1+h)^2 - 1] - [(1)^2 - 1]}{h}$
$= \dfrac{1 + 2h + h^2 - 1 - 0}{h}$
$= \dfrac{2h + h^2}{h}$
$\dfrac{g(1-h) - g(1)}{h} = 2 + h$

53. $\dfrac{g(a+h) - g(a)}{h} = \dfrac{[(a+h)^2 - 1] - (a^2 - 1)}{h}$
$= \dfrac{a^2 + 2ah + h^2 - 1 - a^2 + 1}{h}$
$= \dfrac{2ah + h^2}{h}$
$\dfrac{g(a+h) - g(a)}{h} = 2a + h$

55. $f(x) = 2x$
$f(1) = 2 \cdot 1 = 2$
$h(x) = x - 2$
$h(2) = 2 - 2 = 0$
$g(x) = 3x - 1$
$g(0) = 3 \cdot 0 - 1 = -1$
$g(h[f(1)]) = -1$

57. $g(x) = 3x - 1$
$g(0) = 3 \cdot 0 - 1 = -1$
$h(x) = x - 2$
$h(-1) = -1 - 2 = -3$
$f(x) = 2x$
$f(-3) = 2(-3) = -6$
$f(h[g(0)]) = -6$

59. $h(x) = x - 2$
$f(x - 2) = 2(x - 2) = 2x - 4$
$g(2x - 4) = 3(2x - 4) - 1 = 6x - 12 - 1 = 6x - 13$

61. $f(g(2)) = f(0) = -2$

63. $f(g(-2)) = f(0) = -2$

65. $g(f(0)) = g(-2) = 0$

67. $g(f(-4)) = g(-4) = 6$

Section 9.4

Concept Review 9.4

1. Always true

3. Never true
The function $\{(2, 3), (4, 5), (6, 3)\}$ is not a 1-1 function. The function does not have an inverse.

5. Sometimes true
If a function is a 1-1 function, then the inverse is a function.

Objective 1 Exercises

3. The graph represents a 1-1 function.

5. The graph is not a 1-1 function. It fails the horizontal-line test.

7. The graph is a 1-1 function.

9. The graph is not a 1-1 function. It fails the horizontal- and vertical-line tests.

11. The graph is not a 1-1 function. It fails the horizontal-line test.

13. The graph is not a 1-1 function. It fails the horizontal-line test.

Objective 2 Exercises

17. The inverse of $\{(1, 0), (2, 3), (3, 8), (4, 15)\}$ is $\{(0, 1), (3, 2), (8, 3), (15, 4)\}$.

19. $\{(3, 5), (-3, -5), (2, 5), (-2, -5)\}$ has no inverse because the numbers 5 and -5 would be paired with more than one member of the range.

21. $f(x) = 4x - 8$
$y = 4x - 8$
$x = 4y - 8$
$x + 8 = 4y$
$\frac{1}{4}x + 2 = y$

The inverse function is $f^{-1}(x) = \frac{1}{4}x + 2$.

23. $f(x) = x^2 - 1$ is not a 1-1 function. Therefore, it has no inverse.

25. $f(x) = x - 5$
$y = x - 5$
$x = y - 5$
$x + 5 = y$

The inverse function is $f^{-1}(x) = x + 5$.

27. $f(x) = \frac{1}{3}x + 2$
$y = \frac{1}{3}x + 2$
$x = \frac{1}{3}y + 2$
$x - 2 = \frac{1}{3}y$
$3x - 6 = y$

The inverse function is $f^{-1}(x) = 3x - 6$.

29. $f(x) = -3x - 9$
$y = -3x - 9$
$x = -3y - 9$
$3y = -x - 9$
$y = -\frac{1}{3}x - 3$

The inverse function is $f^{-1}(x) = -\frac{1}{3}x - 3$.

31. $f(x) = \frac{2}{3}x + 4$
$y = \frac{2}{3}x + 4$
$x = \frac{2}{3}y + 4$
$x - 4 = \frac{2}{3}y$
$\frac{3}{2}(x - 4) = y$
$\frac{3}{2}x - 6 = y$

The inverse function is $f^{-1}(x) = \frac{3}{2}x - 6$.

33. $f(x) = -\frac{1}{3}x + 1$
$y = -\frac{1}{3}x + 1$
$x = -\frac{1}{3}y + 1$
$x - 1 = -\frac{1}{3}y$
$-3(x - 1) = y$
$-3x + 3 = y$

The inverse function is $f^{-1}(x) = -3x + 3$.

35. $f(x) = 2x - 5$
$y = 2x - 5$
$x = 2y - 5$
$x + 5 = 2y$
$\frac{1}{2}x + \frac{5}{2} = y$

The inverse function is $f^{-1}(x) = \frac{1}{2}x + \frac{5}{2}$.

37. $f(x) = x^2 + 3$ is not a 1-1 function. Therefore, it has no inverse.

39. $f(g(x)) = f\left(\frac{x}{4}\right)$
$= 4\left(\frac{x}{4}\right)$
$= x$
$g(f(x)) = g(4x)$
$= \frac{4x}{4}$
$= x$

The functions are inverses of each other.

41. $f(h(x)) = f\left(\dfrac{1}{3x}\right)$
$= 3\left(\dfrac{1}{3x}\right)$
$= \dfrac{1}{x}$
$h(f(x)) = h(3x)$
$= \dfrac{1}{3 \cdot 3x}$
$= \dfrac{1}{9x}$
The functions are not inverses of each other.

43. $g(f(x)) = g\left(\dfrac{1}{3}x - \dfrac{2}{3}\right)$
$= 3\left(\dfrac{1}{3}x - \dfrac{2}{3}\right) + 2$
$= x - 2 + 2$
$= x$
$f(g(x)) = f(3x + 2)$
$= \dfrac{1}{3}(3x + 2) - \dfrac{2}{3}$
$= x + \dfrac{2}{3} - \dfrac{2}{3}$
$= x$
The functions are inverses of each other.

45. $f(g(x)) = f(2x + 3)$
$= \dfrac{1}{2}(2x + 3) - \dfrac{3}{2}$
$= x + \dfrac{3}{2} - \dfrac{3}{2}$
$= x$
$g(f(x)) = g\left(\dfrac{1}{2}x - \dfrac{3}{2}\right)$
$= 2\left(\dfrac{1}{2}x - \dfrac{3}{2}\right) + 3$
$= x - 3 + 3$
$= x$
The functions are inverses of each other.

47. The domain of the inverse of f^{-1} is the range of f.

49. For any linear function f and its inverse f^{-1},
$f[f^{-1}(3)] = 3$.

Applying Concepts 9.4

51. If f is a 1-1 function and $f(0) = 5$, then
$f^{-1}(5) = 0$.

53. If f is a 1-1 function and $f(2) = 9$ then
$f^{-1}(9) = 2$.

55. $f(x) = 3x - 5$
From exercise 54,
$f^{-1}(x) = \dfrac{1}{3}x + \dfrac{5}{3}$
$f^{-1}(2) = \dfrac{1}{3}(2) + \dfrac{5}{3}$
$f^{-1}(2) = \dfrac{2}{3} + \dfrac{5}{3}$
$f^{-1}(2) = \dfrac{7}{3}$

57.

59.

61.

Projects and Group Activities

Derivative Investments

1. The interest rate from January 1, 1996 to January 1, 1997 is 5%.
   $10,000 + 0.05(10,000) = 10,000 + 500 = 10,500$

   The interest rate from January 1, 1997 to January 1, 1998 is 6%
   $10,500 + 0.06(10,500) = 10,500 + 630 = 11,130$

   The interest rate from January 1, 1998 to January 1, 1999 is 7%.
   $11,130 + 0.07(11,130) = 11,130 + 779.10$
   $= 11,909.10$

   The interest rate from January 1, 1999 to January 1, 2000 is 8%.
   $1,909.10 + 0.08(11909.10) = 11,909.10 + 954.73$
   $= 12,861.83$

   The interest rate from January 1, 2000 to January 1, 2001 is 9%.
   $12,861.83 + 0.09(12,861.83) = 12,861.83 + 1157.56$
   $= 14,019.39$

   The interest rate from January 1, 2001 to January 1, 2002 is 10%.
   $14,019.39 + 0.10(14019.39) = 14,019.53 + 1401.94$
   $= 15,421.33$

   The value of the traditional investment after 6 years will be $15,421.33.

3. The interest rate from January 1, 1996 to January 1, 1997 is 5%.
   $10,000 + 0.05(10,000) = 10,000 + 500 = 10,500$

   The interest rate from January 1, 1997 to January 1, 1998 is 4%.
   $10,500 + 0.04(10,500) = 10,500 + 420 = 10,920$

   The interest rate from January 1, 1998 to January 1, 1999 is 3%.
   $10,920 + 0.03(10,920) = 10,920 + 327.60$
   $= 11,247.60$

   The interest rate from January 1, 1999 to January 1, 2000 is 2%.
   $11247.60 + 0.02(11,247.60) = 11,247.60 + 224.95$
   $= 11,472.55$

   The interest rate form January 1, 2000 to January 1, 2001 is 1%.
   $11,472.55 + 0.01(11,472.55) = 11,472.55 + 114.73$
   $= 11,587.28$

   The interest rate from January 1, 2001 to January 1, 2002 is 0%.

   The value of the traditional investment after 6 years will be $11,587.28.

5. The general trend of the prime rate during the 1990's was upward. Derivative investments would not have performed as well as traditional investments during this time period.

Properties of Polynomials

1.

   The x-intercepts of the graph are (1.52, 0), (2, 0) and (5.14, 0).

2. The real-number solutions are approximately 1.52, 2 and approximately 5.14.

3.

   The x-intercept of the graph is (2, 0).

4. The real-number solution is 2.

5. One possible conjecture is that a fifth-degree equation can have at most five real-number solutions.

6. All possible fifth-degree equations have five or fewer real-number solutions. The conjecture is satisfied.

The Greatest Integer Function

1.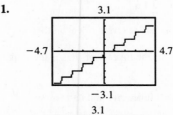

   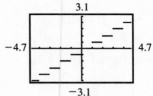

2.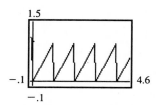

Chapter Review Exercises

1. Yes, the graph is that of a function. It passes the vertical-line test.

2. $f(x) = |x| - 3$
   Domain: $\{x | x \in \text{real numbers}\}$
   Range: $\{y | y \geq -3\}$

3. $f(x) = 3x^3 - 2$
   Domain: $\{x | x \in \text{real numbers}\}$
   Range: $\{y | y \in \text{real numbers}\}$

4. $f(x) = \sqrt{x+4}$
   Domain: $[-4, \infty)$
   Range: $[0, \infty)$

5.

6.

7.

8.

9. $(f+g)(2) = f(2) + g(2)$
   $= [2^2 + 2(2) - 3] + [2^2 - 2]$
   $= [4 + 4 - 3] + [4 - 2]$
   $= 5 + 2$
   $= 7$
   $(f+g)(2) = 7$

10. $(f-g)(-4) = f(-4) - g(-4)$
    $= [(-4)^2 + 2(-4) - 3] - [(-4)^2 - 2]$
    $= [16 - 8 - 3] - [16 - 2]$
    $= 5 - 14$
    $= -9$
    $(f-g)(-4) = -9$

11. $(f \cdot g)(-4) = f(-4) \cdot g(-4)$
    $= [(-4)^2 + 2(-4) - 3] \cdot [(-4)^2 - 2]$
    $= [16 - 8 - 3] \cdot [16 - 2]$
    $= 5 \cdot 14$
    $= 70$
    $(f \cdot g)(-4) = 70$

12. $\left(\dfrac{f}{g}\right)(3) = \dfrac{f(3)}{g(3)}$
    $= \dfrac{3^2 + 2(3) - 3}{3^2 - 2}$
    $= \dfrac{9 + 6 - 3}{9 - 2}$
    $= \dfrac{12}{7}$
    $\left(\dfrac{f}{g}\right)(3) = \dfrac{12}{7}$

13. $f(x) = 3x^2 - 4, \ g(x) = 2x + 1$
    $f[g(x)] = f(2x+1)$
    $= 3(2x+1)^2 - 4$
    $= 3(4x^2 + 4x + 1) - 4$
    $= 12x^2 + 12x + 3 - 4$
    $f[g(x)] = 12x^2 + 12x - 1$

14. $f(x) = x^2 + 4, \ g(x) = 4x - 1$
    $g(0) = 4(0) - 1 = 0 - 1 = -1$
    $f(-1) = (-1)^2 + 4 = 1 + 4 = 5$
    $f[g(0)] = 5$

15. $f(x) = 6x + 8, \ g(x) = 4x + 2$
    $f(-1) = 6(-1) + 8 = -6 + 8 = 2$
    $g(2) = 4(2) + 2 = 8 + 2 = 10$
    $g[f(-1)] = 10$

16. $f(x) = 2x^2 + x - 5, \ g(x) = 3x - 1$
$$g[f(x)] = g(2x^2 + x - 5)$$
$$= 3(2x^2 + x - 5) - 1$$
$$= 6x^2 + 3x - 15 - 1$$
$$g[f(x)] = 6x^2 + 3x - 16$$

17. There are no ordered pairs that have same first component. The relation is a function.

18. The inverse of $\{(-2, 1), (2, 3), (5, -4), (7, 9)\}$ is $\{(1, -2), (3, 2), (-4, 5), (9, 7)\}$

19. A vertical line will intersect the graph at no more than one point. The graph is the graph of a function. A horizontal line will intersect the graph at more than one point. The graph is not the graph of a 1-1 function.

20. A vertical line will intersect the graph at no more than one point. The graph is the graph of a function. A horizontal line will intersect the graph at no more than one point. The graph is the graph of a 1-1 function.

21.
$$f(x) = \frac{1}{2}x + 8$$
$$y = \frac{1}{2}x + 8$$
$$x = \frac{1}{2}y + 8$$
$$x - 8 = \frac{1}{2}y$$
$$2(x - 8) = 2 \cdot \frac{1}{2}y$$
$$2x - 16 = y$$
The inverse function is $f^{-1}(x) = 2x - 16$.

22.
$$f(x) = -6x + 4$$
$$y = -6x + 4$$
$$x = -6y + 4$$
$$x - 4 = -6y$$
$$-\frac{1}{6}x + \frac{2}{3} = y$$
The inverse function is $f^{-1}(x) = -\frac{1}{6}x + \frac{2}{3}$.

23.
$$f(x) = \frac{2}{3}x - 12$$
$$y = \frac{2}{3}x - 12$$
$$x = \frac{2}{3}y - 12$$
$$x + 12 = \frac{2}{3}y$$
$$\frac{3}{2}(x + 12) = y$$
$$\frac{3}{2}x + 18 = y$$
The inverse function is $f^{-1}(x) = \frac{3}{2}x + 18$.

24. $f(g(x)) = f(-4x + 5) = -\frac{1}{4}(-4x + 5) + \frac{5}{4}$
$$= x - \frac{5}{4} + \frac{5}{4} = x$$
$$g(f(x)) = g\left(-\frac{1}{4}x + \frac{5}{4}\right) = -4\left(-\frac{1}{4}x + \frac{5}{4}\right) + 5$$
$$= x - 5 + 5 = x$$
The functions are inverses of each other.

Chapter Test

1. A vertical line intersects the graph at more than one point. The graph is not a function.

2. $(f - g)(2) = f(2) - g(2)$
$$= [2^2 + 2(2) - 3] - [2^3 - 1]$$
$$= [4 + 4 - 3] - [8 - 1]$$
$$= 5 - 7$$
$$= -2$$
$(f - g)(2) = -2$

3. $(f \cdot g)(-3) = f(-3) \cdot g(-3)$
$$= [(-3)^3 + 1][2(-3) - 3]$$
$$= [-27 + 1][-9]$$
$$= 234$$
$(f \cdot g)(-3) = 234$

4. $\left(\dfrac{f}{g}\right)(-2) = \dfrac{f(-2)}{g(-2)}$
$$= \frac{4(-2) - 5}{(-2)^2 + 3(-2) + 4}$$
$$= \frac{-8 - 5}{4 - 6 + 4}$$
$$= -\frac{13}{2}$$
$\dfrac{f}{g}(-2) = \dfrac{-13}{2} = -\dfrac{13}{2}$

5. $(f - g)(-4) = f(-4) - g(-4)$
$$= [(-4)^2 + 4] - [2(-4)^2 + 2(-4) + 1]$$
$$= [16 + 4] - [32 - 8 + 1]$$
$$= 20 - 25$$
$$= -5$$
$(f - g)(-4) = -5$

6. $g(x) = \dfrac{x}{x + 1}$
$$g(3) = \frac{3}{3 + 1} = \frac{3}{4}$$
$$f(x) = 4x + 2$$
$$f\left(\frac{3}{4}\right) = 4\left(\frac{3}{4}\right) + 2$$
$$f[g(3)] = 3 + 2 = 5$$

7.

8.

9. $f(x) = -\sqrt{3-x}$
   The domain is $(-\infty, 3]$.
   The range is $(-\infty, 0]$.

10. $f(x) = \left|\frac{1}{2}x\right| - 2$

    The domain is $\{x | x \in \text{real numbers}\}$.
    The range is $\{y | y \geq -2\}$.

11.

12.

13. $f(x) = \frac{1}{4}x - 4$
    $y = \frac{1}{4}x - 4$
    $x = \frac{1}{4}y - 4$
    $x + 4 = \frac{1}{4}y$
    $4x + 16 = y$

    The inverse of the function is $f^{-1}(x) = 4x + 16$.

14. The inverse of the function of {(2, 6), (3, 5), (4, 4), (5, 3)} is {(6, 2), (5, 3), (4, 4), (3, 5)}.

15. $f[g(x)] = \frac{1}{2}(2x - 4) + 2$
    $= x - 2 + 2$
    $= x$
    $g[f(x)] = 2\left(\frac{1}{2}x + 2\right) - 4$
    $= x + 4 - 4$
    $= x$
    The functions are inverses of each other.

16. $f[g(x)] = f(x - 1)$
    $f[g(x)] = 2(x-1)^2 - 7$
    $= 2(x^2 - 2x + 1) - 7$
    $= 2x^2 - 4x + 2 - 7$
    $= 2x^2 - 4x - 5$

17. $f(x) = \frac{1}{2}x - 3$
    $y = \frac{1}{2}x - 3$
    $x = \frac{1}{2}y - 3$
    $2x = y - 6$
    $y = 2x + 6$
    $f^{-1}(x) = 2x + 6$

18. $f[g(x)] = f\left(\frac{3}{2}x - 3\right)$
    $= \frac{2}{3}\left(\frac{3}{2}x - 3\right) + 3$
    $= x - 2 + 3$
    $= x + 1$
    $g[f(x)] = g\left(\frac{2}{3}x + 3\right)$
    $= \frac{3}{2}\left(\frac{2}{3}x + 3\right) - 3$
    $= x + \frac{9}{2} - 3$
    $= x + \frac{3}{2}$
    The functions are not inverses of each other.

19. $f(x) = x^3 - 3x + 2$
    The domain is $\{x | x \in \text{real numbers}\}$.
    The range is $\{y | y \in \text{real numbers}\}$.

20. The graph does not represent a 1-1 function. It fails the horizontal- and vertical-line tests.

Cumulative Review Exercises

1. $-3a + \left|\dfrac{3a-ab}{3b-c}\right| = -3(2) + \left|\dfrac{3(2)-2(2)}{3(2)-(-2)}\right|$

    $= -6 + \left|\dfrac{6-4}{6+2}\right|$

    $= -6 + \left|\dfrac{2}{8}\right|$

    $= -6 + \left|\dfrac{1}{4}\right|$

    $= -6 + \dfrac{1}{4} = -\dfrac{23}{4}$

2.

3. $\dfrac{3x-1}{6} - \dfrac{5-x}{4} = \dfrac{5}{6}$

    $12\left(\dfrac{3x-1}{6} - \dfrac{5-x}{4}\right) = 12\left(\dfrac{5}{6}\right)$

    $2(3x-1) - 3(5-x) = 2(5)$

    $6x - 2 - 15 + 3x = 10$

    $9x - 17 = 10$

    $9x = 27$

    $x = 3$

4. $4x - 2 < -10$ or $3x - 1 > 8$
    $4x < -8$ $\qquad$ $3x > 9$
    $x < -2$ $\qquad$ $x > 3$
    $\{x | x < -2\}$ $\qquad$ $\{x | x > 3\}$
    $\{x | x < -2\} \cup \{x | x > 3\} = \{x | x < -2 \text{ or } x > 3\}$

5. Vertex: (0, 0)
    Axis of symmetry: $x = 0$

6. $3x - 4y \geq 8$
    $-4y \geq -3x + 8$
    $y \leq \dfrac{3}{4}x - 2$

7. $|8 - 2x| \geq 0$
    $8 - 2x \leq 0$ or $8 - 2x \geq 0$
    $8 \leq 2x$ $\qquad$ $8 \geq 2x$
    $4 \leq x$ $\qquad$ $4 \geq x$
    $\{x | x \geq 4\}$ $\qquad$ $\{x | x \leq 4\}$
    $\{x | x \geq 4\} \cup \{x | x \leq 4\} = \{x | x \in \text{real numbers}\}$

8. $\left(\dfrac{3a^3 b}{2a}\right)^2 \left(\dfrac{a^2}{-3b^2}\right)^3 = \left(\dfrac{3a^2 b}{2}\right)^2 \left(\dfrac{a^2}{-3b^2}\right)^3$

    $= \left(\dfrac{3^2 a^4 b^2}{2^2}\right)\left(\dfrac{a^6}{(-3)^3 b^6}\right)$

    $= \left(\dfrac{9a^4 b^2}{4}\right)\left(\dfrac{a^6}{-27b^6}\right)$

    $= \dfrac{9a^4 b^2 a^6}{4(-27)b^6}$

    $= \dfrac{9a^{10} b^2}{-108b^6}$

    $= -\dfrac{a^{10}}{12b^4}$

9. $\begin{array}{r} 2x^2 + 4x - 1 \\ \hline x - 4 \\ \hline -8x^2 - 16x + 4 \\ 2x^3 + 4x^2 - x \\ \hline 2x^3 - 4x^2 - 17x + 4 \end{array}$

10. $a^4 - 2a^2 - 8 = (a^2)^2 - 2(a^2) - 8$
    $= (a^2 - 4)(a^2 + 2)$
    $= (a+2)(a-2)(a^2 + 2)$

11. $x^3 y + x^2 y^2 - 6xy^3 = xy(x^2 + xy - 6y^2)$
    $= xy(x + 3y)(x - 2y)$

12. $(b+2)(b-5) = 2b + 14$
    $b^2 - 3b - 10 = 2b + 14$
    $b^2 - 5b - 24 = 0$
    $(b-8)(b+3) = 0$
    $b - 8 = 0 \quad b + 3 = 0$
    $b = 8 \quad b = -3$
    The solutions are 8 and −3.

13. $x^2 - 2x > 15$
    $x^2 - 2x - 15 > 0$
    $(x-5)(x+3) > 0$

    $\{x | x < -3 \text{ or } x > 5\}$

14. $\dfrac{x^2+4x-5}{2x^2-3x+1} - \dfrac{x}{2x-1} = \dfrac{(x+5)(x-1)}{(2x-1)(x-1)} - \dfrac{x}{2x-1}$

$\qquad\qquad\qquad\qquad\quad = \dfrac{x+5}{2x-1} - \dfrac{x}{2x-1}$

$\qquad\qquad\qquad\qquad\quad = \dfrac{x+5-x}{2x-1}$

$\qquad\qquad\qquad\qquad\quad = \dfrac{5}{2x-1}$

15. $\qquad\qquad\dfrac{5}{x^2+7x+12} = \dfrac{9}{x+4} - \dfrac{2}{x+3}$

$(x+4)(x+3)\dfrac{5}{(x+4)(x+3)} = (x+4)(x+3)\left[\dfrac{9}{x+4} - \dfrac{2}{x+3}\right]$

$\qquad\qquad\qquad 5 = (x+3)9 - (x+4)2$

$\qquad\qquad\qquad 5 = 9x + 27 - 2x - 8$

$\qquad\qquad\qquad 5 = 7x + 19$

$\qquad\qquad\quad -14 = 7x$

$\qquad\qquad\quad -2 = x$

The solution is -2.

16. $\dfrac{4-6i}{2i} = \dfrac{4-6i}{2i} \cdot \dfrac{i}{i}$

$\qquad\quad = \dfrac{4i - 6i^2}{2i^2}$

$\qquad\quad = \dfrac{4i + 6}{-2}$

$\qquad\quad = -3 - 2i$

17. $m = \dfrac{y_2 - y_1}{x_2 - x_1} = \dfrac{-6-4}{2-(-3)} = \dfrac{-10}{5} = -2$

$y - y_1 = m(x - x_1)$

$y - 4 = -2[x - (-3)]$

$y - 4 = -2(x + 3)$

$y - 4 = -2x - 6$

$\quad y = -2x - 2$

18. The product of the slopes of perpendicular lines is -1.

$\quad 2x - 3y = 6 \qquad\qquad m_1 \cdot m_2 = -1$

$\qquad -3y = -2x + 6 \qquad \dfrac{2}{3} \cdot m_2 = -1$

$\qquad\quad y = \dfrac{2}{3}x - 2 \qquad\qquad m_2 = -\dfrac{3}{2}$

$y - y_1 = m(x - x_1)$

$y - 1 = -\dfrac{3}{2}[x - (-3)]$

$y - 1 = -\dfrac{3}{2}(x + 3)$

$y - 1 = -\dfrac{3}{2}x - \dfrac{9}{2}$

$\quad y = -\dfrac{3}{2}x - \dfrac{7}{2}$

19. $3x^2 = 3x - 1$
$3x^2 - 3x + 1 = 0$
$a = 3, b = -3, c = 1$
$x = \dfrac{-b \pm \sqrt{b^2 - 4ac}}{2a}$
$= \dfrac{-(-3) \pm \sqrt{(-3)^2 - 4(3)(1)}}{2(3)}$
$= \dfrac{3 \pm \sqrt{9 - 12}}{6}$
$= \dfrac{3 \pm \sqrt{-3}}{6}$
$= \dfrac{3 \pm i\sqrt{3}}{6}$
$= \dfrac{1}{2} \pm \dfrac{\sqrt{3}}{6}i$

The solutions are $\dfrac{1}{2} + \dfrac{\sqrt{3}}{6}i$ and $\dfrac{1}{2} - \dfrac{\sqrt{3}}{6}i$.

20. $\sqrt{8x+1} = 2x - 1$
$\left(\sqrt{8x+1}\right)^2 = (2x-1)^2$
$8x + 1 = 4x^2 - 4x + 1$
$0 = 4x^2 - 12x$
$0 = 4x(x - 3)$
$4x = 0 \quad x - 3 = 0$
$x = 0 \quad\ x = 3$
Check:

$\sqrt{8x+1} = 2x - 1$
$\sqrt{8(0)+1} \mid 2(0) - 1$
$\sqrt{1} \mid -1$
$1 \neq -1$

$\sqrt{8x+1} = 2x - 1$
$\sqrt{8(3)+1} \mid 2(3) - 1$
$\sqrt{24+1} \mid 6 - 1$
$\sqrt{25} \mid 5$
$5 = 5$

The solution is 3.

21. $f(x) = 2x^2 - 3$
$a = 2, b = 0, c = -3$
$x = -\dfrac{b}{2a} = \dfrac{-0}{2 \cdot 2} = 0$
$f(0) = 2(0)^2 - 3 = -3$
The minimum value of the function is –3.

22. $f(x) = |3x - 4|$;
domain = {0, 1, 2, 3}
$f(x) = |3x - 4|$
$f(0) = |3(0) - 4| = |0 - 4| = |-4| = 4$
$f(1) = |3(1) - 4| = |3 - 4| = |-1| = 1$
$f(2) = |3(2) - 4| = |6 - 4| = |2| = 2$
$f(3) = |3(3) - 4| = |9 - 4| = |5| = 5$
The range is {1, 2, 4, 5}.

23. {(–3, 0), (–2, 0), (–1, 1), (0, 1)}
Each member of the domain is paired with only one member of the range. The set of ordered pairs is a function.

24. $\sqrt[3]{5x - 2} = 2$
$\left(\sqrt[3]{5x - 2}\right)^3 = 2^3$
$5x - 2 = 8$
$5x = 10$
$x = 2$
The solution is 2.

25. $h(x) = \dfrac{1}{2}x + 4$
$h(2) = \dfrac{1}{2}(2) + 4 = 1 + 4 = 5$
$g(x) = 3x - 5$
$g(5) = 3(5) - 5 = 15 - 5 = 10$
$g(h(2)) = 10$

26. $f(x) = -3x + 9$
$y = -3x + 9$
$x = -3y + 9$
$3y = -x + 9$
$y = -\dfrac{1}{3}x + 3$

The inverse function is $f^{-1}(x) = -\dfrac{1}{3}x + 3$.

27. **Strategy**
Cost per pounds of the mixture: x

	Amount	Cost	Value
$4.50 tea	30	4.50	4.50(30)
$3.60 tea	45	3.60	3.60(45)
Mixture	75	x	75x

The sum of the values before mixing equals the value after mixing.

Solution
$4.50(30) + 3.60(45) = 75x$
$135 + 162 = 75x$
$297 = 75x$
$3.96 = x$

The cost per pound of the mixture is $3.96.

28. Strategy
Pounds of 80% copper alloy: x

	Amount	Percent	Quantity
80%	x	0.80	$0.80x$
20%	50	0.20	$0.20(50)$
40%	$50 + x$	0.40	$0.40(50 + x)$

The sum of the quantities before mixing is equal to the quantity after mixing.

Solution
$$0.80x + 0.20(50) = 0.40(50 + x)$$
$$0.80x + 10 = 20 + 0.40x$$
$$0.40x + 10 = 20$$
$$0.40x = 10$$
$$x = 25$$
25 lb of the 80% copper alloy must be used.

29. Strategy
To find the additional amount of insecticide, write and solve a proportion using x to represent the additional amount of insecticide. Then, $x + 6$ is the total amount of insecticide.

Solution
$$\frac{6}{16} = \frac{x+6}{28}$$
$$\frac{3}{8} = \frac{x+6}{28}$$
$$\frac{3}{8} \cdot 56 = \frac{x+6}{28} \cdot 56$$
$$21 = (x+6)2$$
$$21 = 2x + 12$$
$$9 = 2x$$
$$4.5 = x$$
An additional 4.5 oz of insecticide are required.

30. Strategy
This is a work problem.
Time for the smaller pipe to fill the tank: t
Time for the larger pipe to fill the tank: $t - 8$

	Rate	Time	Part
Smaller pipe	$\frac{1}{t}$	3	$\frac{3}{t}$
Larger pipe	$\frac{1}{t-8}$	3	$\frac{3}{t-8}$

The sum of the parts of the task completed must equal 1.

Solution
$$\frac{3}{t} + \frac{3}{t-8} = 1$$
$$t(t-8)\left(\frac{3}{t} + \frac{3}{t-8}\right) = t(t-8)$$
$$(t-8)3 + 3t = t^2 - 8t$$
$$3t - 24 + 3t = t^2 - 8t$$
$$6t - 24 = t^2 - 8t$$
$$0 = t^2 - 14t + 24$$
$$= (t-2)(t-12)$$
$$t - 2 = 0 \quad t - 12 = 0$$
$$t = 2 \quad\quad t = 12$$
The solution 2 is not possible since the time for the larger pipe would then be a negative number.
$t - 8 = 12 - 8 = 4$
It would take the larger pipe 4 minutes to fill the tank.

31. Strategy
To find the distance:
Write the basic direct variation equation, replace the variables by the given values, and solve for k. Write the direct variation equation, replacing k by its value. Substitute 40 for f and solve for d.

Solution
$$d = kf \quad\quad d = \frac{3}{5}f$$
$$30 = k(50) \quad\quad = \frac{3}{5}(40)$$
$$\frac{3}{5} = k \quad\quad = 24$$
A force of 40 lb will stretch the string 24 in.

32. Strategy
To find the frequency:
Write the basic inverse variation equation, replace the variables by the given values, and solve for k. Write the inverse variation equation, replacing k by its value. Substitute 1.5 for L and solve for f.

Solution
$$f = \frac{k}{L} \quad\quad f = \frac{120}{L}$$
$$60 = \frac{k}{2} \quad\quad = \frac{120}{1.5}$$
$$120 = k \quad\quad = 80$$
The frequency is 80 times per minute.

Chapter 10: Exponential and Logarithmic Functions

Section 10.1

Concept Review 10.1

1. Never true
   The domain is $\{x|x \text{ is a real number}\}$.

3. Never true
   The graph of $f(x) = b^x$ passes through the point $(0, 1)$.

5. Never true
   The range of $f(x) = b^x$ is $\{f(x)|f(x) > 0\}$. The function does not have x-intercepts.

Objective 1 Exercises

3. $f(x) = 3^x$

   a. $f(2) = 3^2 = 9$

   b. $f(0) = 3^0 = 1$

   c. $f(-2) = 3^{-2} = \dfrac{1}{3^2} = \dfrac{1}{9}$

5. $g(x) = 2^{x+1}$

   a. $g(3) = 2^{3+1} = 2^4 = 16$

   b. $g(1) = 2^{1+1} = 2^2 = 4$

   c. $g(-3) = 2^{-3+1} = 2^{-2} = \dfrac{1}{2^2} = \dfrac{1}{4}$

7. $P(x) = \left(\dfrac{1}{2}\right)^{2x}$

   a. $P(0) = \left(\dfrac{1}{2}\right)^{2 \cdot 0} = \left(\dfrac{1}{2}\right)^0 = 1$

   b. $P\left(\dfrac{3}{2}\right) = \left(\dfrac{1}{2}\right)^{2 \cdot \frac{3}{2}} = \left(\dfrac{1}{2}\right)^3 = \dfrac{1}{8}$

   c. $P(-2) = \left(\dfrac{1}{2}\right)^{2(-2)} = \left(\dfrac{1}{2}\right)^{-4} = 2^4 = 16$

9. $G(x) = e^{x/2}$

   a. $G(4) = e^{4/2} = e^2 \approx 7.3891$

   b. $G(-2) = e^{-2/2} = e^{-1} = \dfrac{1}{e} \approx 0.3679$

   c. $G\left(\dfrac{1}{2}\right) = e^{\frac{1}{2}/2} = e^{1/4} = e^{0.25} \approx 1.2840$

11. $H(r) = e^{-r+3}$

    a. $H(-1) = e^{-(-1)+3} = e^{1+3} = e^4 \approx 54.5982$

    b. $H(3) = e^{-3+3} = e^0 = 1$

    c. $H(5) = e^{-5+3} = e^{-2} = \dfrac{1}{e^2} \approx 0.1353$

13. $F(x) = 2^{x^2}$

    a. $F(2) = 2^{2^2} = 2^4 = 16$

    b. $F(-2) = 2^{(-2)^2} = 2^4 = 16$

    c. $F\left(\dfrac{3}{4}\right) = 2^{\left(\frac{3}{4}\right)^2} = 2^{\frac{9}{16}} = \sqrt[16]{2^9} = \sqrt[16]{512} \approx 1.4768$

15. $f(x) = e^{-x^2/2}$

    a. $f(-2) = e^{-(-2)^2/2}$
       $= e^{-4/2}$
       $= e^{-2}$
       $= \dfrac{1}{e^2}$
       ≈ 0.1353

    b. $f(2) = e^{-(2)^2/2} = e^{-4/2} = e^{-2} = \dfrac{1}{e^2} \approx 0.1353$

    c. $f(-3) = e^{-(-3)^2/2} = e^{-9/2} = \dfrac{1}{e^{9/2}} \approx 0.0111$

Objective 2 Exercises

17.

19.

21.

23.

25.

27.

29.

31.

33.
The zero of f is 1.6.

35.
The value of x for which $f(x) = 3$ is 1.1.

37.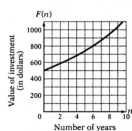
In approximately 9 years the investment will be worth $1000.

39.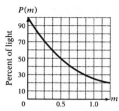
50% of the light reaches 0.5 meter below the surface of the ocean.

41.
The minimum value of f is 0.

43.
The minimum value of f is 0.80.

45. When $n = 100$, $\left(1+\dfrac{1}{n}\right)^n \approx 2.704813829$.

When $n = 1000$, $\left(1+\dfrac{1}{n}\right)^n \approx 2.716923932$.

When $n = 10{,}000$, $\left(1+\dfrac{1}{n}\right)^n \approx 2.718145927$.

When $n = 100{,}000$, $\left(1+\dfrac{1}{n}\right)^n \approx 2.718268237$.

As n increases, $\left(1+\dfrac{1}{n}\right)^n$ becomes closer to e.

47. a.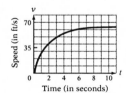

b. At $t = 4$ seconds after the object is dropped, it will be falling at a speed of 55.3 feet per second.

Section 10.2

Concept Review 10.2

1. Always true

3. Never true
   $\dfrac{\log x}{\log y}$ is in simplest form.

5. Always true

7. Always true

Objective 1 Exercises

3. $5^2 = 25$ is equivalent to $\log_5 25 = 2$.

5. $4^{-2} = \dfrac{1}{16}$ is equivalent to $\log_4\left(\dfrac{1}{16}\right) = -2$.

7. $10^y = x$ is equivalent to $\log_{10} x = y$.

9. $a^x = w$ is equivalent to $\log_a w = x$.

11. $\log_3 9 = 2$ is equivalent to $3^2 = 9$.

13. $\log 0.01 = -2$ is equivalent to $10^{-2} = 0.01$.

15. $\ln x = y$ is equivalent to $e^y = x$.

17. $\log_b u = v$ is equivalent to $b^v = u$.

19. $\log_3 81 = x$
$3^x = 81$
$3^x = 3^4$
$x = 4$
$\log_3 81 = 4$

21. $\log_2 128 = x$
$2^x = 128$
$2^x = 2^7$
$x = 7$
$\log_2 128 = 7$

23. $\log 100 = x$
$10^x = 100$
$10^x = 10^2$
$x = 2$
$\log 100 = 2$

25. $\ln e^3 = x$
$3\ln e = x$
$3(1) = x$
$x = 3$
$\ln e^3 = 3$

27. $\log_8 1 = x$
$8^x = 1$
$x = 0$
$\log_8 1 = 0$

29. $\log_5 625 = x$
$5^x = 625$
$5^x = 5^4$
$x = 4$
$\log_5 625 = 4$

31. $\log_3 x = 2$
$3^2 = x$
$9 = x$

33. $\log_4 x = 3$
$4^3 = x$
$64 = x$

35. $\log_7 x = -1$
$7^{-1} = x$
$\dfrac{1}{7} = x$

37. $\log_6 x = 0$
$6^0 = x$
$1 = x$

39. $\log x = 2.5$
$10^{2.5} = x$
$316.23 \approx x$

41. $\log x = -1.75$
$10^{-1.75} = x$
$0.02 \approx x$

43. $\ln x = 2$
$e^2 = x$
$7.39 \approx x$

45. $\ln x = -\dfrac{1}{2}$
$e^{-1/2} = x$
$0.61 \approx x$

Objective 2 Exercises

49. $\log_8(xz) = \log_8 x + \log_8 z$

51. $\log_3 x^5 = 5\log_3 x$

53. $\ln\left(\dfrac{r}{s}\right) = \ln r - \ln s$

55. $\log_3(x^2 y^6) = \log_3 x^2 + \log_3 y^6$
$= 2\log_3 x + 6\log_3 y$

57. $\log_7\left(\dfrac{u^3}{v^4}\right) = \log_7 u^3 - \log_7 v^4$
$= 3\log_7 u - 4\log_7 v$

59. $\log_2(rs)^2 = 2\log_2(rs)$
$= 2(\log_2 r + \log_2 s)$
$= 2\log_2 r + 2\log_2 s$

61. $\log_9 x^2 yz = \log_9 x^2 + \log_9 y + \log_9 z$
$= 2\log_9 x + \log_9 y + \log_9 z$

63. $\ln\left(\dfrac{xy^2}{z^4}\right) = \ln(xy^2) - \ln z^4$
$= \ln x + \ln y^2 - \ln z^4$
$= \ln x + 2\ln y - 4\ln z$

65. $\log_8\left(\dfrac{x^2}{yz^2}\right) = \log_8 x^2 - \log_8(yz^2)$
$= \log_8 x^2 - (\log_8 y + \log_8 z^2)$
$= \log_8 x^2 - \log_8 y - \log_8 z^2$
$= 2\log_8 x - \log_8 y - 2\log_8 z$

67. $\log_7 \sqrt{xy} = \log_7(xy)^{1/2}$
$= \dfrac{1}{2}\log_7(xy)$
$= \dfrac{1}{2}(\log_7 x + \log_7 y)$
$= \dfrac{1}{2}\log_7 x + \dfrac{1}{2}\log_7 y$

69. $\log_2 \sqrt{\dfrac{x}{z}} = \log_2\left(\dfrac{x}{y}\right)^{1/2}$
$= \dfrac{1}{2}\log_2\left(\dfrac{x}{y}\right)$
$= \dfrac{1}{2}(\log_2 x - \log_2 y)$
$= \dfrac{1}{2}\log_2 x - \dfrac{1}{2}\log_2 y$

71. $\ln\sqrt{x^3 y} = \ln(x^3 y)^{1/2}$
$= \dfrac{1}{2}\ln(x^3 y)$
$= \dfrac{1}{2}(\ln x^3 + \ln y)$
$= \dfrac{1}{2}(3\ln x + \ln y)$
$= \dfrac{3}{2}\ln x + \dfrac{1}{2}\ln y$

73. $\log_7\sqrt{\dfrac{x^3}{y}} = \log_7\left(\dfrac{x^3}{y}\right)^{1/2}$
$= \dfrac{1}{2}\log_7\left(\dfrac{x^3}{y}\right)$
$= \dfrac{1}{2}(\log_7 x^3 - \log_7 y)$
$= \dfrac{1}{2}(3\log_7 x - \log_7 y)$
$= \dfrac{3}{2}\log_7 x - \dfrac{1}{2}\log_7 y$

75. $\log_3 x^3 - \log_3 y = \log_3\left(\dfrac{x^3}{y}\right)$

77. $\log_8 x^4 + \log_8 y^2 = \log_8(x^4 y^2)$

79. $3\ln x = \ln x^3$

81. $3\log_5 x + 4\log_5 y = \log_5 x^3 + \log_5 y^4$
$= \log_5(x^3 y^4)$

83. $-2\log_4 x = \log_4(x^{-2}) = \log_4\left(\dfrac{1}{x^2}\right)$

85. $2\log_3 x - \log_3 y + 2\log_3 z$
$= \log_3 x^2 - \log_3 y + \log_3 z^2$
$= \log_3\left(\dfrac{x^2}{y}\right) + \log_3 z^2$
$= \log_3\left(\dfrac{x^2 z^2}{y}\right)$

87. $\log_b x - (2\log_b y + \log_b z)$
$= \log_b x - (\log_b y^2 + \log_b z)$
$= \log_b x - \log_b y^2 z$
$= \log_b\left(\dfrac{x}{y^2 z}\right)$

89. $2(\ln x + \ln y) = 2\ln(xy) = \ln(xy)^2 = \ln(x^2 y^2)$

91. $\dfrac{1}{2}(\log_6 x - \log_6 y) = \dfrac{1}{2}\log_6\left(\dfrac{x}{y}\right)$
$= \log_6\left(\dfrac{x}{y}\right)^{1/2}$
$= \log_6\sqrt{\dfrac{x}{y}}$

93. $2(\log_4 s - 2\log_4 t + \log_4 r)$
$= 2\left(\log_4\dfrac{s}{t^2} + \log_4 r\right)$
$= 2\log_4\left(\dfrac{sr}{t^2}\right)$
$= \log_4\left(\dfrac{sr}{t^2}\right)^2$
$= \log_4\dfrac{s^2 r^2}{t^4}$

95. $\log_5 x - 2(\log_5 y + \log_5 z) = \log_5 x - 2\log_5(yz)$
$= \log_5 x - \log_5(yz)^2$
$= \log_5 x - \log_5 y^2 z^2$
$= \log_5\dfrac{x}{y^2 z^2}$

97. $3\ln t - 2(\ln r - \ln v) = \ln t^3 - 2\ln\left(\dfrac{r}{v}\right)$
$= \ln t^3 - \ln\left(\dfrac{r}{v}\right)^2$
$= \ln t^3 - \ln\left(\dfrac{r^2}{v^2}\right)$
$= \ln\dfrac{t^3}{\frac{r^2}{v^2}}$
$= \ln\dfrac{t^3 v^2}{r^2}$

99. $\frac{1}{2}(3\log_4 x - 2\log_4 y + \log_4 z)$
$= \frac{1}{2}(\log_4 x^3 - \log_4 y^2 + \log_4 z)$
$= \frac{1}{2}\left(\log_4 \frac{x^3}{y^2} + \log_4 z\right)$
$= \frac{1}{2}\log_4\left(\frac{x^3 z}{y^2}\right)$
$= \log_4\left(\frac{x^3 z}{y^2}\right)^{1/2}$
$= \log_4 \sqrt{\frac{x^3 z}{y^2}}$

101. $\ln 4 \approx 1.3863$

103. $\ln\left(\frac{17}{6}\right) = \ln 17 - \ln 6 \approx 1.0415$

105. $\log_8 6 = \frac{\log 6}{\log 8} \approx 0.8617$

107. $\log_5 30 = \frac{\log 30}{\log 5} \approx 2.1133$

109. $\log_3(0.5) = \frac{\log(0.5)}{\log 3} \approx -0.6309$

111. $\log_7(1.7) = \frac{\log 1.7}{\log 7} \approx 0.2727$

113. $\log_5 15 = \frac{\log 15}{\log 5} \approx 1.6826$

115. $\log_{12} 120 = \frac{\log 120}{\log 12} \approx 1.9266$

117. $\log_3(3x-2) = \frac{\log(3x-2)}{\log 3}$

119. $\log_8(4-9x) = \frac{\log(4-9x)}{\log 8}$

121. $5\log_9(6x+7) = 5\frac{\log(67x+7)}{\log 9}$
$= \frac{5}{\log 9}\log(6x+7)$

123. $\log_2(x+5) = \frac{\ln(x+5)}{\ln 2}$

125. $\log_3(x^2+9) = \frac{\ln(x^2+9)}{\ln 3}$

127. $7\log_8(10x-7) = 7\frac{\ln(10x-7)}{\ln 8}$
$= \frac{7}{\ln 8}\ln(10x-7)$

Applying Concepts 10.2

129. $\log_8 x = 3\log_8 2$
$\log_8 x = \log_8 2^3$
$\log_8 x = \log_8 8$
$x = 8$
The solution is 8.

131. $\log_4 x = \log_4 2 + \log_4 3$
$\log_4 x = \log_4(2 \cdot 3)$
$\log_4 x = \log_4 6$
$x = 6$
The solution is 6.

133. $\log_6 x = 3\log_6 2 - \log_6 4$
$\log_6 x = \log_6 2^3 - \log_6 4$
$\log_6 x = \log_6 8 - \log_6 4$
$\log_6 x = \log_6\left(\frac{8}{4}\right)$
$\log_6 x = \log_6 2$
$x = 2$
The solution is 2.

135. $\log x = \frac{1}{3}\log 27$
$\log x = \log 27^{1/3}$
$\log x = \log \sqrt[3]{27}$
$\log x = \log 3$
$x = 3$
The solution is 3.

137. $\log_a a^x = x\log_a a = x(1) = x$

139. **a.** $a^c = b$

b. $\text{antilog}_a(\log_a b) = \text{antilog}_a(c) = b$

141. $S(t) = 8\log_5(6t+2)$
$S(2) = 8\log_5(6 \cdot 2 + 2)$
$= 8\log_5 14$
$= 8\frac{\log 14}{\log 5}$
≈ 13.12

143. $G(x) = -5\log_7(2x+19)$
$G(-3) = -5\log_7[2(-3)+19]$
$= -5\log_7 13$
$= -5\frac{\log 13}{\log 7}$
≈ -6.59

145. $\ln(\ln x) = 1$
$e^1 = \ln x$
$e = \ln x$
$e^e = x$
$15.1543 \approx x$
The solution is 15.1543.

147. $\log\left(\dfrac{x - \sqrt{x^2 - a^2}}{a^2}\right) = -\log\left(x + \sqrt{x^2 - a^2}\right)$

$\log\left(\dfrac{x - \sqrt{x^2 - a^2}}{a^2}\right) = \log\left(x + \sqrt{x^2 - a^2}\right)^{-1}$

$\log\left(\dfrac{x - \sqrt{x^2 - a^2}}{a^2}\right) = \log\dfrac{1}{x + \sqrt{x^2 - a^2}}$

$\dfrac{x - \sqrt{x^2 - a^2}}{a^2} = \dfrac{1}{x + \sqrt{x^2 - a^2}}$

$\dfrac{x - \sqrt{x^2 - a^2}}{a^2} = \dfrac{1}{x + \sqrt{x^2 - a^2}} \cdot \dfrac{x - \sqrt{x^2 - a^2}}{x - \sqrt{x^2 - a^2}}$

$\dfrac{x - \sqrt{x^2 - a^2}}{a^2} = \dfrac{x - \sqrt{x^2 - a^2}}{x^2 - (x^2 - a^2)}$

$\dfrac{x - \sqrt{x^2 - a^2}}{a^2} = \dfrac{x - \sqrt{x^2 - a^2}}{x^2 - x^2 + a^2}$

$\dfrac{x - \sqrt{x^2 - a^2}}{a^2} = \dfrac{x - \sqrt{x^2 - a^2}}{a^2}$

Section 10.3

Concept Review 10.3

1. Never true
The domain of $f(x) = \log_b x$ is $\{x | x > 0\}$.

3. Always true

Objective 1 Exercises

3. $f(x) = \log_4 x$
$y = \log_4 x$

$y = \log_4 x$ is equivalent to $x = 4^y$.

5. $f(x) = \log_3(2x - 1)$
$y = \log_3(2x - 1)$

$y = \log_3(2x - 1)$ is equivalent to $(2x - 1) = 3^y$, $2x = 3^y + 1$, or $x = \dfrac{1}{2}(3^y + 1)$.

7. $f(x) = 3\log_2 x$
$y = 3\log_2 x$
$\dfrac{y}{3} = \log_2 x$

$\dfrac{y}{3} = \log_2 x$ is equivalent to $x = 2^{y/3}$.

9. $f(x) = -\log_2 x$
$y = -\log_2 x$
$-y = \log_2 x$

$-y = \log_2 x$ is equivalent to $x = 2^{-y}$.

11. $f(x) = \log_2(x-1)$
$y = \log_2(x-1)$

$y = \log_2(x-1)$ is equivalent to $(x-1) = 2^y$, or $x = 2^y + 1$.

13. $f(x) = -\log_2(x-1)$
$y = -\log_2(x-1)$
$-y = \log_2(x-1)$

$-y = \log_2(x-1)$ is equivalent to $(x-1) = 2^{-y}$, or $x = 2^{-y} + 1$.

15. $f(x) = \log_2 x - 3$
$y = \log_2 x - 3$
$y = \dfrac{\log x}{\log 2} - 3$
$y = \dfrac{\log x}{0.3010} - 3$

17. $f(x) = -\log_2 x + 2$
$y = -\log_2 x + 2$
$y = -\dfrac{\log x}{\log 2} + 2$
$y = -\dfrac{\log x}{0.3010} + 2$

19. $f(x) = x - \log_2(1-x)$
$y = x - \log_2(1-x)$
$y = x - \dfrac{\log(1-x)}{\log 2}$
$y = x - \dfrac{\log(1-x)}{0.3010}$

21. $f(x) = \dfrac{x}{2} - 2\log_2(x+1)$
$y = \dfrac{x}{2} - \log_2(x+1)^2$
$= \dfrac{x}{2} - \dfrac{\log(x+1)^2}{\log 2}$
$= \dfrac{x}{2} - \dfrac{\log(x+1)^2}{0.3010}$

23. $f(x) = x^2 - 10\ln(x-1)$
$y = x^2 - 10\ln(x-1)$

25.

27.

Applying Concepts 10.3

29. $f(x) = \log_2(x+2)$
$x + 2 > 0$
$x > -2$
The domain is $\{x | x > -2\}$.

31. $f(x) = \ln(x^2 + 4)$
$x^2 + 4 > 0$ for all t.
The domain is $\{x | x \text{ is a real number}\}$.

33. $f(x) = \log_4\left(\dfrac{x}{x+2}\right)$
$\dfrac{x}{x+2} > 0$

```
x       - - - - | - - | + + + + +
x + 2   - - - - | + + | + + + + +
       -5 -4 -3 -2 -1  0  1  2  3  4  5
```

The domain is $\{x | x < -2 \text{ or } x > 0\}$.

35. $f(x) = e^{-x+2}$
$y = e^{-x+2}$
$x = e^{-y+2}$
$\ln x = \ln e^{-y+2}$
$\ln x = (-y+2)\ln e$
$\ln x = (-y+2)(1)$
$\ln x = -y + 2$
$\ln x - 2 = -y$
$-(\ln x - 2) = y$
$f^{-1}(x) = -(\ln x - 2) = -\ln x + 2$

37. $f(x) = \ln(2x) + 3$
$y = \ln(2x) + 3$
$x = \ln(2y) + 3$
$x - 3 = \ln(2y)$
$e^{x-3} = 2y$
$\dfrac{e^{x-3}}{2} = y$
$f^{-1}(x) = \dfrac{e^{x-3}}{2}$

39.
The minimum value of f is 0.8.

41. a. $S = 60 - 7\ln(t+1)$

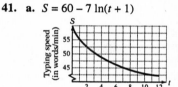

b. After 4 months without typing practice, a typist's proficiency decreases to 49 words per minute.

43. a.

   (graph: Interest rate (as a percent) vs Term (in years))

b. $y = 13$ when $x \approx 8.5$.
A security that has a yield of 13% has a term of 8.5 years.

c. When $x = 30$, $y \approx 12.2$.
The model predicts an interest rate of 12.2%.

Section 10.4

Concept Review 10.4

1. Always true

3. Never true
The logarithm of a negative number is not defined.

5. Always true

Objective 1 Exercises

3. $5^{4x-1} = 5^{x+2}$
$4x - 1 = x + 2$
$3x - 1 = 2$
$3x = 3$
$x = 1$
The solution is 1.

5. $8^{x-4} = 8^{5x+8}$
$x - 4 = 5x + 8$
$-4x - 4 = 8$
$-4x = 12$
$x = -3$
The solution is -3.

7. $5^x = 6$
$\log 5^x = \log 6$
$x \log 5 = \log 6$
$x = \dfrac{\log 6}{\log 5}$
$x \approx 1.1133$
The solution is 1.1133.

9. $12^x = 6$
$\log 12^x = \log 6$
$x \log 12 = \log 6$
$x = \dfrac{\log 6}{\log 12}$
$x \approx 0.7211$
The solution is 0.7211.

11. $\left(\dfrac{1}{2}\right)^x = 3$
$\log\left(\dfrac{1}{2}\right)^x = \log 3$
$x \log\left(\dfrac{1}{2}\right) = \log 3$
$x = \dfrac{\log 3}{\log \frac{1}{2}}$
$x = \dfrac{\log 3}{\log 0.5}$
$x \approx -1.5850$
The solution is -1.5850.

13. $1.5^x = 2$
$\log 1.5^x = \log 2$
$x \log 1.5 = \log 2$
$x = \dfrac{\log 2}{\log 1.5}$
$x \approx 1.7095$
The solution is 1.7095.

15. $10^x = 21$
$\log 10^x = \log 21$
$x \log 10 = \log 21$
$x = \dfrac{\log 21}{\log 10}$
$x \approx 1.3222$
The solution is 1.3222.

17. $2^{-x} = 7$
$\log 2^{-x} = \log 7$
$-x \log 2 = \log 7$
$-x = \dfrac{\log 7}{\log 2}$
$x = -\dfrac{\log 7}{\log 2}$
$x \approx -2.8074$
The solution is -2.8074.

19. $2^{x-1} = 6$
$\log 2^{x-1} = \log 6$
$(x-1) \log 2 = \log 6$
$x - 1 = \dfrac{\log 6}{\log 2}$
$x = \dfrac{\log 6}{\log 1} + 1$
$x \approx 3.5850$
The solution is 3.5850.

21. $3^{2x-1} = 4$
$\log 3^{2x-1} = \log 4$
$(2x-1) \log 3 = \log 4$
$2x - 1 = \dfrac{\log 4}{\log 3}$
$2x = \dfrac{\log 4}{\log 3} + 1$
$x = \dfrac{1}{2}\left(\dfrac{\log 4}{\log 3} + 1\right)$
$x \approx 1.1309$
The solution is 1.1309.

23. $9^x = 3^{x+1}$
$3^{2x} = 3^{x+1}$
$2x = x + 1$
$x = 1$
The solution is 1.

25. $8^{x+2} = 16^x$
$(2^3)^{x+2} = 2^{4x}$
$2^{3x+6} = 2^{4x}$
$3x + 6 = 4x$
$6 = x$
The solution is 6.

27. $5^{x^2} = 21$
$\log 5^{x^2} = \log 21$
$x^2 \log 5 = \log 21$
$x^2 = \dfrac{\log 21}{\log 5}$
$x = \pm\sqrt{\dfrac{\log 21}{\log 5}}$
$x = \pm 1.3754$
The solutions are 1.3754 and -1.3754.

29. $2^{4x-2} = 20$
$\log 2^{4x-2} = \log 20$
$(4x-2) \log 2 = \log 20$
$4x - 2 = \dfrac{\log 20}{\log 2}$
$4x = \dfrac{\log 20}{\log 2} + 2$
$x = \dfrac{1}{4}\left(\dfrac{\log 20}{\log 2} + 2\right)$
$x = 1.5805$
The solution is 1.5805.

31.
$$3^{-x+2} = 18$$
$$\log 3^{-x+2} = \log 18$$
$$(-x+2)\log 3 = \log 18$$
$$-x+2 = \frac{\log 18}{\log 3}$$
$$-x = \frac{\log 18}{\log 3} - 2$$
$$x = -\frac{\log 18}{\log 3} + 2$$
$$x \approx -0.6309$$
The solution is −0.6309.

33.
$$4^{2x} = 100$$
$$\log 4^{2x} = \log 100$$
$$2x \log 4 = 2$$
$$2x = \frac{2}{\log 4}$$
$$x = \frac{1}{2}\left(\frac{2}{\log 4}\right) = \frac{1}{\log 4}$$
$$x \approx 1.6610$$
The solution is 1.6610.

35.
$$2.5^{-x} = 4$$
$$\log 2.5^{-x} = \log 4$$
$$-x \log 2.5 = \log 4$$
$$-x = \frac{\log 4}{\log 2.5}$$
$$x = -\frac{\log 4}{\log 2.5}$$
$$x \approx -1.5129$$
The solution is −1.5129.

37.
$$0.25^x = 0.125$$
$$\log 0.25^x = \log 0.125$$
$$x \log 0.25 = \log 0.125$$
$$x = \frac{\log 0.125}{\log 0.25}$$
$$x = 1.5$$
The solution is 1.5.

39.
$$3^x = 2$$
$$3^x - 2 = 0$$

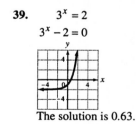

The solution is 0.63.

41.
$$2^x = 2x+4$$
$$2^x - 2x - 4 = 0$$

The solutions are −1.86 and 3.44.

43.
$$e^x = -2x-2$$
$$e^x + 2x + 2 = 0$$

The solution is −1.16.

Objective 2 Exercises

47. $\log_3(x+1) = 2$
Rewrite in exponential form.
$$3^2 = x+1$$
$$9 = x+1$$
$$8 = x$$
The solution is 8.

49. $\log_2(2x-3) = 3$
Rewrite an exponential form.
$$2^3 = 2x-3$$
$$8 = 2x-3$$
$$11 = 2x$$
$$\frac{11}{2} = x$$
The solution is $\frac{11}{2}$.

51. $\log_2(x^2+2x) = 3$
Rewrite in exponential form.
$$2^3 = x^2 + 2x$$
$$8 = x^2 + 2x$$
$$0 = x^2 + 2x - 8$$
$$0 = (x+4)(x-2)$$
$$x+4 = 0 \qquad x-2 = 0$$
$$x = -4 \qquad x = 2$$
The solutions are −4 and 2.

53. $\log_5 \frac{2x}{x-1} = 1$
Rewrite in exponential form.
$$5^1 = \frac{2x}{x-1}$$
$$(x-1)5 = (x-1)\frac{2x}{x-1}$$
$$5x - 5 = 2x$$
$$3x - 5 = 0$$
$$3x = 5$$
$$x = \frac{5}{3}$$
The solution is $\frac{5}{3}$.

55. $\log_7 x = \log_7(1-x)$
Use the fact that if $\log_b u = \log_b v$, then $u = v$.
$$x = 1 - x$$
$$2x = 1$$
$$x = \frac{1}{2}$$
The solution is $\frac{1}{2}$.

57. $\frac{2}{3}\log x = 6$
$\log x^{2/3} = 6$
Rewrite in exponential form.
$$10^6 = x^{2/3}$$
$$(10^6)^{3/2} = (x^{2/3})^{3/2}$$
$$10^9 = x$$
The solution is 1,000,000,000.

59. $\log_2(x-3) + \log_2(x+4) = 3$
$\log_2(x-3)(x+4) = 3$
Rewrite in exponential form.
$$(x-3)(x+4) = 2^3$$
$$x^2 + x - 12 = 8$$
$$x^2 + x - 20 = 0$$
$$(x-4)(x+5) = 0$$
$x - 4 = 0 \qquad x + 5 = 0$
$x = 4 \qquad x = -5$
-5 does not check as a solution. The solution is 4.

61. $\log_3 x + \log_3(x-1) = \log_3 6$
$\log_3 x(x-1) = \log_3 6$
Use the fact that if $\log_b u = \log_b v$, then $u = v$.
$$x(x-1) = 6$$
$$x^2 - x = 6$$
$$x^2 - x - 6 = 0$$
$$(x+2)(x-3) = 0$$
$x + 2 = 0 \qquad x - 3 = 0$
$x = -2 \qquad x = 3$
-2 does not check as a solution. The solution is 3.

63. $\log_2(8x) - \log_2(x^2 - 1) = \log_2 3$
$$\log_2\left(\frac{8x}{x^2-1}\right) = \log_2 3$$
Use the fact that if $\log_b u = \log_b v$, then $u = v$.
$$\frac{8x}{x^2-1} = 3$$
$$(x^2-1)\frac{8x}{x^2-1} = (x^2-1)3$$
$$8x = 3x^2 - 3$$
$$0 = 3x^2 - 8x - 3$$
$$0 = (3x+1)(x-3)$$
$3x + 1 = 0 \qquad x - 3 = 0$
$3x = -1 \qquad x = 3$
$x = -\frac{1}{3}$

$-\frac{1}{3}$ does not check as a solution. The solution is 3.

65. $\log_9 x + \log_9(2x-3) = \log_9 2$
$\log_9 x(2x-3) = \log_9 2$
Use the fact that if $\log_b u = \log_b v$, then $u = v$.
$$x(2x-3) = 2$$
$$2x^2 - 3x = 2$$
$$2x^2 - 3x - 2 = 0$$
$$(2x+1)(x-2) = 0$$
$2x + 1 = 0 \qquad x - 2 = 0$
$2x = -1 \qquad x = 2$
$x = -\frac{1}{2}$

$-\frac{1}{2}$ does not check as a solution. The solution is 2.

67. $\log_8(6x) = \log_8 2 + \log_8(x-4)$
$\log_8(6x) = \log_8 2(x-4)$
Use the fact that if $\log_b u = \log_b v$, then $u = v$.
$$6x = 2(x-4)$$
$$6x = 2x - 8$$
$$4x = -8$$
$$x = -2$$
-2 does not check as a solution. The equation has no solution.

69. $\log_9(7x) = \log_9 2 + \log_9(x^2 - 2)$
$\log_9(7x) = \log_9 2(x^2 - 2)$
Use the fact that if $\log_b u = \log_b v$, then $u = v$.
$7x = 2(x^2 - 2)$
$7x = 2x^2 - 4$
$0 = 2x^2 - 7x - 4$
$0 = (2x+1)(x-4)$
$2x+1 = 0 \qquad x - 4 = 0$
$2x = -1 \qquad x = 4$
$x = -\dfrac{1}{2}$

$-\dfrac{1}{2}$ does not check as a solution. The solution is 4.

71. $\log(x^2 + 3) - \log(x+1) = \log 5$
$\log\left(\dfrac{x^2+3}{x+1}\right) = \log 5$
Use the fact that if $\log_b u = \log_b v$, then $u = v$.
$\dfrac{x^2+3}{x+1} = 5$
$(x+1)\left(\dfrac{x^2+3}{x+1}\right) = (x+1)5$
$x^2 + 3 = 5x + 5$
$x^2 - 5x - 2 = 0$
$x = \dfrac{-(-5) \pm \sqrt{(-5)^2 - 4(1)(-2)}}{2(1)}$
$= \dfrac{5 \pm \sqrt{25+8}}{2}$
$= \dfrac{5 \pm \sqrt{33}}{2}$

The solutions are $\dfrac{5+\sqrt{33}}{2}$ and $\dfrac{5-\sqrt{33}}{2}$.

73. $\log x = -x + 2$
$\log x + x - 2 = 0$

The solution is 1.76.

75. $\log(2x-1) = -x + 3$
$\log(2x-1) + x - 3 = 0$

The solution is 2.42.

77. $\ln(x+2) = x^2 - 3$
$\ln(x+2) - x^2 + 3 = 0$

The solutions are -1.51 and 2.10.

Applying Concepts 10.4

79. $8^{x/2} = 6$
$\log 8^{x/2} = \log 6$
$\dfrac{x}{2} \log 8 = \log 6$
$\dfrac{x}{2} = \dfrac{\log 6}{\log 8}$
$x = \dfrac{2 \log 6}{\log 8}$
$x = 1.7233$
The solution is 1.7233.

81. $5^{3x/2} = 7$
$\log 5^{3x/2} = \log 7$
$\dfrac{3x}{2} \log 5 = \log 7$
$\dfrac{3x}{2} = \dfrac{\log 7}{\log 5}$
$x = \dfrac{2 \log 7}{3 \log 5}$
$x = 0.8060$
The solution is 0.8060.

83. $1.2^{(x/2)-1} = 1.4$
$\log 1.2^{(x/2)-1} = \log 1.4$
$\left(\dfrac{x}{2} - 1\right)(\log 1.2) = \log 1.4$
$\dfrac{x}{2} - 1 = \dfrac{\log 1.4}{\log 1.2}$
$\dfrac{x}{2} = \dfrac{\log 1.4}{\log 1.2} + 1$
$x = 2\left[\dfrac{\log 1.4}{\log 1.2} + 1\right]$
$x = 5.6910$

85. $2^x = 8^y$
$x + y = 4$

$x + y = 4$
$x = -y + 4$

$2^x = 8^y$
$2^x = (2^3)^y$
$2^{-y+4} = 2^{3y}$
$-y + 4 = 3y$
$4 = 4y$
$1 = y$

$x + y = 4$
$x + 1 = 4$
$x = 3$
The solution is (3, 1).

87. $\log(x + y) = 3$
$x = y + 4$

$\log(x + y) = 3$
$\log(y + 4 + y) = 3$
$\log(2y + 4) = 3$
$10^3 = 2y + 4$
$1000 = 2y + 4$
$996 = 2y$
$498 = y$

$x = y + 4 = 498 + 4 = 502$
The solution is (502, 498).

89. $8^{3x} = 4^{2y}$
$x - y = 5$

$x - y = 5$
$x = y + 5$

$8^{3x} = 4^{2y}$
$(2^3)^{3x} = (2^2)^{2y}$
$2^{9x} = 2^{4y}$
$2^{9(y+5)} = 2^{4y}$
$9(y + 5) = 4y$
$9y + 45 = 4y$
$45 = -5y$
$-9 = y$

$x - y = 5$
$x - (-9) = 5$
$x + 9 = 5$
$x = -4$
The solution is (−4, −9).

91. a.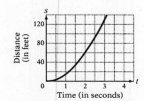

b. It will take the object approximately 2.64 s to fall 100 ft.

Section 10.5

Concept Review 10.5

1. Always true

3. Always true

Objective 1 Exercises

3. **Strategy**
To find the value of the investment, solve the compound interest formula for P. Use $A = 2500$, $n = 7300$, and $i = \dfrac{7.5\%}{365} = \dfrac{0.075}{365} = 0.000205$.

Solution
$P = A(1 + i)^n$
$P = 2500(1 + 0.000205)^{7300}$
$P = 2500(1.000205)^{7300}$
$P \approx 11{,}202.50$
The value of the investment after 20 years is $11,202.50.

5. **Strategy**
To find the amount that needs to be invested, solve the compound interest formula for A. Use $P = 575{,}000$, $n = 36$, and $i = \dfrac{6\%}{12} = \dfrac{0.06}{12} = 0.005$.

Solution
$P = A(1 + i)^n$
$575{,}000 = A(1 + 0.005)^{36}$
$575{,}000 = A(1.005)^{36}$
$\dfrac{575{,}000}{(1.005)^{36}} = A$
$480{,}495.83 = A$
$480,495.83 must be deposited in the account.

7. Strategy
To find the time, solve for t in the exponential decay equation. Use $k = 6$, $A_0 = 30$, and $A = 20$.

Solution
$$A = A_0 \left(\frac{1}{2}\right)^{t/k}$$
$$20 = 30 \left(\frac{1}{2}\right)^{t/6}$$
$$0.\overline{6} = \left(\frac{1}{2}\right)^{t/6}$$
$$0.\overline{6} = (0.5)^{t/6}$$
$$\log(0.\overline{6}) = \log(0.5)^{t/6}$$
$$\log(0.\overline{6}) = \frac{t}{6}\log(0.5)$$
$$\frac{\log(0.\overline{6})}{\log(0.5)} = \frac{t}{6}$$
$$\frac{6\log(0.\overline{6})}{\log(0.5)} = t$$
$$3.510 = t$$

It will take 3.5 hours for the injection to decay to 20 mg.

9. Strategy
To find the half-life, solve for k in the exponential decay equation. Use $A_0 = 25$, $A = 18.95$, and $t = 1$.

Solution
$$A = A_0 \left(\frac{1}{2}\right)^{t/k}$$
$$18.95 = 25 \left(\frac{1}{2}\right)^{1/k}$$
$$0.758 = \left(\frac{1}{2}\right)^{1/k}$$
$$0.758 = (0.5)^{1/k}$$
$$\log(0.758) = \log(0.5)^{1/k}$$
$$\log(0.758) = \frac{1}{k}\log(0.5)$$
$$k\log(0.758) = \log(0.5)$$
$$k = \frac{\log(0.5)}{\log(0.758)}$$
$$k \approx 2.5$$

The half-life is 2.5 years.

11. Strategy
To find the number of weeks, replace P with its given value in the equation and solve for t.

Solution
$$P = 100[1 - (0.75)^t]$$
$$80 = 100[1 - (0.75)^t]$$
$$0.80 = 1 - (0.75)^t$$
$$(0.75)^t = 0.20$$
$$\log(0.75)^t = \log(0.2)$$
$$t\log(0.75) = \log(0.2)$$
$$t = \frac{\log(0.2)}{\log(0.75)}$$
$$t \approx 5.59$$

After 6 weeks the student will make 80% of the welds correctly.

13. Strategy
To find the pH, replace H^+ with its given value and solve for pH.

Solution
$$\text{pH} = -\log(H^+) = -\log(0.045)$$
$$\text{pH} = -\log(4.5 \times 10^{-2})$$
$$= -(\log 4.5 + \log 10^{-2})$$
$$= -[0.6532 + (-2)]$$
$$= 1.3468$$

The pH of the digestive solution is 1.35.

15. Strategy
To find the percent, solve the equation for P. Use $d = 0.5$ and $k = 0.2$.

Solution
$$\log P = -kd$$
$$\log P = -(0.2)(0.5)$$
$$\log P = -0.1$$
$$P = 10^{-0.1}$$
$$P \approx 0.7943$$

79% of the light will pass through the glass.

17. Strategy
To find the thickness needed, solve the given equation for x. Use $I = 0.25 I_0$ and $k = 3.2$.

Solution
$$I = I_0 e^{-kx}$$
$$0.25 I_0 = I_0 e^{-3.2x}$$
$$0.25 = e^{-3.2x}$$
$$\ln 0.25 = \ln(e^{-3.2x})$$
$$\ln 0.25 = -3.2x$$
$$\frac{\ln 0.25}{-3.2} = x$$
$$0.4 = x$$

Use a piece of copper that is 0.4 cm thick.

19. $M = \log A + 3 \log 8t - 2.92$
    $M = \log 30 + 3 \log[8(21)] - 2.92$
    $M \approx 1.47712 + 6.67593 - 2.92$
    $M \approx 5.2$
    The magnitude is approximately 5.2.

21. $W = 0.00031 C^{2.9}$
    $W = 0.00031(534)^{2.9}$
    $W \approx 25{,}190$
    The weight is approximately 25,190 kg.

23. **Strategy**
    To find how many times stronger Japan's earthquake was, use the Richter equation to write a system of equations.
    Solve the equations for the ratio $\dfrac{I_1}{I_2}$.

    Solution
    $8.9 = \log \dfrac{I_1}{I_0}$
    $7.1 = \log \dfrac{I_2}{I_0}$
    $8.9 = \log I_1 - \log I_0$
    $7.1 = \log I_2 - \log I_0$
    $1.8 = \log I_1 - \log I_2$
    $1.8 = \log \dfrac{I_1}{I_2}$
    $\dfrac{I_1}{I_2} = 10^{1.8} \approx 63.10$

    Japan's earthquake was about 63.1 times stronger than San Francisco's earthquake.

25. **Strategy**
    To find the number of decibels, replace I with its given value in the equation and solve for D.

    Solution
    $D = 10(\log I + 16)$
    $= 10[\log(3.2 \times 10^{-10}) + 16]$
    $= 10[\log 3.2 + \log 10^{-10} + 16]$
    $= 10[0.5051 + (-10) + 16]$
    $= 10(6.5051)$
    $= 65.051$
    The number of decibels is 65.

27. **Strategy**
    To find the number of parsecs, use the distance modulus formula to solve for r. Use $M = 5.89$.

    Solution
    $M = 5 \log r - 5$
    $5.89 = 5 \log r - 5$
    $10.89 = 5 \log r$
    $2.178 = \log r$
    $10^{2.178} = r$
    $150.66 \approx r$
    The star is 150.7 parsecs from Earth.

29. **Strategy**
    To find out how many times farther Antares is from Earth than Pollux is, use the distance modulus formula to write a system of equations.
    Let r_1 be the distance from Antares to Earth, and let r_2 be the distance from Pollux to Earth.

    Solution
    $5.4 = 5 \log r_1 - 5$
    $2.7 = 5 \log r_2 - 5$
    $2.7 = 5 \log r_1 - 5 \log r_2$
    $2.7 = 5 \log \dfrac{r_1}{r_2}$
    $0.54 = \log \dfrac{r_1}{r_2}$
    $10^{0.54} = \dfrac{r_1}{r_2}$
    $3.47 \approx \dfrac{r_1}{r_2}$
    $3.47 r_2 = r_1$

    No, Antares is more than twice as far from Earth as Pollux is. Antares is 3.5 times as far from Earth as Pollux is.

31. **Strategy**
    To find the number of barrels, solve the equation for r. Use $T = 20$.

    Solution
    $T = 14.29 \ln(0.00411 r + 1)$
    $20 = 14.29 \ln(0.00411 r + 1)$
    $1.3996 = \ln(0.00411 r + 1)$
    $e^{1.3996} = 0.00411 r + 1$
    $4.0535 \approx 0.00411 r + 1$
    $3.0535 \approx 0.00411 r$
    $742.94 \approx r$
    742.9 billion barrels of oil are necessary to last 20 years.

Applying Concepts 10.5

33. Strategy

To find the time at which the population will be 9×10^6:

Write the exponential growth equation using 1.5×10^6 for A_0, 3×10^6 for A, 3 for t, and 2 for b. Solve for k.

Rewrite the exponential growth equation using 9×10^6 for A, 1.5×10^6 for A_0, 2 for b, and the value of k for k. Solve for t.

Solution
$$A = A_0 b^{kt}$$
$$3 \times 10^6 = (1.5 \times 10^6) 2^{3k}$$
$$2 = 2^{3k}$$
$$2^1 = 2^{3k}$$
$$1 = 3k$$
$$\frac{1}{3} = k$$

$$A = A_0 b^{kt}$$
$$9 \times 10^6 = (1.5 \times 10^6) 2^{(1/3)t}$$
$$6 = 2^{(1/3)t}$$
$$\log 6 = \log 2^{(1/3)t} = \frac{1}{3} t (\log 2)$$
$$\frac{\log 6}{\log 2} = \frac{1}{3} t$$
$$\frac{3 \log 6}{\log 2} = t$$
$$7.7549 = t$$

The population will be 9×10^6 about 8 h after 9 A.M.

The population will be approximately 9×10^6 at 5 P.M.

35. Strategy

To find the annual interest rate, write the compound interest formula using 1000 for A, 1177.23 for P, and 2 for n. Solve for i.

Solution
$$P = A(1+i)^n$$
$$1177.23 = 1000(1+i)^2$$
$$1.17723 = (1+i)^2$$
$$\log 1.17723 = \log(1+i)^2$$
$$\log 1.17723 = 2 \log(1+i)$$
$$0.0709 \approx 2 \log(1+i)$$
$$0.03545 = \log(1+i)$$
$$10^{0.03545} = 1 + i$$
$$10^{0.03545} - 1 = i$$
$$0.08505 = i$$

The annual interest rate is 8.5%.

37.
$$T = 14.29 \ln(0.00411r + 1)$$
$$\frac{T}{14.29} = \ln(0.00411r + 1)$$
$$e^{T/14.29} = 0.00411r + 1$$
$$e^{T/14.29} - 1 = 0.00411r$$
$$\frac{e^{T/14.29} - 1}{0.00411} = r$$

39.
$$y = A 2^{kt}$$
$$\frac{y}{A} = 2^{kt}$$
$$\log_2 \frac{y}{A} = \log_2 2^{kt}$$
$$\log_2 \frac{y}{A} = kt$$
$$\frac{\ln \frac{y}{A}}{\ln 2} = kt$$
$$\ln \frac{y}{A} = (\ln 2)(kt)$$
$$\frac{y}{A} = e^{(\ln 2)(kt)}$$
$$y = A e^{(k \ln 2)t}$$

41. a. $y(t) = At - 16t^2 + \dfrac{A}{k}(M + m - kt)\ln\left(1 - \dfrac{k}{M+m}t\right)$

$y(t) = 8000t - 16t^2 + \dfrac{8000}{250}(8000 + 16,000 - 250t)\ln\left(1 - \dfrac{250}{8000 + 16,000}t\right)$

$y(t) = 8000t - 16t^2 + 32(24,000 - 250t)\ln\left(1 - \dfrac{250}{24,000}t\right)$

When $y = 5280$, $t \approx 14$.
The rocket requires 14 s to reach a height of 1 mile.

b. $v(t) = -32t + A\ln\left(\dfrac{M+m}{M+m-kt}\right)$

$v(14) = -32(14) + 8000\ln\dfrac{8000 + 16,000}{8000 + 16,000 - 250(14)}$

$= -448 + 8000\ln\left(\dfrac{24,000}{20,500}\right)$

≈ 813

The velocity of the rocket is 813 ft/s.

c. $\dfrac{M+m}{M+m-kt} > 0$

Since $M + m > 0$, then
$M + m - kt > 0$
$M + m > kt$
$\dfrac{M+m}{k} > t$
$\dfrac{8000 + 16,000}{250} > t$
$96 > t$

Since t must be greater than 0 and less than 96, the domain is $\{t | 0 < t < 96\}$.

Focus on Problem Solving

1. The sum of $0 + 1 + 2 + 3 + 4 + 5 + 6 + 7 + 8 + 9$ is 45.
   Some of these digits must be used as tens digits.
   Let x = the sum of the tens digits.
   Let S = the sum of the remaining units digits.
   Then $S = 45 - x$.

2. The sum of 10 times the tens digits and the units digits will equal 100.
   $10x + (45 - x) = 100$

3. $10x + (45 - x) = 100$
   $10x + 45 - x = 100$
   $9x = 55$
   $x = \dfrac{55}{9}$

4. x is the sum of integers, so $x = \dfrac{55}{9}$ is not possible. Our assumption that it is possible to use each of the digits 0, 1, 2, 3, 4, 5, 6, 7, 8 and 9 exactly once in such a way that the sum is 100 is not valid.

Projects and Group Activities
Fractals

1. scale factor = $\dfrac{\text{new length}}{\text{old length}}$

   a. scale factor$_{2-1}$ = $\dfrac{2}{1}$ = 2

   b. scale factor$_{3-2}$ = $\dfrac{4}{2}$ = 2

   c. scale factor$_{4-3}$ = $\dfrac{8}{4}$ = 2

2. size ratio = $\dfrac{\text{new size}}{\text{old size}}$

   a. size ratio$_{2-1}$ = $\dfrac{4}{1}$ = 4

   b. size ratio$_{3-2}$ = $\dfrac{16}{4}$ = 4

   c. size ratio$_{4-3}$ = $\dfrac{64}{16}$ = 4

3. a. scale factor = 2

   b. size ratio = 4

4. a. scale factor$_{2-1}$ = $\dfrac{2}{1}$ = 2

   scale factor$_{3-2}$ = $\dfrac{4}{2}$ = 2

   The scale factor is 2.

   b. size ratio$_{2-1}$ = $\dfrac{2^3}{1^3}$ = 8

   size ratio$_{3-2}$ = $\dfrac{4^3}{2^3}$ = 8

   The size ratio is 8.

5. $d = \dfrac{\log(\text{size ratio})}{\log(\text{scale factor})}$

   $d = \dfrac{\log 8}{\log 2} = \dfrac{\log 2^3}{\log 2} = \dfrac{3 \log 2}{\log 2} = 3$

   Thus the cubes are three-dimensional figures.

6. a. scale factor$_{2-1}$ = $\dfrac{2}{1}$ = 2

   scale factor$_{3-2}$ = $\dfrac{4}{2}$ = 2

   b. size ratio$_{2-1}$ = $\dfrac{3}{1}$ = 3

   size ratio$_{3-2}$ = $\dfrac{9}{3}$ = 3

7. $d = \dfrac{\log(\text{size ratio})}{\log(\text{scale factor})}$

   $d = \dfrac{\log 3}{\log 2}$

   $d \approx 1.58$

Chapter Review Exercises

1. $\log_4 16 = x$
   $4^x = 16$
   $4^x = 4^2$
   $x = 2$
   $\log_4 16 = 2$

2. $\dfrac{1}{2}(\log_3 x - \log_3 y) = \dfrac{1}{2}\left(\log_3 \dfrac{x}{y}\right)$
   $= \log_3 \left(\dfrac{x}{y}\right)^{1/2}$
   $= \log_3 \sqrt{\dfrac{x}{y}}$

3. $f(x) = e^{x-2}$
   $f(2) = e^{2-2}$
   $f(2) = e^0$
   $f(2) = 1$

4. $8^x = 2^{x-6}$
   $(2^3)^x = 2^{x-6}$
   $2^{3x} = 2^{x-6}$
   $3x = x - 6$
   $2x = -6$
   $x = -3$
   The solution is −3.

5. $f(x) = \left(\dfrac{2}{3}\right)^x$
   $f(0) = \left(\dfrac{2}{3}\right)^0$
   $f(0) = 1$

6. $\log_3 x = -2$
   $3^{-2} = x$
   $\dfrac{1}{3^2} = x$
   $\dfrac{1}{9} = x$
   The solution is $\dfrac{1}{9}$.

7. $2^5 = 32$ is equivalent to $\log_2 32 = 5$.

8. $\log x + \log(x-4) = \log 12$
$\log x(x-4) = \log 12$
$x(x-4) = 12$
$x^2 - 4x = 12$
$x^2 - 4x - 12 = 0$
$(x-6)(x+2) = 0$
$x - 6 = 0 \qquad x + 2 = 0$
$x = 6 \qquad x = -2$
-2 does not check as a solution. 6 checks as a solution. The solution is 6.

9. $\log_6 \sqrt{xy^3} = \log_6 \sqrt{x}\sqrt{y^3}$
$= \log_6 \sqrt{x} + \log_6 \sqrt{y^3}$
$= \log_6 x^{1/2} + \log_6 y^{3/2}$
$= \frac{1}{2}\log_6 x + \frac{3}{2}\log_6 y$

10. $4^{5x-2} = 4^{3x+2}$
$5x - 2 = 3x + 2$
$2x - 2 = 2$
$2x = 4$
$x = 2$
The solution is 2.

11. $3^{7x+1} = 3^{4x-5}$
$7x + 1 = 4x - 5$
$3x + 1 = -5$
$3x = -6$
$x = -2$
The solution is -2.

12. $f(x) = 3^{x+1}$
$f(-2) = 3^{-2+1} = 3^{-1} = \frac{1}{3}$

13. $\log_2 16 = x$
$2^x = 16$
$2^x = 2^4$
$x = 4$
The solution is 4.

14. $\log_6 2x = \log_6 2 + \log_6(3x-4)$
$\log_6 2x = \log_6 2(3x-4)$
$\log_6 2x = \log_6(6x-8)$
$2x = 6x - 8$
$-4x = -8$
$x = 2$
The solution is 2.

15. $\log_2 5 = \frac{\log 5}{\log 2} = 2.3219$

16. $\log_6 22 = \frac{\log 22}{\log 6} = 1.7251$

17. $4^x = 8^{x-1}$
$(2^2)^x = (2^3)^{x-1}$
$2^{2x} = 2^{3x-3}$
$2x = 3x - 3$
$-x = -3$
$x = 3$
The solution is 3.

18. $f(x) = \left(\frac{1}{4}\right)^x$
$f(-1) = \left(\frac{1}{4}\right)^{-1} = 4$

19. $\log_5 \sqrt{\frac{x}{y}} = \log_5 \frac{\sqrt{x}}{\sqrt{y}}$
$= \log_5 \sqrt{x} - \log_5 \sqrt{y}$
$= \log_5 x^{1/2} - \log_5 y^{1/2}$
$= \frac{1}{2}\log_5 x - \frac{1}{2}\log_5 y$

20. $\log_5 \frac{7x+2}{3x} = 1$
$5^1 = \frac{7x+2}{3x}$
$5 = \frac{7x+2}{3x}$
$15x = 7x + 2$
$8x = 2$
$x = \frac{1}{4}$
The solution is $\frac{1}{4}$.

21. $\log_5 x = 3$
$5^3 = x$
$125 = x$
The solution is 125.

22. $\log x + \log(2x+3) = \log 2$
$\log x(2x+3) = \log 2$
$x(2x+3) = 2$
$2x^2 + 3x = 2$
$2x^2 + 3x - 2 = 0$
$(2x-1)(x+2) = 0$
$2x - 1 = 0 \qquad x + 2 = 0$
$2x = 1 \qquad x = -2$
$x = \frac{1}{2}$
-2 does not check as a solution. $\frac{1}{2}$ checks as a solution. The solution is $\frac{1}{2}$.

23. $3\log_b x - 5\log_b y = \log_b x^3 - \log_b y^5 = \log_b \frac{x^3}{y^5}$

24. $f(x) = 2^{-x-1}$
$f(-3) = 2^{-(-3)-1}$
$f(-3) = 2^{3-1}$
$f(-3) = 2^2 = 4$

25. $\log_3 19 = \dfrac{\log 19}{\log 3} = 2.6801$

26. $3^{x+2} = 5$
$\log 3^{x+2} = \log 5$
$(x+2) \log 3 = \log 5$
$x + 2 = \dfrac{\log 5}{\log 3}$
$x = \dfrac{\log 5}{\log 3} - 2$
$x = -0.5350$
The solution is -0.5350.

27.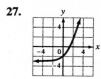

28.

29.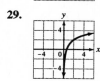

30.

31. 1.0

32. **Strategy**
To find the half-life, solve for k in the exponential decay equation. Use $A_0 = 10$, $A = 9$, and $t = 5$.

Solution
$A = A_0 \left(\dfrac{1}{2}\right)^{t/k}$
$9 = 10 \left(\dfrac{1}{2}\right)^{5/k}$
$0.9 = \left(\dfrac{1}{2}\right)^{5/k}$
$0.9 = (0.5)^{5/k}$
$\log(0.9) = \log(0.5)^{5/k}$
$\log(0.9) = \dfrac{5}{k} \log(0.5)$
$k = \dfrac{5 \log 0.5}{\log 0.9} = 32.89$
The half-life is 33 h.

33. **Strategy**
To find the thickness, solve the equation $\log P = -0.5d$ for d. Use $P = 50\% = 0.5$.

Solution
$\log P = -0.5d$
$\log 0.5 = -0.5d$
$\dfrac{\log 0.5}{-0.5} = d$
$0.602 = d$
The material must be 0.602 cm thick.

Chapter Test

1. $f(x) = \left(\dfrac{3}{4}\right)^x$
$f(0) = \left(\dfrac{3}{4}\right)^0$
$f(0) = 1$

2. $f(x) = 4^{x-1}$
$f(-2) = 4^{-2-1}$
$f(-2) = 4^{-3}$
$f(-2) = \dfrac{1}{4^3}$
$f(-2) = \dfrac{1}{64}$

3. $\log_4 64 = x$
$4^x = 64$
$4^x = 4^3$
$x = 3$

4. $\log_4 x = -2$
$4^{-2} = x$
$\dfrac{1}{4^2} = x$
$\dfrac{1}{16} = x$
The solution is $\dfrac{1}{16}$.

5. $\log_6 \sqrt[3]{x^2 y^5} = \log_6 (x^2 y^5)^{1/3}$
$= \dfrac{1}{3}\log_6(x^2 y^5)$
$= \dfrac{1}{3}(\log_6 x^2 + \log_6 y^5)$
$= \dfrac{1}{3}(2\log_6 x + 5\log_6 y)$
$= \dfrac{2}{3}\log_6 x + \dfrac{5}{3}\log_6 y$

6. $\dfrac{1}{2}(\log_5 x - \log_5 y) = \dfrac{1}{2}\left(\log_5 \dfrac{x}{y}\right)$
$= \log_5 \left(\dfrac{x}{y}\right)^{1/2}$
$= \log_5 \sqrt{\dfrac{x}{y}}$

7. $\log_6 x + \log_6(x-1) = 1$
$\log_6 x(x-1) = 1$
$x(x-1) = 6^1$
$x^2 - x = 6$
$x^2 - x - 6 = 0$
$(x-3)(x+2) = 0$
$x - 3 = 0 \qquad x + 2 = 0$
$x = 3 \qquad x = -2$
−2 does not check as a solution. 3 checks as a solution. The solution is 3.

8. $f(x) = 3^{x+1}$
$f(-2) = 3^{-2+1}$
$f(-2) = 3^{-1}$
$f(-2) = \dfrac{1}{3}$
The solution is $\dfrac{1}{3}$.

9. $3^x = 17$
$\log 3^x = \log 17$
$x \log 3 = \log 17$
$x = \dfrac{\log 17}{\log 3}$
$x = 2.5789$

10. $\log_2 x + 3 = \log_2(x^2 - 20)$
$3 = \log_2(x^2 - 20) - \log_2 x$
$3 = \log_2 \dfrac{(x^2 - 20)}{x}$
$2^3 = \dfrac{x^2 - 20}{x}$
$8 = \dfrac{x^2 - 20}{x}$
$8x = x^2 - 20$
$0 = x^2 - 8x - 20$
$0 = (x-10)(x+2)$
$x - 10 = 0 \qquad x + 2 = 0$
$x = 10 \qquad x = -2$
−2 does not check as a solution. 10 checks as a solution. The solution is 10.

11. $5^{6x-2} = 5^{3x+7}$
$6x - 2 = 3x + 7$
$3x - 2 = 7$
$3x = 9$
$x = 3$
The solution is 3.

12. $4^x = 2^{3x+4}$
$(2^2)^x = 2^{3x+4}$
$2^{2x} = 2^{3x+4}$
$2x = 3x + 4$
$-x = 4$
$x = -4$
The solution is −4.

13. $\log(2x+1) + \log x = \log 6$
$\log x(2x+1) = \log 6$
$x(2x+1) = 6$
$2x^2 + x = 6$
$2x^2 + x - 6 = 0$
$(2x-3)(x+2) = 0$
$2x - 3 = 0 \qquad x + 2 = 0$
$2x = 3 \qquad x = -2$
$x = \dfrac{3}{2}$
−2 does not check as a solution. $\dfrac{3}{2}$ checks as a solution. The solution is $\dfrac{3}{2}$.

14.

15.

16.

17.

18. 1.6

19. Strategy
To find the value, solve the compound interest formula for P. Use $A = 10{,}000$,
$$i = \frac{7.5\%}{12} = \frac{0.075}{12} = 0.00625, \ n = 72.$$

Solution
$P = A(1+i)^n$
$P = 10{,}000(1+0.00625)^{72}$
$\approx 15{,}661$
The value is $15,661.

20. Strategy
To find the half-life, solve for k in the exponential decay equation. Use $A_0 = 40$, $A = 30$, and $t = 10$.

Solution
$$A = A_0\left(\frac{1}{2}\right)^{t/k}$$
$$30 = 40\left(\frac{1}{2}\right)^{10/k}$$
$$0.75 = \left(\frac{1}{2}\right)^{10/k}$$
$$0.75 = (0.5)^{10/k}$$
$$\log 0.75 = \log(0.5)^{10/k}$$
$$\log(0.75) = \frac{10}{k}\log(0.5)$$
$$k = \frac{10\log 0.5}{\log 0.75} \approx 24.09$$
The half-life is 24 h.

Cumulative Review Exercises

1. $4 - 2[x - 3(2-3x) - 4x] = 2x$
$4 - 2[x - 6 + 9x - 4x] = 2x$
$4 - 2[6x - 6] = 2x$
$4 - 12x + 12 = 2x$
$-12x + 16 = 2x$
$-14x = -16$
$x = \dfrac{8}{7}$

The solution is $\dfrac{8}{7}$.

2. $S = 2WH + 2WL + 2LH$
$S - 2WH = 2WL + 2LH$
$S - 2WH = L(2W + 2H)$
$\dfrac{S - 2WH}{2W + 2H} = L$

3. $|2x - 5| \le 3$
$-3 \le 2x - 5 \le 3$
$-3 + 5 \le 2x - 5 + 5 \le 3 + 5$
$2 \le 2x \le 8$
$1 \le x \le 4$
$\{x | 1 \le x \le 4\}$

4. $4x^{2n} + 7x^n + 3 = (4x^n + 3)(x^n + 1)$

5. $x^2 + 4x - 5 \le 0$
$(x+5)(x-1) \le 0$

$\{x | -5 \le x \le 1\}$

6. $\dfrac{1 - \frac{5}{x} + \frac{6}{x^2}}{1 + \frac{1}{x} - \frac{6}{x^2}} = \dfrac{1 - \frac{5}{x} + \frac{6}{x^2}}{1 + \frac{1}{x} - \frac{6}{x^2}} \cdot \dfrac{x^2}{x^2}$
$= \dfrac{x^2 - 5x + 6}{x^2 + x - 6}$
$= \dfrac{(x-2)(x-3)}{(x+3)(x-2)}$
$= \dfrac{x-3}{x+3}$

7. $\dfrac{\sqrt{xy}}{\sqrt{x} - \sqrt{y}} = \dfrac{\sqrt{xy}}{\sqrt{x} - \sqrt{y}} \cdot \dfrac{\sqrt{x} + \sqrt{y}}{\sqrt{x} + \sqrt{y}}$
$= \dfrac{\sqrt{x^2 y} + \sqrt{xy^2}}{(\sqrt{x})^2 - (\sqrt{y})^2}$
$= \dfrac{x\sqrt{y} + y\sqrt{x}}{x - y}$

8. $y\sqrt{18x^5y^4} - x\sqrt{98x^3y^6}$
   $= y\sqrt{2 \cdot 3^2 x^5 y^4} - x\sqrt{2 \cdot 7^2 x^3 y^6}$
   $= y\sqrt{3^2 x^4 y^4 (2x)} - x\sqrt{7^2 x^2 y^6 (2x)}$
   $= y\sqrt{3^2 x^4 y^4}\sqrt{2x} - x\sqrt{7^2 x^2 y^6}\sqrt{2x}$
   $= y \cdot 3x^2 y^2 \sqrt{2x} - x \cdot 7xy^3 \sqrt{2x}$
   $= 3x^2 y^3 \sqrt{2x} - 7x^2 y^3 \sqrt{2x}$
   $= -4x^2 y^3 \sqrt{2x}$

9. $\dfrac{i}{2-i} = \dfrac{i}{2-i} \cdot \dfrac{2+i}{2+i}$
   $= \dfrac{2i + i^2}{4 - i^2}$
   $= \dfrac{2i - 1}{4 + 1}$
   $= \dfrac{-1 + 2i}{5}$
   $= -\dfrac{1}{5} + \dfrac{2}{5}i$

10. $2x - y = 5$
    $-y = -2x + 5$
    $y = 2x - 5$
    $m = 2 \qquad (x_1, y_1) = (2, -2)$
    $y - y_1 = m(x - x_1)$
    $y - (-2) = 2(x - 2)$
    $y + 2 = 2x - 4$
    $y = 2x - 6$
    The equation of the line is $y = 2x - 6$.

11. $(x - r_1)(x - r_2) = 0$
    $\left(x - \dfrac{1}{3}\right)(x - (-3)) = 0$
    $\left(x - \dfrac{1}{3}\right)(x + 3) = 0$
    $x^2 + \dfrac{8}{3}x - 1 = 0$
    $3\left(x^2 + \dfrac{8}{3}x - 1\right) = 3(0)$
    $3x^2 + 8x - 3 = 0$

12. $x^2 - 4x - 6 = 0$
    $x^2 - 4x = 6$
    $x^2 - 4x + 4 = 6 + 4$
    $(x - 2)^2 = 10$
    $\sqrt{(x-2)^2} = \sqrt{10}$
    $x - 2 = \pm\sqrt{10}$
    $x = 2 \pm \sqrt{10}$
    The solutions are is $2 + \sqrt{10}$ and $2 - \sqrt{10}$.

13. $f(x) = x^2 - 3x - 4$
    $f(-1) = (-1)^2 - 3(-1) - 4 = 1 + 3 - 4 = 0$
    $f(0) = 0^2 - 3(0) - 4 = -4$
    $f(1) = 1^2 - 3(1) - 4 = 1 - 3 - 4 = -6$
    $f(2) = 2^2 - 3(2) - 4 = 4 - 6 - 4 = -6$
    $f(3) = 3^2 - 3(3) - 4 = 9 - 9 - 4 = -4$
    The range is $\{-6, -4, 0\}$.

14. $g(x) = 2x - 3$
    $g(0) = 2(0) - 3 = -3$

    $f(x) = x^2 + 2x + 1$
    $f(-3) = (-3)^2 + 2(-3) + 1$
    $\qquad = 9 - 6 + 1$
    $\qquad = 4$
    $f[g(0)] = 4$

15. (1) $\quad 3x - y + z = 3$
    (2) $\quad x + y + 4z = 7$
    (3) $\quad 3x - 2y + 3z = 8$
    Eliminate y.
    Add equation (1) and (2).
    $\quad 3x - y + z = 3$
    $\quad x + y + 4z = 7$
    (4) $\quad 4x + 5z = 10$
    Multiply equation (2) by 2 and add to equation (3).
    $2(x + y + 4z) = 7(2)$
    $3x - 2y + 3z = 8$

    $2x + 2y + 8z = 14$
    $3x - 2y + 3z = 8$
    (5) $\quad 5x + 11z = 22$

    Multiply equation (4) by –5. Multiply equation (5) by 4 and add.
    $-5(4x + 5z) = -5(10)$
    $4(5x + 11z) = 4(22)$

    $-20x - 25z = -50$
    $20x + 44z = 88$

    $19z = 38$
    $z = 2$
    Substitute 2 for z in equation (5).
    $5x + 11(2) = 22$
    $5x + 22 = 22$
    $5x = 0$
    $x = 0$
    Substitute 0 for x and 2 for z in equation (2).
    $x + y + 4z = 7$
    $0 + y + 4(2) = 7$
    $y + 8 = 7$
    $y = -1$
    The solution is $(0, -1, 2)$.

16. (1) $y = -2x - 3$
    (2) $y = 2x - 1$
    Solve by the substitution method.
    $$y = 2x - 1$$
    $$-2x - 3 = 2x - 1$$
    $$-4x - 3 = -1$$
    $$-4x = 2$$
    $$x = -\frac{1}{2}$$
    Substitute into equation (2).
    $$y = 2x - 1$$
    $$y = 2\left(-\frac{1}{2}\right) - 1$$
    $$y = -1 - 1$$
    $$y = -2$$
    The solution is $\left(-\frac{1}{2}, -2\right)$.

17. $f(x) = 3^{-x+1}$
    $f(-4) = 3^{-(-4)+1} = 3^{4+1} = 3^5 = 243$

18. $\log_4 x = 3$
    $4^3 = x$
    $64 = x$
    The solution is 64.

19. $2^{3x+2} = 4^{x+5}$
    $2^{3x+2} = (2^2)^{x+5}$
    $2^{3x+2} = 2^{2x+10}$
    $3x + 2 = 2x + 10$
    $x + 2 = 10$
    $x = 8$
    The solution is 8.

20. $\log x + \log(3x + 2) = \log 5$
    $\log x(3x + 2) = \log 5$
    $x(3x + 2) = 5$
    $3x^2 + 2x = 5$
    $3x^2 + 2x - 5 = 0$
    $(3x + 5)(x - 1) = 0$
    $3x + 5 = 0 \qquad x - 1 = 0$
    $3x = -5 \qquad x = 1$
    $x = -\frac{5}{3}$
    $-\frac{5}{3}$ does not check as a solution. 1 does check as a solution. The solution is 1.

21.
    ```
    ←+─(+─+─)+─+─+─+─+─+→
     -5-4-3-2-1 0 1 2 3 4 5
    ```

22. $\frac{x+2}{x-1} \geq 0$
    ```
    x + 2  - - - -|+ + +|+ + + + +
    x - 1  - - - - - - -|+ + + + +
          ─+─+─+─+─+─+─+─+─+─+─→
          -5-4-3-2-1 0 1 2 3 4 5
    ```
    $\{x \mid x \leq -2 \text{ or } x > 1\}$
    ```
    ←+─+─+─]+─+─+─(+─+─+─+→
     -5-4-3-2-1 0 1 2 3 4 5
    ```

23. $y = -x^2 - 2x + 3$
    $-\frac{b}{2a} = -\frac{(-2)}{2(-1)} = -1$
    $y = -(-1)^2 - 2(-1) + 3 = -1 + 2 + 3 = 4$
    Vertex: $(-1, 4)$
    Axis of symmetry: $x = -1$

24.

25.

26.

27.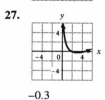

    -0.3

28. Strategy
    Pounds of 25% alloy: x
    Pounds of 50% alloy: $2000 - x$

	Amount	Percent	Quantity
25% alloy	x	0.25	$0.25x$
50% alloy	$2000 - x$	0.50	$0.50(2000 - x)$
40% alloy	2000	0.40	$0.40(2000)$

    The sum of the quantities before mixing equals the quantity after mixing.

    Solution
    $0.25x + 0.50(2000 - x) = 0.40(2000)$
    $0.25x + 1000 - 0.50x = 800$
    $-0.25x + 1000 = 800$
    $-0.25x = -200$
    $x = 800$
    $2000 - x = 2000 - 800 = 1200$
    800 lb of the alloy containing 25% tin and 1200 lb of the alloy containing 50% tin were used.

324 *Chapter 10: Exponential and Logarithmic Functions*

29. Strategy
To find the amount, write and solve an inequality using x to represent the amount of sales.

Solution
$500 + 0.08x \geq 3000$
$0.08x \geq 2500$
$x \geq 31,250$

To earn $3000 or more a month, the sales executive must sell $31,250 or more.

30. Strategy
Time to print the checks when both printers are operating:

	Rate	Time	Part
Old printer	$\frac{1}{30}$	t	$\frac{t}{30}$
New printer	$\frac{1}{10}$	t	$\frac{t}{10}$

The sum of the parts of the task completed by each printer equals 1.

Solution
$$\frac{t}{30} + \frac{t}{10} = 1$$
$$30\left(\frac{t}{30} + \frac{t}{10}\right) = 30(1)$$
$$t + 3t = 30$$
$$4t = 30$$
$$t = 7.5$$

When both printers are operating, it will take 7.5 min to print the checks.

31. Strategy
To find the pressure:
Write the basic inverse variation equation, replace the variables by the given values, and solve for k.

Write the inverse variation equation, replacing k by its value. Substitute 25 for V and solve for P.

Solution
$$P = \frac{k}{V}$$
$$50 = \frac{k}{250}$$
$$12,500 = k$$

$$P = \frac{12,500}{V}$$
$$= \frac{12,500}{25}$$
$$= 500$$

When the volume is 25 ft^3, the pressure is 500 ln/in^2.

32. Strategy
Cost per yard of nylon carpet: n
Cost per yard of wool carpet: w
First purchase:

	Amount	Unit cost	Value
Nylon carpet	45	n	$45n$
Wool carpet	30	w	$30w$

Second purchase:

	Amount	Unit cost	Value
Nylon carpet	25	n	$25n$
Wool carpet	80	w	$80w$

The total of the first purchase was $1170.
The total of the second purchase was $1410.
$45n + 30w = 1170$
$25n + 80w = 1410$

Solution
$45n + 30w = 1170$
$25n + 80w = 1410$

$$25(45n + 30w) = 25(1170)$$
$$-45(25n + 80w) = -45(1410)$$
$$1125n + 750w = 29,250$$
$$-1125n - 3600w = -63,450$$
$$-2850w = -34,200$$
$$w = 12$$

The cost per yard of the wool carpet is $12.

33. Strategy
To find the value of the investment, solve the compound interest formula for P. Use $A = 10,000$, $i = \frac{9\%}{12} = \frac{0.09}{12} = 0.0075$, and $n = 12 \cdot 5 = 60$.

Solution
$$P = A(1+i)^n$$
$$= 10,000(1 + 0.0075)^{60}$$
$$= 10,000(1.0075)^{60}$$
$$\approx 15,657$$

The value of the investment after 5 yr is $15,657.

Chapter 11: Sequences and Series

Section 11.1

Concept Review 11.1

1. Always true

3. Always true

Objective 1 Exercises

3. $a_n = n+1$
   $a_1 = 1+1 = 2$ The first term is 2.
   $a_2 = 2+1 = 3$ The second term is 3.
   $a_3 = 3+1 = 4$ The third term is 4.
   $a_4 = 4+1 = 5$ The fourth term is 5.

5. $a_n = 2n+1$
   $a_1 = 2(1)+1 = 3$ The first term is 3.
   $a_2 = 2(2)+1 = 5$ The second term is 5.
   $a_3 = 2(3)+1 = 7$ The third term is 7.
   $a_4 = 2(4)+1 = 9$ The fourth term is 9.

7. $a_n = 2-2n$
   $a_1 = 2-2(1) = 0$ The first term is 0.
   $a_2 = 2-2(2) = -2$ The second term is -2.
   $a_3 = 2-2(3) = -4$ The third term is -4.
   $a_4 = 2-2(4) = -6$ The fourth term is -6.

9. $a_n = 2^n$
   $a_1 = 2^1 = 2$ The first term is 2.
   $a_2 = 2^2 = 4$ The second term is 4.
   $a_3 = 2^3 = 8$ The third term is 8.
   $a_4 = 2^4 = 16$ The fourth term is 16.

11. $a_n = n^2 + 1$
    $a_1 = 1^2 + 1 = 2$ The first term is 2.
    $a_2 = 2^2 + 1 = 5$ The second term is 5.
    $a_3 = 3^2 + 1 = 10$ The third term is 10.
    $a_4 = 4^2 + 1 = 17$ The fourth term is 17.

13. $a_n = \dfrac{n}{n^2+1}$
    $a_1 = \dfrac{1}{1^2+1} = \dfrac{1}{2}$ The first term is $\dfrac{1}{2}$.
    $a_2 = \dfrac{2}{2^2+1} = \dfrac{2}{5}$ The second term is $\dfrac{2}{5}$.
    $a_3 = \dfrac{3}{3^2+1} = \dfrac{3}{10}$ The third term is $\dfrac{3}{10}$.
    $a_4 = \dfrac{4}{4^2+1} = \dfrac{4}{17}$ The fourth term is $\dfrac{4}{17}$.

15. $a_n = n - \dfrac{1}{n}$
    $a_1 = 1 - \dfrac{1}{1} = 0$ The first term is 0.
    $a_2 = 2 - \dfrac{1}{2} = \dfrac{3}{2}$ The second term is $\dfrac{3}{2}$.
    $a_3 = 3 - \dfrac{1}{3} = \dfrac{8}{3}$ The third term is $\dfrac{8}{3}$.
    $a_4 = 4 - \dfrac{1}{4} = \dfrac{15}{4}$ The fourth term is $\dfrac{15}{4}$.

17. $a_n = (-1)^{n+1} n$
    $a_1 = (-1)^2 (1) = 1$ The first term is 1.
    $a_2 = (-1)^3 (2) = -2$ The second term is -2.
    $a_3 = (-1)^4 (3) = 3$ The third term is 3.
    $a_4 = (-1)^5 (4) = -4$ The fourth term is -4.

19. $a_n = \dfrac{(-1)^{n+1}}{n^2+1}$
    $a_1 = \dfrac{(-1)^2}{1^2+1} = \dfrac{1}{2}$ The first term is $\dfrac{1}{2}$.
    $a_2 = \dfrac{(-1)^3}{2^2+1} = -\dfrac{1}{5}$ The second term is $-\dfrac{1}{5}$.
    $a_3 = \dfrac{(-1)^4}{3^2+1} = \dfrac{1}{10}$ The third term is $\dfrac{1}{10}$.
    $a_4 = \dfrac{(-1)^5}{4^2+1} = -\dfrac{1}{17}$ The fourth term is $-\dfrac{1}{17}$.

21. $a_n = (-1)^n 2^n$
    $a_1 = (-1)^1 \cdot 2^1 = -2$ The first term is -2.
    $a_2 = (-1)^2 \cdot 2^2 = 4$ The second term is 4.
    $a_3 = (-1)^3 \cdot 2^3 = -8$ The third term is -8.
    $a_4 = (-1)^4 \cdot 2^4 = 16$ The fourth term is 16.

23. $a_n = 2\left(\dfrac{1}{3}\right)^{n+1}$
    $a_1 = 2\left(\dfrac{1}{3}\right)^2 = \dfrac{2}{9}$ The first term is $\dfrac{2}{9}$.
    $a_2 = 2\left(\dfrac{1}{3}\right)^3 = \dfrac{2}{27}$ The second term is $\dfrac{2}{27}$.
    $a_3 = 2\left(\dfrac{1}{3}\right)^4 = \dfrac{2}{81}$ The third term is $\dfrac{2}{81}$.
    $a_4 = 2\left(\dfrac{1}{3}\right)^5 = \dfrac{2}{243}$ The fourth term is $\dfrac{2}{243}$.

25. $a_n = 2n - 5$
    $a_{10} = 2(10) - 5 = 15$
    The tenth term is 15.

27. $a_n = \dfrac{n}{n+1}$
    $a_{12} = \dfrac{12}{12+1} = \dfrac{12}{13}$
    The twelfth term is $\dfrac{12}{13}$.

29. $a_n = (-1)^{n-1}(n-1)$
    $a_{25} = (-1)^{24}(25-1) = 24$
    The twenty-fifth term is 24.

31. $a_n = \left(\dfrac{2}{3}\right)^n$
    $a_5 = \left(\dfrac{2}{3}\right)^5 = \dfrac{32}{243}$
    The fifth term is $\dfrac{32}{243}$.

33. $a_n = (n+4)(n+1)$
    $a_7 = (7+4)(7+1) = (11)(8) = 88$
    The seventh term is 88.

35. $a_n = \dfrac{(-1)^{2n}}{n+4}$
    $a_{16} = \dfrac{(-1)^{32}}{16+4} = \dfrac{1}{20}$
    The sixteenth term is $\dfrac{1}{20}$.

37. $a_n = \dfrac{1}{3}n + n^2$
    $a_6 = \dfrac{1}{3}(6) + 6^2 = 2 + 36 = 38$
    The sixth term is 38.

Objective 2 Exercises

41. $\displaystyle\sum_{n=1}^{7}(i+2) = (1+2)+(2+2)+(3+2)+(4+2)+(5+2)+(6+2)+(7+2)$
    $= 3+4+5+6+7+8+9$
    $= 42$

43. $\displaystyle\sum_{n=1}^{7} n = 1+2+3+4+5+6+7 = 28$

45. $\displaystyle\sum_{i=1}^{5}(i^2+1) = (1^2+1)+(2^2+1)+(3^2+1)+(4^2+1)+(5^2+1)$
    $= 2+5+10+17+26$
    $= 60$

47. $\displaystyle\sum_{n=1}^{4} \dfrac{1}{2n} = \dfrac{1}{2(1)} + \dfrac{1}{2(2)} + \dfrac{1}{2(3)} + \dfrac{1}{2(4)} = \dfrac{1}{2} + \dfrac{1}{4} + \dfrac{1}{6} + \dfrac{1}{8} = \dfrac{12+6+4+3}{24} = \dfrac{25}{24}$

49. $\displaystyle\sum_{n=2}^{4} 2^n = 2^2 + 2^3 + 2^4 = 4 + 8 + 16 = 28$

51. $\displaystyle\sum_{i=3}^{6} \dfrac{i+1}{i} = \dfrac{3+1}{3} + \dfrac{4+1}{4} + \dfrac{5+1}{5} + \dfrac{6+1}{6} = \dfrac{4}{3} + \dfrac{5}{4} + \dfrac{6}{5} + \dfrac{7}{6} = \dfrac{80+75+72+70}{60} = \dfrac{297}{60} = \dfrac{99}{20}$

53. $\displaystyle\sum_{i=1}^{5} \dfrac{1}{2i} = \dfrac{1}{2(1)} + \dfrac{1}{2(2)} + \dfrac{1}{2(3)} + \dfrac{1}{2(4)} + \dfrac{1}{2(5)} = \dfrac{1}{2} + \dfrac{1}{4} + \dfrac{1}{6} + \dfrac{1}{8} + \dfrac{1}{10} = \dfrac{60+30+20+15+12}{120} = \dfrac{137}{120}$

55. $\displaystyle\sum_{i=1}^{4}(-1)^{i-1}(i+1) = (-1)^0(1+1)+(-1)^1(2+1)+(-1)^2(3+1)+(-1)^3(4+1)$
$= 2+(-3)+4+(-5) = -2$

57. $\displaystyle\sum_{n=4}^{7}\frac{(-1)^{n-1}}{n-3} = \frac{(-1)^3}{4-3}+\frac{(-1)^4}{5-3}+\frac{(-1)^5}{6-3}+\frac{(-1)^6}{7-3} = -1+\frac{1}{2}-\frac{1}{3}+\frac{1}{4} = \frac{-12+6-4+3}{12} = -\frac{7}{12}$

59. $\displaystyle\sum_{n=1}^{4}\frac{2n}{x} = \frac{2(1)}{x}+\frac{2(2)}{x}+\frac{2(3)}{x}+\frac{2(4)}{x} = \frac{2}{x}+\frac{4}{x}+\frac{6}{x}+\frac{8}{x}$

61. $\displaystyle\sum_{i=1}^{4}\frac{x^i}{i+1} = \frac{x^1}{1+1}+\frac{x^2}{2+1}+\frac{x^3}{3+1}+\frac{x^4}{4+1} = \frac{x}{2}+\frac{x^2}{3}+\frac{x^3}{4}+\frac{x^4}{5}$

63. $\displaystyle\sum_{i=2}^{4}\frac{x^i}{2i-1} = \frac{x^2}{2(2)-1}+\frac{x^3}{2(3)-1}+\frac{x^4}{2(4)-1} = \frac{x^2}{3}+\frac{x^3}{5}+\frac{x^4}{7}$

65. $\displaystyle\sum_{n=1}^{4}x^{2n-1} = x^{2(1)-1}+x^{2(2)-1}+x^{2(3)-1}+x^{2(4)-1} = x+x^3+x^5+x^7$

Applying Concepts 11.1

67. The sequence of the odd natural numbers is expressed by the formula $a_n = 2n-1$.

69. The sequence of the negative odd integers is expressed by the formula $a_n = -2n+1$.

71. The sequence of the positive integers that are divisible by 4 is expressed by the formula $a_n = 4n$.

73. $\displaystyle\sum_{i=1}^{4}\log 2i$
$= \log[2(1)] + \log[2(2)] + \log[2(3)] + \log[2(4)]$
$= \log 2 + \log 4 + \log 6 + \log 8$
$= \log(2 \cdot 4 \cdot 6 \cdot 8)$
$= \log 384$

75. Strategy
Multiply 6 by the number of 6s in the ones, tens, and hundreds places. Then multiply each product by the place value. Add the products and find the hundreds digit.

Solution
$31 \times 6 = 186 \quad\quad 186(1) \;\;= \quad\;\; 186$
$30 \times 6 = 180 \quad\quad 180(10) \;= \quad 1{,}800$
$29 \times 6 = 174 \quad\quad 174(100) = 17{,}400$
$ = \overline{19{,}386}$

The hundreds digit is 3.

77. $a_1 = 1,\ a_2 = 1,\ a_n = a_{n-1}+a_{n-2},\ n \geq 3$
$a_1 = 1$
$a_2 = 1$
$a_3 = a_{3-1}+a_{3-2} = a_2+a_1 = 1+1 = 2$
$a_4 = a_{4-1}+a_{4-2} = a_3+a_2 = 2+1 = 3$
$a_5 = a_{5-1}+a_{5-2} = a_4+a_3 = 3+2 = 5$
The first three terms are 2, 3, and 5.

79. $\dfrac{1}{1}+\dfrac{1}{2}+\dfrac{1}{3}+\ldots+\dfrac{1}{n} = \displaystyle\sum_{i=1}^{n}\dfrac{1}{i}$

Section 11.2

Concept Review 11.2

1. Sometimes true

    The successive terms of the series $\sum_{i=1}^{n}(3-i)$ decreases in value.

3. Sometimes true
   The first term of $-4, -2, 0, 2, \ldots$ is a negative number.

Objective 1 Exercises

3. $d = a_2 - a_1 = 11 - 1 = 10$
   $a_n = a_1 + (n-1)d$
   $a_{15} = 1 + (15-1)(10) = 1 + 14(10) = 1 + 140$
   $a_{15} = 141$

5. $d = a_2 - a_1 = -2 - (-6) = 4$
   $a_n = a_1 + (n-1)d$
   $a_{15} = -6 + (15-1)4 = -6 + (14)4 = -6 + 56$
   $a_{15} = 50$

7. $d = a_2 - a_1 = 7 - 3 = 4$
   $a_n = a_1 + (n-1)d$
   $a_{18} = 3 + (18-1)4 = 3 + 17(4) = 3 + 68$
   $a_{18} = 71$

9. $d = a_2 - a_1 = 0 - \left(-\dfrac{3}{4}\right) = \dfrac{3}{4}$
   $a_n = a_1 + (n-1)d$
   $a_{11} = -\dfrac{3}{4} + (11-1)\dfrac{3}{4}$
   $= -\dfrac{3}{4} + 10\left(\dfrac{3}{4}\right)$
   $= -\dfrac{3}{4} + \dfrac{30}{4}$
   $a_{13} = \dfrac{27}{4}$

11. $d = a_2 - a_1 = \dfrac{5}{2} - 2 = \dfrac{1}{2}$
    $a_n = a_1 + (n-1)d$
    $a_{31} = 2 + (31-1)\dfrac{1}{2} = 2 + 30\left(\dfrac{1}{2}\right) = 2 + 15$
    $a_{31} = 17$

13. $d = a_2 - a_1 = 5.75 - 6 = -0.25$
    $a_n = a_1 + (n-1)d$
    $a_{10} = 6 + (10-1)(-0.25) = 6 + 9(-0.25) = 6 - 2.25$
    $a_{10} = 3.75$

15. $d = a_2 - a_1 = 2 - 1 = 1$
    $a_n = a_1 + (n-1)d$
    $a_n = 1 + (n-1)1$
    $a_n = 1 + n - 1$
    $a_n = n$

17. $d = a_2 - a_1 = 2 - 6 = -4$
    $a_n = a_1 + (n-1)d$
    $a_n = 6 + (n-1)(-4)$
    $a_n = 6 - 4n + 4$
    $a_n = -4n + 10$

19. $d = a_2 - a_1 = \dfrac{7}{2} - 2 = \dfrac{3}{2}$
    $a_n = a_1 + (n-1)d$
    $a_n = 2 + (n-1)\dfrac{3}{2}$
    $a_n = 2 + \dfrac{3}{2}n - \dfrac{3}{2}$
    $a_n = \dfrac{3}{2}n + \dfrac{1}{2}$
    $a_n = \dfrac{3n+1}{2}$

21. $d = a_2 - a_1 = -13 - (-8) = -5$
    $a_n = a_1 + (n-1)d$
    $a_n = -8 + (n-1)(-5)$
    $a_n = -8 - 5n + 5$
    $a_n = -5n - 3$

23. $d = a_2 - a_1 = 16 - 26 = -10$
    $a_n = a_1 + (n-1)d$
    $a_n = 26 + (n-1)(-10)$
    $a_n = 26 - 10n + 10$
    $a_n = -10n + 36$

25. $d = a_2 - a_1 = 11 - 7 = 4$
    $a_n = a_1 + (n-1)d$
    $171 = 7 + (n-1)4$
    $171 = 7 + 4n - 4$
    $171 = 3 + 4n$
    $168 = 4n$
    $n = 42$
    There are 42 terms in the sequence.

27. $d = a_2 - a_1 = \dfrac{5}{3} - \dfrac{1}{3} = \dfrac{4}{3}$
    $a_n = a_1 + (n-1)d$
    $\dfrac{61}{3} = \dfrac{1}{3} + (n-1)\dfrac{4}{3}$
    $\dfrac{61}{3} = \dfrac{1}{3} + \dfrac{4}{3}n - \dfrac{4}{3}$
    $\dfrac{61}{3} = -1 + \dfrac{4}{3}n$
    $\dfrac{64}{3} = \dfrac{4}{3}n$
    There are 16 terms in the sequence.

29. $d = a_2 - a_1 = 8 - 3 = 5$
    $a_n = a_1 + (n-1)d$
    $98 = 3 + (n-1)5$
    $98 = 3 + 5n - 5$
    $98 = 5n - 2$
    $100 = 5n$
    $20 = n$
    There are 20 terms in the sequence.

31. $d = a_2 - a_1 = -3 - 1 = -4$
    $a_n = a_1 + (n-1)d$
    $-75 = 1 + (n-1)(-4)$
    $-75 = 1 - 4n + 4$
    $-75 = 5 - 4n$
    $-80 = -4n$
    $20 = n$
    There are 20 terms in the sequence.

33. $d = a_2 - a_1 = \dfrac{13}{3} - \dfrac{7}{3} = 2$
    $a_n = a_1 + (n-1)d$
    $\dfrac{79}{3} = \dfrac{7}{3} + (n-1)2$
    $\dfrac{79}{3} = \dfrac{7}{3} + 2n - 2$
    $\dfrac{79}{3} = 2n + \dfrac{1}{3}$
    $26 = 2n$
    $13 = n$
    There are 13 terms in the sequence.

35. $d = a_2 - a_1 = 2 - 3.5 = -1.5$
    $a_n = a_1 + (n-1)d$
    $-25 = 3.5 + (n-1)(-1.5)$
    $-25 = 3.5 - 1.5n + 1.5$
    $-25 = 5 - 1.5n$
    $-30 = -1.5n$
    $20 = n$
    There are 20 terms in the sequence.

Objective 2 Exercises

37. $d = a_2 - a_1 = 4 - 2 = 2$
    $a_n = a_1 + (n-1)d$
    $a_{25} = 2 + (25-1)2 = 2 + 24(2) = 50$
    $S_n = \dfrac{n}{2}(a_1 + a_2)$
    $S_{25} = \dfrac{25}{2}(2 + 50) = \dfrac{25}{2}(52) = 650$

39. $d = a_2 - a_1 = 20 - 25 = -5$
    $a_n = a_1 + (n-1)d$
    $a_{22} = 25 + (22-1)(-5)$
    $\phantom{a_{22}} = 25 + 21(-5)$
    $\phantom{a_{22}} = 25 - 105$
    $\phantom{a_{22}} = -80$
    $S_n = \dfrac{n}{2}(a_1 + a_n)$
    $S_{22} = \dfrac{22}{2}[25 + (-80)] = \dfrac{22}{2}(-55) = -605$

41. $d = a_2 - a_1 = \dfrac{11}{4} - 2 = \dfrac{3}{4}$
    $a_n = a_1 + (n-1)d$
    $a_{10} = 2 + (10-1)\dfrac{3}{4} = 2 + 9\left(\dfrac{3}{4}\right) = 2 + \dfrac{27}{4} = \dfrac{35}{4}$
    $S_{10} = \dfrac{n}{2}(a_1 + a_n)$
    $S_{10} = \dfrac{10}{2}\left(2 + \dfrac{35}{4}\right) = \dfrac{10}{2}\left(\dfrac{43}{4}\right) = \dfrac{215}{4}$

43. $a_i = 3i + 4$
    $a_1 = 3(1) + 4 = 7$
    $a_{15} = 3(15) + 4 = 49$
    $S_i = \dfrac{i}{2}(a_1 + a_i)$
    $S_{15} = \dfrac{15}{2}(7 + 49) = \dfrac{15}{2}(56) = 420$

45. $a_n = 1 - 4n$
    $a_1 = 1 - 4(1) = -3$
    $a_{10} = 1 - 4(10) = -39$
    $S_n = \dfrac{n}{2}(a_1 + a_n)$
    $S_{10} = \dfrac{10}{2}(-3 - 39)$
    $\phantom{S_{10}} = \dfrac{10}{2}(-42)$
    $\phantom{S_{10}} = -210$

47. $a_n = 5 - n$
    $a_1 = 5 - 1 = 4$
    $a_{10} = 5 - 10 = -5$
    $S_n = \dfrac{n}{2}(a_1 + a_n)$
    $S_{10} = \dfrac{10}{2}(4 - 5) = \dfrac{10}{2}(-1) = -5$

Objective 3 Exercises

49. Strategy
    To find the number of weeks:
    Write the arithmetic sequence.
    Find the common difference of the arithmetic sequence.
    Use the Formula for the nth Term of an Arithmetic Sequence to find the number of terms in the sequence.

    Solution
    12, 18, 24, ... 60
    $d = a_2 - a_1 = 18 - 12 = 6$
    $a_n = a_1 + (n-1)d$
    $60 = 12 + (n-1)6$
    $60 = 12 + 6n - 6$
    $60 = 6 + 6n$
    $54 = 6n$
    $9 = n$
    In 9 weeks the person will walk 60 min per day.

51. Strategy
To find the total number of seats:
Write the arithmetic sequence.
Find the common difference of the arithmetic sequence.
Use the Formula for the nth Term of an Arithmetic Sequence to find the sum of the sequence.
Use the Formula for the Sum of n Terms of an Arithmetic Sequence to find the sum of the sequence.

Solution
52, 58, 64, ...
$d = a_2 - a_1 = 58 - 52 = 6$
$a_n = a_1 + (n-1)d$
$a_{20} = 52 + (20-1)6 = 52 + 19(6) = 52 + 114 = 166$
$S_n = \dfrac{n}{2}(a_1 + a_n)$
$S_{20} = \dfrac{20}{2}(52 + 166) = 10(218) = 2180$
There are 2180 seats in the theater.

53. Strategy
To find the salary for the ninth month, use the Formula for the nth Term of an Arithmetic Sequence.
To find the total salary, use the Formula for the Sum of n Terms of an Arithmetic Sequence to find the sum of the sequence.

Solution
$a_n = a_1 + (n-1)d$
$a_{10} = 1800 + (10-1)150$
$= 1800 + (9)150$
$= 1800 + 1350$
$= 3150$
$S_n = \dfrac{n}{2}(a_1 + a_n)$
$S_{10} = \dfrac{10}{2}(1800 + 3150) = \dfrac{10}{2}(4800) = 22,275$
The salary for the tenth month is $3150.
The total salary for the ten-month period is $22,275.

55. Strategy
To find a formula for the sequence of distances:
Find the first term of the sequence by finding the circumference of a circle whose radius is equal to the first term of the sequence in Exercise 54, and adding this to twice the length of the straight part of the track, 83.4.
The common difference of the sequence is the circumference of a circle whose radius is equal to the width of a lane, 1.22.
Use the Formula for the nth Term of an Arithmetic Sequence.

Solution
$a_1 = 2\pi(37.11) + 2(83.4)$
$a_1 = 399.85$
$d = 2\pi(1.22)$
$d = 7.66$
$a_n = a_1 + (n-1)d$
$a_n = 399.85 + (n-1)(7.66)$
$a_n = 7.66n + 392.19$

Applying Concepts 11.2

57. $d = a_2 - a_1 = 2 - (-3) = 5$
$a_n = a_1 + (n-1)d$
$= -3 + (n-1)5$
$= -3 + 5n - 5$
$= 5n - 8$
$S_n = \dfrac{n}{2}(a_1 + a_n)$
$116 = \dfrac{n}{2}(-3 + 5n - 8)$
$116 = \dfrac{n}{2}(5n - 11)$
$232 = n(5n - 11)$
$232 = 5n^2 - 11n$
$0 = 5n^2 - 11n - 232$
$0 = (5n + 29)(n - 8)$
$5n + 29 = 0 \qquad n - 8 = 0$
$5n = -29 \qquad n = 8$
$n = -\dfrac{29}{5} \qquad n = 8$
The number of terms must be a natural number, so $-\dfrac{29}{5}$ is not a solution.
8 terms must be added together.

59. Find d in the series in which $a_1 = 9$ and $a_6 = 29$.

$a_n = a_1 + (n-1)d$
$29 = 9 + (6-1)d$
$29 = 9 + 5d$
$20 = 5d$
$4 = d$

d is the same for the series in which $a_4 = 9$ and $a_9 = 29$.

$a_n = a_1 + (n-1)d$
$29 = a_1 + (9-1)4$
$29 = a_1 + 8(4)$
$29 = a_1 + 32$
$-3 = a_1$

The first term is -3.

61. Strategy
To find the sum of the angles:
Write the arithmetic sequence with third term 180°, fourth term 360°, and fifth term 540°.
Find the common difference of the arithmetic sequence.
Find the first term of the arithmetic sequence using the Formula for the nth Term of an Arithmetic Sequence.
Use the Formula for the nth Term of an Arithmetic Sequence to find the 12th term.

Solution
$a_1, a_2, 180, 360, 540, \ldots$

$d = a_4 - a_3 = 360 - 180 = 180$
$a_n = a_1 + (n-1)d$
$a_3 = a_1 + (3-1)180$
$180 = a_1 + 2(180)$
$180 = a_1 + 360$
$-180 = a_1$

$a_n = a_1 + (n-1)d$
$a_{12} = -180 + (12-1)180$
$= -180 + 11(180)$
$= 1800$

The sum of the angles in a dodecagon is 1800°.
The general term is
$a_n = -180 + (n-1)180$
$= -180 + 180n - 180$
$= 180n - 360$
$= 180(n-2)$

The formula for the sum of the angles of an n-sided polygon is $180(n-2)$.

63. The arithmetic sequence of natural numbers that are multiples of 3 is 3, 6, 9, 12, …
Find the common difference:
$d = a_2 - a_1 = 6 - 3 = 3$
Now use the Formula for the nth Term of an Arithmetic Sequence.
$a_n = a_1 + (n-1)d$
$a_n = 3 + (n-1)3$
$a_n = 3 + 3n - 3$
$a_n = 3n$
The formula is $a_n = 3n$.

Section 11.3

Concept Review 11.3

1. Always true

3. Sometimes true
   If r is a negative number, the sum of a geometric series will oscillate.

5. Never true
   $\dfrac{a_{n+1}}{a_n}$ is not a constant ratio.

Objective 1 Exercises

3. $r = \dfrac{a_2}{a_1} = \dfrac{8}{2} = 4$

$a_n = a_1 r^{n-1}$
$a_9 = 2(4)^{9-1} = 2(4)^8 = 2(65{,}536) = 131{,}072$

5. $r = \dfrac{a_2}{a_1} = \dfrac{-4}{6} = -\dfrac{2}{3}$

$a_n = a_1 r^{n-1}$
$a_7 = 6\left(-\dfrac{2}{3}\right)^{7-1} = 6\left(-\dfrac{2}{3}\right)^6 = 6\left(\dfrac{64}{729}\right) = \dfrac{128}{243}$

7. $r = \dfrac{a_2}{a_1} = \dfrac{\sqrt{2}}{1} = \sqrt{2}$

$a_n = a_1 r^{n-1}$
$a_9 = 1(\sqrt{2})^{9-1} = 1(\sqrt{2})^8 = 1(2^4) = 16$

9. $a_n = a_1 r^{n-1}$
$a_4 = 9r^{4-1}$
$\dfrac{8}{3} = 9r^{4-1}$
$\dfrac{8}{3} = 9r^3$
$\dfrac{8}{27} = r^3$
$\dfrac{2}{3} = r$

$a_n = a_1 r^{n-1}$
$a_2 = 9\left(\dfrac{2}{3}\right)^{2-1} = 9\left(\dfrac{2}{3}\right) = 6$
$a_3 = 9\left(\dfrac{2}{3}\right)^{3-1} = 9\left(\dfrac{2}{3}\right)^2 = 9\left(\dfrac{4}{9}\right) = 4$

11. $a_n = a_1 r^{n-1}$
$a_4 = 3r^{4-1}$
$-\dfrac{8}{9} = 3r^{4-1}$
$-\dfrac{8}{9} = 3r^3$
$-\dfrac{8}{27} = r^3$
$-\dfrac{2}{3} = r$
$a_n = a_1 r^{n-1}$
$a_2 = 3\left(-\dfrac{2}{3}\right)^{2-1} = 3\left(-\dfrac{2}{3}\right) = -2$
$a_3 = 3\left(-\dfrac{2}{3}\right)^{3-1} = 3\left(-\dfrac{2}{3}\right)^2 = 3\left(\dfrac{4}{9}\right) = \dfrac{4}{3}$

13. $a_n = a_1 r^{n-1}$
$a_4 = (-3)r^{4-1}$
$192 = (-3)r^{4-1}$
$192 = (-3)r^3$
$-64 = r^3$
$-4 = r$
$a_n = a_1 r^{n-1}$
$a_2 = -3(-4)^{2-1} = -3(-4) = 12$
$a_3 = -3(-4)^{3-1} = -3(-4)^2 = -3(16) = -48$

Objective 2 Exercises

15. $r = \dfrac{a_2}{a_1} = \dfrac{6}{2} = 3$
$S_n = \dfrac{a_1(1-r^n)}{1-r}$
$S_7 = \dfrac{2(1-3^7)}{1-3}$
$= \dfrac{2(1-2187)}{-2}$
$= \dfrac{2(-2186)}{-2}$
$= 2186$

17. $r = \dfrac{a_2}{a_1} = \dfrac{9}{12} = \dfrac{3}{4}$
$S_n = \dfrac{a_1(1-r^n)}{1-r}$
$S_5 = \dfrac{12\left[1-\left(\frac{3}{4}\right)^5\right]}{1-\frac{3}{4}}$
$= \dfrac{12\left(1-\frac{243}{1024}\right)}{\frac{1}{4}}$
$= 48\left(\dfrac{781}{1024}\right)$
$= \dfrac{2343}{64}$

19. $a_i = (2)^i$
$a_1 = (2)^1 = 2$
$a_2 = (2)^2 = 4$
$r = \dfrac{a_2}{a_1} = \dfrac{4}{2} = 2$
$S_i = \dfrac{a_1(1-r^i)}{1-r}$
$S_5 = \dfrac{2(1-2^5)}{1-2} = \dfrac{2(1-32)}{-1} = -2(-31) = 62$

21. $a_i = \left(\dfrac{1}{3}\right)^i$
$a_1 = \left(\dfrac{1}{3}\right)^1 = \dfrac{1}{3}$
$a_2 = \left(\dfrac{1}{3}\right)^2 = \dfrac{1}{9}$
$r = \dfrac{a_2}{a_1} = \dfrac{1}{9} \div \dfrac{1}{3} = \dfrac{1}{9} \cdot \dfrac{3}{1} = \dfrac{1}{3}$
$S_i = \dfrac{a_1(1-r^i)}{1-r}$
$S_5 = \dfrac{\frac{1}{3}\left[1-\left(\frac{1}{3}\right)^5\right]}{1-\frac{1}{3}} = \dfrac{\frac{1}{3}\left(1-\frac{1}{243}\right)}{\frac{2}{3}} = \dfrac{1}{2}\left(\dfrac{242}{243}\right) = \dfrac{121}{243}$

23. $a_i = (4)^i$
$a_1 = 4^1 = 4$
$a_2 = 4^2 = 16$
$r = \dfrac{a_2}{a_1} = \dfrac{16}{4} = 4$
$S_i = \dfrac{a_1(1-r^i)}{1-r}$
$S_5 = \dfrac{4(1-4^5)}{1-4}$
$= \dfrac{4(1-1024)}{1-4}$
$= \dfrac{-4(1023)}{-3}$
$= 1364$

25. $a_i = (7)^i$
$a_1 = 7^1 = 7$
$a_2 = 7^2 = 49$
$r = \dfrac{a_2}{a_1} = \dfrac{49}{7} = 7$
$S_i = \dfrac{a_1(1-r^i)}{1-r}$
$S_4 = \dfrac{7(1-7^4)}{1-7}$
$= \dfrac{7(1-2401)}{-6}$
$= \dfrac{-7(2400)}{-6}$
$= 2800$

27. $a_i = \left(\frac{3}{4}\right)^i$

$a_1 = \left(\frac{3}{4}\right)^1 = \frac{3}{4}$

$a_2 = \left(\frac{3}{4}\right)^2 = \frac{9}{16}$

$r = \frac{a_2}{a_1} = \frac{9}{16} \div \frac{3}{4} = \frac{9}{16} \cdot \frac{4}{3} = \frac{3}{4}$

$S_i = \frac{a_1(1-r^i)}{1-r}$

$S_5 = \frac{\frac{3}{4}\left[1-\left(\frac{3}{4}\right)^5\right]}{1-\frac{3}{4}}$

$= \frac{\frac{3}{4}\left(1-\frac{243}{1024}\right)}{1-\frac{3}{4}}$

$= \frac{\frac{3}{4}\left(\frac{781}{1024}\right)}{\frac{1}{4}}$

$= 3\left(\frac{781}{1024}\right)$

$= \frac{2343}{1024}$

29. $a_i = \left(\frac{5}{3}\right)^i$

$a_1 = \left(\frac{5}{3}\right)^1 = \frac{5}{3}$

$a_2 = \left(\frac{5}{3}\right)^2 = \frac{25}{9}$

$r = \frac{a_2}{a_1} = \frac{25}{9} \div \frac{5}{3} = \frac{25}{9} \cdot \frac{3}{5} = \frac{5}{3}$

$S_i = \frac{a_1(1-r^i)}{1-r}$

$S_4 = \frac{\frac{5}{3}\left[1-\left(\frac{5}{3}\right)^4\right]}{1-\frac{5}{3}}$

$= \frac{\frac{5}{3}\left(1-\frac{625}{81}\right)}{-\frac{2}{3}}$

$= \frac{-\frac{5}{3}\left(\frac{544}{81}\right)}{-\frac{2}{3}}$

$= \frac{5}{2}\left(\frac{544}{81}\right)$

$= \frac{1360}{81}$

Objective 3 Exercises

31. $r = \frac{a_2}{a_1} = \frac{2}{3}$

$S = \frac{a_1}{1-r} = \frac{3}{1-\frac{2}{3}} = \frac{3}{\frac{1}{3}} = 9$

33. $r = \frac{a_2}{a_1} = \frac{-4}{6} = -\frac{2}{3}$

$S = \frac{a_1}{1-r} = \frac{6}{1-\left(-\frac{2}{3}\right)} = \frac{6}{\frac{5}{3}} = \frac{18}{5}$

35. $r = \frac{a_2}{a_1} = \frac{\frac{7}{100}}{\frac{7}{10}} = \frac{1}{10}$

$S = \frac{a_1}{1-r} = \frac{\frac{7}{10}}{1-\frac{1}{10}} = \frac{\frac{7}{10}}{\frac{9}{10}} = \frac{7}{9}$

37. $0.8\overline{88} = 0.8 + 0.08 + 0.008 + \ldots$

$= \frac{8}{10} + \frac{8}{100} + \frac{8}{1000} + \ldots$

$S = \frac{a_1}{1-r} = \frac{\frac{8}{10}}{1-\frac{1}{10}} = \frac{\frac{8}{10}}{\frac{9}{10}} = \frac{8}{9}$

An equivalent fraction is $\frac{8}{9}$.

39. $0.2\overline{22} = 0.2 + 0.02 + 0.002 + \ldots$

$= \frac{2}{10} + \frac{2}{100} + \frac{2}{1000} + \ldots$

$S = \frac{a_1}{1-r} = \frac{\frac{2}{10}}{1-\frac{1}{10}} = \frac{\frac{2}{10}}{\frac{9}{10}} = \frac{2}{9}$

An equivalent fraction $\frac{2}{9}$.

41. $0.45\overline{45} = 0.45 + 0.0045 + 0.000045 + \ldots$

$= \frac{45}{100} + \frac{45}{10,000} + \frac{45}{1,000,000} + \ldots$

$S = \frac{a_1}{1-r} = \frac{\frac{45}{100}}{1-\frac{1}{100}} = \frac{\frac{45}{100}}{\frac{99}{100}} = \frac{45}{99} = \frac{5}{11}$

An equivalent fraction is $\frac{5}{11}$.

43. $0.1\overline{666} = 0.1 + 0.06 + 0.006 + 0.0006 + \ldots$

$= \frac{1}{10} + \frac{6}{100} + \frac{6}{1000} + \frac{6}{10,000} + \ldots$

$S = \frac{a_1}{1-r} = \frac{\frac{6}{100}}{1-\frac{1}{10}} = \frac{\frac{6}{100}}{\frac{9}{10}} = \frac{6}{90} = \frac{1}{15}$

$0.1\overline{666} = \frac{1}{10} + \frac{1}{15} = \frac{5}{30} = \frac{1}{6}$

An equivalent fraction is $\frac{1}{6}$.

Chapter 11: Sequences and Series

ive 4 Exercises

Strategy
Use the Formula for the nth Term of a Geometric Sequence.

Solution
$a_1 = 1, \ r = 3$
$a_n = a_1 r^{n-1}$
$a_n = 1 \cdot 3^{n-1}$
$a_n = 3^{n-1}$

47. Strategy
To find the amount of radioactive material at the beginning of the seventh day, use the Formula for the nth Term of a Geometric Sequence.

Solution
$n = 7, \ a_1 = 500, \ r = \dfrac{1}{2}$
$a_n = a_1 r^{n-1}$
$a_7 = 500\left(\dfrac{1}{2}\right)^{7-1} = 500\left(\dfrac{1}{2}\right)^6 = 500\left(\dfrac{1}{64}\right) = 7.8125$

There will be 7.8125 mg of radioactive material in the sample at the beginning of the seventh day.

49. Strategy
To find the height of the ball on the fifth bounce, use the Formula for the nth Term of a Geometric Sequence. Let a_1 be the height of the ball after the first bounce.

Solution
$n = 5, \ a_1 = 80\% \text{ of } 8 = 6.4, \ r = 80\% = \dfrac{4}{5}$
$a_n = a_1 r^{n-1}$
$a_5 = 6.4\left(\dfrac{4}{5}\right)^{5-1} = 6.4\left(\dfrac{4}{5}\right)^4 = 6.4\left(\dfrac{256}{625}\right) \approx 2.6$

The ball bounces to a height of 2.6 ft on the fifth bounce.

51. Strategy
To find the value of the land in 15 years, use the Formula for the nth Term of a Geometric Sequence. Let a_1 be the value of the land after 1 year.

Solution
$n = 15, \ a_1 = 1.12(15{,}000) = 16{,}800,$
$r = 112\% = 1.12$
$a_n = a_1 r^{n-1}$
$a_{15} = 16{,}800(1.12)^{15-1}$
$= 16{,}800(4.8871123)$
$\approx 82{,}103.49$

The value of the land in 15 years will be $82,103.49.

53. Strategy
To find the value of the house in 30 yr, use the Formula for the nth Term of a Geometric Sequence.

Solution
$n = 30, \ a_1 = 1.05(100{,}000) = 105{,}000,$
$r = 105\% = 1.05$
$a_n = a_1 r^{n-1}$
$a_{30} = 105{,}000(1.05)^{30-1}$
$= (105{,}000)(4.1161356)$
$= 432{,}194.24$

The value of the house in 30 yr will be $432,194.24.

Applying Concepts 11.3

55. $4, -2, 1, \ldots$
$r = -\dfrac{1}{2}$
$a_4 = a_3 r = 1\left(-\dfrac{1}{2}\right) = -\dfrac{1}{2}$
The sequence is geometric. (G)
The next term is $-\dfrac{1}{2}$.

57. $5, 6.5, 8, \ldots$
$d = 1.5$
$a_4 = a_3 + d = 8 + 1.5 = 9.5$
The sequence is arithmetic. (A)
The next term is 9.5.

59. $1, 4, 9, 16, \ldots$
$a_n = n^2$
$a_5 = 5^2 = 25$
The sequence is neither arithmetic nor geometric. (N) The next term is 25.

61. $x^8, x^6, x^4, \ldots$
$r = x^{-2}$
$a_4 = a_3 r = x^4(x^{-2}) = x^2$
The sequence is geometric. (G)
The next term is x^2.

63. $\log x, \ 2\log x, \ 3\log x, \ldots$
$d = \log x$
$a_4 = a_3 + d = 3\log x + \log x = 4\log x$
The sequence is arithmetic. (A)
The next term is $4\log x$.

65. Find r for the geometric series in which $a_1 = 3$ and $a_4 = \dfrac{1}{9}$.

$$a_n = a_1 r^{n-1}$$
$$\frac{1}{9} = 3r^{4-1}$$
$$\frac{1}{9} = 3r^3$$
$$\frac{1}{27} = r^3$$
$$\sqrt[3]{\frac{1}{27}} = \sqrt[3]{r^3}$$
$$\frac{1}{3} = r$$

r is the same for the geometric series in which $a_3 = 3$ and $a_6 = \dfrac{1}{9}$.

$$a_n = a_1 r^{n-1}$$
$$\frac{1}{9} = a_1 \left(\frac{1}{3}\right)^{6-1}$$
$$\frac{1}{9} = a_1 \left(\frac{1}{3}\right)^5$$
$$\frac{1}{9} = a_1 \left(\frac{1}{243}\right)$$
$$27 = a_1$$

The first term is 27.

67.
$$a_n = 2^n$$
$$a_1 = 2^1 = 1$$
$$a_2 = 2^2 = 4$$
$$a_3 = 2^3 = 8$$
$$\vdots$$
$$a_n = 2^n$$
$$a_{n+1} = 2^{n+1}$$

$$b_n = \log a_n$$
$$b_1 = \log a_1 = \log 2$$
$$b_2 = \log a_2 = \log 4 = \log 2^2 = 2 \log 2$$
$$b_3 = \log a_3 = \log 8 = \log 2^3 = 3 \log 2$$
$$\vdots$$
$$b_n = \log a_n = \log 2^n = n \log 2$$
$$b_{n+1} = \log a_{n+1} = \log 2^{n+1} = (n+1) \log 2$$
$$b_{n+1} - b_n = (n+1) \log 2 - n \log 2$$
$$= n \log 2 + \log 2 - n \log 2 = \log 2$$

The common difference is $\log 2$.

69.
$$a_n = 3n - 2$$
$$a_1 = 3(1) - 2 = 1$$
$$a_2 = 3(2) - 2 = 4$$
$$a_3 = 3(3) - 2 = 7$$
$$\vdots$$
$$a_n = 3n - 2$$
$$a_{n+1} = 3(n+1) - 2 = 3n + 3 - 2 = 3n + 1$$
$$b_n = 2^{a_n}$$
$$b_1 = 2^{a_1} = 2^1 = 2$$
$$b_2 = 2^{a_2} = 2^4 = 16$$
$$b_3 = 2^{a_3} = 2^7 = 128$$
$$\vdots$$
$$b_n = 2^{a_n} = 2^{3n-2}$$
$$b_{n+1} = 2^{a_{n+1}} = 2^{3n+1}$$
$$\frac{b^{n+1}}{b^n} = \frac{2^{3n+1}}{2^{3n-2}} = 2^3 = 8$$

The common ratio is 8.

71. a. $R_n = R_1(1.0075)^{n-1}$
$R_{27} = 66.29(1.0075)^{27-1} = 66.29(1.0075)^{26}$
≈ 80.50
$80.50 of the loan is repaid in the twenty-seventh payment.

b. $T = \sum_{k=1}^{n} R_1(1.0075)^{k-1}$

$T = \sum_{k=1}^{20} 66.29(1.0075)^{k-1}$

$T = 66.29(1.0075)^0 + 66.29(1.0075)^1 + 66.29(1.0075)^2 + \ldots + 66.29(1.0075)^{19}$

$a_1 = 66.29(1.0075)^0 = 66.25$

$S_{20} = \dfrac{66.29(1 - 1.0075^{20})}{1 - 1.0075} \approx 1424.65$

The total amount repaid after 20 payments is $1424.65.

c. $5000 - 1424.65 = 3575.35$
The unpaid amount repaid after 20 payments is $3575.35.

Section 11.4

Concept Review 11.4

1. Never true
$0! \cdot 4! = 1 \cdot 24 = 24$

3. Never true
The exponent on the fifth term is 4.

5. Never true

There are $n + 1$ terms in the expansion of $(a + b)^n$.

Objective 1 Exercises

3. $3! = 3 \cdot 2 \cdot 1 = 6$

5. $8! = 8 \cdot 7 \cdot 6 \cdot 5 \cdot 4 \cdot 3 \cdot 2 \cdot 1 = 40{,}320$

7. $0! = 1$

9. $\dfrac{5!}{2!3!} = \dfrac{5 \cdot 4 \cdot 3 \cdot 2 \cdot 1}{(2 \cdot 1)(3 \cdot 2 \cdot 1)} = 10$

11. $\dfrac{6!}{6!0!} = \dfrac{6 \cdot 5 \cdot 4 \cdot 3 \cdot 2 \cdot 1}{(6 \cdot 5 \cdot 4 \cdot 3 \cdot 2 \cdot 1)(1)} = 1$

13. $\dfrac{9!}{6!3!} = \dfrac{9 \cdot 8 \cdot 7 \cdot 6 \cdot 5 \cdot 4 \cdot 3 \cdot 2 \cdot 1}{(6 \cdot 5 \cdot 4 \cdot 3 \cdot 2 \cdot 1)(3 \cdot 2 \cdot 1)} = 84$

15. $\dbinom{7}{2} = \dfrac{7!}{(7-2)!2!} = \dfrac{7!}{5!2!} = \dfrac{7 \cdot 6 \cdot 5 \cdot 4 \cdot 3 \cdot 2 \cdot 1}{(5 \cdot 4 \cdot 3 \cdot 2 \cdot 1)(2 \cdot 1)} = 21$

17. $\dbinom{10}{2} = \dfrac{10}{(10-2)!2!}$
$= \dfrac{10}{8!2!}$
$= \dfrac{10 \cdot 9 \cdot 8 \cdot 7 \cdot 6 \cdot 5 \cdot 4 \cdot 3 \cdot 2 \cdot 1}{(2 \cdot 1)(6 \cdot 5 \cdot 4 \cdot 3 \cdot 2 \cdot 1)}$
$= 45$

19. $\binom{9}{0} = \dfrac{9!}{(9-0)!0!}$

$= \dfrac{9!}{9!0!}$

$= \dfrac{9 \cdot 8 \cdot 7 \cdot 6 \cdot 5 \cdot 4 \cdot 3 \cdot 2 \cdot 1}{(9 \cdot 8 \cdot 7 \cdot 6 \cdot 5 \cdot 4 \cdot 3 \cdot 2 \cdot 1)(1)}$

$= 1$

21. $\binom{6}{3} = \dfrac{6!}{(6-3)!3!} = \dfrac{6!}{3!3!} = \dfrac{6 \cdot 5 \cdot 4 \cdot 3 \cdot 2 \cdot 1}{(3 \cdot 2 \cdot 1)(3 \cdot 2 \cdot 1)} = 20$

23. $\binom{11}{1} = \dfrac{11!}{(11-1)!1!}$

$= \dfrac{11!}{10!1!}$

$= \dfrac{11 \cdot 10 \cdot 9 \cdot 8 \cdot 7 \cdot 6 \cdot 5 \cdot 4 \cdot 3 \cdot 2 \cdot 1}{(10 \cdot 9 \cdot 8 \cdot 7 \cdot 6 \cdot 5 \cdot 4 \cdot 3 \cdot 2 \cdot 1)(1)}$

$= 11$

25. $\binom{4}{2} = \dfrac{4!}{(4-2)!2!} = \dfrac{4!}{2!2!} = \dfrac{4 \cdot 3 \cdot 2 \cdot 1}{(2 \cdot 1)(2 \cdot 1)} = 6$

27. $(x+y)^4$

$= \binom{4}{0}x^4 + \binom{4}{1}x^3 y + \binom{4}{2}x^2 y^2 + \binom{4}{3}xy^3 + \binom{4}{4}y^4$

$= x^4 + 4x^3 y + 6x^2 y^2 + 4xy^3 + y^4$

29. $(x-y)^5 = \binom{5}{0}x^5 + \binom{5}{1}x^4(-y) + \binom{5}{2}x^3(-y)^2 + \binom{5}{3}x^2(-y)^3 + \binom{5}{4}x(-y)^4 + \binom{5}{5}(-y)^5$

$= x^5 - 5x^4 y + 10x^3 y^2 - 10x^2 y^3 + 5xy^4 - y^5$

31. $(2m+1)^4 = \binom{4}{0}(2m)^4 + \binom{4}{1}(2m)^3(1) + \binom{4}{2}(2m)^2(1)^2 + \binom{4}{3}(2m)(1)^3 + \binom{4}{4}(1)^4$

$= 1(16m^4) + 4(8m^3) + 6(4m^2) + 4(2m) + 1(1)$

$= 16m^4 + 32m^3 + 24m^2 + 8m + 1$

33. $(2r-3)^5 = \binom{5}{0}(2r)^5 + \binom{5}{1}(2r)^4(-3) + \binom{5}{2}(2r)^3(-3)^2 + \binom{5}{3}(2r)^2(-3)^3 + \binom{5}{4}(2r)(-3)^4 + \binom{5}{5}(-3)^5$

$= 1(32r^5) + 5(16r^4)(-3) + 10(8r^3)(9) + 10(4r^2)(-27) + 5(2r)(81) + 1(-243)$

$= 32r^5 - 240r^4 + 720r^3 - 1080r^2 + 810r - 243$

35. $(a+b)^{10} = \binom{10}{0}a^{10} + \binom{10}{1}a^9 b + \binom{10}{2}a^8 b^2 + \ldots$

$= a^{10} + 10a^9 b + 45a^8 b^2 + \ldots$

37. $(a-b)^{11} = \binom{11}{0}a^{11} + \binom{11}{1}a^{10}(-b) + \binom{11}{2}a^9(-b)^2 + \ldots$

$= (1)a^{11} + 11a^{10}(-b) + 55a^9 b^2 + \ldots$

$a^{11} - 11a^{10} b + 55a^9 b^2 + \ldots$

39. $(2x+y)^8 = \binom{8}{0}(2x)^8 + \binom{8}{1}(2x)^7 y + \binom{8}{2}(2x)^6 y^2 + \ldots$

$= 1(256x^8) + 8(128x^7)y + 28(64x^6)y^2 + \ldots$

$= 256x^8 + 1024x^7 y + 1792x^6 y^2 + \ldots$

41. $(4x-3y)^8 = \binom{8}{0}(4x)^8 + \binom{8}{1}(4x)^7(-3y) + \binom{8}{2}(4x)^6(-3y)^2 + \ldots$
$= 1(65,536x^8) + 8(16,384x^7)(-3y) + 28(4096x^6)(9y^2) + \ldots$
$= 65,536x^8 - 393,216x^7 y + 1,032,192x^6 y^2 + \ldots$

43. $\left(x+\dfrac{1}{x}\right)^7 = \binom{7}{0}x^7 + \binom{7}{1}x^6\left(\dfrac{1}{x}\right) + \binom{7}{2}x^5\left(\dfrac{1}{x}\right)^2 + \ldots$
$= 1(x^7) + 7x^6\left(\dfrac{1}{x}\right) + 21x^5\left(\dfrac{1}{x^2}\right) + \ldots$
$= x^7 + 7x^5 + 21x^3 + \ldots$

45. $(x^2+3)^5 = \binom{5}{0}(x^2)^5 + \binom{5}{1}(x^2)^4(3) = \binom{5}{2}(x^2)^3(3)^2 + \ldots$
$= 1(x^{10}) + 5(x^8)(3) + 10(x^6)(9) + \ldots$
$= x^{10} + 15x^8 + 90x^6 + \ldots$

47. $n = 7, a = 2x, b = -1, r = 4$
$\binom{7}{4-1}(2x)^{7-4+1}(-1)^{4-1} = \binom{7}{3}(2x)^4(-1)^3 = 35(16x^4)(-1) = -560x^4$

49. $n = 6, a = x^2, b = -y^2, r = 2$
$\binom{6}{2-1}(x^2)^{6-2+1}(-y^2)^{2-1} = \binom{6}{1}(x^2)^5(-y^2) = 6x^{10}(-y^2) = -6x^{10}y^2$

51. $n = 9, a = y, b = -1, r = 5$
$\binom{9}{5-1}y^{9-5+1}(-1)^{5-1} = \binom{9}{4}y^5(-1)^4 = 126y^5(1) = 126y^5$

53. $n = 5, a = n, b = \dfrac{1}{n}, r = 2$
$\binom{5}{2-1}n^{5-2+1}\left(\dfrac{1}{n}\right)^{2-1} = \binom{5}{1}n^4\left(\dfrac{1}{n}\right) = 5n^4\left(\dfrac{1}{n}\right) = 5n^3$

55. $n = 5, a = \dfrac{x}{2}, b = 2, r = 1$
$\binom{5}{1-1}\left(\dfrac{x}{2}\right)^{5-1+1}(2)^{1-1} = \binom{5}{0}\left(\dfrac{x}{2}\right)^5(2)^0 = 1\left(\dfrac{x^5}{32}\right)(1) = \dfrac{x^5}{32}$

Applying Concepts 11.4

57.
```
                    1
                  1   1
                1   2   1
              1   3   3   1
            1   4   6   4   1
          1   5  10  10   5   1
        1   6  15  20  15   6   1
      1   7  21  35  35  21   7   1
```

59. $\dfrac{n!}{(n-1)!} = \dfrac{n(n-1)!}{(n-1)!} = n$

61. $(x^{1/2}+2)^4 = \binom{4}{0}(x^{1/2})^4 + \binom{4}{1}(x^{1/2})^3(2) + \binom{4}{2}(x^{1/2})^2(2^2) + \binom{4}{3}x^{1/2}(2^3) + \binom{4}{4}2^4$
$= x^2 + 4(2x^{3/2}) + 6(4x) + 4(8x^{1/2}) + 16$
$= x^2 + 8x^{3/2} + 24x + 32x^{1/2} + 16$

63. $(1+i)^6 = \binom{6}{0}1^6 + \binom{6}{1}1^5 i + \binom{6}{2}1^4 i^2 + \binom{6}{3}1^3 i^3 + \binom{6}{4}1^2 i^4 + \binom{6}{5}1 i^5 + \binom{6}{6} i^6$
$= 1 + 6i + 15i^2 + 20i^3 + 15i^4 + 6i^5 + i^6$
$= 1 + 6i - 15 - 20i + 15 + 6i - 1 = -8i$

65. $\dfrac{2 \cdot 4 \cdot 6 \cdot 8 \cdots (2n)}{2^n n!} = \dfrac{2(1) \cdot 2(2) \cdot 2(3) \cdot 2(4) \cdots 2(n)}{2^n n!}$
$= \dfrac{2^n(1 \cdot 2 \cdot 3 \cdots n)}{2^n n!}$
$= \dfrac{2^n n!}{2^n n!} = 1$

67. Strategy
To find the coefficient of $a^4 b^2 c^3$ in the expansion of $(a+b+c)^9$, use the formula
$\dfrac{n!}{r! k! (n-r-k)!}$
where $n = 9$, $r = 4$, $k = 2$

Solution
$\dfrac{n!}{r! k! (n-r-k)!} = \dfrac{9!}{4! 2! (9-4-2)!}$
$= \dfrac{9!}{4! 2! 3!} = \dfrac{9 \cdot 8 \cdot 7 \cdot 6 \cdot 5 \cdot 4!}{4! 2 \cdot 6}$
$= \dfrac{9 \cdot 8 \cdot 7 \cdot 6 \cdot 5}{2 \cdot 6} = 1260$

The coefficient is 1260.

Focus on Problem Solving

1. There were no errors on the test.
2. The temperature was at least 30 degrees.
3. The mountain is at most 5000 ft tall.
4. There are more than 5 vacancies for a field trip to New York.
5. Some trees are not tall.
6. Some cats do not chase mice.
7. All flowers do not have red blooms.
8. Some golfers like tennis.
9. All students like math.
10. Some honest people are politicians.
11. Some cars do not have power steering.
12. All televisions are not black and white.

Projects and Group Activities

ISBN and UPC Numbers

1. The first nine digits of the ISBN are 0-895-77692. We are to find the check digit.
   $0(10) + 8(9) + 9(8) + 5(7) + 7(6) + 7(5) + 6(4) + 9(3) + 2(2) + C = 311 + C$
   The last digit of the ISBN is chosen as 8 because $311 + 8 = 319$ and $319 \div 11 = 29$.

2. Is 0-395-12370-4 a possible ISBN?
   $0(10) + 3(9) + 9(8) + 5(7) + 1(6) + 2(5) + 3(4) + 7(3) + 0(2) + 4 = 187$
   $187 \div 11 = 17$, so 0-395-12370-4 is a possible ISBN.

3. Answers may vary.

4. The ISBN number for this book is 0-395-96961-1.

Finding the Proper Dosages

1. **Strategy**
   To find the x-rays:
   Write the formula $A(r) = A_0 e^{-0.2r}$ using $0.40 A_0$ for $A(r)$. Solve for r.

   Solution
   $A(r) = A_0 e^{-0.2r}$
   $0.40 A_0 = A_0 e^{-0.2r}$
   $0.40 = e^{-0.2r}$
   $\ln 0.40 = \ln e^{-0.2r}$
   $\ln 0.40 = -0.2r$
   $\dfrac{\ln 0.40}{-0.2} = r$
   $4.58 \approx r$

   The strength of the x-rays is 4.58 roentgens.

2. **Strategy**
   To find the initial dose:
   Write the formula $f(t) = Ae^{kt}$ where $f(t) = 500$, $k = \dfrac{-\ln 2}{2}$, $t = \dfrac{1}{2}$. Solve for A.

   Solution
   $f(t) = Ae^{kt}$
   $500 = Ae^{[(-\ln 2)/2] \cdot \frac{1}{2}}$
   $500 = Ae^{-(\ln 2)/4}$
   $\dfrac{500}{e^{-\ln 2/4}} = A$
   $594.60 \approx A$

   The initial dose is 594.60 mg.

340 Chapter 11: Sequences and Series

3. **Strategy**
   To find the concentration:
   Write the formula $S = \dfrac{Ae^{kt}(1-e^{nkt})}{1-e^{kt}}$ where
   $A = 0.5$, $k = -\dfrac{\ln 2}{8}$, $t = 4$, $n = 3$. Solve for S.

   Solution
   $S = \dfrac{Ae^{kt}(1-e^{nkt})}{1-e^{kt}}$
   $= \dfrac{0.5e^{[(-\ln 2)/8](4)}(1-e^{3[-(\ln 2)/8]4})}{1-e^{[(-\ln 2)/8](4)}}$
   $= \dfrac{0.5e^{-\ln 2/2}(1-e^{-3\ln 2/2})}{1-e^{-\ln 2/2}}$
   The concentration is 0.7803 mg/ml.

4. **a. Strategy**
   To find the level of medication:
   Write the formula $S = \dfrac{Ae^{kt}(1-e^{nkt})}{1-e^{kt}}$ using
   $A = 50$, $k = \dfrac{-\ln 2}{5}$, $t = 2$, $n = 4$. Solve for S.

   Solution
   $S = \dfrac{50e^{[(-\ln 2)/5](2)}(1-e^{4[(\ln 2/5](2)})}{1-e^{[(-\ln 2)/2](2)}}$
   $= \dfrac{50e^{-2\ln 2/5}(1-e^{-8\ln 2/5})}{1-e^{-2\ln 5/5}}$
   The level of mediation is 104.87 mg.

   b. Strategy
   To find the maintenance dose:
   Write the formula $M = A(1-e^{nkt})$ using
   $A = 50$, $k = \dfrac{-\ln 2}{5}$, $t = 2$, $n = 4$.

   Solution
   $M = A(1-e^{nkt})$
   $= 50(1-e^{4[(-\ln 2)/5](2)})$
   $= 50(1-e^{-8\ln 2/5})$
   $= 33.51$
   The maintenance dose is 33.51 mg.

Chapter Review Exercises

1. $\displaystyle\sum_{i=1}^{4} 3x^i = 3x + 3x^2 + 3x^3 + 3x^4$

2. $d = a_2 - a_1 = -8 - (-5) = -3$
   $a_n = a_1 + (n-1)d$
   $-50 = -5 + (n-1)(-3)$
   $-50 = -5 - 3n + 3$
   $-50 = -2 - 3n$
   $-48 = -3n$
   $16 = n$
   There are 16 terms.

3. $r = \dfrac{a_2}{a_1} = \dfrac{4\sqrt{2}}{4} = \sqrt{2}$
   $a_n = a_1 r^{n-1}$
   $a_7 = 4(\sqrt{2})^{7-1}$
   $= 4(\sqrt{2})^6$
   $= 4(8)$
   $= 32$

4. $r = \dfrac{a_2}{a_1} = \dfrac{3}{4}$
   $S = \dfrac{a_1}{1-r} = \dfrac{4}{1-\frac{3}{4}} = \dfrac{4}{\frac{1}{4}} = 16$

5. $\dbinom{9}{3} = \dfrac{9!}{(9-3)!3!}$
   $= \dfrac{9!}{6!3!}$
   $= \dfrac{9 \cdot 8 \cdot 7 \cdot 6 \cdot 5 \cdot 4 \cdot 3 \cdot 2 \cdot 1}{(6 \cdot 5 \cdot 4 \cdot 3 \cdot 2 \cdot 1)(3 \cdot 2 \cdot 1)}$
   $= 84$

6. $a_n = \dfrac{8}{n+2}$
   $a_{14} = \dfrac{8}{14+2} = \dfrac{8}{16} = \dfrac{1}{2}$

7. $d = a_2 - a_1 = -4 - (-10) = 6$
   $a_n = a_1 + (n-1)d$
   $a_{10} = -10 + (10-1)6$
   $= -10 + 9(6)$
   $= -10 + 54$
   $= 44$

8. $d = a_2 - a_1 = -19 - (-25) = 6$
   $a_n = a_1 + (n-1)d$
   $a_{18} = -25 + (18-1)6$
   $= -25 + (17)6$
   $= -25 + 102$
   $= 77$
   $S_n = \dfrac{n}{2}(a_1 + a_n)$
   $S_{18} = \dfrac{18}{2}(-25 + 77) = 9(52) = 468$

9. $r = \dfrac{a_2}{a_1} = \dfrac{12}{-6} = -2$
   $S_n = \dfrac{a_1(1-r^n)}{1-r}$
   $S_5 = \dfrac{-6[1-(-2)^5]}{1-(-2)}$
   $= \dfrac{-6[1-(-32)]}{3}$
   $= \dfrac{-6(33)}{3}$
   $= -66$
   $n = 9$, $a = 3x$, $b = y$, $r = 7$

10. $\dfrac{8!}{4!4!} = \dfrac{8\cdot 7\cdot 6\cdot 5\cdot 4\cdot 3\cdot 2\cdot 1}{(4\cdot 3\cdot 2\cdot 1)(4\cdot 3\cdot 2\cdot 1)} = 70$

11. $n = 9,\ a = 3x,\ b = y,\ r = 7$
$$\binom{n}{r-1}a^{n-r+1}b^{r-1}$$
$$\binom{9}{7-1}(3x)^{9-7+1}y^{7-1} = \binom{9}{6}(3x)^3 y^6$$
$$= 84(27x^3 y^6)$$
$$= 2268 x^3 y^6$$

12. $\displaystyle\sum_{n=1}^{n}(3n+1)$
$= [3(1)+1] + [3(2)+1] + [3(3)+1] + [3(4)+1]$
$= 4 + 7 + 10 + 13 = 34$

13. $a_n = \dfrac{n+1}{n}$
$a_6 = \dfrac{6+1}{6} = \dfrac{7}{6}$

14. $d = a_2 - a_1 = 9 - 12 = -3$
$a_n = a_1 + (n-1)d$
$a_n = 12 + (n-1)(-3)$
$a_n = 12 - 3n + 3$
$a_n = -3n + 15$

15. $r = \dfrac{a_2}{a_1} = \dfrac{2}{6} = \dfrac{1}{3}$
$a_n = a_1 r^{n-1}$
$a_5 = 6\left(\dfrac{1}{3}\right)^{5-1} = 6\left(\dfrac{1}{3}\right)^4 = 6\left(\dfrac{1}{81}\right) = \dfrac{2}{27}$

16. $0.23\overline{3} = 0.02 + 0.03 + 0.003 + 0.0003 + \ldots$
$= \dfrac{2}{10} + \dfrac{3}{100} + \dfrac{3}{1000} + \dfrac{3}{10{,}000} + \ldots$
$S = \dfrac{a_1}{1-r} = \dfrac{\frac{3}{100}}{1-\frac{1}{10}} = \dfrac{\frac{3}{100}}{\frac{9}{10}} = \dfrac{1}{30}$
$0.23\overline{3} = \dfrac{2}{10} + \dfrac{1}{30} = \dfrac{7}{30}$
An equivalent fraction is $\dfrac{7}{30}$.

17. $d = a_2 - a_1 = -16 - (-13) = -3$
$a_n = a_1 + (n-1)d$
$a_{35} = -13 + (35-1)(-3)$
$= -13 + (34)(-3)$
$= -13 - 102$
$= -115$

18. $S_n = \dfrac{a_1(1-r^n)}{1-r}$
$S_6 = \dfrac{1\left[1-\left(\frac{3}{2}\right)^6\right]}{1-\frac{3}{2}} = \dfrac{1-\frac{729}{64}}{-\frac{1}{2}} = \dfrac{-\frac{665}{64}}{-\frac{1}{2}} = \dfrac{665}{32}$

19. $d = a_2 - a_1 = 12 - 5 = 7$
$a_n = a_1 + (n-1)d$
$\quad = 5 + (21-1)7$
$\quad = 5 + 20(7)$
$\quad = 5 + 140 = 145$
$S_n = \dfrac{n}{2}(a_1 + a_n)$
$S_{21} = \dfrac{21}{2}(5+145) = \dfrac{21}{2}(150) = 1575$

20. $n = 7,\ a = x,\ b = -2y,\ r = 4$
$$\binom{n}{r-1}a^{n-r+1}b^{r-1}$$
$$\binom{7}{4-1}x^{7-4+1}(-2y)^{4-1} = \binom{7}{3}x^4(-2y)^3$$
$$= 35x^4(-8y^3)$$
$$= -280x^4 y^3$$

21. $d = a_2 - a_1 = 7 - 1 = 6$
$a_n = a_1 + (n-1)d$
$121 = 1 + (n-1)6$
$121 = 1 + 6n - 6$
$121 = 6n - 5$
$126 = 6n$
$21 = n$
There are 21 terms in the sequence.

22. $r = \dfrac{a_2}{a_1} = \dfrac{\frac{3}{4}}{\frac{3}{8}} = 2$
$a_n = a_1 r^{n-1}$
$a_8 = \dfrac{3}{8}(2)^{8-1} = \dfrac{3}{8}(2)^7 = \dfrac{3}{8}(128) = 48$

23. $\displaystyle\sum_{i=1}^{5} 2i = 2(1) + 2(2) + 2(3) + 2(4) + 2(5)$
$= 2 + 4 + 6 + 8 + 10 = 30$

24. $r = \dfrac{a_2}{a_1} = \dfrac{4}{1} = 4$
$S_n = \dfrac{a_1(1-r^n)}{1-r}$
$S_5 = \dfrac{1(1-4^5)}{1-4} = \dfrac{1(1-1024)}{-3} = \dfrac{-1023}{-3} = 341$

25. $5! = 5\cdot 4\cdot 3\cdot 2\cdot 1 = 120$

26. $n = 6,\ a = x,\ b = -4,\ r = 3$
$$\binom{n}{r-1}a^{n-r+1}b^{r-1}$$
$$\binom{6}{3-1}x^{6-3+1}(-4)^{3-1} = \binom{6}{2}x^4(-4)^2$$
$$= 15x^4(16) = 240x^4$$

27. $d = a_2 - a_1 = 3 - (-2) = 5$
$a_n = a_1 + (n-1)d$
$a_{30} = -2 + (30-1)5$
$= -2 + (29)(5)$
$= -2 + 145$
$= 143$

28. $d = a_2 - a_1 = 21 - 25 = -4$
$a_n = a_1 + (n-1)d$
$a_{25} = 25 + (25-1)(-4)$
$= 25 + (24)(-4)$
$= 25 - 96 = -71$
$S_n = \frac{n}{2}(a_1 + a_n)$
$S_{25} = \frac{25}{2}[25 + (-71)] = \frac{25}{2}(-46) = -575$

29. $a_n = \frac{(-1)^{2n-1} n}{n^2 + 2}$
$a_5 = \frac{(-1)^{2(5)-1} \cdot 5}{(5)^2 + 2} = \frac{(-1)^9 \cdot 5}{25 + 2} = \frac{-5}{27}$
The fifth term is $-\frac{5}{27}$.

30. $\sum_{i=1}^{4} 2x^{i-1} = 2x^{1-1} + 2x^{2-1} + 2x^{3-1} + 2x^{4-1}$
$= 2 + 2x + 2x^2 + 2x^3$

31. $0.\overline{23} = 0.23 + 0.0023 + 0.000023 + \ldots$
$= \frac{23}{100} + \frac{23}{10,000} + \frac{23}{1,000,000} + \ldots$
$S = \frac{a_1}{1-r} = \frac{\frac{23}{100}}{1 - \frac{1}{100}} = \frac{\frac{23}{100}}{\frac{99}{100}} = \frac{23}{99}$
An equivalent fraction is $\frac{23}{99}$.

32. $r = \frac{a_2}{a_1} = \frac{-1}{4} = -\frac{1}{4}$
$S = \frac{a_1}{1-r} = \frac{4}{1-\left(-\frac{1}{4}\right)} = \frac{4}{\frac{5}{4}} = \frac{16}{5}$

33. $a_n = 2(3)^n$
$a_1 = 2(3)^1 = 6$
$a_2 = 2(3)^2 = 18$
$r = \frac{a_2}{a_1} = \frac{18}{6} = 3$
$S_n = \frac{a_1(1-r^n)}{1-r}$
$S_5 = \frac{6(1-3^5)}{1-3} = \frac{6(1-243)}{-2} = -3(-242) = 726$

34. $n = 11, \ a = x, \ b = -2y, \ r = 8$
$\binom{n}{r-1} a^{n-r+1} b^{r-1}$
$\binom{11}{8-1}(x)^{11-8+1}(-2y)^{8-1} = \binom{11}{7} x^4 (-2y)^7$
$= 330 x^4 (-128 y^7)$
$= 42,240 x^4 y^7$

35. $a_n = \left(\frac{1}{2}\right)^n$
$a_1 = \left(\frac{1}{2}\right)^1 = \frac{1}{2}$
$a_2 = \left(\frac{1}{2}\right)^2 = \frac{1}{4}$
$r = \frac{a_2}{a_1} = \frac{\frac{1}{4}}{\frac{1}{2}} = \frac{1}{4} \cdot \frac{2}{1} = \frac{1}{2}$
$S_n = \frac{a_1(1-r^n)}{1-r}$
$= \frac{\frac{1}{2}\left(1 - \left(\frac{1}{2}\right)^8\right)}{1 - \frac{1}{2}}$
$= \frac{\frac{1}{2}\left(1 - \frac{1}{256}\right)}{\frac{1}{2}}$
$= \frac{255}{256}$
≈ 0.996

36. $r = \frac{a_2}{a_1} = \frac{\frac{4}{3}}{2} = \frac{2}{3}$
$S = \frac{a_1}{1-r} = \frac{2}{1 - \frac{2}{3}} = \frac{2}{\frac{1}{3}} = 6$

37. $0.6\overline{3} = 0.6 + 0.03 + 0.003 + \ldots$
$= \frac{6}{10} + \frac{3}{100} + \frac{3}{1000} + \ldots$
$S = \frac{a_1}{1-r} = \frac{\frac{3}{100}}{1 - \frac{1}{10}} = \frac{\frac{3}{100}}{\frac{9}{10}} = \frac{3}{100} \cdot \frac{10}{9} = \frac{1}{30}$
$0.6\overline{33} = \frac{6}{10} + \frac{1}{30} = \frac{19}{30}$

38. $(x-3y^2)^5 = \binom{5}{0}x^5 + \binom{5}{1}x^4(-3y^2)^1 + \binom{5}{2}x^3(-3y^2)^2 + \binom{5}{3}x^2(-3y^2)^3 + \binom{5}{4}x^1(-3y^2)^4 + \binom{5}{5}(-3y^2)^5$
$= x^5 + 5x^4(-3y^2) + 10x^3(9y^9) + 10x^2(-27y^6) + 5x(81y^8) + 1(-243y^{10})$
$= x^5 - 15x^4y^2 + 90x^3y^4 - 270x^2y^6 + 405xy^8 - 243y^{10}$

39. $d = a_2 - a_1 = 2 - 8 = -6$
$a_n = a_1 + (n-1)d$
$-118 = 8 + (n-1)d$
$-118 = 8 - 6n + 6$
$-118 = 14 - 6n$
$-132 = -6n$
There are 22 terms in the sequence.

40. $\dfrac{12!}{5!\,8!} = \dfrac{12\cdot 11\cdot 10\cdot 9\cdot 8\cdot 7\cdot 6\cdot 5\cdot 4\cdot 3\cdot 2\cdot 1}{(5\cdot 4\cdot 3\cdot 2\cdot 1)(8\cdot 7\cdot 6\cdot 5\cdot 4\cdot 3\cdot 2\cdot 1)} = 99$

41. $\sum_{i=1}^{5} \dfrac{(2x)^i}{i} = \dfrac{(2x)^1}{1} + \dfrac{(2x)^2}{2} + \dfrac{(2x)^3}{3} + \dfrac{(2x)^4}{4} + \dfrac{(2x)^5}{5}$
$= 2x + \dfrac{4x^2}{2} + \dfrac{8x^3}{3} + \dfrac{16x^4}{4} + \dfrac{32x^5}{5}$
$= 2x + 2x^2 + \dfrac{8}{3}x^3 + 4x^4 + \dfrac{32}{5}x^5$

42. $\sum_{n=1}^{4} \dfrac{(-1)^{n-1}n}{n+1} = \dfrac{(-1)^{1-1}\cdot 1}{1+1} + \dfrac{(-1)^{2-1}\cdot 2}{2+1} + \dfrac{(-1)^{3-1}\cdot 3}{3+1} + \dfrac{(-1)^{4-1}\cdot 4}{4+1}$
$= \dfrac{1}{2} + \left(-\dfrac{2}{3}\right) + \dfrac{3}{4} + \left(-\dfrac{4}{5}\right) = \dfrac{30 - 40 + 45 - 48}{60}$
$= -\dfrac{13}{60}$

43. Strategy
To find the total salary for the nine-month period:
Write the arithmetic sequence.
Find the common difference of the arithmetic sequence.
Use the Formula for the nth Term of an Arithmetic Sequence to find the ninth term.
Use the Formula for the Sum of n Terms of an Arithmetic Sequence to find the sum of nine terms of the sequence.

Solution
$1200, $1240, $1280, ...
$d = a_2 - a_1 = 1240 - 1200 = 40$
$a_n = a_1 + (n-1)d$
$a_9 = 1200 + (9-1)40 = 1200 + 320 = 1520$
$S_n = \dfrac{n}{2}(a_1 + a_n)$
$S_9 = \dfrac{9}{2}(1200 + 1520) = \$12,240$
The total salary for the nine-month period is $12,240.

44. Strategy
To find the temperature of the spa after 8 hours, use the Formula for the nth Term of a Geometric Sequence.

Solution
$n = 8,\ a_1 = 102(0.95) = 96.9,\ r = 0.95$
$a_n = a_1 r^{n-1}$
$a_8 = 96.9(0.95)^7 \approx 67.7$
The temperature is 67.7° F.

Chapter Test

1. $a_n = \dfrac{6}{n+4}$

   $a_{14} = \dfrac{6}{14+4} = \dfrac{6}{18} = \dfrac{1}{3}$

   The fourteenth term is $\dfrac{1}{3}$.

2. $a_n = \dfrac{n-1}{n}$

   $a_9 = \dfrac{9-1}{9} = \dfrac{8}{9}$

   The ninth term is $\dfrac{8}{9}$.

   $a_{10} = \dfrac{10-1}{10} = \dfrac{9}{10}$

   The tenth term is $\dfrac{9}{10}$.

3. $\displaystyle\sum_{n=1}^{4}(2n+3)$
   $= [2(1)+3]+[2(2)+3]+[2(3)+3]+[2(4)+3]$
   $= 5+7+9+11$
   $= 32$

4. $\displaystyle\sum_{i=1}^{4} 2x^{2i} = 2x^{2(1)} + 2x^{2(2)} + 2x^{2(3)} + 2x^{2(4)}$
   $= 2x^2 + 2x^4 + 2x^6 + 2x^8$

5. $d = a_2 - a_1 = -16-(-12) = -16+12 = -4$
   $a_n = a_1 + (n-1)d$
   $a_{28} = -12 + (28-1)(-4) = -12 + 27(-4) = -120$

6. $d = a_2 - a_1 = -1-(-3) = -1+3 = 2$
   $a_n = a_1 + (n-1)d$
   $a_n = -3 + (n+1)(2) = -3 + 2n - 2$
   $a_n = 2n - 5$

7. $d = a_2 - a_1 = 3 - 7 = -4$
   $a_n = a_1 + (n-1)d$
   $-77 = 7 + (n-1)d$
   $-77 = 7 - 4n + 4$
   $-77 = 11 - 4n$
   $-88 = -4n$
   $22 = n$

8. $d = a_2 - a_1 = -33 - (-42) = -33 + 42 = 9$
   $a_n = a_1 + (n-1)d$
   $a_{15} = -42 + (15-1)9 = -42 + 14(9) = 84$
   $S_n = \dfrac{n}{2}(a_1 + a_n)$
   $S_{15} = \dfrac{15}{2}(-42+84) = 315$

9. $d = a_2 - a_1 = 2-(-4) = 2+4 = 6$
   $a_n = a_1 + (n-1)d$
   $a_{24} = -4 + (24-1)6 = -4 + 23(6) = 134$
   $S_n = \dfrac{n}{2}(a_1 + a_n)$
   $S_{24} = \dfrac{24}{2}(-4+134) = 1560$

10. $\dfrac{10!}{5!5!} = \dfrac{10\cdot 9\cdot 8\cdot 7\cdot 6\cdot 5\cdot 4\cdot 3\cdot 2\cdot 1}{(5\cdot 4\cdot 3\cdot 2\cdot 1)(5\cdot 4\cdot 3\cdot 2\cdot 1)} = 252$

11. $r = \dfrac{a_2}{a_1} = \dfrac{-4\sqrt{2}}{2} = -\sqrt{2}$
    $a_n = 4r^{n-1}$
    $a_{10} = 4(-\sqrt{2})^{10-1} = 4\cdot(-\sqrt{2})^9 = -64\sqrt{2}$

12. $r = \dfrac{a_2}{a_1} = \dfrac{3}{5}$
    $a_n = a_1 r^{n-1}$
    $a_5 = 5\cdot\left(\dfrac{3}{5}\right)^{5-1} = 5\cdot\left(\dfrac{3}{5}\right)^4 = 5\cdot\dfrac{81}{625} = \dfrac{81}{125}$

13. $r = \dfrac{a_2}{a_1} = \dfrac{\frac{3}{4}}{1} = \dfrac{3}{4}$
    $S_n = \dfrac{a_1(1-r^n)}{1-r}$
    $S_5 = \dfrac{1\left(1-\left(\frac{3}{4}\right)^5\right)}{1-\frac{3}{4}} = \dfrac{1-\frac{243}{1024}}{\frac{1}{4}} = \dfrac{\frac{781}{1024}}{\frac{1}{4}} = \dfrac{781}{256}$

14. $r = \dfrac{a_2}{a_1} = \dfrac{10}{-5} = -2$
    $S_n = \dfrac{a_1(1-r^n)}{1-r}$
    $S_5 = \dfrac{-5(1-(-2)^5)}{1-(-2)}$
    $= \dfrac{-5(1-(-32))}{3}$
    $= \dfrac{-5(33)}{3}$
    $= -55$

15. $r = \dfrac{a_2}{a_1} = \dfrac{1}{2}$
    $S_n = \dfrac{a_1}{1-r} = \dfrac{2}{1-\frac{1}{2}} = \dfrac{2}{\frac{1}{2}} = 4$

16. $0.2\overline{3} = 0.2 + 0.03 + 0.003 + \ldots$
    $= \dfrac{2}{10} + \dfrac{3}{100} + \dfrac{3}{1000} + \ldots$
    $S = \dfrac{a_1}{1-r} = \dfrac{\frac{3}{100}}{1-\frac{1}{10}} = \dfrac{\frac{3}{100}}{\frac{9}{10}} = \dfrac{3}{90} = \dfrac{1}{30}$
    $0.23\overline{3} = \dfrac{2}{10} + \dfrac{1}{30} = \dfrac{7}{30}$
    An equivalent fraction is $\dfrac{7}{30}$.

17. $\binom{11}{4} = \dfrac{11!}{(11-4)!4!}$
$= \dfrac{11!}{7!4!}$
$= \dfrac{11 \cdot 10 \cdot 9 \cdot 8 \cdot 7 \cdot 6 \cdot 5 \cdot 4 \cdot 3 \cdot 2 \cdot 1}{(7 \cdot 6 \cdot 5 \cdot 4 \cdot 3 \cdot 2 \cdot 1)(4 \cdot 3 \cdot 2 \cdot 1)}$
$= 330$

18. $n = 8,\ a = 3x,\ b = -y,\ r = 5$
$\binom{n}{r-1} a^{n-r+1} b^{r-1}$
$\binom{8}{5-1}(3x)^{8-5+1}(-y)^{5-1} = \binom{8}{4}(3x)^4(-y)^4$
$= 70(81x^4)(y^4)$
$= 5670 x^4 y^4$

19. **Strategy**
To find how much material was in stock after the shipment on October 1:
Write the arithmetic sequence.
Find the common difference of the arithmetic sequence.
Use the Formula for the nth Term of an Arithmetic Sequence to find the tenth term.

Solution
7500, 6950, 6400, ...
$d = a_2 - a_1 = 6950 - 7500 = -550$
$a_n = a_1 + (n-1)d$
$a_{10} = 7500 + (10-1)(-550) = 2550$
The inventory after the October 1 shipment was 2550 yd.

20. **Strategy**
To find the amount of radioactive material at the beginning of the fifth day, use the Formula for the nth Term of a Geometric Sequence.

Solution
$n = 5,\ a_1 = 320,\ r = \dfrac{1}{2}$
$a_n = a_1 r^{n-1}$
$a_5 = 320\left(\dfrac{1}{2}\right)^{5-1} = 320\left(\dfrac{1}{2}\right)^4 = 20$
There will be 20 mg of radioactive material in the sample at the beginning of the fifth day.

Cumulative Review Exercises

1. $\dfrac{4x^2}{x^2 + x - 2} - \dfrac{3x-2}{x+2}$
$= \dfrac{4x^2}{(x+2)(x-1)} - \dfrac{3x-2}{x+2} \cdot \dfrac{x-1}{x-1}$
$= \dfrac{4x^2 - (3x^2 - 5x + 2)}{(x+2)(x-1)}$
$= \dfrac{x^2 + 5x - 2}{(x+2)(x-1)}$

2. $2x^6 + 16 = 2(x^6 + 8)$
$= 2((x^2)^3 + 2^3)$
$= 2(x^2 + 2)(x^4 - 2x^2 + 4)$

3. $\sqrt{2y}(\sqrt{8xy} - \sqrt{y}) = \sqrt{16xy^2} - \sqrt{2y^2}$
$= \sqrt{16y^2(x)} - \sqrt{y^2(2)}$
$= 4y\sqrt{x} - y\sqrt{2}$

4. $\left(\dfrac{x^{-\frac{3}{4}} x^{\frac{3}{2}}}{x^{-\frac{5}{2}}}\right)^{-8} = \dfrac{x^6 x^{-12}}{x^{20}} = \dfrac{x^{-6}}{x^{20}} = \dfrac{1}{x^{26}}$

5. $5 - \sqrt{x} = \sqrt{x+5}$
$(5-\sqrt{x})^2 = (\sqrt{x+5})^2$
$25 - 10\sqrt{x} + x = x + 5$
$-10\sqrt{x} = -20$
$\sqrt{x} = 2$
$(\sqrt{x})^2 = 2^2$
$x = 4$

Check:
$5 - \sqrt{x} = \sqrt{x+5}$	
$5 - \sqrt{4}$	$\sqrt{4+5}$
$5 - 2$	$\sqrt{9}$

$3 = 3$
The solution is 4.

6. $2x^2 - x + 7 = 0$
$a = 2,\ b = -1,\ c = 7$
$x = \dfrac{-b \pm \sqrt{b^2 - 4ac}}{2a}$
$= \dfrac{-(-1) \pm \sqrt{(-1)^2 - 4(2)(7)}}{2(2)}$
$= \dfrac{1 \pm \sqrt{1 - 56}}{4}$
$= \dfrac{1 \pm \sqrt{-55}}{4}$
$= \dfrac{1}{4} \pm \dfrac{\sqrt{55}}{4} i$

The solutions are $\dfrac{1}{4} + \dfrac{\sqrt{55}}{4}i$ and $\dfrac{1}{4} - \dfrac{\sqrt{55}}{4}i$.

346 *Chapter 11: Sequences and Series*

7. (1) $3x - 3y = 2$
    (2) $6x - 4y = 5$
    Eliminate x.
    Multiply equation (1) by -2 and add equation (2).
    $(-2)(3x - 3y) = (-2)(2)$
    $6x - 4y = 5$
    $-6x + 6y = -4$
    $6x - 4y = 5$
    $2y = 1$
    $y = \dfrac{1}{2}$

    Substitute $\dfrac{1}{2}$ for y in equation (2).
    $6x - 4\left(\dfrac{1}{2}\right) = 5$
    $6x - 2 = 5$
    $6x = 7$
    $x = \dfrac{7}{6}$

    The solution is $\left(\dfrac{7}{6}, \dfrac{1}{2}\right)$.

8. $2x - 1 > 3$ or $1 - 3x > 7$
    $2x > 4$ $\quad\quad -3x > 6$
    $x > 2$ $\quad\quad\quad x < -2$
    $\{x | x > 2\}$ $\quad \{x | x < -2\}$
    $\{x | x < -2 \text{ or } x > 2\}$

9. $\begin{vmatrix} -3 & 1 \\ 4 & 2 \end{vmatrix} = -3(2) - 4(1) = -6 - 4 = -10$

10. $\log_5 \sqrt{\dfrac{x}{y}} = \log_5 \left(\dfrac{x}{y}\right)^{\frac{1}{2}} = \dfrac{1}{2} \log_5 \left(\dfrac{x}{y}\right) = \dfrac{1}{2}(\log_5 x - \log_5 y) = \dfrac{1}{2} \log_5 x - \dfrac{1}{2} \log_5 y$

11. $4^x = 8^{x-1}$
    $(2^2)^x = (2^3)^{x-1}$
    $2^{2x} = 2^{3(x-1)}$
    $2x = 3(x - 1)$
    $2x = 3x - 3$
    $-x = -3$
    $x = 3$

12. $a_n = n(n - 1)$
    $a_5 = 5(5 - 1) = 5(4) = 20$ The fifth term is 20.
    $a_6 = 6(6 - 1) = 6(5) = 30$ The sixth term is 30.

13. $\displaystyle\sum_{n=1}^{7} (-1)^{n-1}(n+2) = (-1)^{1-1}(1+2) + (-1)^{2-1}(2+2) + (-1)^{3-1}(3+2) + (-1)^{4-1}(4+2) + (-1)^{5-1}(5+2)$
    $+ (-1)^{6-1}(6+2) + (-1)^{7-1}(7+2)$
    $= (-1)^0(3) + (-1)^1(4) + (-1)^2(5) + (-1)^3(6) + (-1)^4(7) + (-1)^5(8) + (-1)^6(9)$
    $= 3 - 4 + 5 - 6 + 7 - 8 + 9 = 6$

14. (1) $x + 2y + z = 3$
(2) $2x - y + 2z = 6$
(3) $3x + y - z = 5$
To eliminate y add equations (2) and (3).
(2) $2x - y + 2z = 6$
(3) $3x + y - z = 5$
(4) $5x + z = 11$
To eliminate y from equations (1) and (2), multiply equation (2) by 2 and add it to equation (1).
$x + 2y + z = 3$
$2(2x - y + 2z) = 2(6)$

$x + 2y + z = 3$
$4x - 2y + 4z = 12$
(5) $5x + 5z = 15$
Eliminate x from equations (4) and (5) by multiplying equation (4) by -1 and adding to equation (5).
$-1(5x + z) = (-1)(11)$
$5x + 5z = 15$

$-5x - z = -11$
$5x + 5z = 15$

$4z = 4$
$z = 1$
Solve for x by substituting 1 for z in equation (5).
$5x + 5z = 15$
$5x + 5(1) = 15$
$5x + 5 = 15$
$5x = 10$
$x = 2$
Solve for y by substituting 1 for z and 2 for x in equation (3).
$3x + y - z = 5$
$3(2) + y - 1 = 5$
$6 + y - 1 = 5$
$5 + y = 5$
$y = 0$
The solution is $(2, 0, 1)$.

15. $\log_6 x = 3$
$6^3 = x$
$216 = x$
The solution is 216.

16. $(4x^3 - 3x + 5) \div (2x + 1)$

$$\begin{array}{r} 2x^2 - x - 1 \\ 2x+1\overline{)4x^3 + 0x^2 - 3x + 5} \\ \underline{4x^3 + 2x^2} \\ -2x^2 - 3x \\ \underline{-2x^2 - x} \\ -2x + 5 \\ \underline{-2x - 1} \\ 6 \end{array}$$

The solution is $2x^2 - x - 1 + \dfrac{6}{2x+1}$.

17. $g(x) = -3x + 4$
$g(1 + h) = -3(1 + h) + 4$
$= -3 - 3h + 4 = -3h + 1$

18. $f(a) = \dfrac{a^3 - 1}{2a + 1}$
The domain is $\{0, 1, 2\}$.
$f(0) = \dfrac{0^3 - 1}{2(0) + 1} = \dfrac{-1}{1} = -1$
$f(1) = \dfrac{1^3 - 1}{2(1) + 1} = \dfrac{0}{3} = 0$
$f(2) = \dfrac{2^3 - 1}{2(2) + 1} = \dfrac{8 - 1}{5} = \dfrac{7}{5}$
The range is $\{-1, 0, \dfrac{7}{5}\}$.

19. $3x - 2y = -4$
$-2y = -3x - 4$
$y = \dfrac{3}{2}x + 2$

20. $2x - 3y < 9$
$-3y < -2x + 9$
$y > \dfrac{2}{3}x - 3$

21. Strategy

Time required for the older computer: t
Time required for the new computer: $t-16$

	Rate	Time	Part
Older computer	$\frac{1}{t}$	15	$\frac{15}{t}$
New computer	$\frac{1}{t-16}$	15	$\frac{15}{t-16}$

The sum of the part of the task completed by the older computer and the part of the task completed by the new computer is 1.

$$\frac{15}{t} + \frac{15}{t-16} = 1$$

Solution

$$\frac{15}{t} + \frac{15}{t-16} = 1$$

$$t(t-16)\left(\frac{15}{t} + \frac{15}{t-16}\right) = 1(t(t-16))$$

$$15(t-16) + 15t = t^2 - 16t$$

$$15t - 240 + 15t = t^2 - 16t$$

$$0 = t^2 - 46t + 240$$

$$0 = (t-40)(t-6)$$

$t - 40 = 0 \quad t - 6 = 0$
$t = 40 \quad t = 6$

$t = 6$ does not check as a solution since
$t - 16 = 6 - 16 = -10$.
$t - 16 = 40 - 16 = 24$

The new computer takes 24 min to complete the payroll.
The old computer takes 40 min to complete the payroll.

22. Strategy

Rate of boat in calm water: x
Rate of current: y

	Rate	Time	Distance
With current	$x+y$	2	$2(x+y)$
Against current	$x-y$	3	$3(x-y)$

The distance traveled with the current is 15 mi.
The distance traveled against the current is 15 mi.
$2(x+y) = 15$
$3(x-y) = 15$

Solution

$2(x+y) = 15 \quad \frac{1}{2} \cdot 2(x+y) = \frac{1}{2} \cdot 15$
$3(x-y) = 15 \quad \frac{1}{3} \cdot 3(x-y) = \frac{1}{3} \cdot 15$

$ x + y = 7.5$
$ x - y = 5$
$ 2x = 12.5$
$ x = 6.25$

$x + y = 7.5$
$6.25 + y = 7.5$
$y = 1.25$

The rate of the boat in calm water is 6.25 mph.
The rate of the current is 1.25 mph.

23. To find the half-life, solve for k in the exponential decay equation.
Use $A_0 = 80$, $A = 55$, $t = 30$.

Solution

$$A = A_0\left(\frac{1}{2}\right)^{t/k}$$

$$55 = 80\left(\frac{1}{2}\right)^{30/k}$$

$$0.6875 = (0.5)^{30/k}$$

$$\log 0.6875 = \frac{30}{k} \log(0.5)$$

$$k = \frac{30 \log 0.5}{\log 0.6875}$$

$$\approx 55.49$$

The half-life is approximately 55 days.

24. Strategy
To find the total number of seats in the 12 rows of the theater.
Write the arithmetic sequence.
Find the common difference of the arithmetic sequence.
Use the Formula for the nth Term of an Arithmetic Sequence to find the 12th term.
Use the Formula for the Sum of n Terms of an Arithmetic Sequence to find the sum of the 12 terms of the sequence.

Solution
62, 74, 86, ...
$d = a_2 - a_1 = 74 - 62 = 12$
$a_n = a_1 + (n-1)d$
$a_{12} = 62 + (12-1)12 = 62 + 132 = 194$
$S_n = \dfrac{n}{2}(a_1 + a_n)$
$S_{12} = \dfrac{12}{2}(62 + 194) = 1536$
The total number of seats in the theater is 1536.

25. Strategy
To find the height of the ball on the fifth bounce, use the Formula for the nth Term of a Geometric Sequence.

Solution
$n = 5, \ a_1 = 80\% \text{ of } 10 = 8, \ r = 80\% = 0.8$
$a_n = a_1 r^{n-1}$
$a_5 = 8(0.8)^{5-1}$
$\quad = 8(0.4096) = 3.2768$
The height of the ball on the fifth bounce is 3.3 ft.

Chapter 12: Conic Sections

Section 12.1

Concept Review 12.1

1. Sometimes true
   The graph of the parabola $y = x^2 - 2$ is the graph of a function. The graph of the parabola $x = y^2 - 2$ is not the graph of a function.

3. Sometimes true
   A parabola may have one intercept, two intercepts, or no intercepts.

5. Sometimes true
   The axis of symmetry of the parabola $y = x^2 + 2x - 3$ is the line $x = -1$.

Objective 1 Exercises

3. $y = x^2 - 2x - 4$
   $-\dfrac{b}{2a} = -\dfrac{-2}{2(1)} = 1$
   $y = 1^2 - 2(1) - 4 = -5$
   Vertex: $(1, -5)$
   Axis of symmetry: $x = 1$

5. $y = -x^2 + 2x - 3$
   $-\dfrac{b}{2a} = -\dfrac{2}{2(-1)} = 1$
   $y = -(1)^2 + 2(1) - 3 = -2$
   Vertex: $(1, -2)$
   Axis of symmetry: $x = 1$

7. $x = y^2 + 6y + 5$
   $-\dfrac{b}{2a} = -\dfrac{6}{2(1)} = -3$
   $x = (-3)^2 + 6(-3) + 5 = -4$
   Vertex: $(-4, -3)$
   Axis of symmetry: $y = -3$

9. $y = 2x^2 - 4x + 1$
   $-\dfrac{b}{2a} = -\dfrac{-4}{2(2)} = 1$
   $y = 2(1)^2 - 4(1) + 1 = -1$
   Vertex: $(1, -1)$
   Axis of symmetry: $x = 1$

11. $y = x^2 - 5x + 4$
    $-\dfrac{b}{2a} = -\dfrac{-5}{2(1)} = \dfrac{5}{2}$
    $y = \left(\dfrac{5}{2}\right)^2 - 5\left(\dfrac{5}{2}\right) + 4 = -\dfrac{9}{4}$
    Vertex: $\left(\dfrac{5}{2}, -\dfrac{9}{4}\right)$
    Axis of symmetry: $x = \dfrac{5}{2}$

13. $x = y^2 - 2y - 5$
    $-\dfrac{b}{2a} = -\dfrac{-2}{2(1)} = 1$
    $x = 1^2 - 2(1) - 5 = -6$
    Vertex: $(-6, 1)$
    Axis of symmetry: $y = 1$

15. $y = -3x^2 - 9x$
    $-\dfrac{b}{2a} = -\dfrac{-9}{2(-3)} = -\dfrac{3}{2}$
    $y = -3\left(-\dfrac{3}{2}\right)^2 - 9\left(-\dfrac{3}{2}\right) = \dfrac{27}{4}$
    Vertex: $\left(-\dfrac{3}{2}, \dfrac{27}{4}\right)$
    Axis of symmetry: $x = -\dfrac{3}{2}$

17. $x = -\frac{1}{2}y^2 + 4$

$-\frac{b}{2a} = -\frac{0}{2(-\frac{1}{2})} = 0$

$x = -\frac{1}{2}(0)^2 + 4 = 4$

Vertex: (4, 0)
Axis of symmetry: $y = 0$

19. $x = \frac{1}{2}y^2 - y + 1$

$-\frac{b}{2a} = -\frac{-1}{2(\frac{1}{2})} = 1$

$x = \frac{1}{2}(1)^2 - 1 + 1 = \frac{1}{2}$

Vertex: $\left(\frac{1}{2}, 1\right)$
Axis of symmetry: $y = 1$

21. $y = \frac{1}{2}x^2 + 2x - 6$

$-\frac{b}{2a} = -\frac{2}{2(\frac{1}{2})} = -2$

$y = \frac{1}{2}(-2)^2 + 2(-2) - 6 = -8$

Vertex: (-2, -8)
Axis of symmetry: $x = -2$

Applying Concepts 12.1

23. $y = x^2 - 4x - 2$

$-\frac{b}{2a} = -\frac{-4}{2(1)} = \frac{4}{2} = 2$

$y = x^2 - 4x - 2$
$= (2)^2 - 4(2) - 2$
$= 4 - 8 - 2$
$= -6$

a is positive. The minimum value of y is -6.
The domain is all real numbers.
The range is all real numbers greater than or equal to -6.

25. $y = -x^2 + 2x - 3$

$-\frac{b}{2a} = -\frac{2}{2(-1)} = -\frac{2}{-2} = 1$

$y = -x^2 + 2x - 3$
$= -(1)^2 + 2(1) - 3$
$= -1 + 2 - 3$
$= -2$

a is negative. The maximum value of y is -2.
The domain is all real numbers. The range is all real numbers less than or equal to -2.

27. $x = y^2 + 6y - 5$

$-\frac{b}{2a} = -\frac{6}{2(1)} = -\frac{6}{2} = -3$

$x = y^2 + 6y - 5$
$= (-3)^2 + 6(-3) - 5$
$= 9 - 18 - 5$
$= -14$

a is positive. The minimum value of x is -14.
The domain is all real numbers greater than or equal to -14. The range is all real numbers.

29. $x = -y^2 - 2y + 6$

$-\frac{b}{2a} = -\frac{-2}{2(-1)} = -\frac{-2}{-2} = -1$

$x = -y^2 - 2y + 6$
$= -(-1)^2 - 2(-1) + 6$
$= -1 + 2 + 6$
$= 7$

a is negative. The maximum value of x is 7.
The domain is all real numbers less than or equal to 7. The range is all real numbers.

31. $y = 2x^2 - 4x + 1$
$a = 2,\ b = -4,\ c = 1$

$x = -\frac{b}{2a} = -\frac{-4}{2(2)} = 1$

$y = 2(1)^2 - 4(1) + 1$
$= 2 - 4 + 1$
$= -1$

The vertex is $(1, -1)$. The axis of symmetry is $x = 1$. Since a is positive, the parabola opens up. The focus is $\frac{1}{4a} = \frac{1}{4(2)} = \frac{1}{8}$ units from the vertex on the line $x = 1$. The focus is

$\left(1, -1 + \frac{1}{8}\right) = \left(1, -\frac{7}{8}\right)$.

352 Chapter 12: Conic Sections

33. $x = \frac{1}{2}y^2 + y - 2$

   $a = \frac{1}{2}$, $b = 1$, $c = -2$

   $y = -\frac{b}{2a} = -\frac{1}{2(\frac{1}{2})} = -1$

   $x = \frac{1}{2}y^2 + y - 2$

   $= \frac{1}{2}(-1)^2 + (-1) - 2$

   $= \frac{1}{2} - 1 - 2$

   $= -\frac{5}{2}$

   The vertex is $\left(-\frac{5}{2}, -1\right)$. The axis of symmetry is $y = -1$. Since a is positive, the parabola opens to the right. The focus is $\frac{1}{4a} = \frac{1}{4(\frac{1}{2})} = \frac{1}{2}$ units from the vertex on the line $y = -1$. The focus is $\left(-\frac{5}{2} + \frac{1}{2}, -1\right) = (-2, -1)$.

35. **a.** The parabola passes through the point whose coordinates are (3.79, 100). Because the parabola opens to the right and its vertex is at the origin, the equation of the parabola is of the form $x = ay^2$.

   $x = ay^2$
   $3.79 = a(100)^2$ • Substitute (3.79, 100).
   $0.000379 = a$ • Solve for a.

   The equation of the mirror is $x = 0.000379y^2$ or $x = \frac{1}{2639}y^2$.

   b. The equation is valid over the interval $0 \le x \le 3.79$, or [0, 3.79].

Section 12.2

Concept Review 12.2

1. Sometimes true
   The center of the circle given by the equation $(x - 3)^2 + (y + 1)^2 = 4$ is $(3, -1)$.

3. Always true

5. Never true
   The square of the radius of a circle cannot be a negative value.

Objective 1 Exercises

3.

5.

7.

9.

11. $(x - h)^2 + (y - k)^2 = r^2$
    $(x - 2)^2 + [y - (-1)]^2 = 2^2$
    $(x - 2)^2 + (y + 1)^2 = 4$

13. $(x_1, y_1) = (1, 2)$, $(x_2, y_2) = (-1, 1)$

    $d = \sqrt{(x_2 - x_1)^2 + (y_2 - y_1)^2}$
    $= \sqrt{(-1 - 1)^2 + (1 - 2)^2}$
    $= \sqrt{(-2)^2 + (-1)^2}$
    $= \sqrt{4 + 1}$
    $= \sqrt{5}$

    $(x - h)^2 + (y - k)^2 = r^2$
    $[x - (-1)]^2 + (y - 1)^2 = \left(\sqrt{5}\right)^2$
    $(x + 1)^2 + (y - 1)^2 = 5$

15. The endpoints of the diameter are $(-1, 4)$ and $(-5, 8)$. The center of the circle is the midpoint of the diameter.
$(x_1, y_1) = (-1, 4)$ $(x_2, y_2) = (-5, 8)$
$x_m = \dfrac{x_1 + x_2}{2}$ $y_m = \dfrac{y_1 + y_2}{2}$
$= \dfrac{-1 + (-5)}{2}$ $= \dfrac{4 + 8}{2}$
$= -3$ $= 6$

The center of the circle is $(-3, 6)$. The radius of the circle is the length of the segment connecting the center of the circle $(-3, 6)$ to an endpoint of the diameter (use either $(-1, 4)$ or $(-5, 8)$).
$r = \sqrt{(x_1 - x_m)^2 + (y_1 - y_m)^2}$
$r = \sqrt{(-1-(-3))^2 + (4-6)^2}$
$r = \sqrt{4+4}$
$r = \sqrt{8}$

Write the equation of the circle with center $(-3, 6)$ and radius $\sqrt{8}$.
$(x+3)^2 + (y-6)^2 = 8$

17. The endpoints of the diameter are $(-4, 2)$ and $(0, 0)$. The center of the circle is the midpoint of the diameter.
$(x_1, y_1) = (-4, 2)$ $(x_2, y_2) = (0, 0)$
$x_m = \dfrac{x_1 + x_2}{2}$ $y_m = \dfrac{y_1 + y_2}{2}$
$= \dfrac{-4 + 0}{2}$ $= \dfrac{2 + 0}{2}$
$= -2$ $= 1$

The center of the circle is $(-2, 1)$. The radius of the circle is the length of the segment connecting the center of the circle $(-2, 1)$ to an endpoint of the diameter (use either $(-4, 2)$ or $(0, 0)$).
$r = \sqrt{(x_1 - x_m)^2 + (y_1 - y_m)^2}$
$r = \sqrt{(-4-(-2))^2 + (2-1)^2}$
$r = \sqrt{4+1}$
$r = \sqrt{5}$

Write the equation of the circle with center $(-2, 1)$ and radius $\sqrt{5}$.
$(x+2)^2 + (y-1)^2 = 5$

Objective 2 Exercises

19. $x^2 + y^2 - 2x + 4y - 20 = 0$
$(x^2 - 2x) + (y^2 + 4y) = 20$
$(x^2 - 2x + 1) + (y^2 + 4y + 4) = 20 + 1 + 4$
$(x-1)^2 + (y+2)^2 = 25$
Center: $(1, -2)$
Radius: 5

21. $x^2 + y^2 + 6x + 8y + 9 = 0$
$(x^2 + 6x) + (y^2 + 8y) = -9$
$(x^2 + 6x + 9) + (y^2 + 8y + 16) = -9 + 9 + 16$
$(x+3)^2 + (y+4)^2 = 16$
Center: $(-3, -4)$
Radius: 4

23. $x^2 + y^2 - x + 4y + \dfrac{13}{4} = 0$
$(x^2 - x) + (y^2 + 4y) = -\dfrac{13}{4}$
$\left(x^2 - x + \dfrac{1}{4}\right) + (y^2 + 4y + 4) = -\dfrac{13}{4} + \dfrac{1}{4} + 4$
$\left(x - \dfrac{1}{2}\right)^2 + (y+2)^2 = 1$
Center: $\left(\dfrac{1}{2}, -2\right)$
Radius: 1

25. $x^2 + y^2 - 6x + 4y + 4 = 0$
$(x^2 - 6x) + (y^2 + 4y) = -4$
$(x^2 - 6x + 9) + (y^2 + 4y + 4) = -4 + 9 + 4$
$(x-3)^2 + (y+2)^2 = 9$
Center: $(3, -2)$
Radius: 3

Applying Concepts 12.2

27. $(x_1, y_1) = (3, 0)$ $(x_2, y_2) = (0, 0)$
$r = \sqrt{(x_2 - x_1)^2 + (y_2 - y_1)^2}$
$= \sqrt{(0-3)^2 + (0-0)^2} = \sqrt{9} = 3$
$(x-h)^2 + (y-k)^2 = r^2$
$(x-3)^2 + (y-0)^2 = 3^2$
$(x-3)^2 + y^2 = 9$

354 Chapter 12: Conic Sections

29. If the circle lies in quadrant II, has a radius of 1, and is tangent to both axes, then it must pass through the points (0, 1), (–1, 0), (–2, 1), and (–1, 2). The center must be (–1, 1).

$$(x-h)^2 + (y-k)^2 = r^2$$
$$[x-(-1)]^2 + (y-1)^2 = 1^2$$
$$(x+1)^2 + (y-1)^2 = 1$$

Section 12.3

Concept Review 12.3

1. Never true
   A vertical line will intersect the graph of an ellipse at more than one point. By the vertical-line test, the graph of an ellipse is not the graph of a function.

3. Always true

5. Sometimes true
   The hyperbola $\dfrac{y^2}{4} - \dfrac{x^2}{16} = 1$ has no x-intercepts.

Objective 1 Exercises

1. x-intercepts: (2, 0) and (–2, 0)
   y-intercepts: (0, 3) and (0, –3)

3. x-intercepts: (5, 0) and (–5, 0)
   y-intercepts: (0, 3) and (0, –3)

5. x-intercepts: (6, 0) and (–6, 0)
   y-intercepts: (0, 4) and (0, –4)

7. x-intercepts: (3, 0) and (–3, 0)
   y-intercepts: (0, 5) and (0, –5)

9. x-intercepts: $(2\sqrt{3}, 0)$ and $(-2\sqrt{3}, 0)$
   y-intercepts: (0, 2) and (0, –2)

11. x-intercepts: (6, 0) and (–6, 0)
    y-intercepts: (0, 3) and (0, –3)

Objective 2 Exercises

15. Axis of symmetry: x-axis
    Vertices: (3, 0) and (–3, 0)
    Asymptotes: $y = \dfrac{4}{3}x$ and $y = -\dfrac{4}{3}x$

17. Axis of symmetry: y-axis
    Vertices: (0, 4) and (0, –4)
    Asymptotes: $y = \dfrac{4}{3}x$ and $y = -\dfrac{4}{3}x$

19. Axis of symmetry: x-axis
    Vertices: (2, 0) and (–2, 0)
    Asymptotes: $y = \dfrac{5}{2}x$ and $y = -\dfrac{5}{2}x$

21. Axis of symmetry: y-axis
    Vertices: (0, 5) and (0, –5)
    Asymptotes: $y = \dfrac{5}{3}x$ and $y = -\dfrac{5}{3}x$

23. Axis of symmetry: x-axis
Vertices: $(5, 0)$ and $(-5, 0)$
Asymptotes: $y = \frac{4}{5}x$ and $y = -\frac{4}{5}x$

25. Axis of symmetry: y-axis
Vertices: $(0, 4)$ and $(0, -4)$
Asymptotes: $y = 2x$ and $y = -2x$

27. Axis of symmetry: x-axis
Vertices: $(5, 0)$ and $(-5, 0)$
Asymptotes: $y = \frac{3}{5}x$ and $y = -\frac{3}{5}x$

Applying Concepts 12.3

29. The graph of $4x^2 + y^2 = 16$ is an ellipse.
$$4x^2 + y^2 = 16$$
$$\frac{4x^2 + y^2}{16} = \frac{16}{16}$$
$$\frac{4x^2}{16} + \frac{y^2}{16} = 1$$
$$\frac{x^2}{4} + \frac{y^2}{16} = 1$$
x-intercepts: $(2, 0)$ and $(-2, 0)$
y-intercepts: $(0, 4)$ and $(0, -4)$

31. The graph of $y^2 - 4x^2 = 16$ is a hyperbola.
$$y^2 - 4x^2 = 16$$
$$\frac{y^2 - 4x^2}{16} = \frac{16}{16}$$
$$\frac{y^2}{16} - \frac{4x^2}{16} = 1$$
$$\frac{y^2}{16} - \frac{x^2}{4} = 1$$
Axis of symmetry: y-axis
Vertices: $(0, 4)$ and $(0, -4)$
Asymptotes: $y = 2x$ and $y = -2x$

33. The graph of $9x^2 - 25y^2 = 225$ is a hyperbola.
$$9x^2 - 25y^2 = 225$$
$$\frac{9x^2 - 25y^2}{225} = \frac{225}{225}$$
$$\frac{9x^2}{225} - \frac{25y^2}{225} = 1$$
$$\frac{x^2}{25} - \frac{y^2}{9} = 1$$
Axis of symmetry: x-axis
Vertices: $(5, 0)$ and $(-5, 0)$
Asymptotes: $y = \frac{3}{5}x$ and $y = -\frac{3}{5}x$

35. a. $\dfrac{x^2}{a^2}+\dfrac{y^2}{b^2}=1$

$\dfrac{x^2}{18^2}+\dfrac{y^2}{4.5^2}=1$

$\dfrac{x^2}{324}+\dfrac{y^2}{20.25}=1$

b. Major axis = 36 AU
36 AU ÷ 2 = 18 AU
18 AU = 18(92,960,000 mi)
       = 1,673,280,000 mi

$\sqrt{a^2-b^2} = \sqrt{324-20.25}$
$= \sqrt{303.75}$
≈ 17.42842506

17.42842506 AU
= 17.42842506(92,960,000 mi)
≈ 1,620,146,393 mi
1,673,280,000 + 1,620,146,393
≈ 3,293,400,000
The distance from the Sun to the point at the aphelion is about 3,293,400,000 mi.

c. $18\text{ AU} - \sqrt{a^2-b^2}$ AU
= 1,673,280,000 mi − 1,620,146,393 mi
= 53,133,607 ≈ 53,100,000
The distance from the Sun to the point at the perihelion is about 53,100,000 mi.

37. a. $\dfrac{x^2}{a^2}+\dfrac{y^2}{b^2}=1$

$\dfrac{x^2}{1.52^2}+\dfrac{y^2}{1.495^2}=1$

$\dfrac{x^2}{2.310}+\dfrac{y^2}{2.235}=1$

b. Major axis = 3.04 AU
3.04 AU ÷ 2 = 1.52 AU
1.52 AU = 1.52(92,960,000 mi)
        = 141,299,200 mi

$\sqrt{a^2-b^2} = \sqrt{2.310-2.235}$
$= \sqrt{0.075}$
≈ 0.2738612788

0.2738612788 AU
= 0.2738612788(92,960,000 mi)
≈ 25,488,144 mi
141,299,200 + 25,458,144 = 168,757,344
                         ≈ 166,800,000
The aphelion is about 166,800,000 mi.

c. $1.52\text{ AU} - \sqrt{a^2-b^2}$ AU
= 141,299,200 mi − 25,458,144 mi
≈ 115,800,000
The perihelion is about 115,800,000 mi.

Section 12.4

Concept Review 12.4

1. Never true
Two ellipses with centers at the origin may intersect at four points or they may not intersect.

3. Sometimes true
A straight line may intersect a parabola at one point or two points, or the line may not intersect the parabola.

5. Always true

Objective 1 Exercises

3. (1) $y = x^2 - x - 1$
(2) $y = 2x + 9$
Use the substitution method.
$y = x^2 - x - 1$
$2x + 9 = x^2 - x - 1$
$0 = x^2 - 3x - 10$
$0 = (x-5)(x+2)$
$x - 5 = 0 \quad x + 2 = 0$
$x = 5 \quad\quad x = -2$
Substitute into equation (2).
$y = 2x + 9 \quad\quad y = 2x + 9$
$y = 2(5) + 9 \quad y = 2(-2) = 9$
$y = 10 + 9 \quad\quad y = -4 + 9$
$y = 19 \quad\quad\quad y = 5$
The solutions are (5, 19) and (−2, 5).

5. (1) $y^2 = -x + 3$
(2) $x - y = 1$
Solve equation (2) for x.
$x - y = 1$
$x = y + 1$
Use the substitution method.
$y^2 = -x + 3$
$y^2 = -(y+1) + 3$
$y^2 = -y - 1 + 3$
$y^2 = -y + 2$
$y^2 + y - 2 = 0$
$(y+2)(y-1) = 0$
$y + 2 = 0 \quad y - 1 = 0$
$y = -2 \quad\quad y = 1$
Substitute into equation (2).
$x - y = 1 \quad\quad x - y = 1$
$x - (-2) = 1 \quad x - 1 = 1$
$x + 2 = 1 \quad\quad x = 2$
$x = -1$
The solutions are (−1, −2) and (2, 1).

7. (1) $y^2 = 2x$
(2) $x + 2y = -2$
Solve equation (2) for x.
$x + 2y = -2$
$x = -2y - 2$
Use the substitution method.
$y^2 = 2x$
$y^2 = 2(-2y - 2)$
$y^2 = -4y - 4$
$y^2 + 4y + 4 = 0$
$(y + 2)(y + 2) = 0$
$y + 2 = 0$
$y = -2$
The solution is a double root.
Substitute into equation (2).
$x + 2y = -2$
$x + 2(-2) = -2$
$x - 4 = -2$
$x = 2$
The solution is $(2, -2)$.

9. (1) $x^2 + 2y^2 = 12$
(2) $2x - y = 2$
Solve equation (2) for y.
$2x - y = 2$
$-y = -2x + 2$
$y = 2x - 2$
Use the substitution method.
$x^2 + 2y^2 = 12$
$x^2 + 2(2x - 2)^2 = 12$
$x^2 + 2(4x^2 - 8x + 4) = 12$
$x^2 + 8x^2 - 16x + 8 = 12$
$9x^2 - 16x - 4 = 0$
$(x - 2)(9x + 2) = 0$
$x - 2 = 0 \quad 9x + 2 = 0$
$x = 2 \qquad 9x = -2$
$\qquad\qquad x = -\dfrac{2}{9}$
Substitute into equation (2).
$2x - y = 2 \qquad\qquad 2x - y = 2$
$2(2) - y = 2 \qquad 2\left(-\dfrac{2}{9}\right) - y = 2$
$4 - y = 2 \qquad\qquad -\dfrac{4}{9} - y = 2$
$-y = -2 \qquad\qquad\quad -y = \dfrac{22}{9}$
$y = 2 \qquad\qquad\qquad\quad y = -\dfrac{22}{9}$
The solutions are $(2, 2)$ and $\left(-\dfrac{2}{9}, -\dfrac{22}{9}\right)$.

11. (1) $x^2 + y^2 = 13$
(2) $x + y = 5$
Solve equation (2) for y.
$x + y = 5$
$y = -x + 5$
Use the substitution method.
$x^2 + y^2 = 13$
$x^2 + (-x + 5)^2 = 13$
$x^2 + x^2 - 10x + 25 = 13$
$2x^2 - 10x + 12 = 0$
$2(x^2 - 5x + 6) = 0$
$2(x - 3)(x - 2) = 0$
$x - 3 = 0 \quad x - 2 = 0$
$x = 3 \qquad x = 2$
Substitute into equation (2).
$x + y = 5 \quad x + y = 5$
$3 + y = 5 \quad 2 + y = 5$
$y = 2 \qquad y = 3$
The solutions are $(3, 2)$ and $(2, 3)$.

13. (1) $4x^2 + y^2 = 12$
(2) $y = 4x^2$
Use the substitution method.
$4x^2 + y^2 = 12$
$4x^2 + (4x^2)^2 = 12$
$4x^2 + 16x^4 = 12$
$16x^4 + 4x^2 - 12 = 0$
$4(4x^4 + x^2 - 3) = 0$
$4(4x^2 - 3)(x^2 + 1) = 0$
$4x^2 - 3 = 0 \qquad x^2 + 1 = 0$
$4x^2 = 3 \qquad\qquad x^2 = -1$
$x^2 = \dfrac{3}{4} \qquad\qquad x = \pm\sqrt{-1}$
$x = \pm\dfrac{\sqrt{3}}{2}$
Substitute the real number solutions into equation (2).
$y = 4x^2 \qquad\qquad\qquad y = 4x^2$
$y = 4\left(\dfrac{\sqrt{3}}{2}\right)^2 \qquad y = 4\left(-\dfrac{\sqrt{3}}{2}\right)^2$
$y = 4\left(\dfrac{3}{4}\right) \qquad\qquad y = 4\left(\dfrac{3}{4}\right)$
$y = 3 \qquad\qquad\qquad\quad y = 3$
The solutions are $\left(\dfrac{\sqrt{3}}{2}, 3\right)$ and $\left(-\dfrac{\sqrt{3}}{2}, 3\right)$.

15. (1) $y = x^2 - 2x - 3$
    (2) $y = x - 6$
    Use the substitution method.
    $$y = x^2 - 2x - 3$$
    $$x - 6 = x^2 - 2x - 3$$
    $$0 = x^2 - 3x + 3$$
    $$x = \frac{-b \pm \sqrt{b^2 - 4ac}}{2a}$$
    $$= \frac{-(-3) \pm \sqrt{(-3)^2 - 4(1)(3)}}{2(1)}$$
    $$= \frac{3 \pm \sqrt{9 - 12}}{2}$$
    $$= \frac{3 \pm \sqrt{-3}}{2}$$
    Since the discriminant is less than zero, the equation has two complex number solutions. Therefore, the system of equations has no real number solution.

17. (1) $3x^2 - y^2 = -1$
    (2) $x^2 + 4y^2 = 17$
    Use the addition method.
    Multiply equation (1) by 4.
    $$12x^2 - 4y^2 = -4$$
    $$x^2 + 4y^2 = 17$$
    $$13x^2 = 13$$
    $$x^2 = 1$$
    $$x = \pm\sqrt{1} = \pm 1$$
    Substitute into equation (2).
    $$x^2 + 4y^2 = 17 \qquad x^2 + 4y^2 = 17$$
    $$1^2 + 4y^2 = 17 \qquad (-1)^2 + 4y^2 = 17$$
    $$1 + 4y^2 = 17 \qquad 1 + 4y^2 = 17$$
    $$4y^2 = 16 \qquad 4y^2 = 16$$
    $$y^2 = 4 \qquad y^2 = 4$$
    $$y = \pm\sqrt{4} \qquad y = \pm\sqrt{4}$$
    $$y = \pm 2 \qquad y = \pm 2$$
    The solutions are (1, 2), (1, –2), (–1, 2), and (–1, –2).

19. (1) $2x^2 + 3y^2 = 30$
    (2) $x^2 + y^2 = 13$
    Use the addition method.
    Multiply equation (2) by –2.
    $$2x^2 + 3y^2 = 30$$
    $$-2x^2 - 2y^2 = -26$$
    $$y^2 = 4$$
    $$y = \pm\sqrt{4} = \pm 2$$
    Substitute into equation (2).
    $$x^2 + y^2 = 13 \qquad x^2 + y^2 = 13$$
    $$x^2 + 2^2 = 13 \qquad x^2 + (-2)^2 = 13$$
    $$x^2 + 4 = 13 \qquad x^2 + 4 = 13$$
    $$x^2 = 9 \qquad x^2 = 9$$
    $$x = \pm\sqrt{9} \qquad x = \pm\sqrt{9}$$
    $$x = \pm 3 \qquad x = \pm 3$$
    The solutions are (3, 2), (3, –2), (–3, 2), and (–3, –2).

21. (1) $y = 2x^2 - x + 1$
    (2) $y = x^2 - x + 5$
    Use the substitution method.
    $$y = 2x^2 - x + 1$$
    $$x^2 - x + 5 = 2x^2 - x + 1$$
    $$0 = x^2 - 4$$
    $$0 = (x + 2)(x - 2)$$
    $$x + 2 = 0 \qquad x - 2 = 0$$
    $$x = -2 \qquad x = 2$$
    Substitute into equation (2).
    $$y = x^2 - x + 5 \qquad y = x^2 - x + 5$$
    $$y = (-2)^2 - (-2) + 5 \qquad y = 2^2 - 2 + 5$$
    $$y = 4 + 2 + 5 \qquad y = 4 - 2 + 5$$
    $$y = 11 \qquad y = 7$$
    The solutions are (2, 7) and (–2, 11).

23. (1) $2x^2 + 3y^2 = 24$
    (2) $x^2 - y^2 = 7$
    Use the addition method.
    Multiply equation (2) by 3.
    $$2x^2 + 3y^2 = 24$$
    $$3x^2 - 3y^2 = 21$$
    $$5x^2 = 45$$
    $$x^2 = 9 = \pm\sqrt{9} = \pm 3$$
    Substitute into equation (2).
    $$x^2 - y^2 = 7 \qquad x^2 - y^2 = 7$$
    $$3^2 - y^2 = 7 \qquad (-3)^2 - y^2 = 7$$
    $$9 - y^2 = 7 \qquad 9 - y^2 = 7$$
    $$-y^2 = -2 \qquad -y^2 = -2$$
    $$y^2 = 2 \qquad y^2 = 2$$
    $$y = \pm\sqrt{2} \qquad y = \pm\sqrt{2}$$
    The solutions are $(3, \sqrt{2})$, $(3, -\sqrt{2})$, $(-3, \sqrt{2})$, and $(-3, -\sqrt{2})$.

25. (1) $x^2 + y^2 = 36$
(2) $4x^2 + 9y^2 = 36$
Use the addition method.
Multiply equation (1) by −4.
$-4x^2 - 4y^2 = -144$
$4x^2 + 9y^2 = 36$
$5y^2 = -108$
$y^2 = -\dfrac{108}{5}$
$y = \pm\sqrt{-\dfrac{108}{5}}$

The system of equations has no real number solution.

27. (1) $11x^2 - 2y^2 = 4$
(2) $3x^2 + y^2 = 15$
Use the addition method.
Multiply equation (2) by 2.
$11x^2 - 2y^2 = 4$
$6x^2 + 2y^2 = 30$
$17x^2 = 34$
$x^2 = 2$
$x = \pm\sqrt{2}$
Substitute into equation (2).

$3x^2 + y^2 = 15$ $3x^2 + y^2 = 15$
$3(\sqrt{2})^2 + y^2 = 15$ $3(-\sqrt{2})^2 + y^2 = 15$
$3(2) + y^2 = 15$ $3(2) + y^2 = 15$
$6 + y^2 = 15$ $6 + y^2 = 15$
$y^2 = 9$ $y^2 = 9$
$y = \pm\sqrt{9}$ $y = \pm\sqrt{9}$
$y = \pm 3$ $y = \pm 3$

The solutions are $(\sqrt{2}, 3)$, $(\sqrt{2}, -3)$, $(-\sqrt{2}, 3)$, and $(-\sqrt{2}, -3)$.

29. (1) $2x^2 - y^2 = 7$
(2) $2x - y = 5$
Solve equation (2) for y.
$2x - y = 5$
$-y = -2x + 5$
$y = 2x - 5$
Use the substitution method.
$2x^2 - y^2 = 7$
$2x^2 - (2x - 5)^2 = 7$
$2x^2 - (4x^2 - 20x + 25) = 7$
$2x^2 - 4x^2 + 20x - 25 = 7$
$-2x^2 + 20x - 32 = 0$
$-2(x^2 - 10x + 16) = 0$
$-2(x - 2)(x - 8) = 0$
$x - 2 = 0 \quad x - 8 = 0$
$x = 2 \quad\quad x = 8$
Substitute into equation (2).

$2x - y = 5$ $2x - y = 5$
$2(2) - y = 5$ $2(8) - y = 5$
$4 - y = 5$ $16 - y = 5$
$-y = 1$ $-y = -11$
$y = -1$ $y = 11$

The solutions are $(2, -1)$ and $(8, 11)$.

31. (1) $y = 3x^2 + x - 4$
(2) $y = 3x^2 - 8x + 5$
Use the substitution method.
$y = 3x^2 + x - 4$
$3x^2 - 8x + 5 = 3x^2 + x - 4$
$-9x = -9$
$x = 1$
Substitute into equation (1).
$y = 3x^2 + x - 4$
$y = 3(1)^2 + (1) - 4$
$y = 3(1) + 1 - 4$
$y = 3 + 1 - 4$
$y = 0$
The solution is $(1, 0)$.

Applying Concepts 12.4

33. $y = 2^x$
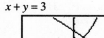
$x + y = 3$

The solution is $(1.000, 2.000)$.

35. $y = \log_2 x$
$\dfrac{x^2}{9} + \dfrac{y^2}{1} = 1$

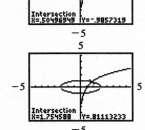

The approximate solutions are (1.755, 0.811) and (0.505, −0.986).

37. $y = -\log_3 x$
$x + y = 4$

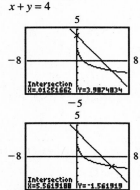

The approximate solutions are (0.013, 3.987) and (5.562, −1.562).

Section 12.5

Concept Review 12.5

1. Always true

3. Never true
   $0 + 0 > 4$ is not a true statement. The point (0, 0) is not a solution of the inequality.

5. Sometimes true
   If the inequality includes the symbol ≤ or ≥, the solution set will include the boundary.

Objective 1 Exercises

1. $y \le x^2 - 4x + 3$
   Substitute the point (0, 0) into the inequality.
   $0 \le 0^2 - 4(0) + 3$
   $0 \le 3$ true
   The point (0, 0) should be in the shaded region.

3. $(x-1)^2 + (y+2)^2 \le 9$
   Substitute the point (0, 0) into the inequality.
   $(0-1)^2 + (0+2)^2 \le 9$
   $-1^2 + 2^2 \le 9$
   $3 \le 9$ true
   The point (0, 0) should be in the shaded region.

5. $(x+3)^2 + (y-2)^2 \ge 9$
   Substitute the point (0, 0) into the inequality.
   $(0+3)^2 + (0-2)^2 \ge 9$
   $3^2 + (-2)^2 \ge 9$
   $13 \ge 9$ true
   The point (0, 0) should be in the shaded region.

7. $\dfrac{x^2}{16} + \dfrac{y^2}{25} < 1$
   Substitute the point (0, 0) into the inequality.
   $\dfrac{0^2}{16} + \dfrac{0^2}{25} < 1$
   $0 < 1$ true
   The point (0, 0) should be in the shaded region.

Section 12.5 **361**

9. $\dfrac{x^2}{25} - \dfrac{y^2}{9} \leq 1$

Substitute the point (0, 0) into the inequality.

$\dfrac{0^2}{25} - \dfrac{0^2}{9} \leq 1$

$0 \leq 1$ true

The point (0, 0) should be in the shaded region.

11. $\dfrac{x^2}{4} + \dfrac{y^2}{16} \geq 1$

Substitute the point (0, 0) into the inequality.

$\dfrac{0^2}{4} + \dfrac{0^2}{16} \geq 1$

$0 \geq 1$ false

The point (0, 0) should not be in the shaded region.

13. $y \leq x^2 - 2x + 3$

Substitute the point (0, 0) into the inequality.

$0 \leq 0^2 - 2(0) + 3$

$0 \leq 3$ true

The point (0, 0) should be in the shaded region.

15. $\dfrac{y^2}{9} - \dfrac{x^2}{16} \leq 1$

Substitute the point (0, 0) into the inequality.

$\dfrac{0^2}{9} - \dfrac{0^2}{16} \leq 1$

$0 \leq 1$ true

The point (0, 0) should be in the shaded region.

17. $\dfrac{x^2}{9} + \dfrac{y^2}{1} \leq 1$

Substitute the point (0, 0) into the inequality.

$\dfrac{0^2}{9} + \dfrac{0^2}{1} \leq 1$

$0 \leq 1$ true

The point (0, 0) should be in the shaded region.

19. $(x-1)^2 + (y+3)^2 \leq 25$

Substitute the point (0, 0) into the inequality.

$(0-1)^2 + (0+3)^2 \leq 25$

$(-1)^2 + 3^2 \leq 25$

$10 \leq 25$ true

The point (0, 0) should be in the shaded region.

21. $\dfrac{y^2}{25} - \dfrac{x^2}{4} \leq 1$

Substitute the point (0, 0) into the inequality.

$\dfrac{0^2}{25} - \dfrac{0^2}{4} \leq 1$

$0 \leq 1$ true

The point (0, 0) should be in the shaded region.

23. $\dfrac{x^2}{25} + \dfrac{y^2}{9} \leq 1$

Substitute the point (0, 0) into the inequality.

$\dfrac{0^2}{25} + \dfrac{0^2}{9} \leq 1$

$0 \leq 1$ true

The point (0, 0) should be in the shaded region.

Objective 2 Exercises

27. $y \le x^2 - 4x + 4$
$y + x > 4$

29. $x^2 + y^2 < 16$
$y > x + 1$

31. $\dfrac{x^2}{4} + \dfrac{y^2}{16} \le 1$
$y \le -\dfrac{1}{2}x + 2$

33. $x \ge y^2 - 3y + 2$
$y \ge 2x - 2$

35. $x^2 + y^2 < 25$
$\dfrac{x^2}{9} + \dfrac{y^2}{36} < 1$

37. $x^2 + y^2 > 4$
$x^2 + y^2 < 25$

Applying Concepts 12.5

39. $y > x^2 - 3$
$y < x + 3$
$x \le 0$

41. $x^2 + y^2 < 3$
$x > y^2 - 1$
$y \ge 0$

43. $\dfrac{x^2}{4} + \dfrac{y^2}{1} \le 4$
$x^2 + y^2 \le 4$
$x \ge 0$
$y \le 0$

45. $y > 2^x$
$x + y < 4$

47. $y \ge \log_2 x$
$x^2 + y^2 < 9$

49. $y < 3^{-x}$

$\dfrac{x^2}{4} - \dfrac{y^2}{1} \geq 1$

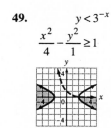

Focus on Problem Solving

1. Label the coins 1, 2, 3, 4, 5, 6, 7, and 8.
   Set coins 1 and 2 aside. Put coins 3, 4, and 5 on one side of the balance and coins 6, 7, and 8 on the other side the balance. If the set balances, then coin 1 or coin 2 is the lighter coin. You can determine which by using the balance a second time.
   If the set does not balance, choose the lighter set of coins. Assume the lighter set includes coins 6, 7, and 8.
   Choose two coins to put on the balance (say coin 6 and coin 7). If the coins do not balance, you have found the lighter coin. If coins 6 and 7 are equal in weight, then coin 8 is the lighter coin. Thus the lighter coin can be found in two weighings.

2. The sequence is 1, 1, 2, 3, 5, 8, 13, ...
   The third term is the sum of the two preceding terms. $1 + 1 = 2$
   The fourth term is the sum of the two preceding terms. $1 + 2 = 3$
   The fifth term is the sum of the two preceding terms. $2 + 3 = 5$
   The sixth term is the sum of the two preceding terms. $3 + 5 = 8$
   The seventh term is the sum of the two preceding terms. $5 + 8 = 13$

   If the pattern holds, then the eighth term is $8 + 13 = 21$.

3.

6	7	2
1	5	9
8	3	4

4. To find the price of the larger pizza so that the price per square inch is the same, multiply the price by the ratio of the area of the larger pizza to that of the smaller pizza.
   $$C = 5.00\left(\dfrac{A^2}{A^1}\right) = 5.00\left(\dfrac{\pi(9)^2}{\pi(4.5)^2}\right) = 5.00(4) = 20$$
   The larger pizza will cost $20.

5. Use the weights 1 g, 2 g, and 4 g to weigh 7 g.
   Use the weights 4 g and 8 g to weigh 12 g.

6. To get from A to B requires 7 moves to the right and 5 moves upward. There are 12! moves to get from A to B. However, 7 moves to the right are the same and 5 moves upward are the same. Thus the number of paths from A to B will be $\dfrac{12!}{7!5!}$.
   $$\dfrac{12!}{7!5!} = \dfrac{12 \cdot 11 \cdot 10 \cdot 9 \cdot 8 \cdot 7 \cdot 6 \cdot 5 \cdot 4 \cdot 3 \cdot 2 \cdot 1}{7 \cdot 6 \cdot 5 \cdot 4 \cdot 3 \cdot 2 \cdot 1 \cdot 5 \cdot 4 \cdot 3 \cdot 2 \cdot 1} = 792$$
   There are 792 paths in going from A to B by moving right or up along one of the grid lines.

7. The checkerboard has 79 squares. Each domino will cover two squares, and the dominos cannot overlap. Thus the checkerboard cannot be covered with dominos.

Projects and Group Activities

1. $\dfrac{x^2}{25} + \dfrac{y^2}{49} = 1$

   $\dfrac{y^2}{49} = 1 - \dfrac{x^2}{25}$

   $y^2 = 49\left(1 - \dfrac{x^2}{25}\right)$

   $y = \pm 7\sqrt{1 - \dfrac{x^2}{25}}$

2. $\dfrac{x^2}{4} + \dfrac{y^2}{64} = 1$

   $\dfrac{y^2}{64} = 1 - \dfrac{x^2}{4}$

   $y^2 = 64\left(1 - \dfrac{x^2}{4}\right)$

   $y = \pm 8\sqrt{1 - \dfrac{x^2}{4}}$

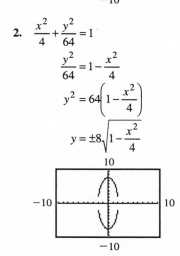

364   Chapter 12: Conic Sections

3. $\dfrac{x^2}{16} - \dfrac{y^2}{4} = 1$

$\dfrac{x^2}{16} - 1 = \dfrac{y^2}{4}$

$4\left(\dfrac{x^2}{16} - 1\right) = y^2$

$\pm 2\sqrt{\dfrac{x^2}{16} - 1} = y$

4. $\dfrac{x^2}{9} - \dfrac{y^2}{36} = 1$

$\dfrac{x^2}{9} - 1 = \dfrac{y^2}{36}$

$36\left(\dfrac{x^2}{9} - 1\right) = y^2$

$\pm 6\sqrt{\dfrac{x^2}{9} - 1} = y$

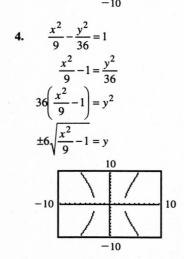

Chapter Review Exercises

1. $y = x^2 - 4x + 8$

$-\dfrac{b}{2a} = -\dfrac{-4}{2(1)} = \dfrac{4}{2} = 2$

$y = 2^2 - 4(2) + 8$
$= 4$
Vertex: (2, 4)
Axis of symmetry: $x = 2$

2. $y = -x^2 + 7x - 8$

$-\dfrac{b}{2a} = -\dfrac{7}{2(-1)} = \dfrac{7}{2}$

$y = -\left(\dfrac{7}{2}\right)^2 + 7\left(\dfrac{7}{2}\right) - 8$

$= -\dfrac{49}{4} + \dfrac{49}{2} - 8$

$= \dfrac{17}{4}$

Vertex: $\left(\dfrac{7}{2}, \dfrac{17}{4}\right)$

Axis of symmetry: $x = \dfrac{7}{2}$

3. $y = -2x^2 + x - 2$

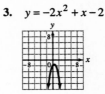

4. $x = 2y^2 - 6y + 5$

5. $(x_1, y_1) = (2, -1) \quad (x_2, y_2) = (-1, 2)$

$r = \sqrt{(x_2 - x_1)^2 + (y_2 - y_1)^2}$
$= \sqrt{(-1 - 2)^2 + (2 - (-1))^2}$
$= \sqrt{(-3)^2 + 3^2}$
$= \sqrt{9 + 9} = \sqrt{18}$

$(x - h)^2 + (y - k)^2 = r^2$
$[x - (-1)]^2 + (y - 2)^2 = (\sqrt{18})^2$
$(x + 1)^2 + (y - 2)^2 = 18$

6. $(x - h)^2 + (y - k)^2 = r^2$
$(x - (-1))^2 + (y - 5)^2 = 6^2$
$(x + 1)^2 + (y - 5)^2 = 36$

7. $(x + 3)^2 + (y + 1)^2 = 1$

Center: $(-3, -1)$
Radius: 1

8. $x^2 + (y - 2)^2 = 9$

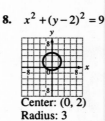

Center: $(0, 2)$
Radius: 3

9. $(x_1, y_1) = (4, 6) \quad (x_2, y_2) = (0, -3)$

$r = \sqrt{(x_2 - x_1)^2 + (y_2 - y_1)^2}$
$= \sqrt{(0 - 4)^2 + (-3 - 6)^2}$
$= \sqrt{(-4)^2 + (-9)^2}$
$= \sqrt{16 + 81} = \sqrt{97}$

$(x - h)^2 + (y - k)^2 = r^2$
$(x - 0)^2 + (y - (-3))^2 = (\sqrt{97})^2$
$x^2 + (y + 3)^2 = 97$

10.
$$x^2 + y^2 + 4x - 2y = 4$$
$$(x^2 + 4x) + (y^2 - 2y) = 4$$
$$(x^2 + 4x + 4) + (y^2 - 2y + 1) = 4 + 4 + 1$$
$$(x + 2)^2 + (y - 1)^2 = 9$$

11. $\dfrac{x^2}{1} + \dfrac{y^2}{9} = 1$
x-intercepts: $(1, 0)$ and $(-1, 0)$
y-intercepts: $(0, 3)$ and $(0, -3)$

12. $\dfrac{x^2}{25} + \dfrac{y^2}{9} = 1$
x-intercepts: $(5, 0)$ and $(-5, 0)$
y-intercepts: $(0, 3)$ and $(0, -3)$

13. $\dfrac{x^2}{25} - \dfrac{y^2}{1} = 1$
Axis of symmetry: x-axis
Vertices: $(5, 0)$ and $(-5, 0)$
Asymptotes: $y = \dfrac{1}{5}x$ and $y = -\dfrac{1}{5}x$

14. $\dfrac{y^2}{16} - \dfrac{x^2}{9} = 1$
Axis of symmetry: y-axis
Vertices: $(0, 4)$ and $(0, -4)$
Asymptotes: $y = \dfrac{4}{3}x$ and $y = -\dfrac{4}{3}x$

15. (1) $y = x^2 + 5x - 6$
(2) $y = x - 10$
Use the substitution method.
$$x - 10 = x^2 + 5x - 6$$
$$0 = x^2 + 4x + 4$$
$$0 = (x + 2)^2$$
$x + 2 = 0 \quad x + 2 = 0$
$x = -2 \quad x = -2$
Substitute into equation (2).
$y = -2 - 10 \quad y = -2 - 10$
$y = -12 \quad y = -12$
The solution is $(-2, -12)$

16. (1) $2x^2 + y^2 = 19$
(2) $3x^2 - y^2 = 6$
Use the addition method.
$$2x^2 + y^2 = 19$$
$$3x^2 - y^2 = 6$$
$$5x^2 = 25$$
$$x^2 = 5$$
$$x = \pm\sqrt{5}$$
Substitute into equation (1).
$2x^2 + y^2 = 19 \qquad 2x^2 + y^2 = 19$
$2(\sqrt{5})^2 + y^2 = 19 \qquad 2(-\sqrt{5})^2 + y^2 = 19$
$10 + y^2 = 19 \qquad 10 + y^2 = 19$
$y^2 = 9 \qquad y^2 = 9$
$y = \pm\sqrt{9} \qquad y = \pm\sqrt{9}$
$y = \pm 3 \qquad y = \pm 3$
The solutions are $(\sqrt{5}, 3)$, $(\sqrt{5}, -3)$, $(-\sqrt{5}, 3)$, and $(-\sqrt{5}, -3)$.

17. (1) $x = 2y^2 - 3y + 1$
    (2) $3x - 2y = 0$
    Use the substitution method.
    $3x - 2y = 0$
    $3x = 2y$
    $x = \frac{2}{3}y$

    $\frac{2}{3}y = 2y^2 - 3y + 1$
    $2y = 6y^2 - 9y + 3$
    $0 = 6y^2 - 11y + 3$
    $0 = (2y - 3)(3y - 1)$
    $0 = 2y - 3 \quad 0 = 3y - 1$
    $2y = 3 \quad \quad 3y = 1$
    $y = \frac{3}{2} \quad \quad y = \frac{1}{3}$
    Substitute into equation (2).
    $3x - 2\left(\frac{3}{2}\right) = 0 \quad 3x - 2\left(\frac{1}{3}\right) = 0$
    $3x - 3 = 0 \quad \quad 3x - \frac{2}{3} = 0$
    $3x = 3 \quad \quad \quad 3x = \frac{2}{3}$
    $x = 1 \quad \quad \quad \quad x = \frac{2}{9}$

    The solutions are $\left(1, \frac{3}{2}\right)$ and $\left(\frac{2}{9}, \frac{1}{3}\right)$.

18. (1) $y^2 = 2x^2 - 3x + 6$
    (2) $y^2 = 2x^2 + 5x - 2$
    Use the addition method.
    Multiply equation (2) by -1.
    $y^2 = 2x^2 - 3x + 6$
    $-y^2 = -2x^2 - 5x + 2$
    $0 = -8x + 8$
    $8x = 8$
    $x = 1$
    Substitute into equation (1).
    $y^2 = 2x^2 - 3x + 6$
    $y^2 = 2(1)^2 - 3(1) + 6$
    $y^2 = 2 - 3 + 6$
    $y^2 = 5$
    $y = \pm\sqrt{5}$
    The solutions are $(1, \sqrt{5})$ and $(1, -\sqrt{5})$.

19. $(x-2)^2 + (y+1)^2 \le 16$

20. $\frac{x^2}{9} - \frac{y^2}{16} < 1$

21. $y \ge -x^2 - 2x + 3$

22. $\frac{x^2}{16} + \frac{y^2}{4} > 1$

23. (1) $y \ge x^2 - 4x + 2$
    (2) $y \le \frac{1}{3}x - 1$
    Write equation (1) in standard form.
    $y \ge (x^2 - 4x + 4) - 4 + 2$
    $y \ge (x-2)^2 - 2$

24. $\frac{x^2}{25} + \frac{y^2}{16} \le 1$
    $\frac{y^2}{4} - \frac{x^2}{4} \ge 1$

25. $\frac{x^2}{9} + \frac{y^2}{1} \ge 1$
    $\frac{x^2}{4} - \frac{y^2}{1} \le 1$

26. $\dfrac{x^2}{16}+\dfrac{y^2}{4}<1$

$x^2+y^2>9$

Chapter Test

1. $y=-x^2+6x-5$

$-\dfrac{b}{2a}=\dfrac{-6}{2(-1)}=\dfrac{-6}{-2}=3$

Axis of symmetry: $x=3$

2. $y=-x^2+3x-2$

$-\dfrac{b}{2a}=\dfrac{-3}{2(-1)}=\dfrac{3}{2}$

$y=-\left(\dfrac{3}{2}\right)^2+3\left(\dfrac{3}{2}\right)-2=-\dfrac{9}{4}+\dfrac{9}{2}-2=\dfrac{1}{4}$

Vertex: $\left(\dfrac{3}{2},\dfrac{1}{4}\right)$

3. $y=\dfrac{1}{2}x^2+x-4$

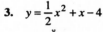

4. $x=y^2-y-2$

5. $(x-h)^2+(y-k)^2=r^2$

$[x-(-3)]^2+[y-(-3)]^2=4^2$

$(x+3)^2+(y+3)^2=16$

6. (1) $x^2+2y^2=4$
   (2) $x+y=2$

Use the substitution method.

$x+y=2$
$y=2-x$
$x^2+2(2-x)^2=4$
$x^2+2(4-4x+x^2)=4$
$x^2+8-8x+2x^2=4$
$3x^2-8x+4=0$
$(3x-2)(x-2)=0$

$3x-2=0 \qquad x-2=0$
$x=\dfrac{2}{3} \qquad x=2$

Substitute into equation (2).

$x+y=2 \qquad x+y=2$
$\dfrac{2}{3}+y=2 \qquad 2+y=2$
$y=\dfrac{4}{3} \qquad y=0$

The solutions are $\left(\dfrac{2}{3},\dfrac{4}{3}\right)$ and $(2, 0)$.

7. (1) $x=3y^2+2y-4$
   (2) $x=y^2-5y$

Use the addition method.
Multiply equation (2) by -1.

$x=3y^2+2y-4$
$-x=-y^2+5y$
$0=2y^2+7y-4$
$0=(2y-1)(y+4)$

$2y-1=0 \qquad y+4=0$
$y=\dfrac{1}{2} \qquad y=-4$

Substitute into equation (2).

$x=y^2-5y \qquad x=y^2-5y$
$x=\left(\dfrac{1}{2}\right)^2-5\left(\dfrac{1}{2}\right) \qquad x=(-4)^2-5(-4)$
$x=\dfrac{1}{4}-\dfrac{5}{2} \qquad x=16+20$
$x=-\dfrac{9}{4} \qquad x=36$

The solutions are $\left(-\dfrac{9}{4},\dfrac{1}{2}\right)$ and $(36, -4)$.

8. (1) $x^2 - y^2 = 24$
(2) $2x^2 + 5y^2 = 55$
Use the addition method.
Multiply equation (1) by –2.
$$-2(x^2 - y^2) = -2 \cdot 24$$
$$2x^2 + 5y^2 = 55$$
$$7y^2 = 7$$
$$y^2 = 1$$
$$y = \pm\sqrt{1}$$
$$y = \pm 1$$
Substitute into equation (1).

$x^2 - y^2 = 24 \quad\quad x^2 - y^2 = 24$
$x^2 - (1)^2 = 24 \quad x^2 - (-1)^2 = 24$
$x^2 - 1 = 24 \quad\quad x^2 - 1 = 24$
$x^2 = 25 \quad\quad\quad x^2 = 25$
$x = \pm\sqrt{25} \quad\quad x = \pm\sqrt{25}$
$x = \pm 5 \quad\quad\quad x = \pm 5$

The solutions are (5, 1), (–5, 1), (5, –1), and (–5, –1).

9. $(x_1, y_1) = (2, 4) \quad (x_2, y_2) = (-1, -3)$
$$r = \sqrt{(x_2 - x_1)^2 + (y_2 - y_1)^2}$$
$$= \sqrt{(-1 - 2)^2 + (-3 - 4)^2}$$
$$= \sqrt{(-3)^2 + (-7)^2}$$
$$= \sqrt{9 + 49}$$
$$= \sqrt{58}$$
$$(x - h)^2 + (y - k)^2 = r^2$$
$$[x - (-1)]^2 + [y - (-3)]^2 = (\sqrt{58})^2$$
$$(x + 1)^2 + (y + 3)^2 = 58$$

10. $(x - 2)^2 + (y + 1)^2 = 9$

11. $(x - h)^2 + (y - k)^2 = r^2$
$(x - (-2))^2 + (y - 4)^2 = 3^2$
$(x + 2)^2 + (y - 4)^2 = 9$

12. $(x_1, y_1) = (2, 5) \quad (x_2, y_2) = (-2, 1)$
$$d = \sqrt{(x_2 - x_1)^2 + (y_2 - y_1)^2}$$
$$= \sqrt{(-2 - 2)^2 + (1 - 5)^2}$$
$$= \sqrt{(-4)^2 + (-4)^2}$$
$$= \sqrt{32}$$
$$(x - h)^2 + (y - k)^2 = r^2$$
$$(x - (-2))^2 + (y - 1)^2 = \sqrt{32}$$
$$(x + 2)^2 + (y - 1)^2 = 32$$

13. $x^2 + y^2 - 4x + 2y + 1 = 0$
$(x^2 - 4x) + (y^2 + 2y) = -1$
$(x^2 - 4x + 4) + (y^2 + 2y + 1) = -1 + 4 + 1$
$(x - 2)^2 + (y + 1)^2 = 4$
Center: (2, –1)
Radius: 2

14. $\dfrac{y^2}{25} - \dfrac{x^2}{16} = 1$
Axis of symmetry: y-axis
Vertices: (0, 5) and (0, –5)
Asymptotes: $y = \dfrac{5}{4}x$ and $y = -\dfrac{5}{4}x$

15. $\dfrac{x^2}{9} - \dfrac{y^2}{4} = 1$
Axis of symmetry: x-axis
Vertices: (3, 0) and (–3, 0)
Asymptotes: $y = \dfrac{2}{3}x$ and $y = -\dfrac{2}{3}x$

16. $\dfrac{x^2}{16} + \dfrac{y^2}{4} = 1$
x-intercepts: (4, 0) and (–4, 0)
y-intercepts: (0, 2) and (0, –2)

17. $\dfrac{x^2}{16} - \dfrac{y^2}{25} < 1$

18. $x^2 + y^2 < 36$
    $x + y > 4$

19. $\dfrac{x^2}{25} + \dfrac{y^2}{4} \le 1$

20. $\dfrac{x^2}{25} - \dfrac{y^2}{16} \ge 1$
    $x^2 + y^2 \le 9$

The solution sets of these inequalities do not intersect, so the system has no real number solution.

Final Exam

1. $12 - 8[3-(-2)]^2 \div 5 - 3 = 12 - 8[5]^2 \div 5 - 3$
   $= 12 - 8(25) \div 5 - 3$
   $= 12 - 200 \div 5 - 3$
   $= 12 - 40 - 3$
   $= -31$

2. $\dfrac{a^2 - b^2}{a - b} = \dfrac{3^2 - (-4)^2}{3 - (-4)}$
   $= \dfrac{9 - 16}{3 + 4} = \dfrac{-7}{7}$
   $= -1$

3. $5 - 2[3x - 7(2 - x) - 5x] = 5 - 2[3x - 14 + 7x - 5x]$
   $= 5 - 2[5x - 14]$
   $= 5 - 10x + 28$
   $= 33 - 10x$

4. $\dfrac{3}{4}x - 2 = 4$
   $\dfrac{3}{4}x = 6$
   $\dfrac{4}{3} \cdot \dfrac{3}{4}x = \dfrac{4}{3} \cdot 6$
   $x = 8$
   The solution is 8.

5. $\dfrac{2 - 4x}{3} - \dfrac{x - 6}{12} = \dfrac{5x - 2}{6}$
   $12\left(\dfrac{2 - 4x}{3} - \dfrac{x - 6}{12}\right) = 12\left(\dfrac{5x - 2}{6}\right)$
   $4(2 - 4x) - (x - 6) = 2(5x - 2)$
   $8 - 16x - x + 6 = 10x - 4$
   $14 - 17x = 10x - 4$
   $-27x = -18$
   $x = \dfrac{2}{3}$
   The solution is $\dfrac{2}{3}$.

6. $8 - |5 - 3x| = 1$
   $-|5 - 3x| = -7$
   $|5 - 3x| = 7$
   $5 - 3x = 7$ $5 - 3x = -7$
   $-3x = 2$ $-3x = -12$
   $x = -\dfrac{2}{3}$ $x = 4$
   The solutions are $-\dfrac{2}{3}$ and 4.

7. $|2x + 5| < 3$
   $-3 < 2x + 5 < 3$
   $-3 - 5 < 2x < 3 - 5$
   $-8 < 2x < -2$
   $\dfrac{1}{2}(-8) < \dfrac{1}{2}(2x) < \dfrac{1}{2}(-2)$
   $-4 < x < -1$
   $\{x | -4 < x < -1\}$

8. $2 - 3x < 6$ and $2x + 1 > 4$
   $-3x < 4$ $2x > 3$
   $x > -\dfrac{4}{3}$ $x > \dfrac{3}{2}$
   $\{x | x > -\dfrac{4}{3}\} \cap \{x | x > \dfrac{3}{2}\} = \{x | x > \dfrac{3}{2}\}$

9. $3x - 2y = 6$
$-2y = -3x + 6$
$y = \dfrac{3}{2}x - 3$
$m_1 = \dfrac{3}{2}$
$m_1 \cdot m_2 = -1$
$\dfrac{3}{2} m_2 = -1$
$m_2 = -\dfrac{2}{3}$ $(x_1, y_1) = (-2, 1)$
$y - y_1 = m(x - x_1)$
$y - 1 = -\dfrac{2}{3}(x - (-2))$
$y - 1 = -\dfrac{2}{3}(x + 2)$
$y - 1 = -\dfrac{2}{3}x - \dfrac{4}{3}$
$y = -\dfrac{2}{3}x - \dfrac{1}{3}$

The equation of the line is $y = -\dfrac{2}{3}x - \dfrac{1}{3}$.

10. $2a[5 - a(2 - 3a) - 2a] + 3a^2$
$= 2a[5 - 2a + 3a^2 - 2a] + 3a^2$
$= 2a[5 - 4a + 3a^2] + 3a^2$
$= 10a - 8a^2 + 6a^3 + 3a^2$
$= 6a^3 - 5a^2 + 10a$

11. $\dfrac{3}{2 + i} = \dfrac{3}{2 + i} \cdot \dfrac{2 - i}{2 - i}$
$= \dfrac{6 - 3i}{4 - i^2}$
$= \dfrac{6 - 3i}{4 + 1}$
$= \dfrac{6 - 3i}{5}$
$= \dfrac{6}{5} - \dfrac{3}{5}i$

12. $(x - r_1)(x - r_2) = 0$
$\left(x - \left(-\dfrac{1}{2}\right)\right)(x - 2) = 0$
$\left(x + \dfrac{1}{2}\right)(x - 2) = 0$
$x^2 - \dfrac{3}{2}x - 1 = 0$
$2\left(x^2 - \dfrac{3}{2}x - 1\right) = 0$
$2x^2 - 3x - 2 = 0$

13. $8 - x^3 y^3 = 2^3 - (xy)^3$
$= (2 - xy)(4 + 2xy + x^2 y^2)$

14. $x - y - x^3 + x^2 y = x - y - x^2(x - y)$
$= 1(x - y) - x^2(x - y)$
$= (x - y)(1 - x^2)$
$= (x - y)(1 - x)(1 + x)$

15. $\begin{array}{r} x^2 - 2x - 3 \\ 2x - 3 \overline{\smash{)}2x^3 - 7x^2 + 0x + 4} \\ \underline{2x^3 - 3x^2} \\ -4x^2 + 0x \\ \underline{-4x^2 + 6x} \\ -6x + 4 \\ \underline{-6x + 9} \\ -5 \end{array}$

$x^2 - 2x - 3 - \dfrac{5}{2x - 3}$

16. $\dfrac{x^2 - 3x}{2x^2 - 3x - 5} \div \dfrac{4x - 12}{4x^2 - 4} = \dfrac{x^2 - 3x}{2x^2 - 3x - 5} \cdot \dfrac{4x^2 - 4}{4x - 12}$
$= \dfrac{x(x - 3)}{(2x - 5)(x + 1)} \cdot \dfrac{4(x + 1)(x - 1)}{4(x - 3)}$
$= \dfrac{x(x-3)4(x+1)(x-1)}{(2x-5)(x+1)4(x-3)} = \dfrac{x(x - 1)}{2x - 5}$

17. $\dfrac{x - 2}{x + 2} - \dfrac{x + 3}{x - 3} = \dfrac{x - 2}{x + 2} \cdot \dfrac{x - 3}{x - 3} - \dfrac{x + 3}{x - 3} \cdot \dfrac{x + 2}{x + 2}$
$= \dfrac{x^2 - 5x + 6 - (x^2 + 5x + 6)}{(x + 2)(x - 3)}$
$= \dfrac{x^2 - 5x + 6 - x^2 - 5x - 6}{(x + 2)(x - 3)}$
$= -\dfrac{10x}{(x + 2)(x - 3)}$

18. $\dfrac{\dfrac{3}{x} + \dfrac{1}{x+4}}{\dfrac{1}{x} + \dfrac{3}{x+4}} = \dfrac{\dfrac{3}{x} + \dfrac{1}{x+4}}{\dfrac{1}{x} + \dfrac{3}{x+4}} \cdot \dfrac{x(x + 4)}{x(x + 4)}$
$= \dfrac{3(x + 4) + x}{x + 4 + 3x} = \dfrac{3x + 12 + x}{4x + 4}$
$= \dfrac{4x + 12}{4x + 4} = \dfrac{4(x + 3)}{4(x + 1)} = \dfrac{x + 3}{x + 1}$

19.
$$\frac{5}{x-2} - \frac{5}{x^2-4} = \frac{1}{x+2}$$
$$(x+2)(x-2)\left(\frac{5}{x-2} - \frac{5}{(x+2)(x-2)}\right) = (x+2)(x-2)\frac{1}{x+2}$$
$$5(x+2) - 5 = x-2$$
$$5x + 10 - 5 = x - 2$$
$$5x + 5 = x - 2$$
$$4x = -7$$
$$x = -\frac{7}{4}$$

The solution is $-\frac{7}{4}$.

20.
$$a_n = a_1 + (n-1)d$$
$$a_n - a_1 = (n-1)d$$
$$\frac{a_n - a_1}{n-1} = d$$

21. $\left(\frac{4x^2y^{-1}}{3x^{-1}y}\right)^{-2}\left(\frac{2x^{-1}y^2}{9x^{-2}y^2}\right)^3 = \frac{4^{-2}x^{-4}y^2}{3^{-2}x^2y^{-2}} \cdot \frac{2^3x^{-3}y^6}{9^3x^{-6}y^6}$

$= 4^{-2} \cdot 3^{-(-2)} \cdot x^{-4-2}y^{2-(-2)} \cdot 2^3 9^{-3} x^{-3-(-6)}y^{6-6}$

$= 4^{-2} \cdot 3^2 x^{-6} y^4 \cdot 2^3 \cdot 9^{-3} x^3 y^0$

$= \frac{9x^{-3}y^4 \cdot 8}{16 \cdot 729} = \frac{y^4}{162x^3}$

22. $\left(\frac{3x^{2/3}y^{1/2}}{6x^2y^{4/3}}\right)^6 = \frac{3^6 x^4 y^3}{6^6 x^{12} y^8}$

$= \frac{729 x^{4-12} y^{3-8}}{46656}$

$= \frac{1x^{-8}y^{-5}}{64} = \frac{1}{64x^8y^5}$

23. $x\sqrt{18x^2y^3} - y\sqrt{50x^4y}$
$= x\sqrt{3^2x^2y^2(2y)} - y\sqrt{5^2x^4(2y)}$
$= 3x^2y\sqrt{2y} - 5x^2y\sqrt{2y} = -2x^2y\sqrt{2y}$

24. $\frac{\sqrt{16x^5y^4}}{\sqrt{32xy^7}} = \sqrt{\frac{16x^5y^4}{32xy^7}}$

$= \sqrt{\frac{x^4}{2y^3}}$

$= \sqrt{\frac{x^4}{y^2(2y)}}$

$= \frac{x^2}{y}\sqrt{\frac{1}{2y}} \cdot \sqrt{\frac{2y}{2y}}$

$= \frac{x^2}{y}\sqrt{\frac{1 \cdot 2y}{(2y)^y}}$

$= \frac{x^2\sqrt{2y}}{2y^2}$

25. $2x^2 - 3x - 1 = 0$
$a = 2, b = -3, c = -1$

$x = \dfrac{-b \pm \sqrt{b^2 - 4ac}}{2a}$

$= \dfrac{-(-3) \pm \sqrt{(-3)^2 - 4(2)(-1)}}{2(2)}$

$= \dfrac{3 \pm \sqrt{9+8}}{4} = \dfrac{3 \pm \sqrt{17}}{4}$

The solutions are $\dfrac{3+\sqrt{17}}{4}$ and $\dfrac{3-\sqrt{17}}{4}$.

26. $x^{2/3} - x^{1/3} - 6 = 0$
$(x^{1/3})^2 - x^{1/3} - 6 = 0$
Let $u = x^{1/3}$.
$u^2 - u - 6 = 0$
$(u-3)(u+2) = 0$

$u - 3 = 0 \qquad u + 2 = 0$
$u = 3 \qquad u = -2$
$x^{1/3} = 3 \qquad x^{1/3} = -2$
$(x^{1/3})^3 = 3^3 \qquad (x^{1/3})^3 = (-2)^3$
$x = 27 \qquad x = -8$

The solutions are 27 and -8.

27. $(x_1, y_1) = (3, -2), (x_2, y_2) = (1, 4)$

$m = \dfrac{y_2 - y_1}{x_2 - x_1} = \dfrac{4-(-2)}{1-3} = \dfrac{6}{-2} = -3$

$y - y_1 = m(x - x_1)$
$y - (-2) = -3(x - 3)$
$y + 2 = -3x + 9$
$y = -3x + 7$

The equation of the line is $y = -3x + 7$.

28. $\dfrac{2}{x} - \dfrac{2}{2x+3} = 1$

$x(2x+3)\left(\dfrac{2}{x} - \dfrac{2}{2x+3}\right) = x(2x+3)(1)$

$2(2x+3) - 2x = 2x^2 + 3x$
$4x + 6 - 2x = 2x^2 + 3x$
$2x + 6 = 2x^2 + 3x$
$0 = 2x^2 + x - 6$
$0 = (2x - 3)(x + 2)$

$2x - 3 = 0 \qquad x + 2 = 0$
$x = \dfrac{3}{2} \qquad x = -2$

The solutions are $\dfrac{3}{2}$ and -2.

29. (1) $3x - 2y = 1$
    (2) $5x - 3y = 3$
Eliminate y.
Multiply equation (1) by -3 and equation (2) by 2.
Add the two new equations.

$-3(3x - 2y) = -3(1) \qquad -9x + 6y = -3$
$2(5x - 3y) = 2(3) \qquad 10x - 6y = 6$
$\qquad\qquad\qquad\qquad\qquad x = 3$

Substitute 3 for x in equation (1).
$3x - 2y = 1$
$3(3) - 2y = 1$
$9 - 2y = 1$
$-2y = -8$
$y = 4$

The solution is (3, 4).

30. $\begin{vmatrix} 3 & 4 \\ -1 & 2 \end{vmatrix} = 3(2) - (-1)(4)$
$= 6 + 4$
$= 10$

31. $\log_3 x - \log_3 (x - 3) = \log_3 2$

$\log_3 \left(\dfrac{x}{x-3}\right) = \log_3 2$

Use the fact that if $\log_b u = \log_b v$, then $u = v$.

$\dfrac{x}{x-3} = 2$

$(x - 3) \cdot \dfrac{x}{x-3} = 2 \cdot (x - 3)$

$x = 2x - 6$
$-x = -6$
$x = 6$

The solution is 6.

32. $\displaystyle\sum_{i=1}^{5} 2y^i = 2y^1 + 2y^2 + 2y^3 + 2y^4 + 2y^5$
$= 2y + 2y^2 + 2y^3 + 2y^4 + 2y^5$

33. $0.5\overline{1} = 0.5 + 0.01 + 0.001 + 0.0001 + \cdots$

$= \dfrac{5}{10} + \dfrac{1}{100} + \dfrac{1}{1000} + \dfrac{1}{10{,}000} + \cdots$

$r = \dfrac{a_2}{a_1} = \dfrac{\frac{1}{1000}}{\frac{1}{100}} = \dfrac{1}{10}$

$S = \dfrac{a_1}{1-r} = \dfrac{\frac{1}{100}}{1-\frac{1}{10}} = \dfrac{\frac{1}{100}}{\frac{9}{10}} = \dfrac{1}{100} \cdot \dfrac{10}{9} = \dfrac{1}{90}$

$0.5\overline{1} = \dfrac{5}{10} + \dfrac{1}{90} = \dfrac{46}{90} = \dfrac{23}{45}$

34. $n = 9, a = x, b = -2y, r = 3$

$\dbinom{9}{3-1} x^{9-3+1}(-2y)^{3-1} = \dbinom{9}{2} x^7 (-2y)^2$

$= 36x^7 \cdot 4y^2$
$= 144x^7 y^2$

35. (1) $x^2 - y^2 = 4$
(2) $x + y = 1$
Solve equation (2) for y and substitute into equation (1).

$$x + y = 1$$
$$y = -x + 1$$
$$x^2 - y^2 = 4$$
$$x^2 - (-x+1)^2 = 4$$
$$x^2 - (x^2 - 2x + 1) = 4$$
$$x^2 - x^2 + 2x - 1 = 4$$
$$2x = 5$$
$$x = \frac{5}{2}$$

Substitute $\frac{5}{2}$ for x into equation (2).

$$\frac{5}{2} + y = 1$$
$$y = -\frac{3}{2}$$

The solution is $\left(\frac{5}{2}, -\frac{3}{2}\right)$.

36. $f(x) = \frac{2}{3}x - 4$

$$y = \frac{2}{3}x - 4$$
$$x = \frac{2}{3}y - 4$$
$$x + 4 = \frac{2}{3}y$$
$$\frac{3}{2}(x+4) = \frac{3}{2} \cdot \frac{2}{3}y$$
$$\frac{3}{2}x + 6 = y$$
$$f^{-1}(x) = \frac{3}{2}x + 6$$

37. $2(\log_2 a - \log_2 b) = 2\log_2 \frac{a}{b}$
$$= \log_2 \left(\frac{a}{b}\right)^2 = \log_2 \frac{a^2}{b^2}$$

38.
x-intercept   y-intercept
$2x - 3y = 9$ $2x - 3y = 9$
$2x - 3(0) = 9$ $2(0) - 3y = 9$
$2x = 9$ $-3y = 9$
$x = \frac{9}{2}$ $y = -3$
                    $(0, -3)$
$\left(\frac{9}{2}, 0\right)$

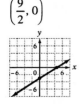

39. $3x + 2y > 6$
$$2y > -3x + 6$$
$$y > -\frac{3}{2}x + 3$$

40. $f(x) = -x^2 + 4$
$$-\frac{b}{2a} = -\frac{0}{2(-1)} = 0$$
$$f(x) = -0^2 + 4 = 4$$
Vertex: $(0, 4)$
Axis of symmetry: $x = 0$

41. $\frac{x^2}{16} + \frac{y^2}{4} = 1$
x-intercepts: $(4, 0)$ and $(-4, 0)$
y-intercepts: $(0, 2)$ and $(0, -2)$

42. $f(x) = \log_2 (x+1)$
$$y = \log_2 (x+1)$$
$$2^y = x + 1$$
$$2^y - 1 = x$$

43. $f(x) = x + 2^{-x}$

-2 and 1.7

44. $f(x) = \ln x$
$g(x) = \ln(x+3)$

Chapter 12: Conic Sections

45. Strategy
To find the range of scores on the 5th test, write and solve a compound inequality using x to represent the 5th test.

Solution
$70 \leq$ average of the 5 test scores ≤ 79
$$79 \leq \frac{64 + 58 + 82 + 77 + x}{5} \leq 79$$
$$70 \leq \frac{281 + x}{5} \leq 79$$
$$350 \leq 281 + x \leq 395$$
$$69 \leq x \leq 114$$
The range of scores is 69 or better.

46. Strategy
Average speed of jogger: x
Average speed of cyclist: $2.5x$

	Rate	Time	Distance
Jogger	x	2	$2x$
Cyclist	$2.5x$	2	$2(2.5x)$

• The distance traveled by the cyclist is 24 more miles than the distance traveled by the jogger.
$2x + 24 = 2(2.5x)$

Solution
$2x + 24 = 2(2.5x)$
$2x + 24 = 5x$
$24 = 3x$
$8 = x$
$2(2.5x) = 5x = 5(8) = 40$
The cyclist traveled 40 mi.

47. Strategy
Amount invested at 8.5%: x
Amount invested at 6.4%: $12000 - x$

	Amount	Rate	Interest
8.5%	x	0.085	$0.085x$
6.4%	$12,000 - x$	0.064	$0.064(12,000 - x)$

The sum of the interest earned by the two investments is $936.
$0.085x + 0.064(12,000 - x) = 936$

Solution
$0.085x + 0.064(12,000 - x) = 936$
$0.085x + 768 - 0.064x = 936$
$0.021x + 768 = 936$
$0.021x = 168$
$x = 8000$
$12,000 - x = 4000$
The amount invested at 8.5% is $8000.
The amount invested at 6.4% is $4000.

48. Strategy
The width of the rectangle: x
The length of the rectangle: $3x - 1$
Use the formula for the area of a rectangle $(A = LW)$ if the area is 140 ft^2.

Solution
$A = L \cdot W$
$140 = (3x - 1)x$
$140 = 3x^2 - x$
$0 = 3x^2 - x - 140$
$0 = (3x + 20)(x - 7)$
$3x + 20 = 0 \qquad x - 7 = 0$
$x = -\frac{20}{3} \qquad x = 7$

The solution $-\frac{20}{3}$ does not check because the width cannot be negative.
$3x - 1 = 3(7) - 1 = 20$
The length of the rectangle is 20 ft and the width is 7 ft.

49. Strategy
To find the number of additional shares, write and solve a proportion using x to represent the additional number of shares.

Solution
$$\frac{300}{486} = \frac{300 + x}{810}$$
$$(810 \cdot 486) \cdot \frac{300}{486} = \frac{300 + x}{810} \cdot (810 \cdot 486)$$
$$300(810) = (300 + x)486$$
$$243,000 = 145,800 + 486x$$
$$97,200 = 486x$$
$$200 = x$$
The number of additional shares to be purchased is 200.

50. Strategy
Rate of car: x
Rate of plane: $7x$

	Distance	Rate	Time
Car	45	x	$\frac{45}{x}$
Plane	1050	$7x$	$\frac{1050}{7x}$

The total time traveled is $3\frac{1}{4}$ h.
$$\frac{45}{x} + \frac{1050}{7x} = 3\frac{1}{4}$$

Solution
$$\frac{45}{x} + \frac{1050}{7x} = 3\frac{1}{4}$$
$$\frac{45}{x} + \frac{150}{x} = \frac{13}{4}$$
$$\frac{195}{x} = \frac{13}{4}$$
$$4x\left(\frac{195}{x}\right) = 4x\left(\frac{13}{4}\right)$$
$$780 = 13x$$
$$60 = x$$
$$7x = 6(60) = 420$$
The rate of the plane is 420 mph.

51. Strategy
To find the distance of the object has fallen, substitute 75 ft/s for v in the formula and solve for d.

Solution
$$v = \sqrt{64d}$$
$$75 = \sqrt{64d}$$
$$75^2 = (\sqrt{64d})^2$$
$$5625 = 64d$$
$$87.89 \approx d$$
The distance traveled is 88 ft.

52. Strategy
Rate traveled during the first 360 mi: x
Rate traveled during the next 300 mi: $x + 30$.

	Distance	Rate	Time
First part of trip	360	x	$\frac{360}{x}$
Second part of trip	300	$x+30$	$\frac{300}{x+30}$

The total time traveled during the trip was 5 h.
$$\frac{360}{x} + \frac{300}{x+30} = 5$$

Solution
$$\frac{360}{x} + \frac{300}{x+30} = 5$$
$$x(x+30)\left(\frac{360}{x} + \frac{300}{x+30}\right) = 5x(x+30)$$
$$360(x+30) + 300x = 5(x^2 + 30x)$$
$$360x + 10800 + 300x = 5x^2 + 150x$$
$$660x + 10800 = 5x^2 + 150x$$
$$0 = 5x^2 - 510x - 10800$$
$$0 = 5(x^2 - 102x - 2160)$$
$$0 = (x+18)(x-120)$$
$$x + 18 = 0 \qquad x - 120 = 0$$
$$x = -18 \qquad x = 120$$
The solution -18 does not check because the rate cannot be negative.
The rate of the plane for the first 360 mi was 120 mph.

53. Strategy
To find the intensity:
Write the basic inverse variation equation, replace the variable by the given values, and solve for k. Write the inverse variation equation, replacing k by its value. Substitute 4 for d and solve for L.

Solution
$$L = \frac{k}{d^2}$$
$$8 = \frac{k}{(20)^2}$$
$$8 \cdot 400 = k$$
$$3200 = k$$
$$L = \frac{3200}{d^2}$$
$$L = \frac{3200}{4^2}$$
$$L = \frac{3200}{16} = 200$$
The intensity is 200 lumens.

54. Strategy
Rate of the boat in calm water: x
Rate of the current: y

	Rate	Time	Distance
With current	$x + y$	2	$2(x + y)$
Against current	$x - y$	3	$3(x - y)$

The distance traveled with the current is 30 mi.
The distance traveled against the current is 30 mi.
$2(x + y) = 30$
$3(x - y) = 30$

Solution

$2(x+y) = 30 \qquad \frac{1}{2} \cdot 2(x+y) = \frac{1}{2} \cdot 30$
$3(x-y) = 30 \qquad \frac{1}{3} \cdot 3(x-y) = \frac{1}{3} \cdot 30$
$\qquad \qquad \qquad \qquad x + y = 15$
$\qquad \qquad \qquad \qquad x - y = 10$

$\qquad \qquad \qquad \qquad 2x = 25$
$\qquad \qquad \qquad \qquad x = 12.5$

$x + y = 15$
$12.5 + y = 15$
$\qquad y = 2.5$

The rate of the boat in calm water is 12.5 mph.
The rate of the current is 2.5 mph.

55. Strategy
The find the value of the investment after two years, solve the compound interest formula for P.
Use $A = 4000$, $n = 24$,
$i = \dfrac{9\%}{12} = \dfrac{0.09}{12} = 0.0075$.

Solution
$P = A(1+i)^n$
$P = 4000(1 + 0.0075)^{24}$
$P = 4000(1.0075)^{24}$
$P \approx 4785.65$
The value of the investment is $4785.65.

56. Strategy
To find the value of the house in 20 years, use the formula for the nth term of a geometric sequence.

Solution
$n = 20$, $a_1 = 1.06(80,000) = 84,800$, $r = 1.06$
$a_n = a_1(r)^{n-1}$
$a_{20} = 84,800(1.06)^{20-1}$
$\quad = 84,800(1.06)^{19}$
$\quad \approx 256,571$
The value of the house will be $256,571.